# Innovative Management and Business Practices in Asia

Patricia Ordoñez de Pablos
*The University of Oviedo, Spain*

Xi Zhang
*Tianjin University, China*

Kwok Tai Chui
*The Open University of Hong Kong, Hong Kong*

A volume in the Advances in Human Resources Management and Organizational Development (AHRMOD) Book Series

IGI Global
DISSEMINATOR OF KNOWLEDGE

Published in the United States of America by
IGI Global
Business Science Reference (an imprint of IGI Global)
701 E. Chocolate Avenue
Hershey PA, USA 17033
Tel: 717-533-8845
Fax: 717-533-8661
E-mail: cust@igi-global.com
Web site: http://www.igi-global.com



Library of Congress Cataloging-in-Publication Data

Names: Ordóñez de Pablos, Patricia, 1975- editor. | Zhang, Xi, 1982- editor. | Chui, Kwok Tai, 1991- editor.
Title: Innovative management and business practices in Asia / Patricia Ordoñez de Pablos, Xi Zhang, and Kwok Tai Chui, editors.
Description: Hershey : Business Science Reference, 2019. | Includes bibliographical references and index. | Summary: “”This book contains research that enhances understanding and collaboration in business, management, and technology in Asia for the present and in the future”--Provided by publisher”-- Provided by publisher.
Identifiers: LCCN 2019031694 (print) | LCCN 2019031695 (ebook) | ISBN 9781799815662 (hardcover) | ISBN 9781799815679 (paperback) | ISBN 9781799815686 (ebook)
Subjects: LCSH: Entrepreneurship--Asia. | Management--Asia. | Technological innovations--Asia--Management.
Classification: LCC HB615 .I5647 2019 (print) | LCC HB615 (ebook) | DDC 338/.04095--dc23
LC record available at https://lccn.loc.gov/2019031694
LC ebook record available at https://lccn.loc.gov/2019031695

This book is published in the IGI Global book series Advances in Human Resources Management and Organizational Development (AHRMOD) (ISSN: 2327-3372; eISSN: 2327-3380)

British Cataloguing in Publication Data
A Cataloguing in Publication record for this book is available from the British Library.

# Advances in Human Resources Management and Organizational Development (AHRMOD) Book Series

Patricia Ordóñez de Pablos
Universidad de Oviedo, Spain

ISSN:2327-3372
EISSN:2327-3380

## Mission

A solid foundation is essential to the development and success of any organization and can be accomplished through the effective and careful management of an organization's human capital. Research in human resources management and organizational development is necessary in providing business leaders with the tools and methodologies which will assist in the development and maintenance of their organizational structure.

The **Advances in Human Resources Management and Organizational Development (AHRMOD) Book Series** aims to publish the latest research on all aspects of human resources as well as the latest methodologies, tools, and theories regarding organizational development and sustainability. The **AHRMOD Book Series** intends to provide business professionals, managers, researchers, and students with the necessary resources to effectively develop and implement organizational strategies.

## Coverage

- Personnel Policies
- E-Human Resources Management
- Change Management
- Recruitment Process
- Personnel Retention
- Disputes Resolution
- Coaching and Mentoring
- Work-Life Balance
- Upward Feedback
- Workplace Culture

IGI Global is currently accepting manuscripts for publication within this series. To submit a proposal for a volume in this series, please contact our Acquisition Editors at Acquisitions@igi-global.com or visit: http://www.igi-global.com/publish/.

The Advances in Human Resources Management and Organizational Development (AHRMOD) Book Series (ISSN 2327-3372) is published by IGI Global, 701 E. Chocolate Avenue, Hershey, PA 17033-1240, USA, www.igi-global.com. This series is composed of titles available for purchase individually; each title is edited to be contextually exclusive from any other title within the series. For pricing and ordering information please visit http://www.igi-global.com/book-series/advances-human-resources-management-organizational/73670. Postmaster: Send all address changes to above address. 

## Titles in this Series

*For a list of additional titles in this series, please visit: https://www.igi-global.com/book-series/advances-human-resources-management-organizational/73670*

***Handbook of Research on Positive Organizational Behavior for Improved Workplace Performance***
Elif Baykal (Istanbul Medipol University, Turkey)
Business Science Reference • © 2020 • 445pp • H/C (ISBN: 9781799800583) • US $295.00

***Analyzing Workplace Deviance in Modern Organizations***
Naman Sharma (Entrepreneurship Development Institute of India, India)
Business Science Reference • © 2020 • 354pp • H/C (ISBN: 9781522599968) • US $205.00

***Recent Advances in the Roles of Cultural and Personal Values in Organizational Behavior***
Zlatko Nedelko (University of Maribor, Slovenia) and Maciej Brzozowski (Poznan University of Economics and Business, Poland)
Business Science Reference • © 2020 • 420pp • H/C (ISBN: 9781799810131) • US $225.00

***Handbook of Research on Social and Organizational Dynamics in the Digital Era***
Efosa C. Idemudia (Arkansas Tech University, USA)
Business Science Reference • © 2020 • 667pp • H/C (ISBN: 9781522589334) • US $295.00

***Corporate Governance Models and Applications in Developing Economies***
Otuo Serebour Agyemang (University of Cape Coast, Ghana) Abraham Ansong (University of Cape Coast, Ghana) and Ben Kwame Agyei-Mensah (Solbridge International School of Business, South Korea)
Business Science Reference • © 2020 • 330pp • H/C (ISBN: 9781522596073) • US $195.00

***Macro and Micro-Level Issues Surrounding Women in the Workforce Emerging Research and Opportunities***
Başak Uçanok Tan (Istanbul Bilgi University, Turkey)
Business Science Reference • © 2020 • 309pp • H/C (ISBN: 9781522591634) • US $195.00

***Cases on Learning Design and Human Performance Technology***
Jill Stefaniak (University of Georgia, USA)
Business Science Reference • © 2020 • 362pp • H/C (ISBN: 9781799800545) • US $195.00

***Organizational Culture and Achieving Business Excellence Emerging Research and Opportunities***
Rassel Kassem (Abu Dhabi Judicial Department, UAE) and Mian M. Ajmal (Abu Dhabi University, UAE)
Business Science Reference • © 2019 • 143pp • H/C (ISBN: 9781522584131) • US $165.00

IGI Global
DISSEMINATOR OF KNOWLEDGE

701 East Chocolate Avenue, Hershey, PA 17033, USA
Tel: 717-533-8845 x100 • Fax: 717-533-8661
E-Mail: cust@igi-global.com • www.igi-global.com

# Editorial Advisory Board



# List of Reviewers

# Table of Contents

# Detailed Table of Contents



Strategic Alliances (SAs) have been formed by firms to attain strategic objectives which alone firms were unable to secure. Typically, firms formed SAs to gain market share or to seek innovation and research and development capabilities form the partnering firm. In the extant literature, market seeking SAs have been viewed as defensive while innovation and research and development capabilities seeking SAs as an aggressive strategy. In this research, the authors conducted a study of 165 strategic alliances cases in the Indian business scenario to study the motives of SAs. This study classified SAs formed by the top 50 firms listed on India's National Stock Exchange (NSE) (NIFTY50 index) into different categories. Inferences and findings have been drawn using content analysis of the available dataset. It was observed that, out of the 165 unique cases of SAs, a substantial number of SAs could be classified as 'business alliances' focusing on the market seeking from the alliance partner.



This empirical study investigates the symmetric and asymmetric dynamic correlations and volatility linkages between ASEAN-5 and the Indian equity markets. Granger causality test results reveal that bidirectional causal relation between the pairs of India-Indonesia, and India-Singapore. However, India-Phillippines and India-Thailand have a unidirectional causal relationship. Variance decomposition results show that India's equity market volatility contributes moderate fluctuations in the variance of Indonesia, Phillippines, Singapore Thailand. Finally, the Markov regime transition probabilities show that the high transition probabilities of p11 and p22 for India-Malaysia, India-Philippines and India-Thailand indicates a high degree of regime stability. The study on financial integration provides important inputs to investors in sharing risk internationally since restrictions on investment are removed. This study provides an essential insight to policymakers, portfolio managers, domestic and international investors, risk analysts and financial researchers in an emerging market.

Innovating products or services are not an emerging issue rather a reality in today's marketplace. Organizations are relentlessly chasing innovation to compete with their rivals. Considering these phenomena, literature from the different management studies suggest that to being mechanistic it would better to become organic in its operation. The organic structure of the firm also highlights teamwork as a priority to become successful. This study illustrates the effects of team dynamics and mediating effects of organizational learning on organizational innovativeness. By proposing the conceptual model, the chapter presents different practical implications to the practitioners, researchers, and academicians connected to the industry and can be a source for future research Bangladesh and other developing economies to develop different insights.



Manufacturing plays a substantial role in the economic development of any country because of its multiplier impact on the growth of value addition. Currently, industry 4.0 requires manufacturers to deliver highly customized products without compromising on quality at a reduced life cycle. The objective of this study was to find out a solution for the optimum operation of manufacturing firms. By applying resource-based view, dynamic capability, and effectuation theory, this study has proposed an integrated framework of the organizational network, entrepreneurial bricolage, strategic agility and business performance in the context of the industry 4.0. Moreover, the positive effect of the organizational networks on the strategic agility ultimately improves the business performance of manufacturers. Furthermore, strategic agility is also claimed to play its role as mediator between organizational networks and business performance.



Facing a severely competitive environment and unique consumer expectations, cultivating consumer loyalty seems crucial for every firm. Rather than to cultivate, marketers find it more difficult to sustain a loyal consumer base. Understanding loyalty drives is of utmost importance now, and it is the main concern of this study. Faced with lack of structural management approaches in reference to relationship practices, this chapter provides a cohesive understanding of loyalty drivers in application of acquisition,

Venture capital (VC) provides a platform to empowered individuals with financial constraints to transform their ideas into business models and attain commercial success. This article reviewed the growth and trends of VC industry across various regions such as the United States of America (USA), Europe, China, and India. Initially, VC firms flourished and developed in the USA and still it harbors the largest VC industry. From the USA, VC firms spread to Europe and then much later to emerging economies like China and India. Although the VC ecosystem had started late in China, it had registered higher growth when compared to Europe in terms of VC investment. China has become the second largest VC market. It was backed by government initiatives, vast market opportunities, and home-grown technology firm investments. India has started observing growth in VC space later than China but had ample opportunities to allow for a surge in VC activities.



This chapter presents a recent survey on the development and accessibility for finance of small and medium-size enterprises (SMEs) in Mongolia. The survey covers a sample of 1922 SMEs from Ulaanbaatar, the capital city, and 21 provinces. We find that banks and local government administrations are the most supportive institutions for SME development. Political instability, corruption, and labour supply, a high lending rate, short maturity loans, and service fees are perceived as the major obstacles that SMEs face in the business environment. Our results also suggest that SMEs in Mongolia are less likely to have access to external finance because of tight credit condition, potentially explaining the lack of SME growth. Implementing country-specific reform strategy for SME development covering key building blocks is needed to promote SME financial inclusion and facilitate SMEs to contribute to the economic growth.

# Preface

## INTRODUCTION

Asia comprises 30% of the world's land area, contributed by about 50 countries and 60% of world population (Asia Population 2019, 2019). About 60% of the Asia's population are from China and India. Today's information era allows everybody contributes to the economy in a fair opportunity by proposing and realizing their innovations on management and business practices, which is not mainly based on the population size of the countries. Innovation does not only restricted by creating something totally new but also ideas in optimizing the process, design, practice, marketing and management. There should be a compromise between different parties that a trade-off solution of the optimization problem is usually recommended. This can be explained by there is usually no best solution that satisfies perfectly for everybody.

One of the key benefits of innovation is to facilitate and promote economic growth locally and internationally. Put it this way, innovation can lead to higher productivity, which is equivalent to the same input will generate a greater output. More goods and services are produced as a result of productivity rise. In other words, the economy grows. Nowadays, the prevalence of data-driven approach has been increased for applying innovations, as people (particularly senior management) have witnessed the effectiveness of the approach. Business data is being collected and further analyzed to produce valuable insights in the form of action items. Evaluation will be made on the items and may result in fine-tuning based on users' feedback so that innovation takes full effect.

## AIMS AND SCOPE OF THE BOOK

The book offers a valuable collection of 15 chapters exploring and discussing the state of the art, emerging topics, challenges and success factors in business, management and innovation in Asia. The book emphasizes empirical studies and case studies in Bangladesh, India, Indonesia, Malaysia, Mongolia and Pakistan.

The book is an international platform to bring together academics, researchers, lecturers, decision makers, policy makers, and practitioners to share new theories and conceptual frameworks, research findings, and case studies, to enhance understanding, visibility and collaboration in management, business, innovation and technology in Asia now. It also discusses emerging trends with potential impact on companies, economies and societies in the Asian region.

## TARGET AUDIENCE

Professors in academia, deans, heads of departments, director of masters, students (undergraduate and postgraduate level), politicians, policy makers, corporate heads of firms, senior general managers, managing directors, libraries, etc.

## MAIN TOPICS DISCUSSED

The book discusses new and emerging topics that are key to understand business, management, innovation and disruption technology in Asia, from equity market integration in ASEAN region and access to finance of SMEs in Mongolia to IoT and knowledge management in SMEs in Pakistan and social networks and social media in Bangladesh, among other interesting topics.

## CONTENTS OF THE BOOK

The first chapter (Chapter 1) titled "Study of Motives of Indian Strategic Alliances - Marketing or Innovation Seeking?" (authors: Vihang Vivek Jumle and Som Sekhar Bhattacharyya) state that "strategic alliances (SAs) have been formed by firms to attain strategic objectives which alone firms were unable to secure. Typically, firms formed SAs to gain market share or to seek innovation and research and development capabilities form the partnering firm. In the extant literature, market seeking SAs have been viewed as defensive while innovation and research and development capabilities seeking SAs as an aggressive strategy. In this research, the authors conducted a study of 165 strategic alliances cases in the Indian business scenario to study the motives of SAs. This study classified SAs formed by the top 50 firms listed on India's National Stock Exchange (NSE) (NIFTY50 index) into different categories. Inferences and findings have been drawn using content analysis of the available dataset. It was observed that, out of the 165 unique cases of SAs, a substantial number of SAs could be classified as 'business alliances' focusing on the market seeking from the alliance partner."

Chapter 2, titled "An Empirical Investigation on Equity Market Integration of ASEAN-India" (authors: Swamy Perumandla and Padma Kurisetti) presents the results of an empirical study that explores "the symmetric and asymmetric dynamic correlations and volatility linkages between ASEAN-5 & Indian equity markets. Granger causality test results reveal that bidirectional causal relation between the pairs of India-Indonesia, and India-Singapore. However, India-Philippines and India-Thailand have a unidirectional causal relationship. Variance decomposition results show that India's equity market volatility contributes moderate fluctuations in the variance of Indonesia, Phillippines, Singapore Thailand. Finally, Markov regime transition probabilities show that the high transition probabilities of p11 and p22 for India-Malaysia, India-Philippines and India-Thailand indicates a high degree of regime stability. The study on Financial integration provides important inputs to investors in sharing risk internationally since restrictions on investment are removed. This study provides an essential insight to policymakers, portfolio managers, domestic and international investors, risk analysts and financial researchers in an emerging market."

Chapter 3, titled "Do the Team Dynamics Influence the Organizations to be Innovative?" (authors: Ikramul Hasan, MD. Nazmul Islam and Mohammad Ashraful Ferdous Chowdhury) proposes that "innovating product or service is not an emerging issue rather a reality in today's marketplace. Organizations are relentlessly making their effort to confirm innovation to compete with their rivals. Considering these phenomena literature from the different management studies suggest that to being mechanistic it would better to become organic in its operation. The organic structure of the firm also highlights teamwork as a priority to become successful. This study illustrates the effects of team dynamics and mediating effects of organizational learning on organizational innovativeness. By proposing the conceptual model the paper present different practical implications to the practitioners, researchers, and academician connected to the industry and can be a source for future research Bangladesh and other developing economies to develop different insights."

Chapter 4, titled "Enhancing Business Performance of Pakistani Manufacturing Firms via Strategic Agility in Industry 4.0 Era: The Role of Entrepreneurial Bricolage as Moderator" (authors: Qaisar Iqbal, Noor Hazlina Ahmad, Heru Kurnianto Tjahjono, Adeel Nasim, Muhammad Mustafa Muqaddis and Majang Palupi) suggest that "manufacturing plays a substantial role in the economic development of any country because of its multiplier impact on the growth of value addition. Currently, industry 4.0 requires manufacturers to deliver highly customized products without compromising on quality at a reduced life cycle. The objective of this study was to find out a solution for the optimum operation of manufacturing firms. By applying resource-based view, dynamic capability, and effectuation theory, this study has proposed an integrated framework of the organizational network, entrepreneurial bricolage, strategic agility and business performance in the context of the industry 4.0. Moreover, the positive effect of the organizational networks on the strategic agility ultimately improves the business performance of manufacturers. Furthermore, strategic agility is also claimed to play its role as mediator between organizational networks and business performance."

Chapter 5, titled "Drivers of Consumer Loyalty in Wellness Sector" (authors: Shivani Saini and Jagwinder Singh) observe that "facing a severely competitive environment and unique consumer expectations, cultivating consumer loyalty seems crucial for every firm. Rather than cultivate marketers finds it more difficult to sustain a loyal consumer base. Understanding loyalty drivers is utmost importance now, and it is the main concern of this study. Faced with lack of structural management approaches in reference to relationship practices, this chapter provides a cohesive understanding of loyalty drivers in application of acquisition, retention, and experience strategies. The study involved a survey of 200 consumers of wellness firms. A confirmatory factor analysis has been employed to evaluate consumers' priority among acquisition, retention, and experience-based marketing strategies. The study offers valuable insights to wellness firms to fine-tune their CRM programmes in view of traditional and modern practices. As experience is very complex while acquisition and retention are highly imitable, firms need to embrace it with clarity to select the right elements to gain consumer loyalty."

Chapter 6, titled "Inter-relational Dynamics of various HR aspects in High Altitude Illness Attrition" (authors: Ajeya Jha, Ajay Dheer, Vijay Kumar Mehta and Saibal Kumar Saha) state that "the adverse health effects of high altitude are of considerable importance since they may seriously interfere with working efficiency of an organization that is actively involved with inescapable duties. The objective of the current study is to explore inter-relational dynamics of various HR aspects in HAIA. The HR aspects included are job delay, poor team, motivation, less leave, high working hours, poor decision making, personal stress, family stress, personal discomfort, uncertainty, poor relations, health, accidents, quality and performance. decision-making trial and evaluation laboratory have been to explore the interrelation dynamics of various factors of HR. The results indicate that that personal stress has the highest impact priority which is followed by poor performance, poor team and motivation. Uncertainty, less leave and high working hours has the least impact priority. It is also found that high working hours, less leave and poor health are the major causes whereas decrease in motivation, poor quality of work are the major effects."

Chapter 7, titled "Green Architecture of Malay Traditional Houses Exhibition: Mini Malaysia and ASEAN Cultural Park (MMACP) Melaka" (authors: Chew Boon Cheong, Mohd Syaiful Rizal Abd Hamid and Saifuddin Isa) affirm that "mini Malaysia and ASEAN Cultural Park (MMACP) is one of the tourism spots in Melaka which showcases the Malay traditional houses from thirteen states in Malaysia. Most of the visitors viewing the houses depicting the culture of each states. Contrarily we revisited the park with a curiosity and we would like to know, "How far these Malay traditional houses portray the green architecture?" We designed this research in an exploratory, looking the Malay traditional houses found in MMACP and trace for their green architectural criteria. Through observation, we analysed the data from our field notes, photos, description found on the park's storyboards and explanations from the tour guides. We can conclude that the Malay traditional houses are green buildings that portray green architecture. From this research, we appreciate our forefathers' skills and wisdom, to build the houses that harmonises with the environment."

Chapter 8, titled "Open Innovation Challenges and Coopetition-Based Open-Innovation Empirical Evidence from Malaysia" (authors: Jawad Iqbal and Waseem Ul Hameed) states that "in the highly competitive environment, the innovation trends have been changed and companies are moving towards the open innovation model rather than to follow close or traditional innovation model. Therefore, this chapter demonstrated various determinants of open innovation. For this purpose, a survey was carried out among Malaysian small and medium-sized enterprises (SMEs). The outcomes of the survey highlighted that, the success of open innovation model is based on five major elements, namely, 1) motivating spillovers, 2) incorporation of external knowledge, 3) intellectual property management, 4) maximization of internal innovation, and 5) financial constraints. These five elements are the major challenges for companies while adopting open innovation model. More specifically, the phenomenon of coopetition-based open-innovation is emerging rapidly among the companies. Now a day, by following the open innovation activities, competitors are collaborating with each other rather than to compete."

Chapter 9, titled "How is it Different from Conventional Learning?: The Growing Trend of Corporate Universities in Indonesia" (authors: Hary Febriansyah, Haifa Labdhagati and Widi Galih Anggara) state that "in the current Indonesian business scene, corporate universities have become fashionable. Since their first emergence in the 2010s, Indonesian corporate universities have shown that they are more than just dressed-up learning centers operating under a new name. As prominent organizations in Indonesia start to build their own learning initiatives adopting the model of the corporate university, success stories of the new approach radiate. Bringing an example of a developing country of Indonesia, this chapter attempts to discuss the trends, challenges, and practices of corporate university. This chapter also examines best practices from Indonesian corporate universities and how they differ from the conventional university and their learning techniques differ from traditional learning methods."

Chapter 10, titled "Health Sustainability and Socialization Agents Role on Organ Donation: A Malaysian Youth Case" (authors: Pushpavalli Maniam and Izzal Asnira Zolkepli) analyses that "how sustainable health can be encouraged through the role of socialization amongst Youth in Malaysia in the context of decision-making process towards becoming an organ donor. This is imperative for the nation especially the policymaker and health business sector in designing strategy pertaining to health issues. The concept of the 'duality of structure' is used as a starting point to link between health sustainability and health communication on organ donation in congregating intergenerational equity by uncovering the structural properties or conditions which either enable or constrain the future of health initiatives. Therefore, this article is aimed to focus on the values and norms commonly transferred by the socialization agents regarding the behavioral development of the potential donor. Data was gathered using self-administered questionnaires from 162 youth. The findings illustrate that the supportive influence from agents of socialization affected towards certain degree on the behavioral formation on becoming an organ donor."

Chapter 11, titled "The Role of Narrative Element in Gamification Towards Value Co-creation: A Case of Mobile App Users in Malaysia" (authors: Cheah Wen Kit and Izzal Asnira Zolkepli) discusses "the role of narrative element in gamification towards value co-creation in crowdsourcing application system. The discourse addresses the gap of knowledge to understand the user motivation and experience to co-create value in a gamified system. Value co-creation is an interactive engagement process that refers to the act of collaborating with a group of intended consumers through crowdsourcing approach. As the decentralisation of the web enables participation of the people to shape the future based on their contributions, understanding Internet users' motivation and experience to co-create value is crucial in ensuring that the initiatives are reciprocated by the intended parties. As gamification has been widely utilised in numerous contexts in order to encourage users to contribute their resources of knowledge and skills, the effectiveness of its elements, namely narrative, remains questionable."

Chapter 12, titled "Nexus between Social Network, Social Media Use, and Loneliness: A Case Study of University Students, Bangladesh" (authors: Bezon Kumar and Md. Aminul Islam) investigates "how real-life social network and social media use are related to loneliness among university students in Bangladesh. To carry out this paper, primary data and several methods are used. This chapter uses Lubben Social Network Scale and UCLA loneliness scale to measure the level of real-life social network and loneliness, respectively. Besides Pearson's partial correlation matrix is used to find out the correlation between social network, social media use, and loneliness. The study finds that students are averagely engaged in real-life social network and moderately lonely. The study also finds a significantly positive relationship between social media (Facebook) use and loneliness, and a significantly negative relationship between real life social network and loneliness. This chapter calls for the students be careful in using social media and be engaged more in real life social network to avoid loneliness."

Chapter 13, titled "A Comparative Historical Analysis of the Evolution of Venture Capital Industry in the Economic Regions of United States of America, Europe and China with that of India" (by Som Sekhar Bibhash Laik, Divya Sharma and Tirthankar Bose) states that "venture capital (VC) provided a platform empowered individuals having financial constraints to transform their ideas into business models and attain commercial success. This article reviewed the growth and trends of VC industry across various regions such as United States of America (USA), Europe, China and India. Initially, VC firms had flourished and developed in USA and still it harbored the largest VC industry. From USA the VC firms had spread in Europe and then much later to emerging economies like China & India. Although VC ecosystem had started late in China it had registered higher growth as compared to Europe in terms of VC investment. China had become the second largest VC market. It was backed up by government initiatives, vast market opportunities and home-grown technology firm investments. India had started observing growth in VC space a bit later than China but had ample opportunities to register surge in VC activities."

Chapter 14, titled "Development and Access to Finance of Small and Medium-Sized Enterprises in Mongolia" (authors: Gan-Ochir Doojav, Davaajargal Luvsannyam, Bilguun Sukhbaatar, Bilguunzul Sodnomdarjaa, Tsolmon Otgonbat, Khuslen Batmunkh, Munkhbayar Gantumur and Elbegjargal Enkh-Amgalan) provides "a recent survey on development and access to finance of small and medium-size enterprises (SMEs) in Mongolia. The survey covers a sample of 1922 SMEs from Ulaanbaatar (capital city) and 21 provinces. We find that banks and local government administration are the most supportive institution for SME development. Political instability, corruption and labour supply, high lending rate, short maturity loans, service fees are perceived as the major obstacles that SMEs face in their business environments. Our results also suggest that SMEs in Mongolia are less likely to have access to external finance because of tight credit condition, potentially explaining the lack of SMEs' growth. Implementing country-specific reform strategy for SME development covering key building blocks is needed to promote SME financial inclusion and facilitate SMEs to contribute the economic growth."

Finally the last chapter of the book, Chapter 15, titled "The Role of Internet of Things, Knowledge Management and Open Innovation in SMEs Sustainability" (authors: Muhammad Imran, Jawad Iqbal and Hassan Mujtaba Nawaz Saleem) examines "the relationship between internet of things and knowledge management; knowledge management and open innovation; open innovation and SMEs sustainability. The relationship between the constructs developed and discuss on the behalf of past studies. The present chapter found that internet of things is playing the important role in knowledge generation and management, further, knowledge management is very important for open innovation environment in SMEs. Moreover, the open innovation sustains the SMEs performance. In respect of implications, the owner / managers of SMEs should consider the Internet of things, knowledge management, and open innovation capabilities during the decision making for SMEs sustainability. Moreover, this is a process framework which brings the effect of one variable to other variables. However, the future studies should empirically validate the proposed research framework."

After the summaries of the chapters included in the book, it is important to thank authors for their interest in participation in this book as well as the members of the Editorial Advisory Board and reviewers for the time and efforts they invested in choosing the best collection of chapters for this book. Thank you very much to all of you.

*October 2019*

*Patricia Ordóñez de Pablos*
*The University of Oviedo, Spain*

*Xi Zhang*
*Tianjin University, China*

*Kwok Tai Chui*
*The Open University of Hong Kong, Hong Kong*

## REFERENCES

Asia Population. (2019), World Population Review. Retrieved from http://worldpopulationreview.com/continents/asia-population/

# Chapter 1
# Study of Motives of Indian Strategic Alliances:
## Marketing or Innovation Seeking?

**Vihang Vivek Jumle**
*Thadomal Shahani Engineering College, India*

**Som Sekhar Bhattacharyya**
https://orcid.org/0000-0003-3259-0319
*National Institute of Industrial Engineering, India*

## ABSTRACT

*Strategic Alliances (SAs) have been formed by firms to attain strategic objectives which alone firms were unable to secure. Typically, firms formed SAs to gain market share or to seek innovation and research and development capabilities form the partnering firm. In the extant literature, market seeking SAs have been viewed as defensive while innovation and research and development capabilities seeking SAs as an aggressive strategy. In this research, the authors conducted a study of 165 strategic alliances cases in the Indian business scenario to study the motives of SAs. This study classified SAs formed by the top 50 firms listed on India's National Stock Exchange (NSE) (NIFTY50 index) into different categories. Inferences and findings have been drawn using content analysis of the available dataset. It was observed that, out of the 165 unique cases of SAs, a substantial number of SAs could be classified as 'business alliances' focusing on the market seeking from the alliance partner.*

## INTRODUCTION

A Strategic Alliance (SAs) is a business agreement between two or more firms brought about to pursue mutually important objectives (Hyder & Eriksson, 2005; Bhattacharyya, 2018). Firms have formed SAs to gain an advantage in the market which individual firms alone could not attain (Kogut, 1991). Hence, multiple firms have strategically collaborated on various aspect of their business to achieve objectives that all partner firms considered important for their growth (Padula & Dagnino, 2007; Bhattacharyya,

DOI: 10.4018/978-1-7998-1566-2.ch001

2019a). Generally, firms that have entered a SA, tended to maintain their independence (Hyder & Eriksson, 2005; Padula & Dagnino, 2007). SA could be national or international depending on the partner firms' business objectives (Bhattacharyya & Shaik, 2009). SAs could be equity-based, where equity is shared between the partner firms, or non-equity based (Bhattacharyya, 2018). The literature has classified SAs in different ways. Koza and Lewin (2000) classified SAs as Learning, Business & Hybrid based on their objectives. Kale and Singh (2009) classified SAs based on the strategic choices the partner firms made. It was noted by Chan, Kensinger, Keown & Martin, (1997) that SAs tended to add a good amount of value to the partner firms and it was observed that some firms were keen to learn the tacit knowledge capability resident in the other partner. Ironically, Das & Teng, (2000) observed that around 70% SAs failed to fulfil the set strategic objectives but despite this high rate of alliances failure, companies have kept on investing company resources and management bandwidth in forming new alliances (Hyder & Eriksson, 2005; Bhattacharyya, 2019a). Therefore, a lot of literature has revolved around justifying this failure rate. Koza & Lewin, (2000) proposed a mismatch in expectations as a reason for alliance failure. While Dyer, Kale & Singh, (2004) proposed forming alliances instead of acquisitions and vice-versa as a potential reason. The literature also discussed various methods that could be used to form successful SAs. Homberg & Cummings, (2009) provided literature on selecting appropriate partners. While Zeneldin & Bredonlow, (2003) provided literature on understanding partner synergies. Zeneldin & Bredonlow, (2003) and Bhattacharyya & Shaik, (2009) observed that similar studies concerning Indian SAs were limited in number. Given this dearth of studies in Indian alliances concerning the role of SAs, the authors were interested in empirically exploring this research domain. The grouped list of alliances was termed as a strategic alliance portfolio (Yamakawa, Yang & Lin, 2011) and the authors were interested in studying the changes in portfolios of Indian SAs with a change in dynamics of the firms' business objectives. The focus of this research was on Indian SAs. This study was centric to SAs that have been formed by the top 50 companies of the National Stock Exchange's 'Nifty 50' index. The purpose of this study was:

1. To study alliance portfolios of NIFTY 50 over a course of time (here 5 years, 2013 - 2017) based on the dynamics of firms' business objectives.
2. To distinctly classify the formed SAs into different types based on the literature provided by Hitt, Ireland & Hoskisson (1997) and Koza & Lewin, (2000).

## LITERATURE SURVEY

The literature review was used to develop a comprehensive understanding of SAs for this study. Discussions provided by the literature allowed the authors to form a contextual framework for their study. Out of the forty-six studies referred by the authors, some particular studies played a significant role in forming the basis of this research. The discussion provided by Koza & Lewin (2000) provided the authors with a clear picture of the issues in the management of SAs. Moreover, the framework provided by Koza & Lewin (2000) and Hitt, Ireland & Hoskisson (1997) aided the authors in similarly classifying Indian SAs. Dyer, Kale & Singh (2004) discussed biases in forming SAs which allowed the authors to better understand the motives behind SAs formed by Indian firms. Additionally, Dyer, Kale & Singh (2004) also discussed the factors that influenced SA decisions and provided literature on the difference between SAs and acquisitions. The literature provided by Lee, Tim & Tan (2000) on SAs between small and large enterprises helped the authors to identify strategic objectives behind such SAs in the Indian scenario.

Other studies that allowed the authors to expand their understanding of the topic include Hartmann (2018) that highlighted the importance of interface managers in SA, Robson et al. (2019) that demonstrated how SA develop partnership capabilities such that they complement each other, Li et al. (2019) that showed alliance behaviour in technology conglomerates, Lechler et al. (2019) that studied how firms collaborated within assessment sharing strategic alliances (ASSAs) to manage supplies, Ho et al. (2019) that explained why international strategic alliances work in some cases and fail in others, Elia et al. (2019) that showed how cultural differences arising out of SA caused innovation levels to suffer, and Klus et al. (2019) that particularly studies SA between banks and fintech firms.

A summarized version of the complete literature has been provided in table 1.

The authors also studied literature which was recently produced. A summarised version of all those studies mentioned below.

Hartmann (2018) suggested alliance contracts should touch upon a wide range of issues. A good amount of time should be given to understand the alliance partner. A specialist should be assigned to deal with the partnership, ideally. Strategic alliances should plan on an entry strategy and exit strategy beforehand. Li et al. (2019) showed that technology conglomerates tended to form more strategic alliances and were likely to fill for more patents than other industry alliances. These technology conglomerates shared a significant amount of their knowledge and had high cross-fertilisation of ideas. Lechler et al. (2019) conducted a multi-level case study and analyses to discover that information asymmetry and goal conflict could be mitigated by collaborating with ASSAs and hence multi-level suppliers' compliance with corporate sustainability standards was fostered. Ho et al. (2019) tried to discover the reason behind some firms' ability to partner with foreign firms. The authored studied knowledge ambiguity and the negative effects of knowledge ambiguity on knowledge acquisition in international alliances. Ho et al. (2019) proved that knowledge ambiguity does affect foreign SAs and provides a list of methods on how firms can work around it. Elia et. al (2019) studied how diversity affected strategic alliances and showed that innovativeness was muted in subsidiaries of multinational corporations because it involved the merger of cultures, one culture from the home locations and the other from the organization. The authors studied 161 SAs in the biotech industry to come to this conclusion. Robson et al. (2019) showed how SA tended to develop resources complimentary to each other and enhanced their partnership capabilities. Robson et al. (2019) also stated that trust had an inverted U shape relationship with performance and resource complementarity drove trust and performance. Finally, Klus et. al (2019) used exploratory research designs and conducted semi-structured interviews to fill the voids and throw light on the motivation of banks and fintech firms to form alliances. Klus et al. (2019) discovered that banks were interested in benefiting from rapid innovation without being a part of the development process whereas fintech firms demanded resources and the know-how to scale their business in the highly regulated financial industry.

## DATA COLLECTION AND ANALYSIS

The authors studied the NSE 50 index and listed all the 50 firms. The authors through a study of annual reports, media releases, website information of these 50 firms, shortlisted the alliances between the five years period of 2013-2017. The authors found that 33 firms of NIFTY50 firms had reported about forming SAs in their annual report. There was a total of 165 incidents of unique SAs. The authors then extracted the text data from the annual reports, media releases and website information and content analysed the

*Table 1. Analysis of the literature review*

| S.No. | Author(s) | Finding(s) |
|---|---|---|
| 1 | Bronder & Pritzi (1992) | The four phases of SAs: strategic decision for an alliance, alliance configuration, partner selection, alliance management, were discussed. |
| 2 | Chan, Kensinger, Keown & Martin (1997) | The production of positive wealth effect by SAs for combined partner firms with no evidence of wealth transfer between partners was stated. |
| | | Horizontal alliances added more value when the alliance involved the transfer and/or pooling of technical knowledge, compared with marketing alliances was stated. |
| | | It was stated that firms entering SAs exhibited superior operating performance compared to their industry peers. |
| 3 | Das & Teng (1998) | The two forms of risks: Relational & Performance were identified. |
| | | The four types of resources: Financial, Technical, Physical & Managerial were identified. |
| 4 | Koza & Lewin (2000) | Issues in management processes of SAs were discussed. A framework to aid in the selection of an alliance was presented. |
| | | Reasons that lead companies into entering SAs were discussed. |
| | | Failure to understand and articulate the strategic objective of the alliance was highlighted as the main reason for SA failures. |
| | | Failure to realize the relationship and uneven expectations by both parties were discussed as a reason that often led to alliance failures. |
| | | "Exploration or Exploitation" was stated as a reason to enter SA. |
| 5 | Fey & Beamish (2000) | Conflicts were an outcome of inter-party joint ventures was stated. |
| | | Inter-party alliances between similar firms were less likely to face conflicts compared to dissimilar firms was stated. |
| 6 | Kumar & Andersen (2000) | "Meaning and Legitimacy" as a concept was introduced. Cultural clashes between partnering management were explained. |
| | | Various meanings associated with the managerial domain were studied. Association of these meanings with a specific form of legitimacy was discussed. |
| | | Three forms of legitimacy: Pragmatic, Moral and Cognitive were stated. |
| 7 | Whipple & Frankel (2000) | The paper surveyed sources of character-based trust in strategic alliances. |
| 8 | Lee, Lim & Tan (2000) | The conditions & criteria for strategic alliances between small & medium enterprises and large corporations were discussed. |
| 9 | Bower (2001) | The paper discussed the paradox concerning the wide presence of M&A and failure of managers to pick the right form of M&A for their projects. |
| | | Five common reasons for M&A and suggested strategies that can be used by the management were discussed. |
| 10 | Pett & Dibrell (2001) | Stages of strategic alliances as Exploration, recurrent, relational, outcome were discussed. |
| 11 | Elmulti & Kathawala (2001) | Salient features of SAs were discussed. |
| 12 | Kauser & Shaw (2003) | International SAs and the impact of behavioural aspects on these SAs were discussed. |
| 13 | Cavusgil, Calantone & Zhao (2003) | Close and frequent interaction amongst participating firms led to the transfer of knowledge was stated. |
| 14 | Dyre, Kale & Singh (2004) | Managers' behaviour to frequently interchange alliances and acquisitions for their projects was discussed. |
| | | The paper stated that distinct strategies should be used for distinct situations. |
| | | The companies' biases towards forming alliances, when in fact acquisitions were needed, because it formed alliances in the past was discussed. |
| | | The factors that should influence the decision making when acquisitions and alliances were concerned were discussed. |
| 15 | Chung, Luo & Wagner (2006) | The literature on firm size, the formation of strategic alliances and knowledge industries was discussed. |
| | | SAs in large and small consulting firms were compared. It was discussed if small firms had an advantage over large firms. |
| | | It was concluded that most alliances between big and small firms were formed for knowledge transfer. |
| 16 | Butler (2007) | The problems which occurred for European defence manufacturers in cross-border SAs were discussed. |
| 17 | Chen & Wu (2008) | The mechanism for partner selection by using relative weights for set criteria in SAs was discussed. |
| 18 | Joia & Malheiros (2009) | The positive impact of strategic alliances on the intellectual capital of a company was discussed. |
| | | The correlation between intangible assets and strategic alliances was discussed. |
| 19 | Homberg & Cummings (2009) | Gaps in existing partner selection literature that discussed high alliance requirement and high failure rate observed due to improper partner selection, were filled. |

continued on following page

Table 1. Continued

| S.No. | Author(s) | Finding(s) |
|---|---|---|
| 20 | Kale & Singh (2009) | The paradox of needing more alliances and the high failure rate of these alliances was identified. |
| | | The usefulness of taking a portfolio approach to alliances was discussed. |
| 21 | Butler (2010) | The different methods adopted by the UK and non-UK defence industry managers for decision-making were discussed. |
| 22 | Cummings & Homberg (2012) | Gaps in existing partner selection literature that discussed high alliance requirement and high failure rate observed due to improper partner selection, were filled. |
| | | Four critical success factors: Learning-related; Task-related; Partner related; Risk factors were identified. |
| 23 | Tavallaei, Hosseinalipour & Mohebifar (2015) | The existing literature on critical success factors was discussed. |
| 24 | Margherita Russo1 & Maurizio Cesarani1 (2017) | The paper presented phases of alliances as formation, operational and evaluation phase |
| 25 | Islam, Hossain & Mia (2017) | Regression model to study causality between four different factors of SAs was discussed. |
| | | It was concluded that SAs improved product development and scarce resources can be better utilized by forming alliances. |
| | | The paper argued that strategic alliances tend to have a positive relationship with innovation but the correlation was relatively weak. |

text data (Weber, 1990). Thematic content analysis was done (Weber, 1990). Thematic content analysis (Hsieh & Shannon, 2005) helped the authors to seek answers to the research questions posed mainly:

1. What are the strategic objectives of the formation of SAs, i.e. R&D/Innovation seeking or Market seeking?
2. What were the classifications of SA as per the literature provided by Koza & Lewin (2000) and Hitt, Ireland & Hoskisson (1997)?

Table 2, presents the SAa data of National Stock Exchange 50 firms which contained data of 33 parent firms and year wise alliance partners.

The authors were interested in classifying the alliances formed by Indian firms into two categories: innovation-seeking (Narula & Hagedoorn, 1999; Kotabe & Scott 1995; Tian & Johnston, 2004; Deeds & Hill, 1996; Gerwin, 2004) and market seeking (Bucklin & Sengupta1993; Varadarajan & Cunningham, 1995; Dev, Klein & Fisher, 1996; Beverland & Bretherton, 2001). An alliance whose business objective was to seek innovation (such as enhancing business operations, development of a new product, etc), was termed as an innovation-seeking alliance (Gerwin & Ferris, 2004; Kotabe & Scott 1995; Tian & Johnston, 2004; Deeds & Hill, 1996; Gerwin, 2004). Whereas, an alliance whose business objective was to seek market growth (such as increase in revenue, increase in sales, increase in manufacture of product, etc) was termed as market seeking alliance (Simonin, 1999; Bucklin & Sengupta1993; Varadarajan & Cunningham, 1995; Dev, Klein & Fisher, 1996; Beverland & Bretherton, 2001; Park, Chen & Gallagher, 2002). This categorization has been useful in outlining the Indian SAs scenario. The authored have marked if a SA fell under the "innovation-seeking" or "market seeking" category. The authors have also mentioned the reason as to why a particular SA was classified of a particular type. In some cases, it was also observed that SAs had overlapping features basis which they have been classified under both categories. Some firms formed many SAs however there was very little information available in the public domain to conclude their categories. In such cases, the authored have left the cells blank with a hyphen.". The complete list of alliance and their classification can be found in the table -3. A remark column in the table provided an explanation behind these classifications as understood by the authors.

*Table 2. Strategic alliances data of National Stock Exchange 50 firms*

<table>
<tr><th rowspan="2">S.No.</th><th rowspan="2">Company</th><th colspan="5">Years</th></tr>
<tr><th>2017</th><th>2016</th><th>2015</th><th>2014</th><th>2013</th></tr>
<tr><td rowspan="2">1</td><td rowspan="2">ACC Ltd.</td><td colspan="3">OneIndia BSC Pvt. Ltd.</td><td colspan="2" rowspan="2"></td></tr>
<tr><td>Aakaash Manufacturing Company Pvt. Ltd.</td><td colspan="2"></td></tr>
<tr><td rowspan="4">2</td><td rowspan="4">Ambuja Cements Ltd.</td><td colspan="5">Wardha Vaalley Coal Field Pvt. Ltd. (Joint Operation), Counto Microfine Products Pvt. Ltd.</td></tr>
<tr><td colspan="2">Alcon Cement Company Pvt. Ltd. (Associate of Subsidiary), Asian Concretes and Cements Pvt. Ltd. (Associate of Subsidiary)</td><td colspan="3">Siam City Cement Public Company Ltd.</td></tr>
<tr><td colspan="2">MP AMRL (Semaria) Coal Company Ltd. (Joint Venture of Subsidiary), MP AMRL (Bicharpur) Coal Company Ltd. (Joint Venture of Subsidiary), MP AMRL (Marki Barka) Coal Company Ltd. (Joint Venture of Subsidiary), MP AMRL (Morga) Coal Company Ltd. (Joint Venture of Subsidiary)</td><td>OneIndia BSC Pvt. Ltd.</td><td></td><td rowspan="2"></td></tr>
<tr><td></td><td colspan="3">Aakaash Manufacturing Company Pvt. Ltd. (Associate of Subsidiary)</td></tr>
<tr><td rowspan="2">3</td><td rowspan="2">Ashok Leyland Ltd.</td><td colspan="5">Ashley Alteams India Ltd., Ashok Leyland John Deere Construction Equipment Company Pvt. Ltd., Automotive Infotronics Ltd., Nissan Ashok Leyland Powertrain Ltd., Nissan Ashok Leyland Technologies Ltd.</td></tr>
<tr><td>Ashok Leyland Nissan Vehicles Ltd., Hinduja Tech Ltd.</td><td colspan="3"></td><td>Ashok Leyland Nissan Vehicles Ltd.</td></tr>
<tr><td rowspan="2">4</td><td rowspan="2">Aurobindo Pharma Ltd.</td><td colspan="5">Novagen Pharma (Pty) Ltd., South Africa (Joint venture of a subsidiary)</td></tr>
<tr><td>Eugia Pharma Specialities Ltd., Tergene Biotech Pvt. Ltd., India (w.e.f. April 1 2015)</td><td colspan="2"></td><td colspan="2">Zao Auros Pharma, Russia (Joint venture of a subsidiary) (Closed during the year without any operations)</td></tr>
<tr><td>5</td><td>Bank of Baroda</td><td colspan="5">India First Life Insurance Company Ltd., India International Bank (Malaysia) Bhd., India Infradebt Ltd. Associates, Baroda Uttar Pradesh Gramin Bank, Baroda Rajasthan Kshetriya Gramin Bank, Baroda Gujarat Gramin Bank, Baroda Pioneer Asset Management Company Ltd., Indo Zambia Bank Ltd., Baroda Pioneer Trustee Company Pvt. Ltd.</td></tr>
<tr><td rowspan="3">6</td><td rowspan="3">Bharat Heavy Electricals Ltd.</td><td colspan="5">Powerplant Performance Improvement Ltd., BHEL-GE Gas Turbine Services Pvt. Ltd., NTPC-BHEL Power Projects Pvt. Ltd., Raichur Power Corporation Ltd., Dada Dhuniwale Khandwa Power Ltd.</td></tr>
<tr><td colspan="3"></td><td colspan="2">Latur Power Company Ltd.</td></tr>
<tr><td colspan="4"></td><td>Udangudi Power Corporation Ltd.</td></tr>
<tr><td rowspan="2">7</td><td rowspan="2">Cadila Healthcare Ltd.</td><td colspan="5">Zydus Hospira Oncology Pvt. Ltd., Bayer Zydus Pharma Pvt. Ltd., Zydus Takeda Healthcare Pvt. Ltd.</td></tr>
<tr><td colspan="3"></td><td colspan="2">Zydus BSV Pharma Pvt. Ltd.</td></tr>
<tr><td rowspan="3">8</td><td rowspan="3">Container Corporation of India Ltd.</td><td colspan="5">Star Track Terminals Pvt. Ltd., Albatross Inland Ports Pvt. Ltd., Gateway Terminals India Pvt. Ltd., Himalayan Terminals Pvt. Ltd. (Foreign Joint Venture), India Gateway Terminal Pvt. Ltd., Container Gateway Ltd., Allcargo Logistics Park Pvt. Ltd., CMA-CGM Logistics Park (Dadri) Pvt. Ltd., HALCON</td></tr>
<tr><td colspan="4">TCI-CONCOR Multimodal Solutions Pvt. Ltd. (formerly known as Infinite Logistics Solutions Pvt. Ltd.)</td><td rowspan="2">CONYK Cartrac Pvt. Ltd, Infinite Logistics Solutions Pvt. Ltd.</td></tr>
<tr><td colspan="3">Angul Sukinda Railway Ltd.</td><td></td></tr>
<tr><td rowspan="3">9</td><td rowspan="3">Cummins India Ltd.</td><td colspan="5">Cummins Research and Technology India Ltd. (CRTI), Valvoline Cummins Ltd. (VCL)</td></tr>
<tr><td colspan="2">Cummins Sales & Service Ltd.</td><td colspan="3">Cummins SVAM Sales & Service Ltd. (CSSSL)</td></tr>
<tr><td colspan="3">Cummins Generator Technologies India Pvt. Ltd. (CGT)</td><td colspan="2"></td></tr>
<tr><td rowspan="2">10</td><td rowspan="2">Dabur India Ltd.</td><td colspan="5">Forum 1 Aviation Ltd.</td></tr>
<tr><td colspan="3"></td><td>Dabon International Ltd</td><td></td></tr>
<tr><td>11</td><td>General Insurance Corporation of India</td><td>5th Reinsurance Company</td><td>4th Reinsurance Company</td><td>3rd Reinsurance Company</td><td>2nd Reinsurance Company</td><td>1st Reinsurance Company</td></tr>
<tr><td>12</td><td>Godrej Consumer Products Ltd.</td><td colspan="2">Godrej Easy IP Holdings (FZC) (Dubai)</td><td colspan="3"></td></tr>
<tr><td>13</td><td>Havells India Ltd.</td><td colspan="5">Jiangsu Havells Sylvania Lighting Co., Ltd.</td></tr>
<tr><td>14</td><td>Hindustan Zinc Ltd.</td><td colspan="5">Madanpur South Coal Company Ltd.</td></tr>
</table>

continued on following page

Table 2. Continued

<table>
<tr><th rowspan="2">S.No.</th><th rowspan="2">Company</th><th colspan="5">Years</th></tr>
<tr><th>2017</th><th>2016</th><th>2015</th><th>2014</th><th>2013</th></tr>
<tr><td>15</td><td>ICICI Prudential Life Insurance Company Ltd.</td><td colspan="5">ICICI Bank Ltd. and Prudential Corporation Holdings Ltd.</td></tr>
<tr><td>16</td><td>Idea Cellular Ltd.</td><td colspan="5">Indus Towers Ltd.</td></tr>
<tr><td rowspan="2">17</td><td rowspan="2">JSW Steel Ltd.</td><td colspan="3">Vijayanagar Minerals Pvt. Ltd., Rohne Coal Company Pvt. Ltd., JSW Severfield Structures Ltd., Gourangdih Coal Ltd., GEO Steel LLC, JSW Structural Metal Decking Ltd., JSW MI Steel Service Center Pvt. Ltd., JSW Vallabh Tinplate Pvt. Ltd.</td><td rowspan="2"></td><td rowspan="2">Vijayanagar Minerals Pvt. Ltd., Rohne Coal Company Pvt. Ltd., JSW Severfield Structures Ltd., Gourangdih Coal Ltd., JSW Structural Metal Decking Ltd., JSW MI Steel Service Center Pvt. Ltd., JSW Vallabh Tinplate Pvt. Ltd.</td></tr>
<tr><td>Accialtalia S.p.A.</td><td colspan="2">MJSJ Coal Ltd., Toshiba JSW Power System Pvt. Ltd.</td></tr>
<tr><td>18</td><td>L&T Finance Holdings Ltd.</td><td>Metro Tunnelling Group</td><td colspan="4"></td></tr>
<tr><td rowspan="2">19</td><td rowspan="2">Marico Ltd.</td><td colspan="2">Bellezimo Professionale Products Pvt. Ltd</td><td colspan="3"></td></tr>
<tr><td>Zed Lifestyle Pvt. Ltd</td><td colspan="4"></td></tr>
<tr><td rowspan="2">20</td><td rowspan="2">Motherson Sumi Systems Ltd.</td><td colspan="3">Kyungshin Industrial Motherson Ltd., Calsonic Kansei Motherson Auto Products Ltd., Eissmann SMP Automotive Interieur Slovensko s.r.o (through SMP Deutschland GmbHGlobal Environment Management (FZC) (held by MSSL Mauritius Holdings Ltd.) (Included Global Environment Management Australia Pty Ltd.</td><td rowspan="2">Sumitomo Wiring Systems Ltd., Japan Kyungshin Corporation, Korea Woco Franz Josef Wolf Holding GmbH, Germany Calsonic Kansei Corporation, Japan E-Compost Pty. Ltd., Australia Dremotech GmbH & Co. KG., Germany Cross Motorsport Systems AG Blanos Partners S.L. Changshu Automobile Interior Decoration Co., Ltd. Ningbo Huaxiang Electronic Co., Ltd. Eissmann Automotive Slovensko s.r.o.</td><td rowspan="2">Kyungshin Industrial Motherson Ltd. Calsonic Kansei Motherson Auto Products Ltd. Ningbo SMR Huaxiang Automotive Mirrors Co. Ltd. (through SMR) (Includes Chongqing SMR Huaxiang Automotive Products Ltd.) Celulosa Fabril S.A. (CEFA) (through SMP automotive technology Iberica S.L) (Including Modulos Ribera Alto S.LU.) Eissmann SMP Automotive Interieur Slovensko s.r.o (through SMP Deutschland GmbHGlobal Environment Management (FZC) (held by MSSL Mauritius Holdings Ltd.) (Included Global Environment Management Australia Pty Ltd.</td></tr>
<tr><td>Ningbo SMR Huaxiang Automotive Mirrors Co. Ltd. (through SMR) (Includes Chongqing SMR Huaxiang Automotive Products Ltd.), Celulosa Fabril S.A. (CEFA) (through SMP automotive technology Iberica S. L) (Including Modulos Ribera Alto S.LU.)</td><td colspan="2">Samvardhana Motherson Nippisun Technology Ltd., Celulosa Fabril (Cefa) Modulos Rivera Alta S.L.U.</td></tr>
<tr><td rowspan="3">21</td><td rowspan="3">NHPC Ltd.</td><td colspan="5">National High-Power Test Laboratory (P) Ltd</td></tr>
<tr><td></td><td colspan="4">National Power Exchange Ltd</td></tr>
<tr><td colspan="4">Chenab Valley Power Projects Pvt. Ltd</td><td></td></tr>
<tr><td rowspan="2">22</td><td rowspan="2">NMDC Ltd.</td><td colspan="5">Kopano-NMDC Minerals (Proprietary) Ltd., Johannesburg, South Africa</td></tr>
<tr><td></td><td>NMDC-CMDC Ltd., RAIPUR (CG) Jharkhand National Mineral Development Corporation Ltd. NMDC-SAIL Ltd., Bastar Railway Pvt. Ltd., Chhattisgarh Mining Ventures Ltd.</td><td colspan="3"></td></tr>
<tr><td>23</td><td>Oil India Ltd.</td><td></td><td colspan="3">Beas Rovuma Energy Mozambique Ltd Suntera Nigeria 205 Ltd.</td><td>Numaligarh Refinery Ltd. (NRL) Brahmaputra Cracker and Polymer Ltd. (BCPL) Suntera Nigeria 205 Ltd. Duliajan Numaligarh Pipeline (DNP) Ltd.</td></tr>
</table>

continued on following page

Table 2. Continued

<table>
<tr><th rowspan="2">S.No.</th><th rowspan="2">Company</th><th colspan="5">Years</th></tr>
<tr><th>2017</th><th>2016</th><th>2015</th><th>2014</th><th>2013</th></tr>
<tr><td rowspan="2">24</td><td rowspan="2">Petronet LNG Ltd.</td><td colspan="5">Adani Petronet (Dahej) Port Pvt. Ltd</td></tr>
<tr><td>India LNG Transport Co (No. 4) Pvt. Ltd (ILT4)</td><td colspan="4"></td></tr>
<tr><td>25</td><td>Pidilite Industries Ltd.</td><td colspan="2">Plus Call Technical Services L.L.C.</td><td colspan="2"></td><td>Building Envelope Systems India Ltd.</td></tr>
<tr><td rowspan="3">26</td><td rowspan="3">Piramal Enterprises Ltd.</td><td colspan="3">Shrilekha Financial Services</td><td></td><td></td></tr>
<tr><td></td><td colspan="4">Allergan India Pvt. Ltd.</td></tr>
<tr><td>Convergence Chemicals Pvt. Ltd.</td><td>Zebra Management Services Pvt. Ltd.</td><td>Novus Cloud Solutions Pvt. Ltd.</td><td colspan="2"></td></tr>
<tr><td>27</td><td>Power Finance Corporation Ltd.</td><td colspan="5">National Power Exchange Ltd. Energy Efficiency Services Ltd.</td></tr>
<tr><td>28</td><td>Punjab National Bank</td><td colspan="5">Everest Bank Ltd., Kathmandu, Nepal</td></tr>
<tr><td>29</td><td>Rural Electrification Corporation Ltd.</td><td colspan="5">Energy Efficiency Services Ltd.</td></tr>
<tr><td>30</td><td>SBI Life Insurance Company Ltd.</td><td colspan="5">BNP Paribas Cardif</td></tr>
<tr><td rowspan="5">31</td><td rowspan="5">Steel Authority of India Ltd.</td><td colspan="5">SAIL Bansal Service Centre Ltd., Mjunction Services Ltd., Bhilai Jaypee Cement Ltd., S&T Mining Company Pvt. Ltd., SAIL- SAIL Kobe Iron India Pvt. Ltd., TMTSAL SAIL JV Ltd., SAL SAIL JVC Ltd., N.E. Steel & Galvanising Pvt. Ltd., NMDC SAIL Ltd., Bastar Railway Pvt. Ltd.</td></tr>
<tr><td>NTPC-SAIL Power Company Ltd., Bokaro Power Supply Company Pvt. Ltd., SAIL & MOIL Ferro Alloys Pvt. Ltd., International Coal Ventures Pvt. Ltd., SAIL-SCI Shipping Pvt. Ltd., SAIL SCL Kerala Ltd., RITES Bengal Wagon Industry Pvt. Ltd., SAIL-Bengal Alloy Castings Pvt. Ltd., North Bengal Dolomite Ltd. Romelt-SAIL (India) Ltd.</td><td colspan="4">UEC-SAIL Information Technology Ltd.</td></tr>
<tr><td colspan="3">Prime Gold-SAIL JVC Ltd.</td><td colspan="2"></td></tr>
<tr><td colspan="4">VSL SAIL JVC Ltd., Abhinav-SAIL JVC Ltd.</td><td></td></tr>
<tr><td colspan="2"></td><td colspan="3">Bokaro Jaypee Cement Ltd.</td></tr>
<tr><td>32</td><td>Sun TV Network Ltd.</td><td>South Asia FM Ltd., Digital Radio (Mumbai) Broadcasting Ltd., Asia Radio Broadcast Pvt. Ltd., Pioneer Radio Training Services Pvt. Ltd., Digital Radio (Kolkata) Broadcasting Ltd., Digital Radio (Delhi) Broadcasting Ltd., Optimum Media Services Pvt. Ltd., South Asia Multimedia Pvt. Ltd.</td><td colspan="4"></td></tr>
</table>

*Table 3. Classification of alliances based on 'innovation seeking' and 'market seeking'*

| Company | Year(s) | Alliance | Innovation Seeking | Market Seeking | Remark |
|---|---|---|---|---|---|
| ACC Ltd. | 2015 - 2017 | OneIndia BSC Pvt. Ltd. | ✓ | ☒ | The alliance was formed to discover new ways of improving business operations. |
| | 2017 | Aakaash Manufacturing Company Pvt. Ltd. | ☒ | ✓ | ALCON provided the alliance partner with grinding, packaging and production facility in Surat and Goa. |
| Ambuja Cements Ltd. | 2013 - 2017 | Wardha Valley Coal Field Pvt. Ltd. (Joint Operation) | ☒ | ✓ | The alliance produced peat for the partners. |
| | | Counto Microfine Products Pvt. Ltd. | ☒ | ✓ | ALCON provided the alliance partner with grinding, packaging and production facility in Surat and Goa. |
| | 2016 - 2017 | MP AMRL (Semaria) Coal Company Ltd. (Joint Venture of Subsidiary) | - | | - |
| | | MP AMRL (Bicharpur) Coal Company Ltd. (Joint Venture of Subsidiary) | | | |
| | | MP AMRL (Marki Barka) Coal Company Ltd. (Joint Venture of Subsidiary) | | | |
| | | MP AMRL (Morga) Coal Company Ltd. (Joint Venture of Subsidiary) | | | |
| | 2013 - 2015 | Siam City Cement Public Company Ltd., Thailand (JV of fellow Subs) | ☒ | ✓ | The alliance was formed to pool services and production facilities between the partners. |
| | 2015 | OneIndia BSC Pvt. Ltd. | ✓ | ☒ | The alliance was formed to discover new ways of improving business operations. |
| | 2015 - 2017 | Alcon Cement Company Pvt. Ltd. (Associate of Subsidiary) | - | | - |
| | | Asian Concretes and Cements Pvt. Ltd. (Associate of Subsidiary) | | | |
| Ashok Leyland Ltd. | 2013 - 2017 | Ashley Alteams India Ltd. | ☒ | ✓ | The alliance was a foundry and it manufactured light cast metal for Ashok Leyland Ltd. |
| | | Ashok Leyland John Deere Construction Equipment Company Pvt. Ltd. | ✓ | ✓ | John Deere manufactured land moving vehicles and it wanted to expand to Indian markets. The alliance was formed to provide John Deere with market intelligence for its product. |
| | | Automotive Infotronics Ltd. | ☒ | ✓ | The alliance manufactured general-purpose machinery. |
| | | Nissan Ashok Leyland Powertrain Ltd. | ✓ | ✓ | Nissan manufactured lightweight commercial vehicles and it wanted to expand to Indian markets. The alliance was formed to provide John Deere with market intelligence for its product. |
| | | Nissan Ashok Leyland Technologies Ltd. | ✓ | ✓ | |
| | 2013, 2017 | Ashok Leyland Nissan Vehicles Ltd. | ✓ | ✓ | |
| | 2017 | Hinduja Tech Ltd. | ☒ | ✓ | The alliance was formed to provided digital solutions to the partners. |
| Aurobindo Pharma Ltd. | 2013 - 2017 | Novagen Pharma (Pty) Ltd., South Africa (Joint venture of a subsidiary) | ✓ | ✓ | The alliance helped Aurobindo Pharma Ltd. sell its product in South Africa. |
| | 2017 | Eugia Pharma Specialities Ltd. | - | | |
| | | Tergene Biotech Pvt. Ltd., India | ✓ | ☒ | The alliance provided R&D facility and expertise to the partners. |
| | 2013 - 2014 | Zao Auros Pharma, Russia (Joint venture of a subsidiary) (Closed during the year without any operations) | - | | |

continued on following page

Table 3. Continued

| Company | Year(s) | Alliance | Innovation Seeking | Market Seeking | Remark |
|---|---|---|---|---|---|
| Bank of Baroda | 2013 - 2017 | India First Life Insurance Company Ltd. | x | ✓ | Large partners collaborated to form an alliance and provided one insurance service (youngest in India) to be a dominant market player. |
| | | India International Bank (Malaysia) Bhd. | x | ✓ | Large partners collaborated to provide non-primary banking services (trade financing, forex, etc) to be a dominant market player. |
| | | India Infradebt Ltd. Associates | x | ✓ | Large partners collaborated to provide infrastructure financing services to be a dominant market player. |
| | | Baroda Uttar Pradesh Gramin Bank | - | | |
| | | Baroda Rajasthan Kshetriya Gramin Bank | | | |
| | | Baroda Gujarat Gramin Bank | | | |
| | | Baroda Pioneer Asset Management Company Ltd. | | | |
| | | Indo Zambia Bank Ltd. | | | |
| | | Baroda Pioneer Trustee Company Pvt. Ltd. | | | |
| Bharat Heavy Electricals Ltd. | 2013 - 2017 | Powerplant Performance Improvement Ltd. | ✓ | x | The alliance was formed to improve the performance of power plants. |
| | | BHEL-GE Gas Turbine Services Pvt. Ltd. | x | ✓ | The alliance provided servicing for gas turbines. |
| | | NTPC-BHEL Power Projects Pvt. Ltd. | ✓ | ✓ | The alliance acted as an administrator for the partner's work. The alliance performed all activities from project planning to post-deployment services. |
| | | Raichur Power Corporation Ltd. | ✓ | ✓ | The alliance was formed to pool resources in order to set up a thermal power plant. |
| | | Dada Dhuniwale Khandwa Power Ltd. | - | | |
| | 2013 - 2014 | Latur Power Company Ltd. | ✓ | ✓ | The alliance was formed to pool resources in order to set up a thermal power plant. |
| | 2013 | Udangudi Power Corporation Ltd. | - | | |
| Cadila Healthcare Ltd. | 2013 - 2017 | Zydus Hospira Oncology Pvt. Ltd. | ✓ | x | The alliance provided a platform for partners to collectively research on tumour fighting medicines. |
| | | Bayer Zydus Pharma Pvt. Ltd. | x | ✓ | The alliance was formed to share partners' expertise on their existing service. |
| | | Zydus Takeda Healthcare Pvt. Ltd. | ✓ | ✓ | The alliance was formed to provide a partner with R&D and production facility. |
| | 2013- 2014 | Zydus BSV Pharma Pvt. Ltd. | - | | |

continued on following page

Table 3. Continued

| Company | Year(s) | Alliance | Innovation Seeking | Market Seeking | Remark |
|---|---|---|---|---|---|
| Container Corporation of India Ltd. | 2013 - 2017 | Star Track Terminals Pvt. Ltd. | ✓ | x | The alliance performed CSR activities for one partner. |
| | | Albatross Inland Ports Pvt. Ltd. | x | ✓ | The alliance aided in operations of the services provided by Container Corporation of India Ltd. |
| | | Gateway Terminals India Pvt. Ltd. | x | ✓ | |
| | | Himalayan Terminals Pvt. Ltd. (Foreign Joint Venture) | - | | |
| | | India Gateway Terminal Pvt. Ltd. | x | ✓ | The alliance aided in operations of the services provided by Container Corporation of India Ltd. |
| | | Container Gateway Ltd. | x | ✓ | |
| | | Allcargo Logistics Park Pvt. Ltd. | x | ✓ | |
| | | CMA-CGM Logistics Park (Dadri) Pvt. Ltd. | x | ✓ | |
| | | HALCON | x | ✓ | |
| | 2014 - 2017 | TCI-CONCOR Multimodal Solutions Pvt. Ltd. | x | ✓ | |
| | 2015 - 2017 | Angul Sukinda Railway Ltd. | x | ✓ | |
| | 2013 | CONYK Cartrac Pvt. Ltd, | x | ✓ | |
| | | Infinite Logistics Solutions Pvt. Ltd. | x | ✓ | |
| Cummins India Ltd. | 2013 - 2017 | Cummins Research and Technology India Ltd. (CRTI) | ✓ | x | The alliance was formed to facilitate R & D initiatives. |
| | | Valvoline Cummins Ltd. (VCL) | x | ✓ | The alliance was formed to manufacture a product using parents' resources. |
| | 2016 - 2017 | Cummins Sales & Service Ltd. | x | ✓ | The alliance was formed to provide services using partners' resources. |
| | 2015 - 2017 | Cummins Generator Technologies India Pvt. Ltd. (CGT) | x | ✓ | |
| | 2013 - 2015 | Cummins SVAM Sales & Service Ltd. (CSSSL) | - | | |
| Dabur India Ltd. | 2013 - 2017 | Forum 1 Aviation Ltd. | | | |
| | 2014 | Dabon International Ltd | ✓ | ✓ | The alliance was formed to increase Dabur's market presence. |
| General Insurance Corporation of India | 2017 | 5th Reinsurance Company | - | | |
| | 2016 | 4th Reinsurance Company | | | |
| | 2015 | 3rd Reinsurance Company | | | |
| | 2014 | 2nd Reinsurance Company | | | |
| | 2013 | 1st Reinsurance Company | | | |
| Godrej Consumer Products Ltd. | 2016 - 2017 | Godrej Easy IP Holdings (FZC) (Dubai) | - | | |
| Havells India Ltd. | 2013 - 2017 | Jiangsu Havells Sylvania Lighting Co. Ltd. | x | ✓ | The alliance was formed to collectively produce for product units. |
| Hindustan Zinc Ltd. | 2013 - 2017 | Madanpur South Coal Company Ltd. | x | ✓ | The alliance was formed to increase product market reach. |
| ICICI Prudential Life Insurance Company Ltd. | 2013 - 2017 | ICICI Bank Ltd. and Prudential Corporation Holdings Ltd. (Itself a JV) | x | ✓ | The alliance was formed to increase product market reach. |
| Idea Cellular Ltd. | 2013 - 2017 | Indus Towers Ltd. | x | ✓ | The alliance was formed to provide Idea Cellular Ltd. with resources. |

continued on following page

Table 3. Continued

| Company | Year(s) | Alliance | Innovation Seeking | Market Seeking | Remark |
|---|---|---|---|---|---|
| JSW Steel Ltd. | 2013, 2015-2017 | Vijayanagar Minerals Pvt. Ltd. | x | ✓ | JSW had metal production facility while VMPL had ore extraction facility. The alliance was formed to pool resources. |
| | | Rohne Coal Company Pvt. Ltd. | x | ✓ | JSW had a metal production facility, while VMPL had coal extraction facility. The alliance was formed to pool resources. |
| | | JSW Severfield Structures Ltd. | x | ✓ | JSW had a metal production facility, while JSWSSL has a metal fabrication facility. The alliance was formed to pool resources. |
| | | Gourangdih Coal Ltd. | x | ✓ | JSW had a metal production facility, while VMPL had coal extraction facility. The alliance was formed to pool resources. |
| | | GEO Steel L.L.C. | x | ✓ | JSW had a metal production facility, while VMPL had metal decking facility. The alliance was formed to pool resources. |
| | | JSW Structural Metal Decking Ltd. | x | ✓ | JSW had a metal production facility, while VMPL had metal decking facility. The alliance was formed to pool resources. |
| | | JSW MI Steel Service Center Pvt. Ltd. | x | ✓ | JSW had a metal production facility, while VMPL had metal distribution facility. The alliance was formed to pool resources. |
| | | JSW Vallabh Tinplate Pvt. Ltd. | x | ✓ | JSW had a metal production facility, while VMPL has a metal galvanizing facility. The alliance was formed to pool resources. |
| | 2017 | AcciaItalia S.p. A | - | | |
| | 2015 - 2016 | MJSJ Coal Ltd. | | | |
| | | Toshiba JSW Power System Pvt. Ltd. | | | |
| L&T Finance Holdings Ltd. | 2017 | Metro Tunnelling Group | ✓ | ✓ | The alliance was formed by five partners to create a novel metro service. |
| Marico Ltd. | 2016 - 2017 | Bellezimo Professionale Products Pvt. Ltd. | x | ✓ | The alliance was formed to provide resources to the partner and improve salon distribution channel. |
| | 2017 | Zed Lifestyle Pvt. Ltd. | x | ✓ | |
| NHPC Ltd. | 2013 - 2017 | National High-Power Test Laboratory (P) Ltd. | ✓ | x | The alliance was formed by 5 partners to establish a testing facility. |
| | 2013 - 2016 | National Power Exchange Ltd. | - | | |
| | 2014 - 2017 | Chenab Valley Power Projects Pvt. Ltd. | x | ✓ | The alliance was formed to collectively execute hydroelectric products. |
| NMDC Ltd. | 2013 - 2017 | Kopano-NMDC Minerals (Proprietary) Ltd., Johannesburg, South Africa | x | ✓ | The alliance was formed to provide coal to NMDC Ltd. |
| | 2016 | NMDC-CMDC Ltd., Raipur (CG) | x | ✓ | The alliance was formed to allow partners to collectively manufacture the product. |
| | | Jharkhand National Mineral Development Corporation Ltd. | x | ✓ | The alliance facilitated the mining operations. |
| | | RANCHI NMDC-SAIL Ltd., Hyderabad | x | ✓ | The alliance facilitated the casting of metals. |
| | | Bastar Railway Pvt. Ltd., Raipur | x | ✓ | The alliance allowed the partners to collectively lay railway tracks. |
| | | Chhattisgarh Mining Ventures Ltd., Chhattisgarh | | | |

continued on following page

Table 3. Continued

| Company | Year(s) | Alliance | Innovation Seeking | Market Seeking | Remark |
|---|---|---|---|---|---|
| Oil India Ltd. | 2014 - 2016 | Beas Rovuma Energy Mozambique Ltd. | x | ✓ | The alliance set up a gas field. |
| | 2013 | Numaligarh Refinery Ltd. (NRL) | | | |
| | | Brahmaputra Cracker and Polymer Ltd. (BCPL) | x | ✓ | The alliance provided a petrochemical complex. |
| | | Suntera Nigeria 205 Ltd. | - | | |
| | | Duliajan Numaligarh Pipeline (DNP) Ltd. | x | ✓ | The alliance provided pipeline complex. |
| Petronet LNG Ltd. | 2013 - 2017 | Adani Petronet (Dahej) Port Pvt. Ltd. | x | ✓ | The alliance allowed the partners to collectively manufacture products. |
| | 2017 | India LNG Transport Co (No 4) Pvt. Ltd (ILT4) | - | | |
| Pidilite Industries Ltd. | 2016 - 2017 | Plus Call Technical Services L.L.C. | | | |
| | 2013 | Building Envelope Systems India Ltd. | | | |
| Piramal Enterprises Ltd. | 2015 - 2017 | Shrilekha Financial Services | ✓ | ✓ | The alliance administered financial services for Piramal. |
| | 2013 - 2016 | Allergan India Pvt. Ltd. | ✓ | ✓ | The alliance administered pharmaceutical business for Piramal. |
| | 2017 | Convergence Chemicals Pvt. Ltd. | ✓ | ✓ | The alliance administered chemical manufacturing for Piramal. |
| | 2016 | Zebra Management Services Pvt. Ltd. | ✓ | ✓ | The alliance administered tax advisory services for Piramal. |
| | 2015 | Novus Cloud Solutions Pvt. Ltd. | ✓ | ✓ | The alliance administered IT services for Piramal. |
| Power Finance Corporation Ltd. | 2013 - 2017 | National Power Exchange Ltd. Energy Efficiency Services Ltd. | x | ✓ | The partners financed the alliance's product. |
| Punjab National Bank | 2013 - 2017 | Everest Bank Ltd., Kathmandu, Nepal | x | ✓ | The alliance allowed the partners to conduct international business. |
| Rural Electrification Corporation Ltd. | 2013 - 2017 | Energy Efficiency Services Ltd. | x | ✓ | The partners financed the alliance's product. |
| SBI Life Insurance Company Ltd. | 2013 - 2017 | BNP Paribas Cardif | x | ✓ | The alliance allowed the partners to conduct international business. |

continued on following page

Table 3. Continued

| Company | Year(s) | Alliance | Innovation Seeking | Market Seeking | Remark |
|---|---|---|---|---|---|
| Steel Authority of India Ltd. | 2013 - 2017 | SAIL Bansal Service Centre Ltd. | x | ✓ | The alliance allowed the partners to collectively cast metal. |
| | | Mjunction Services Ltd. | x | ✓ | The alliance increased market outreach. |
| | | Bhilai Jaypee Cement Ltd. | x | ✓ | The alliance allowed the partners to expand the business into the cement industry. |
| | | S&T Mining Company Pvt. Ltd. | x | ✓ | The alliance allowed the partners to collectively mine for metal. |
| | | SAIL- SAIL Kobe Iron India Pvt. Ltd. | - | | |
| | | TMTSAL SAIL JV Ltd. | | | |
| | | SAL SAIL JVC Ltd. | | | |
| | | N.E. Steel & Galvanising Pvt. Ltd. | | | |
| | | NMDC SAIL Ltd., Bastar Railway Pvt. Ltd. | | | |
| | 2017 | NTPC-SAIL Power Company Ltd. | x | ✓ | The alliance allowed the partners to collectively administer thermal plants. |
| | | Bokaro Power Supply Company Pvt. Ltd. | x | ✓ | |
| | | SAIL & MOIL Ferro Alloys Pvt. Ltd. | x | ✓ | |
| | | International Coal Ventures Pvt. Ltd. | x | ✓ | |
| | | SAIL-SCI Shipping Pvt. Ltd. | x | ✓ | |
| | | SAIL SCL Kerala Ltd. | x | ✓ | |
| | | RITES Bengal Wagon Industry Pvt. Ltd. | - | | |
| | | SAIL-Bengal Alloy Castings Pvt. Ltd. | | | |
| | | North Bengal Dolomite Ltd. Romelt-SAIL (India) Ltd. | | | |
| | 2013 - 2016 | UEC-SAIL Information Technology Ltd. | - | | |
| | 2014 - 2017 | VSL SAIL JVC Ltd. | x | ✓ | The alliance allowed the partners to collectively manufacture iron and steel. |
| | | , Abhinav-SAIL JVC Ltd. | - | | |
| | 2013 - 2015 | Bokaro Jaypee Cement Ltd. | | | |
| Sun TV Network Ltd. | 2017 | South Asia FM Ltd. | | | |
| | | Digital Radio (Mumbai) Broadcasting Ltd. | x | ✓ | The alliance was formed to expand market reach. |
| | | Asia Radio Broadcast Pvt. Ltd. | x | ✓ | |
| | | Pioneer Radio Training Services Pvt. Ltd. | x | ✓ | |
| | | Digital Radio (Kolkata) Broadcasting Ltd., | x | ✓ | |
| | | Digital Radio (Delhi) Broadcasting Ltd. | x | ✓ | |
| | | Optimum Media Services Pvt. Ltd. | x | ✓ | |
| | | South Asia Multimedia Pvt. Ltd. | x | ✓ | |

*Table 4. Classification of SAs based on Hitt, Ireland & Hoskisson (1997) & Koza & Lewin (2000)*

| Company | Year(s) | Alliance | Market Cycle | Potential Main Reason for Strategic Alliance by Market Type | Relationship between partners | Control Mechanism |
|---|---|---|---|---|---|---|
| ACC Ltd. | 2015 - 2017 | OneIndia BSC Pvt. Ltd. | Fast Cycle | Maintain market leadership | Learning Alliance | Behaviour and Process |
| | 2017 | Aakaash Manufacturing Company Pvt. Ltd. | Standard Cycle | Pool resources for very large capital projects | Business Alliance | Output |
| Ambuja Cements Ltd. | 2013 - 2017 | Wardha Vaalley Coal Field Pvt. Ltd. (Joint Operation) | Lack of data for processing information | | | |
| | | Counto Microfine Products Pvt. Ltd. | Standard Cycle | Pool resources for very large capital projects | Business Alliance | Output |
| | 2016 - 2017 | MP AMRL (Semaria) Coal Company Ltd. (Joint Venture of Subsidiary) | Lack of data for processing information | | | |
| | | MP AMRL (Bicharpur) Coal Company Ltd. (Joint Venture of Subsidiary) | | | | |
| | | MP AMRL (Marki Barka) Coal Company Ltd. (Joint Venture of Subsidiary) | | | | |
| | | MP AMRL (Morga) Coal Company Ltd. (Joint Venture of Subsidiary) | | | | |
| | 2013 - 2015 | Siam City Cement Public Company Ltd., Thailand (JV of fellow Subs) | Standard Cycle | Pool resources for very large capital projects | Business Alliance | Output |
| | 2015 | OneIndia BSC Pvt. Ltd. | Fast Cycle | Maintain market leadership | Learning Alliance | Behaviour and Process |
| | 2015 - 2017 | Alcon Cement Company Pvt. Ltd. (Associate of Subsidiary) | Lack of data for processing information | | | |
| | | Asian Concretes and Cements Pvt. Ltd. (Associate of Subsidiary) | | | | |
| Ashok Leyland Ltd. | 2013 - 2017 | Ashley Alteams India Ltd. | Standard Cycle | Pool resources for very large capital projects | Business Alliance | Output |
| | | Ashok Leyland John Deere Construction Equipment Company Pvt. Ltd. | Slow Cycle | Gain access to restricted market | Hybrid Alliance | Behaviour and Process & Output |
| | | Automotive Infotronics Ltd. | Lack of data for processing information | | | |
| | | Nissan Ashok Leyland Powertrain Ltd. | Slow Cycle | Gain access to restricted market | Hybrid Alliance | Behaviour and Process & Output |
| | | Nissan Ashok Leyland Technologies Ltd. | | | | |
| | 2013, 2017 | Ashok Leyland Nissan Vehicles Ltd. | Slow Cycle | | | |
| | 2017 | Hinduja Tech Ltd. | Standard Cycle | Gain access to complementary resources | Business Alliance | Output |
| Aurobindo Pharma Ltd. | 2013 - 2017 | Novagen Pharma (Pty) Ltd., South Africa (Joint venture of a subsidiary) | Slow Cycle | Gain access to restricted market | | |
| | 2017 | Eugia Pharma Specialities Ltd. | Lack of data for processing information | | | |
| | | Tergene Biotech Pvt. Ltd., India | Standard Cycle | Gain access to complementary resources | Learning Alliance | Behaviour and Process |
| | 2013 - 2014 | Zao Auros Pharma, Russia (Joint venture of a subsidiary) (Closed during the year without any operations) | Lack of data for processing information | | | |

continued on following page

Table 4. Continued

| Company | Year(s) | Alliance | Market Cycle | Potential Main Reason for Strategic Alliance by Market Type | Relationship between partners | Control Mechanism |
|---|---|---|---|---|---|---|
| Bank of Baroda | 2013 - 2017 | India First Life Insurance Company Ltd. | Fast Cycle | Speed up new goods or service entry | Business Alliance | Output |
| | | India International Bank (Malaysia) Bhd. | | | | |
| | | India Infradebt Ltd. Associates | | | | |
| | | Baroda Uttar Pradesh Gramin Bank | Lack of data for processing information | | | |
| | | Baroda Rajasthan Kshetriya Gramin Bank | | | | |
| | | Baroda Gujarat Gramin Bank | | | | |
| | | Baroda Pioneer Asset Management Company Ltd. | | | | |
| | | Indo Zambia Bank Ltd. | | | | |
| | | Baroda Pioneer Trustee Company Pvt. Ltd. | | | | |
| Bharat Heavy Electricals Ltd. | 2013 - 2017 | Powerplant Performance Improvement Ltd. | Fast Cycle | Gain access to complementary resources | Business Alliance | Output |
| | | BHEL-GE Gas Turbine Services Pvt. Ltd. | | Speed up new goods or service entry | | |
| | | NTPC-BHEL Power Projects Pvt. Ltd. | Standard Cycle | Pool resources for very large capital projects | Hybrid Alliance | Behaviour and Process & Output |
| | | Raichur Power Corporation Ltd. | | | | |
| | | Dada Dhuniwale Khandwa Power Ltd. | Lack of data for processing information | | | |
| | 2013 - 2014 | Latur Power Company Ltd. | Standard Cycle | Pool resources for very large capital projects | Hybrid Alliance | Behaviour and Process & Output |
| | 2013 | Udangudi Power Corporation Ltd. | Lack of data for processing information | | | |
| Cadila Healthcare Ltd. | 2013 - 2017 | Zydus Hospira Oncology Pvt. Ltd. | Standard Cycle | Gain access to complementary resources | Hybrid Alliance | Behaviour and Process & Output |
| | | Bayer Zydus Pharma Pvt. Ltd. | | Meet competitive challenge by other competitors | Business Alliance | Output |
| | | Zydus Takeda Healthcare Pvt. Ltd. | Standard Cycle/Fast Cycle | Gain access to complementary resources/ Share risky R&D expenses | Hybrid Alliance | Behaviour and Process & Output |
| | 2013- 2014 | Zydus BSV Pharma Pvt. Ltd. | Lack of data for processing information | | | |

continued on following page

Table 4. Continued

| Company | Year(s) | Alliance | Market Cycle | Potential Main Reason for Strategic Alliance by Market Type | Relationship between partners | Control Mechanism |
|---|---|---|---|---|---|---|
| Container Corporation of India Ltd. | 2013 - 2017 | Star Track Terminals Pvt. Ltd. | Standard Cycle | Learn new business techniques | Learning Alliance | Behaviour and Process |
| | | Albatross Inland Ports Pvt. Ltd. | Fast Cycle/Standard Cycle | Maintain market leadership/ Overcome trade barriers | Business Alliance | Output |
| | | Gateway Terminals India Pvt. Ltd. | | | | |
| | | Himalayan Terminals Pvt. Ltd. (Foreign Joint Venture) | Lack of data for processing information | | | |
| | | India Gateway Terminal Pvt. Ltd. | Fast Cycle/Standard Cycle | Maintain market leadership/ Overcome trade barriers | Business Alliance | Output |
| | | Container Gateway Ltd. | | | | |
| | | Allcargo Logistics Park Pvt. Ltd. | | | | |
| | | CMA-CGM Logistics Park (Dadri) Pvt. Ltd. | | | | |
| | | HALCON | | | | |
| | 2014 - 2017 | TCI-CONCOR Multimodal Solutions Pvt. Ltd. | | | | |
| | 2015 - 2017 | Angul Sukinda Railway Ltd. | | | | |
| | 2013 | CONYK Cartrac Pvt. Ltd. | | | | |
| | | Infinite Logistics Solutions Pvt. Ltd. | | | | |
| Cummins India Ltd. | 2013 - 2017 | Cummins Research and Technology India Ltd. (CRTI) | Fast Cycle | Share risky R&D expenses | Learning Alliance | Behaviour and Process |
| | | Valvoline Cummins Ltd. (VCL) | Fast Cycle | Speed up new goods or service entry | Business Alliance | Output |
| | 2016 - 2017 | Cummins Sales & Service Ltd. | | | | |
| | 2015 - 2017 | Cummins Generator Technologies India Pvt. Ltd. (CGT) | | | | |
| | 2013 - 2015 | Cummins SVAM Sales & Service Ltd. (CSSSL) | | | | |
| Dabur India Ltd. | 2013 - 2017 | Forum 1 Aviation Ltd. | Standard Cycle | Gain access to complementary resources | Business Alliance | Output |
| | 2014 | Dabon International Ltd | Slow Cycle | Establish a franchise in a new market | Hybrid Alliance | Behaviour and Process & Output |
| General Insurance Corporation of India | 2017 | 5th Reinsurance Company | Lack of data for processing information | | | |
| | 2016 | 4th Reinsurance Company | | | | |
| | 2015 | 3rd Reinsurance Company | | | | |
| | 2014 | 2nd Reinsurance Company | | | | |
| | 2013 | 1st Reinsurance Company | | | | |
| Godrej Consumer Products Ltd. | 2016 - 2017 | Godrej Easy IP Holdings (FZC) (Dubai) | | | | |
| Havells India Ltd. | 2013 - 2017 | Jiangsu Havells Sylvania Lighting Co. Ltd. | Standard Cycle | Meet competitive challenge by other competitors | Business Alliance | Output |
| Hindustan Zinc Ltd. | 2013 - 2017 | Madanpur South Coal Company Ltd. | Slow Cycle | Gain access to a restricted market | Business Alliance | Output |
| ICICI Prudential Life Insurance Company Ltd. | 2013 - 2017 | ICICI Bank Ltd. and Prudential Corporation Holdings Ltd. (Itself a JV) | Lack of data for processing information | | | |

continued on following page

Table 4. Continued

| Company | Year(s) | Alliance | Market Cycle | Potential Main Reason for Strategic Alliance by Market Type | Relationship between partners | Control Mechanism |
|---|---|---|---|---|---|---|
| Idea Cellular Ltd. | 2013 - 2017 | Indus Towers Ltd. | Standard Cycle | Pool resources for very large capital projects | Business Alliance | Output |
| JSW Steel Ltd. | 2013, 2015-2017 | Vijayanagar Minerals Pvt. Ltd. | | | | |
| | | Rohne Coal Company Pvt. Ltd. | | | | |
| | | JSW Severfield Structures Ltd. | | | | |
| | | Gourangdih Coal Ltd. | | | | |
| | | GEO Steel L.L.C. | | | | |
| | | JSW Structural Metal Decking Ltd. | | | | |
| | | JSW MI Steel Service Center Pvt. Ltd. | | | | |
| | | JSW Vallabh Tinplate Pvt. Ltd. | | | | |
| | 2017 | AcciaItalia S.p.A | Lack of data for processing information | | | |
| | 2015 - 2016 | MJSJ Coal Ltd. | | | | |
| | | Toshiba JSW Power System Pvt. Ltd. | | | | |
| L&T Finance Holdings Ltd. | 2017 | Metro Tunnelling Group | Standard Cycle | Gain access to complementary resources | Hybrid Alliance | Behaviour and Process & Output |
| Marico Ltd. | 2016 - 2017 | Bellezimo Professionale Products Pvt. Ltd. | Standard Cycle | Gain access to complementary resources | Business Alliance | Output |
| | 2017 | Zed Lifestyle Pvt. Ltd. | | | | |
| NHPC Ltd. | 2013 - 2017 | National High-Power Test Laboratory (P) Ltd. | Fast Cycle | Share risky R&D expenses | Learning Alliance | Behaviour and Process |
| | 2013 - 2016 | National Power Exchange Ltd. | Lack of data for processing information | | | |
| | 2014 - 2017 | Chenab Valley Power Projects Pvt. Ltd. | Standard Cycle | Pool resources for very large capital projects | Business Alliance | Output |
| NMDC Ltd. | 2013 - 2017 | Kopano-NMDC Minerals (Proprietary) Ltd., Johannesburg, South Africa | | | | |
| | 2016 | NMDC-CMDC Ltd. (CG) | Fast Cycle | Maintain market leadership | | |
| | | Jharkhand National Mineral Development Corporation Ltd. | Standard Cycle | Pool resources for very large capital projects | | |
| | | NMDC-SAIL Ltd. | | | | |
| | | Bastar Railway Pvt. Ltd. | | Learn new business techniques | | |
| | | Chhattisgarh Mining Ventures Ltd. | Lack of data for processing information | | | |

continued on following page

Table 4. Continued

| Company | Year(s) | Alliance | Market Cycle | Potential Main Reason for Strategic Alliance by Market Type | Relationship between partners | Control Mechanism |
|---|---|---|---|---|---|---|
| Oil India Ltd. | 2014 - 2016 | Beas Rovuma Energy Mozambique Ltd Suntera Nigeria 205 Ltd. | Standard Cycle | Pool resources for very large capital projects | Business Alliance | Output |
| | 2013 | Numaligarh Refinery Ltd. (NRL) | Lack of data for processing information | | | |
| | | Brahmaputra Cracker and Polymer Ltd. (BCPL) | Standard Cycle | Pool resources for very large capital projects | Business Alliance | Output |
| | | Suntera Nigeria 205 Ltd. | Lack of data for processing information | | | |
| | | Duliajan Numaligarh Pipeline (DNP) Ltd. | Standard Cycle | Pool resources for very large capital projects | Business Alliance | Output |
| Petronet LNG Ltd. | 2013 - 2017 | Adani Petronet (Dahej) Port Pvt. Ltd. | Fast Cycle | Maintain market leadership | | |
| | 2017 | India LNG Transport Co (No 4) Pvt. Ltd.(ILT4) | Lack of data for processing information | | | |
| Pidilite Industries Ltd. | 2016 - 2017 | Plus Call Technical Services L.L.C. | | | | |
| | 2013 | Building Envelope Systems India Ltd. | | | | |
| Piramal Enterprises Ltd. | 2015 - 2017 | Shrilekha Financial Services | Standard Cycle | Learn new business techniques | Hybrid Alliance | Behaviour and Process & Output |
| | 2013 - 2016 | Allergan India Pvt. Ltd. | | | | |
| | 2017 | Convergence Chemicals Pvt. Ltd. | | | | |
| | 2016 | Zebra Management Services Pvt. Ltd. | | | | |
| | 2015 | Novus Cloud Solutions Pvt. Ltd. | | | | |
| Power Finance Corporation Ltd. | 2013 - 2017 | National Power Exchange Ltd. | | Pool resources for very large capital projects | Business Alliance | Output |
| Punjab National Bank | 2013 - 2017 | Everest Bank Ltd., Kathmandu, Nepal | Slow Cycle | Gain access to restricted market | | |
| Rural Electrification Corporation Ltd. | 2013 - 2017 | Energy Efficiency Services Ltd. | Standard Cycle | Pool resources for very large capital projects | | |
| SBI Life Insurance Company Ltd. | 2013 - 2017 | BNP Paribas Cardif | Slow Cycle | Gain access to restricted market | | |

continued on following page

Table 4. Continued

| Company | Year(s) | Alliance | Market Cycle | Potential Main Reason for Strategic Alliance by Market Type | Relationship between partners | Control Mechanism |
|---|---|---|---|---|---|---|
| Steel Authority of India Ltd. | 2013 - 2017 | SAIL Bansal Service Centre Ltd. | Standard Cycle | Pool resources for very large capital projects | Business Alliance | Output |
| | | Mjunction Services Ltd. | Standard Cycle | Gain access to restricted market | Business Alliance | Output |
| | | Bhilai Jaypee Cement Ltd. | Standard Cycle | Gain access to restricted market | Business Alliance | Output |
| | | S&T Mining Company Pvt. Ltd. | Standard Cycle | Pool resources for very large capital projects | | |
| | | SAIL- SAIL Kobe Iron India Pvt. Ltd. | Lack of data for processing information | | | |
| | | TMTSAL SAIL JV Ltd. | | | | |
| | | SAL SAIL JVC Ltd. | | | | |
| | | N.E. Steel & Galvanising Pvt. Ltd. | | | | |
| | | NMDC SAIL Ltd. | | | | |
| | 2017 | NTPC-SAIL Power Company Ltd. | Standard Cycle | Pool resources for very large capital projects | Business Alliance | Output |
| | | Bokaro Power Supply Company Pvt. Ltd. | | | | |
| | | SAIL & MOIL Ferro Alloys Pvt. Ltd. | | | | |
| | | International Coal Ventures Pvt. Ltd. | | | | |
| | | SAIL-SCI Shipping Pvt. Ltd. | | | | |
| | | SAIL SCL Kerala Ltd. | | | | |
| | | RITES Bengal Wagon Industry Pvt. Ltd. | Lack of data for processing information | | | |
| | | SAIL-Bengal Alloy Castings Pvt. Ltd. | Standard Cycle | Pool resources for very large capital projects | Business Alliance | Output |
| | | North Bengal Dolomite Ltd. Romelt-SAIL (India) Ltd. | Lack of data for processing information | | | |
| | 2013 - 2016 | UEC-SAIL Information Technology Ltd. | Standard Cycle | Pool resources for very large capital projects | Business Alliance | Output |
| | 2014 - 2017 | VSL SAIL JVC Ltd. | Lack of data for processing information | | | |
| | | , Abhinav-SAIL JVC Ltd. | | | | |
| | 2013 - 2015 | Bokaro Jaypee Cement Ltd. | | | | |
| Sun TV Network Ltd. | 2017 | South Asia FM Ltd. | | | | |
| | | Digital Radio (Mumbai) Broadcasting Ltd. | Fast Cycle | Maintain market leadership | Business Alliance | Output |
| | | Asia Radio Broadcast Pvt. Ltd. | | | | |
| | | Pioneer Radio Training Services Pvt. Ltd. | | | | |
| | | Digital Radio (Kolkata) Broadcasting Ltd. | | | | |
| | | Digital Radio (Delhi) Broadcasting Ltd. | | | | |
| | | Optimum Media Services Pvt. Ltd. | | | | |
| | | South Asia Multimedia Pvt. Ltd. | | | | |

continued on following page

Table- 4, has been provided to present the classification of SAs based on the literature provided by Hitt, Ireland & Hoskisson (1997) and Koza & Lewin (2000). In the presented table, each alliance has been studied and classified under a certain category. The literature provided by Hitt, Ireland & Hoskisson (1997) was used to categorize alliances under a certain 'Market Cycle' and 'Market Type'. Whereas, the literature provided by Koza & Lewin (2000) was used to categorize alliances under 'Relationship between partners' and 'Control Mechanism'. An alliance, based on Hitt, Ireland & Hoskisson (1997) can be classified as either Slow, Standard or Fast cycle under Market cycle. Whereas, an alliance, based on Koza & Lewin (2000) can be classified as learning, business or hybrid alliances under 'relationship between partners', and as 'behaviour and process, output or both' under 'control mechanism'. Some firms formed many SAs however there was very little information available in the public domain to conclude their categories. In such cases, the authored have mentioned "lack of data for processing information". It was also observed that some SAs had overlapping categories and in such cases, both of them have been mentioned in the table. Complete categorization of the SAs has been provided in table 4. This is based upon authors understanding.

## DISCUSSION

This research was majorly centric to the literature provided by Hitt, Ireland & Hoskisson (1997) and Koza & Lewin (2000). The concept discussed by Hitt, Ireland & Hoskisson (1997) on 'Market Cycles' and 'potential major reason for SA' was directly adopted and explicitly used in this research. The authors seconded to this methodology and subjected Indian SAs to understand the Indian SAs scenario. The discussion provided by Koza & Lewin (2000) equally played an important role in broadening the scope of this research. The authors seconded to the methodology proposed by Koza & Lewin (2000) and used the proposed method to classify Indian SAs into different types of alliances. The authors were optimistic and convinced with the final classifications and drew conclusions based on all the results.

## CONCLUSION

This research studied 165 unique cases of alliances formed by 38 companies. 96 alliances were classified. Based on the literature provided by Hitt, Ireland & Hoskisson (1997), 65 (67.7%) of the cases were classified under 'Standard cycle'. 34 (35.4%) of the cases were classified under 'Fast cycle'. 9 (9.4%) and were classified as 'Slow cycle'. Based on the literature provided by Koza & Lewin (2000), 6 (6.25%) of the cases were classified as 'Learning Alliance'. 74 (77%) of the cases were classified as 'Business alliance'. 16 (16.6%) of the cases were classified as 'Hybrid alliance'. 24 (25%) alliances sought innovation. 88 (91.6%) alliances sought marketing. It was observed that many Indian firms relied on SAs for meeting their business objectives. 'Pool resources for very large capital projects' was observed to be the prominent reason to form an alliance. The Indian scenario tended to form many business alliances, aimed to increase the manufacturing of products or improving business operations. It could be speculated that India, being a developing country, the focused on increasing business output by forming business alliances as compared to learning alliances. It was also observed that some firms kept their alliances' portfolio unchanged throughout the period (example, Punjab National Bank) while some firms rapidly changed their alliances' portfolio. (example, Steel Authority of India Ltd.). India as an emerging economy represents both chaos (Bhattacharyya, Rangarajan & Vyas, 2011) and opportunities, which require special strategic attention (Saran& Guo,2005; Bhattacharyya, 2011). Strategic alliances and Mergers and Acquisition represent a way of competing in Indian business landscape (Miller et al., 2005; Bhattacharyya, 2018; Kumar & Bansal, 2008; Bhattacharyya, 2019b; Chand & Katou, 2012). It was observed that Indian SAs tended to be more centric to 'Market Seeking' as against 'Innovative Seeking'. Moreover, it was also observed that the formed SAs have, in some cases, remained the same over the time period, while some cases saw a constant revision in the formed SAs. This phenomenon aligned with individual firms' business motives and long-term objectives. This work in extending a conversation whether the Indian SAs were focussing on market seeking behaviour or towards seeking innovation so as to be globally a pioneer not a laggard.

## REFERENCES

Arifio, A. (2003). Measures of Strategic Alliances Performance: An Analysis of Construct Validity. *Journal of International Business Studies*, *34*(1), 66–79. doi:10.1057/palgrave.jibs.8400005

Bas, T. K., & Teng, B. S. (2000). A Resource-Based Theory of Strategic Alliances. *Journal of Management*, *26*(1), 31–61. doi:10.1177/014920630002600105

Beverland, M., & Bretherton, P. (2001). The uncertain search for opportunities: Determinants of strategic alliances. *Qualitative Market Research*, *4*(2), 88–99. doi:10.1108/13522750110388581

Bhattacharyya, S. S. (2011). Reflections on strategic insights for winning in the complex emerging market of India. *International Journal of Business Excellence*, *4*(1), 15–43. doi:10.1504/IJBEX.2011.037247

Bhattacharyya, S. S. (2018). A holistic perspective on strategic alliances for Indian managers. *Journal of Services Research*, *18*(1).

Bhattacharyya, S. S. (2019a). Development of an Integrated Model of Strategic Alliance. *Indian Journal of Industrial Relations*, *54*(3).

Bhattacharyya, S. S. (2019b). Mergers and Acquisition by Indian Firms: Managerial Perspectives on Challenges and Solutions. In Dynamic Perspectives on Globalization and Sustainable Business in Asia (pp. 41-55). Hershey, PA: IGI Global.

Bhattacharyya, S. S., Rangarajan, R., & Vyas, K. G. (2011). Reflections on mapping chaos in the business organisational landscape. *International Journal of Business Innovation and Research*, *6*(1), 76–116. doi:10.1504/IJBIR.2012.044258

Bhattacharyya, S. S., & Shaik, J. (2009). Topic–strategic alliances: Perspectives from Indian life insurance industry. *International Journal of Indian Culture and Business Management*, *3*(1), 98–120. doi:10.1504/IJICBM.2010.029531

Bower, J. L. (2001). *Not All M&As Are Alike – and That Matters* (pp. 92–101). Harvard Business School Publishing.

Bronder, C., & Pritzi, R. (1992). Developing strategic alliances: A conceptual framework for successful co-operation. *European Management Journal*, *10*(4), 412–421. doi:10.1016/0263-2373(92)90005-O

Bucklin, L. P., & Sengupta, S. (1993). Organizing successful co-marketing alliances. *Journal of Marketing, 57*(2), 32–46. doi:10.1177/002224299305700203

Butler, C. J. (2008). Problems in global strategic alliance management for European defence manufacturing firms. *Management Decision, 46*(2), 330–341. doi:10.1108/00251740810854186

Butler, C. J. (2010). Internal and lateral communication in strategic alliance decision making. *Management Decision, 48*(5), 698–712. doi:10.1108/00251741011043885

Canzaniello, A., Hartmann, E., & Fifka, M. S. (2017). Intra-industry strategic alliances for managing sustainability-related supplier risks: Motivation and outcome. *International Journal of Physical Distribution & Logistics Management, 47*(5), 387–409. doi:10.1108/IJPDLM-01-2016-0034

Cavusgil, S. T., Calantone, R. J., & Zhao, Y. (2003). Tacit knowledge transfer and firm innovation capability. *Journal of Business and Industrial Marketing, 18*(1), 6–21. doi:10.1108/08858620310458615

Chan, S. H., Kensinger, J. W., Keown, A. J., & Martin, J. D. (1997). Do strategic alliances create value? *Journal of Financial Economics, 46*(2), 199–221. doi:10.1016/S0304-405X(97)00029-9

Chand, M., & Katou, A. A. (2012). Strategic determinants for the selection of partner alliances in the Indian tour operator industry: A cross-national study. *Journal of World Business, 47*(2), 167–177. doi:10.1016/j.jwb.2011.04.003

Chen, S. H., Lee, H. T., & Wu, Y. F. (2008). Applying ANP approach to partner selection for strategic alliance. *Management Decision, 46*(3), 449–465. doi:10.1108/00251740810863889

Chung, Q. B., Luo, W., & Wagner, W. P. (2006). Strategic alliance of small firms in knowledge industries: A management consulting perspective. *Business Process Management Journal, 12*(2), 206–233. doi:10.1108/14637150610657549

Cummings, J. L., & Holmberg, S. R. (2012). Best-fit alliance partners: The use of critical success factors in a comprehensive partner selection process. *Long Range Planning, 45*(2-3), 136–159. doi:10.1016/j.lrp.2012.01.001

Das, T. K., & Teng, B. S. (1998). Resource and risk management in the strategic alliance making process. *Journal of Management, 24*(1), 21–42. doi:10.1177/014920639802400103

Das, T. K., & Teng, B. S. (2001). Trust, control and risk in strategic alliances: An integrated framework. *Organization Studies, 22*(2), 251–283. doi:10.1177/0170840601222004

Deeds, D. L., & Hill, C. W. (1996). Strategic alliances and the rate of new product development: An empirical study of entrepreneurial biotechnology firms. *Journal of Business Venturing, 11*(1), 41–55. doi:10.1016/0883-9026(95)00087-9

Dev, C. S., Klein, S., & Fisher, R. A. (1996). A market-based approach for partner selection in marketing alliances. *Journal of Travel Research, 35*(1), 11–17. doi:10.1177/004728759603500103

Doz, Y. L. (1996). The evolution of cooperation in strategic alliance: Initial conditions or learning processes? *Strategic Management Journal, 17*(1), 55–83.

Dyre, J. H., Kale, P., & Singh, H. (2014). When to Ally and When to Acquire. *Harvard Business Review*, 3–10.

Elia, S., Messeni Petruzzelli, A., & Piscitello, L. (2019). The impact of cultural diversity on innovation performance of MNC subsidiaries in strategic alliances. *Journal of Business Research, 98*, 204–213. doi:10.1016/j.jbusres.2019.01.062

Elmuti, D., & Kathawala, Y. (2001). An overview of strategic alliances. *Management Decision, 39*(3), 205–217. doi:10.1108/EUM0000000005452

Fey, C. F., & Beamish, P. W. (2000). Joint venture conflict: The case of Russian international joint ventures. *International Business Review, 9*(2), 139–162. doi:10.1016/S0969-5931(99)00034-7

Gerwin, D. (2004). Coordinating new product development in strategic alliances. *Academy of Management Review, 29*(2), 241–257. doi:10.5465/amr.2004.12736085

Gerwin, D., & Ferris, J. S. (2004). Organizing new product development projects in strategic alliances. *Organization Science, 15*(1), 22–37. doi:10.1287/orsc.1030.0052

Grunwald, R., & Kieser, A. (2007). Learning to Reduce Interorganizational Learning: An Analysis of Architectural Product Innovation in Strategic Alliances. *Journal of Product Innovation Management, 24*(4), 369–391. doi:10.1111/j.1540-5885.2007.00256.x

Hagedoorn, J., & Duysters, G. (2002). External sources of innovative capabilities: The preference for strategic alliances or mergers and acquisitions. *Journal of Management Studies, 39*(2), 167–188. doi:10.1111/1467-6486.00287

Hamel, G. (1991). Competition for competence and interpartner learning within international strategic alliances. *Strategic Management Journal, 12*(1), 83–103. doi:10.1002mj.4250120908

Hartmann, A. M. (2018). *Negotiating for Strategic Alliances*. In *The Palgrave Handbook of Cross-Cultural Business Negotiation* (pp. 53–70). Palgrave.

Hillman, A. J., Withers, M. C., & Collins, B. J. (2009). Resource dependence theory: A review. *Journal of Management, 35*(6), 1404–1427. doi:10.1177/0149206309343469

Hitt, M. A., Ireland, R. D., & Hoskisson, R. E. (1997). *Strategic Management: Competitiveness and Globalization* (2nd ed.). St. Paul, MN: West Publishing Co.

Ho, M. H.-W., Ghauri, P. N., & Kafouros, M. (2019). Knowledge Acquisition in International Strategic Alliances: The Role of Knowledge Ambiguity. *Management International Review, 59*(3), 439–463. doi:10.100711575-019-00383-w

Hoffmann, W. H., & Scholsser, R. (2001). Success factors of strategic alliances in small and medium-sized enterprises – An empirical survey. *Long Range Planning, 34*(3), 357–381. doi:10.1016/S0024-6301(01)00041-3

Holmberg, S. R., & Cummings, J. L. (2009). Building successful strategic alliances. *Long Range Planning, 42*(6), 164–193. doi:10.1016/j.lrp.2009.01.004

Hsieh, H. F., & Shannon, S. E. (2005). Three approaches to qualitative content analysis. *Qualitative Health Research, 15*(9), 1277–1288. doi:10.1177/1049732305276687 PMID:16204405

Hyder, A. S., & Eriksson, L. T. (2005). Success is not enough: The spectacular rise and fall of a strategic alliance between two multinationals. *Industrial Marketing Management, 34*(8), 783–796. doi:10.1016/j.indmarman.2004.12.008

Ireland, R. D., Hitt, M. A., & Vaidyanath, D. (2002). Alliance Management as a source of competitive advantage. *Journal of Management, 28*(3), 413–446. doi:10.1177/014920630202800308

Islam, M., Hossain, A. T., & Mia, L. (2018). Role of strategic alliance and innovation on organizational sustainability. *Benchmarking: An International Journal, 25*(5), 1581–1596. doi:10.1108/BIJ-12-2016-0188

Joia, L. A., & Malheiros, R. (2009). Strategic alliances and the intellectual capital of firms. *Journal of Intellectual Capital, 10*(4), 539–558. doi:10.1108/14691930910996634

Joshi, A. M., & Nerkar, A. (2011). When do strategic alliances inhibit innovation by firms? Evidence from patent pools in the global optical disc industry. *Strategic Management Journal, 32*(11), 1139–1160. doi:10.1002mj.929

Kale, P., & Singh, H. (2009). Managing Strategic Alliances: What do we know now, and where do we go from here? *The Academy of Management Perspectives, 23*(3), 45–62. doi:10.5465/amp.2009.43479263

Kale, P., Singh, H., & Perlmutter, H. (2000). Learning and protection of proprietary assets in strategic alliances: Building relational capital. *Strategic Management Journal, 21*(3), 217–237. doi:10.1002/(SICI)1097-0266(200003)21:3<217::AID-SMJ95>3.0.CO;2-Y

Kauser, S., & Shaw, V. (2004). The influence of behavioural and organisational characteristics on the success of international strategic alliances. *International Marketing Review, 21*(1), 17–52. doi:10.1108/02651330410522934

Klus, M. A., Lohwasser, T. S., Holotiuk, F., & Moormann, J. (2019). Strategic Alliances between Banks and Fintechs for Digital Innovation: Motives to Collaborate and Types of Interaction. *The Journal of Entrepreneurial Finance, 21*(1).

Kogut, B. (1991). Joint ventures and the option to expand and acquire. *Management Science, 37*(1), 13–34. doi:10.1287/mnsc.37.1.19

Kotabe, M., & Scott Swan, K. (1995). The role of strategic alliances in high-technology new product development. *Strategic Management Journal, 16*(8), 621–636. doi:10.1002mj.4250160804

Koza, M., & Lewin, A. (2000). Managing Partnerships and Strategic Alliances: Raising the Odds of Success. *European Management Journal, 18*(2), 146–151. doi:10.1016/S0263-2373(99)00086-9

Kumar, R., & Andersen, P. H. (2000). Inter firm diversity and the management of meaning in international strategic alliances. *International Business Review, 9*(2), 237–252. doi:10.1016/S0969-5931(99)00038-4

Kumar, S., & Bansal, L. K. (2008). The impact of mergers and acquisitions on corporate performance in India. *Management Decision, 46*(10), 1531–1543. doi:10.1108/00251740810920029

Larsson, R., Bengtsson, L., Henriksson, K., & Sparks, J. (1998). The interorganizational learning dilemma: Collective knowledge development in strategic alliances. *Organization Science*, *9*(3), 285–305. doi:10.1287/orsc.9.3.285

Larsson, R., Bengtsson, L., Henriksson, K., & Sparks, J. (1998). The Interorganizational Learning Dilemma: Collective Knowledge Development in Strategic Alliances. *Organization Science*, *9*(3), 285–305. doi:10.1287/orsc.9.3.285

Lawrence, P., & ul-Haq, R. (1998). Qualitative research into strategic alliances. *Qualitative Market Research*, *1*(1), 15–24. doi:10.1108/13522759810197569

Lechler, S., Canzaniello, A., & Hartmann, E. (2019). Assessment sharing intra-industry strategic alliances: Effects on sustainable supplier management within multi-tier supply chains. *International Journal of Production Economics*, 217, 1–52.

Lee, D., & Husk, K. K. (2014). Diversity in alliance portfolio and performance outcomes: A Meta-Analysis. *Journal of Management*, *43*(5), 1–26.

Lee, K. S., Lim, G. H., & Tan, J. (2000). Feasibility pf strategic alliance as an entry strategy into markets dominated by major competitors. *Journal of Small Business and Enterprise Development*, *7*(1), 43–57. doi:10.1108/EUM0000000006804

Li, K., Qiu, J., & Wang, J. (2019). Technology Conglomeration, Strategic Alliances, and Corporate Innovation. *Management Science*, 65(11), 1–26.

Luo, X., & Deng, L. (2009). Do Birds of a Feather Flock Higher? The Effects of Partner Similarity on Innovation in Strategic Alliances in Knowledge-Intensive Industries. *Journal of Management Studies*, *46*(6), 1005–1030. doi:10.1111/j.1467-6486.2009.00842.x

Miles, R. E., & Snow, C. C. (1992). Causes of failure in network organization. *California Management Review*, *34*(4), 53–72. doi:10.2307/41166703

Miller, M. G., Fitzgerald, S. P., Murrell, K. L., Preston, J., & Ambekar, R. (2005). Appreciative inquiry in building a transcultural strategic alliance: The case of a biotech alliance between a US multinational and an Indian family business. *The Journal of Applied Behavioral Science*, *41*(1), 91–110. doi:10.1177/0021886304273060

Narula, R., & Hagedoorn, J. (1999). Innovating through strategic alliances: Moving towards international partnerships and contractual agreements. *Technovation*, *19*(5), 283–294. doi:10.1016/S0166-4972(98)00127-8

Nielsen, B. B., & Nielsen, S. (2009). Learning and Innovation in International Strategic Alliances: An Empirical Test of the Role of Trust and Tacitness. *Journal of Management Studies*, *46*(6), 1031–1056. doi:10.1111/j.1467-6486.2009.00840.x

Padula, G., & Dagnino, G. B. (2007). Untangling the rise of coopetition: The intrusion of competition in a cooperative game structure. *International Studies of Management & Organization*, *37*(2), 32–52. doi:10.2753/IMO0020-8825370202

Park, S. H., Chen, R., & Gallagher, S. (2002). Firm resources as moderators of the relationship between market growth and strategic alliances in semiconductor start-ups. *Academy of Management Journal*, *45*(3), 527–545.

Pett, T. L., & Dibrell, C. C. (2001). A process model of global strategic alliance formation. *Business Process Management Journal*, *7*(4), 349–364. doi:10.1108/14637150110401042

Robson, M. J., Katsikeas, C. S., Schlegelmilch, B. B., & Pramböck, B. (2018). Alliance capabilities, interpartner attributes, and performance outcomes in international strategic alliances. *Journal of World Business*, *54*(2), 137–153. doi:10.1016/j.jwb.2018.12.004

Russo, M., & Cesarami, M. (2017). Strategic alliances success factors: A literature review on alliance lifecycle. *International Journal of Business Administration*, *8*(3), 1–9. doi:10.5430/ijba.v8n3p1

Saran, A., & Guo, C. (2005). Competing in the global marketplace: The case of India and China. *Business Horizons*, *48*(2), 135–142. doi:10.1016/j.bushor.2004.10.010

Shi, W., & Prescott, J. E. (2011). Sequence patterns of firms' acquisition and alliance behaviour and their performance implications. *Journal of Management Studies*, *48*(5), 1044–1070. doi:10.1111/j.1467-6486.2010.00953.x

Simonin, B. L. (1999). Transfer of marketing know-how in international strategic alliances: An empirical investigation of the role and antecedents of knowledge ambiguity. *Journal of International Business Studies*, *30*(3), 463–490. doi:10.1057/palgrave.jibs.8490079

Simonin, B. L. (1999). Ambiguity and the process of knowledge transfer in strategic alliances. *Strategic Management Journal*, *20*(7), 595–623. doi:10.1002/(SICI)1097-0266(199907)20:7<595::AID-SMJ47>3.0.CO;2-5

Spekman, R. E., Forbes, T. M., Isabella, L. A., & MacAvoy, T. C. (1998). Alliance management: A view from the past and a look to the future. *Journal of Management Studies*, *35*(6), 747–772. doi:10.1111/1467-6486.00118

Thomas, J. B., & Trevino, L. K. (1993). Information processing in strategic alliance building: A multiple-case approach. *Journal of Management Studies*, *30*(5), 779–814. doi:10.1111/j.1467-6486.1993.tb00326.x

Tian Xie, F., & Johnston, W. J. (2004). Strategic alliances: Incorporating the impact of e-business technological innovations. *Journal of Business and Industrial Marketing*, *19*(3), 208–222. doi:10.1108/08858620410531342

Todeva, E., & Knoke, D. (2005). Strategic alliances & model of collaboration. *Management Decision*, *43*(1), 1–22. doi:10.1108/00251740510572533

Varadarajan, P. R., & Cunningham, M. H. (1995). Strategic alliances: A synthesis of conceptual foundations. *Journal of the Academy of Marketing Science*, *23*(4), 282–296. doi:10.1177/009207039502300408

Varra, E., Kleymann, B., & Seristo, H. (2004). Strategies as discursive constructions: The case of airline alliances. *Journal of Management Studies*, *41*(1), 1–35. doi:10.1111/j.1467-6486.2004.00419.x

Weber, R. P. (1990). Basic content analysis. Thousand Oaks, CA: Sage. doi:10.4135/9781412983488

Whipple, J. M., & Frankel, R. (2000). Strategic alliance success factor. *The Journal of Supply Chain Management, 36*(2), 21–28. doi:10.1111/j.1745-493X.2000.tb00248.x

Yamakawa, Y., Yang, H., & Lin, Z. J. (2011). Exploration versus exploitation in alliance portfolio: Performance implications of organizational, strategic, and environmental fit. *Research Policy, 40*(2), 287–296. doi:10.1016/j.respol.2010.10.006

Zineldin, M., & Bredenlow, T. (2003). Strategic alliance: synergies and challenges: A case of strategic outsourcing relationships "SOUR". *International Journal of Physical Distribution & Logistics Management, 33*(5), 449–464. doi:10.1108/09600030310482004

# Chapter 2
# An Empirical Investigation on Equity Market Integration of ASEAN-India

**Swamy Perumandla**
*National Institute of Technology, Warangal, India*

**Padma Kurisetti**
*National Institute of Technology, Warangal, India*

## ABSTRACT

*This empirical study investigates the symmetric and asymmetric dynamic correlations and volatility linkages between ASEAN-5 and the Indian equity markets. Granger causality test results reveal that bidirectional causal relation between the pairs of India-Indonesia, and India-Singapore. However, India-Phillippines and India-Thailand have a unidirectional causal relationship. Variance decomposition results show that India's equity market volatility contributes moderate fluctuations in the variance of Indonesia, Phillippines, Singapore Thailand. Finally, the Markov regime transition probabilities show that the high transition probabilities of p11 and p22 for India-Malaysia, India-Philippines and India-Thailand indicates a high degree of regime stability. The study on financial integration provides important inputs to investors in sharing risk internationally since restrictions on investment are removed. This study provides an essential insight to policymakers, portfolio managers, domestic and international investors, risk analysts and financial researchers in an emerging market.*

## INTRODUCTION

Financial development and globalization are significantly dependent on integrated stock markets around the world. Stock market integration plays a vital role in emerging markets' development. The degree of stockmarket interdependence and integration can explain the accessability of capital flows across the international capital markets to firms. The higher degree of interdependence and integration provides higher access to international capital markets to firms with a lower cost of equity. It also provides an

DOI: 10.4018/978-1-7998-1566-2.ch002

exposure to local markets from global and regional shocks. The degree of market comovements is a critical factor for assessing the diversification opportunities across national and international financial markets. Few studies revealed that the market comovements are strongly influenced by the international trade channel(Frankel and Rose, 1998; Pentecôte et al., 2015) and financial market integration (Baele et al., 2004; Wälti, 2011; Aloui et al., 2011). Dewandaru, Masih, and Masih (2016) studied the ontagion amongst the Asia-Pacific equity markets (Japan, Hong Kong and Australia) during twelve major crises around the world. They found that shocks were transmitted via excessive linkages, with the Asian crisis as the most influential in relation to a sudden stop. Finally, they found low co-movements in the short run, suggesting a partial convergence across the markets. Chowdhury, Haque, and Islam (2017) Due to increased globalization and economic integration in the global economy, contagion effects have been considered an important matter for the investors and policymakers. Dewandaru, Masih, and Masih (2018) prior to the subprime crisis, contagion effects generated short-term shocks and their study provided an evidence of increasing short-run and long-run stock market integration of Economic and Monetary Union (EMU). They found evidence of structural breaks in the volatility of time series for the majority of markets.

ASEAN-India trade and investment relations have been growing gradually. ASEAN is India's fourth largest trading partner. According to the Asian Development Bank Report (2018), Singapore, India, Indonesia and Malaysia are listed among Asia's top 10 FDI recipients in 2017. Singapore Foreign Domestic Investment(FDI) Receipts stands at the US $ 62.0 billion, followed by India at the US $ 39.9 billion, Indonesia at US$ 23.1 billion and Malaysia at US$ 9.5 billion. India's trade with ASEAN stands at US$ 81.33 billion which is 10.6% of India's overall trade. Investment flows are also substantial in both the ways, with ASEAN accounting for approximately 18.28% of investment flows into India since 2000. FDI inflows into India from ASEAN between April 2000 to March 2018 was about the US $ 68.91 billion, while FDI outflows from India to ASEAN countries, from April 2007 to March 2015 was about US$38.672 billion, as per data maintained by the Department of External Affairs (DEA). The India-ASEAN Free Trade Agreement (FTA) which was signed on 13 August 2009 after six years of negotiation was a benchmark to mark how India-ASEAN relation has taken to a higher level. In September 2014 India signed an agreement on India-ASEAN FTA in Services and Investments with all the ASEAN members which came into force in April 2015. Hence, there will be huge scope for investing in financial markets in ASEAN-India. It is also essential to investigate and examine the interrelationship between ASEAN-Indian financial markets, for asset allocation and diversifying portfolio risk.

On the other hand, Southeast Asian nations have endorsed far-reaching commitments to economic advancement and financial integration in the region. Areas covered under the region's financial integration framework including financial services, payments & settlement, and capital markets. Moreover, the ASEAN trading link was launched in 2012 to integrate equity markets in Malaysia, Singapore, and Thailand. Hence, it is essential to understand the financial integration among ASEAN countries. The investigation of the cross-market time-varying relations of return and volatility transmission is crucial to investors, fund managers and policymakers. Strong interdependence across markets around the world may effectively lead to high exposure to contagious effects. In this paper study attempt to evaluate the degree of market interdependence and shock transmission between stock markets of India and selected ASEAN-5 countries for portfolio investment and public policy purpose.

This study empirically investigates the relationships between ASEAN 5-India equity markets. This study intends to analyze the time-varying co-movements and degree of financial integration of ASEAN 5- India by using the symmetric and asymmetric versions of Dynamic Conditional Correlation - Generalized Autoregressive Conditional Heteroskedasticity (DCC-GARCH and ADCC-EGARCH) model.

On the other side, this study also identifies the causal relationships between the ASEAN-India equity markets. Pairwise Granger Causality Test will be conducted to know the direction of causation, such as uni-directional or bi-directional between the all possible pairs. The empirical results of this study will highlight dynamic volatility and correlation linkages which are essential while diversifying the investments across ASEAN-Indian markets. Variance decomposition analysis and Markov regime switching model also used to analyse the data. The integration relation between ASEAN-5 and Indian stock markets is characterized as a Markov two-regime process, where regimes are stable and generally persistent.

The remaining paper is organized as follows - section 2 Reviews the Literature, section 3. Methods and Materials, section 4. Describes the Data and Preliminary analysis, section 5 Presents the Empirical Results and Discussions Section 6 Explains Future Research Directions, Finally, 7 Conclusion.

## 2. REVIEW OF LITERATURE

Section 2 presents an extensive review of literature related to various aspects which are crucial and relevant in the context of the present study. Mensah & Premaratne (2018) studied the integration of ASEAN banking sector stocks using a Dynamic Conditional Correlation (DCC) GARCH framework. The study revealed that the evidence of rising correlation in market movements over time, both internal and external to ASEAN, although correlation coefficients remain low to moderate. Komatsubara et al. (2017) investigated the dynamics of integration in East Asian equity markets between 1995 and 2013 using a smooth-transition correlation GARCH model. Results revealed that there is an increasing integration among ASEAN markets caused by correlation increases in after trading hours. Chevallier et al. (2018) studied the market integration and financial linkages among stock markets in Pacific basin countries. This study revealed that the cross-market linkages in the Pacific Basin region are time-varying and have become stronger over time, with large increases in the level of shock spillover. Jonathan A. Batten et al. (2019) studied on Time-Varying Energy and Stock Market Integration in Asia Estimation results reveal a significant energy-related equity risk premium during the high integration regime. Teng et al. (2013) investigated on Financial Market Integration of ASEAN-5 with China and India. Results revealed that ASEAN-5's stock markets were highly integrated with the BSE and DJII, except for the Philippines. Boubakri & Guilaumin (2015) employed GARCH models to assess the dynamics of regional financial integration among East Asian countries. Beirne et al. (2013) found that the volatility in emerging markets (ASEAN-5, China and India) during the turbulent period is relatively lower than the mature markets. Balli et al. (2014) examined the effects of local and global shocks on the sector indices and national returns of the Association of Southeast Asian Nations (ASEAN) by using the univariate AR-GARCH model. They found that regional and global shocks have different influences on the ASEAN-wide sector and national equity indices.

Previously some researchers studied the integration of markets globally in other regions also. Kenourgios & Samitas (2011) examined the Equity market integration in emerging Balkan Markets using Asymmetric Generalized Dynamic Conditional Correlation (AG-DCC) multivariate GARCH model of Cappiello et al. Results show that stock market dependence is heightened, supporting the herding behaviour during the 2008 stock market crash period. (Majdoub & Mansour, 2014) studied Islamic equity market integration; Virk and Javed (2017) investigated European equity market integration; Gil-Alana et al. (2018) studied Africa's emerging equity markets and global markets integration; Alotaibi & Mishra (2017) studied Financial Integration for GCC Stock Markets time-varying market integration using DCC-GARCH methodology;) Abdennadher & Hellara (2018) studied the interdependencies in terms of stock market volatility and to assess the impact of Global Financial Crisis (GFC) on these interdependencies. They found the causality during both the tranquil and crisis period. These additional linkages during crisis periods in excess of those that arise during non-crisis periods contributes significantly in amplifying the international transmission of volatility and the risk of contagion. Perumandla and Kurisetti (2018) studied the integration between Indian commodity and equity markets using DCC-GARCH models.

Our study contributes to the strand of literature investigating time-varying financial integration by employing DCC & ADCC-GARCH models and Markov 2 state regime switching model. Study on Financial integration provides important inputs to investors in sharing risk internationally since restrictions on investment are removed (Obstfeld, 1994; Wright, 2005; Gourinchas & Jeanne, 2006; Bekaert, Harvey, and Lundblad, 2006). Hence, the study focuses on the integration of equity markets between ASEAN-5 & India.

## 3. METHODS AND METERIALS

### DCC and ADCC GARCH

The main purpose of the study is to analyze the dynamic relations of ASEAN-5 and India. Hence, the study employs Dynamic Conditional Correlation [DCC] -GARCH model introduced by Engle (2002) is a generalization of the CCC model, which allows the correlation matrix to vary over time rather than requiring them to be constant. Cappiello et al. (2006) introduced the asymmetric version of DCC GARCH to address the impact of asymmetric information on the time-varying correlations. In the first step, the study reveals the univariate parameters in two ways [standard and exponential]. At the second step, time-varying correlations were estimated by relying on lagged values of residuals and covariance matrices. In the present study both symmetric and asymmetric version (DCC - GARCH (Engle, 2002) and ADCC - GARCH (Cappiello et al., 2006) of modelling time-varying correlations were used. The Covariance matrix in DCC GARCH (Engle, 2002) defined as:

$$H_t = D_t R_t D_t \tag{1}$$

*Ht* is the conditional covariance matrix. $D_t$ is the k×k diagonal matrix of time-varying standard deviations from univariate GARCH models with $\left[\sigma_{i,t}^2\right]^{1/2}$ on the $i_{th}$ diagonal

$$D_t = \begin{bmatrix} \sqrt{\sigma_{c,t}^2} & 0 \\ 0 & \sqrt{\sigma_{e,t}^2} \end{bmatrix} \tag{2}$$

$R_t$ is the time varying correlation matrix.

$$R_t = \begin{bmatrix} \varepsilon_{cc,t} & \varepsilon_{ce,t} \\ \varepsilon_{ec,t} & \varepsilon_{ee,t} \end{bmatrix} = \begin{bmatrix} 1 & \varepsilon_{ce,t} \\ \varepsilon_{ec,t} & 1 \end{bmatrix} \tag{3}$$

Further, R has to be definite positive, and all the parameters should be equal to or less than one. In order to ensure this $R_t$ has been modeled as

$$= Q^{*-1}_{ce,t} Q_{ce,t} Q^{*-1}_{ce,t} \tag{4}$$

Where

$$Q_{ce,t} = [1-\theta_1-\theta_2] \cdot Q^* + \theta_1\left[\varepsilon_{c,t-1}\varepsilon_{e,t-1}\right] + \theta_2\left[Q_{ce,t-1}\right] \tag{5}$$

Where $Q_{ce,t}$ is the unconditional variance between series and i and j and follows a GARCH process, Q* is the unconditional covariance between the series estimated in step 1 and the scalar parameters $\theta_1$ and $\theta_2$ are non-negative and satisfy $\theta_1+\theta_2 < 1$.

Following the methodology of Engle [2002], the parameters $\theta_1$ and $\theta_2$ are estimated by maximizing the log-likelihood function. The log-likelihood function can be expressed as:

$$= -\frac{1}{2}\sum_{t-1}^{T}\left(k \log 2\pi\right) + 2\log([D_t]) + \log([R_t]) + \varepsilon_t' R_t^{-1} \varepsilon_t) \tag{6}$$

As the above model DCC model does not allow for asymmetries and asset specific news impact parameter, the modified model of Cappiello et al. [2006] for incorporating the asymmetrical effect and asset specific news impact can be written as:

$$Q_{ce,t} = [1-\theta_1-\theta_2]\left[Q^- - \theta_3.\right]\xi_t^- + \theta_1\left[\varepsilon_{c,t}\varepsilon_{e,t-1}\right] + \theta_2\left[Q_{ce,t-1}\right] + \theta_3\left(\varphi_{t-1}\varphi'_{t-1}\right) \tag{7}$$

Where

$\overline{\xi_t} = \mathrm{E}[\overline{\varphi_{ct}\varphi_{et}}] +$ and $\overline{\varphi_{ct}} = (I\overline{\varphi_{ct}}\ [<0]\ \overline{c\varepsilon_{ct}})$

Here θ3 is the asymmetric term which captures periods where both commodity and stock market experience bad news making $\left[\varphi_{ct}\varphi_{et}\right] = I_t$. This model is estimated using Maximum Likelihood techniques based on BFGS optimization algorithm. The study adopt student-t multivariate distribution of the time series returns, which is more suitable and gives better estimation results.

## Granger Causality

After analyzing the correlation among the pairs, study continued to find the causal relationships between the portfolio pairs, to know which country returns cause other country returns. For this purpose, study used pairwise Granger causality test. Granger causality test suggested by Granger (1969). To test the direction of causation, bi-direction or uni-direction between all the possible pairs employed bivariate regressions of the form:

$$y_t = \alpha_0 + \alpha_1 y_{t-1} + ... + \alpha_1 y_{t-1} + \beta_1 x_{t-1} + ...\beta_1 x_{-1} + \varepsilon_t \quad (8)$$

$$x_t = \alpha_0 + \alpha_1 x_{t-1} + ... + \alpha_1 x_{t-1} + \beta_1 y_{t-1} + ...\beta_1 y_{-1} + u_t \quad (9)$$

$$\beta_1 = \beta_2 = .... = \beta_t = 0$$

For each equation, the null hypothesis is that x does not Granger-cause $y$ in the first regression and that $y$ does not Granger-cause $x$ in the second regression.

## Variance Decomposition Analysis

The study used variance decomposition analysis to assess to what extent shocks to a certain market are explained by another market in India-ASEAN-5 countries. Variance decomposition tends to show the percentage of forecast error variance for each of the market selected that may attribute to its own shocks and to fluctuations in other markets. Information from this analysis provide some further evidence on the patterns of linkages among India-ASEAN-5 equity markets. This analysis enhances the insights on how markets react to system-wide shocks and how these responses over time span. In a statistical sense, if a variable explains most of its own shock, then it does not allow variances of other variables to contribute to it hence, it is said to be relatively exogenous.

## Markov 2 State Regime Switching Model

Markov Regime Switching Model with 2 Regimes

The number of states (or regimes) is 2, i.e. $St \in \Omega = \{1, 2\}$. This implies that e.g. the log returns of a financial time series are drawn from 2 distinct normal distributions, depending on what state the HMM is currently in. This would give us the following model to work with:

$$Yt = \mu 1 + \varepsilon t \text{ for state 1} \quad (10)$$

$$Yt = \mu 2 + \varepsilon t \text{ for state 2} \quad (11)$$

Where

$\varepsilon t \sim N(0, \sigma 1 2)$ for state 1 (12)

$\varepsilon t \sim N(0, \sigma 2 2)$ for state 2 (13)

This means that when the state of the HMM for time t is 1, then the expectation of the dependent variable is $\mu 1$ and the variance of the innovations is $\sigma 1 2$, etc. Since the underlying Markov chain is hidden one cannot observe what state the HMM is in directly, but only deduce its operation through the observed behaviour of $Yt$ . In order to attain the probability law governing the observed data $Yt$ a probabilistic model of what causes the change from state $St$ = i to state $St$ = j is required. This can be specified using the transition probabilities of an N state HMM (Hamilton, 2005); the switching of the states of the underlying HMM is a stochastic process itself.

## 4. THE DATA AND CERTAIN STYLIZED FACTS

The study considers ASEAN-5 and Indian stock market indices for analysis. The ASEAN-5 countries are, Indonesia, Malaysia, Phillippines, Singapore and Thailand. The stock market indices of selected ASEAN-5 countries are 1) Jakarta Stock Exchange Composite Stock Price Index (JKSE) of Indonesia. 2) Kuala Lumpur Stock Exchange Composite Price index (KLCI) of Malaysia, 3) The Philippines Stock Exchange Composite Price index (PSEi) of the Phillippines, 4) Straits Times Stock Exchange Index of Singapore (STI) of Singapore and 5) Thai Composite stock market index (SETI) of Thailand. The Indian stock market index Nifty 50 is considered for the study . The time span of the study runs from Jan 1st, 2012 to December 31st, 2018 and the data is obtained from investing.com. The selected time span of the study is based on the availability of data and on the othehand ASEAN-India trade and investment relations have been growing gradually in the selected time span. The daily data is considered for the analysis purpose. It is customary to calculate the return of an asset as the logarithmic value of the ratio of two consecutive prices. The continuously compounded daily returns are computed using the following logarithmic filter:

$$r_{i,t} = In\left(\frac{p_{i,t}}{p_{i,t-1}}\right) \quad (14)$$

The descriptive statistics reveal the characteristics of the data at a preliminary stage. Table No.1 presented results.

Table 1 reports the summary statistics of selected ASEAN 5 -India's equity indices of daily natural log returns data set. As evident from Table 1, The results indicate that India's Nifty 50 index offers the highest mean returns (0.049%) followed by the Philippines (0.030%), Indonesia (0.028%). However, Thailand (0.0074%) index offers the lowest mean returns followed by Malaysia (0.0063%). The Phillippines index is most risky, as approximated by a standard deviation of (1.00%) followed by Indonesia (0.095%),

*Table 1. Descriptive statistics of selected sectoral indices*

| | India | Indonesia | Malaysia | Phillippines | Singapore | Thailand |
|---|---|---|---|---|---|---|
| Mean | 0.000492 | 0.000281 | 0.0000639 | 0.000306 | 0.000241 | 0.0000740 |
| Std. Dev. | 0.009104 | 0.009562 | 0.005523 | 0.010085 | 0.008575 | 0.007292 |
| Skewness | -0.309692 | -0.280746 | -0.393334 | -0.200499 | -0.261357 | -0.200795 |
| Kurtosis | 5.371617 | 7.471148 | 6.374513 | 4.998757 | 6.920487 | 5.032191 |
| Jarque-Bera | 433.5918(0)*** | 1465.448(0)*** | 866.4464(0)*** | 299.9124(0)*** | 1128.935(0)*** | 309.6724(0)*** |
| ARCH | 3.705404(0.05)** | 46.26541(0)*** | 66.58457(0)*** | 148.7248(0)*** | 40.99788(0)*** | 25.58917(0)*** |
| Observations | 1732 | 1732 | 1732 | 1732 | 1732 | 1732 |

Source: Researcher's own computation. The figures in the parenthesis are probabilities. Notes: JB stat and ARCH stat are the statistics testing for normal distribution and ARCH effect, respectively. Heteroskedasticity test is used to check the presence of ARCH effects. * Denotes rejections of null hypothesis at 10% significance level. ** Denotes rejections of null hypothesis at 5% significance level. *** Denotes rejections of null hypothesis at 1% significance level.

*Table 2. Stationarity of the India & ASEAN-5 Indices*

| | ADF | | PP | |
|---|---|---|---|---|
| | With Constant & Trend | Without Constant & Trend | With Constant & Trend | Without Constant & Trend |
| India | -38.9675(0)*** | -38.8837(0)*** | -38.9073(0)*** | -38.8661(0)*** |
| Indonesia | -26.5866(0)*** | -26.559(0)*** | -39.6806(0)*** | -39.628(0)*** |
| Malaysia | -38.4232(0)*** | -38.4101(0)*** | -38.3207(0)*** | -38.3144(0)*** |
| Phillippines | -23.8382(0)*** | -23.7248(0)*** | -38.3946(0)*** | -38.3092(0)*** |
| Singapore | -39.7555(0)*** | -39.7039(0)*** | -39.7596(0)*** | -39.7026(0)*** |
| Thailand | -41.3857(0)*** | -41.3813(0)*** | -41.5142(0)*** | -41.5181(0)*** |

Source: Researcher's own computation. The figures in the parenthesis are probabilities *** Denotes rejections of null hypothesis at 1% significance level.

India (0.091%). Further, to arrive at the distribution of the indices returns, skewness and kurtosis are calculated. The excess kurtosis of all indices is also noticed from the fat tail property of financial time series. Hence extreme returns are more likely. The Returns of all series are negatively skewed suggests a higher likelihood of experiencing losses than gain in a given period and the kurtosis is much higher than 3 for all the indices. It is an indication of deviation of the series from a normal distribution, implying the presence of extreme movements in either direction, which is further confirmed by the Jarque–Bera (JB) statistics. JB statistics validate that, series are not normally distributed. The chi-square statistics of ARCH test also reject the null hypothesis of no ARCH effect, which motivates us to employ GARCH specification to capture the dynamics of conditional variance of assets. To establish the dynamic nature of the correlation and integration between ASEAN-India equity market returns, time series properties of asset returns and specific diagnostic tests need to be carried out.

*Figure 1. Time series plots of the india and selected ASEAN-5 indices with daily prices*

Plot 1: India-Indonesia

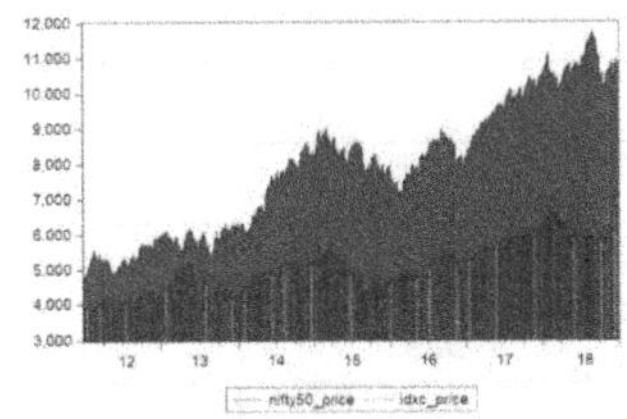

Plot 2: India-Malaysia

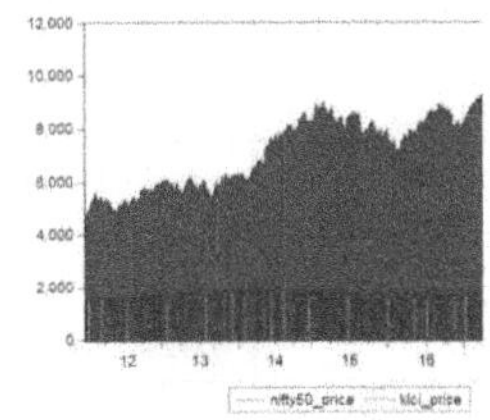

Plot 3: India-Phillippines

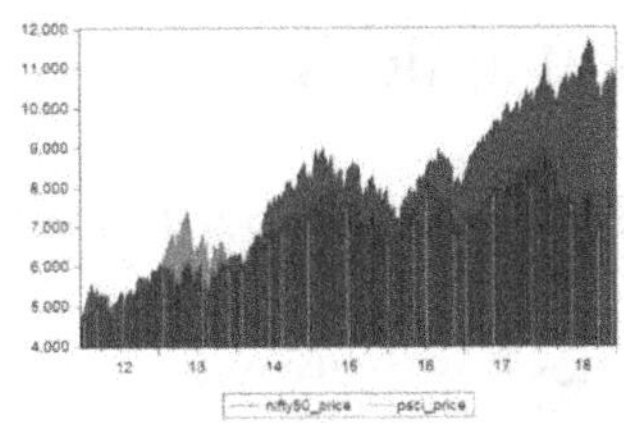

Plot 4: India-Singapore

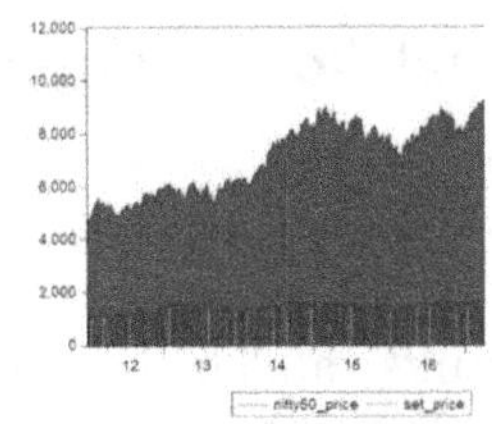

Plot 5: India-Thailand

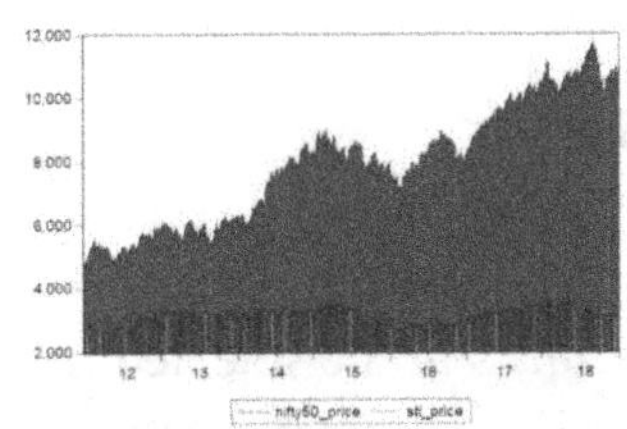

The stationarity of the data sets is examined by applying two commonly used methods i.e., Augumented Dickey and Fuller (1979) and Phillips & Perron (1988) tests. The analysis is carried on data by taking natural logarithms of all selected sectoral indices.

The results from Table.2 show that all indices are stationary at levels with respect to constant& trend and without constant. The results revealed that the null hypothesis is rejected at the 1% level of significance. Results with the constant & trend as well as without constant & trend indicating that all series are stationary at level i.e., I(0).

Sectoral Indices Price Serices Graphs:

Figure 1 provides the time series plots of India and selected ASEAN indices daily prices. plot no 1 to plot 5 depicts the Sectoral Indices Price Series Graphs.

*Table 3. Unconditional correlations of the selected sectoral indices*

| | India | Indonesia | Malaysia | Phillippines | Singapore | Thailand |
|---|---|---|---|---|---|---|
| India | 1 | | | | | |
| Indonesia | 0.390486 | 1 | | | | |
| Malaysia | 0.026612 | 0.009439 | 1 | | | |
| Phillippines | 0.298288 | 0.442831 | 0.00435 | 1 | | |
| Singapore | 0.420081 | 0.424376 | 0.035638 | 0.337389 | 1 | |
| Thailand | 0.453619 | 0.439294 | 0.042587 | 0.365555 | 0.43954 | 1 |

## 4.1. Unconditional Correlations of the Selected Indices

The analysis of static, as well as dynamic correlation, are important for portfolio choice or selection.

Table 3 presents the unconditional correlation results of the selected indices. The static or unconditional correlation results validate that India's return correlates positively with all other selected ASEAN countries. India returns highly correlate with Thailand followed by Singapore. However, India returns low correlate with Malaysia. Phillippines and Singapore high correlates with Indonesia. It is worth mentioning that the correlation analysis is unconditional and static hence, it fails to capture the effects of different unforeseen events. The static unconditional correlation represents the long-run average. Hence, the researcher further analysed the selected indices with dynamic conditional correlations.

## 5. EMPIRICAL RESULTS AND DISCUSSIONS

In this Section, the researcher present and discuss the results of the univariate GARCH and EGARCH to know the persistence of volatility shocks on the daily returns in the selected indices. The researcher investigated the time-varying comovements by using bivariate parameter estimates of symmetric and asymmetric versions of DCC-GARCH & ADCC-EGARCH. Granger Causality test to test the direction of causation, bi-direction or uni-direction between all the possible pairs by using bivariate regressions. The study used variance decomposition analysis to assess to what extent shocks to a certain sector are explained by another sector in the equity market. Finally, the Markov regime switching model is used to analyze the integration and regime stability conditions of India-ASEAN-5 countries.

In order to examine the dynamic relationship between India-ASEAN-5 and inter ASEAN-5 returns, the researcher first estimates the univariate GARCH(1,1) processes. Table 4 reports the results for mean and variance equations of the GARCH model estimations. The study first checks the stability conditions of the GARCH model by imposing the constraints $\omega > 0$, $0 \leq \alpha$, $0 \leq \beta$, $\alpha + \beta < 1$. All of the estimated parameters in the GARCH models satisfy the stability conditions therefore reserchers proceed to draw some inferences.

The degree of persistence ($\alpha + \beta$) in each period shows that the volatility shocks are very persistent which implies a fair amount of persistence of volatility shocks on the daily returns. Large values of the ARCH and GARCH parameters influence the conditional volatility in different ways. A high ARCH parameter implies that the effects of a shock are more pronounced in the subsequent period. Therefore,

*Table 4. Univariate parameter estimates of GARCH & EGARCH*

| Panel-A GARCH Parameter estimates | | | | | | |
|---|---|---|---|---|---|---|
| | ω | α | β | | AIC | SC |
| India | 0.00072(0.0001)*** | 0.049467(0)*** | 0.936642(0)*** | | -6.67995 | -6.6642 |
| Indonesia | 0.000642(0.0002)*** | 0.091244(0)*** | 0.889916(0)*** | | -6.71618 | -6.700427 |
| Malaysia | 0.000174(0.0972)* | 0.102175(0)*** | 0.87291(0)*** | | -7.77608 | -7.760327 |
| Philippines | 0.000587(0.0049)*** | 0.087571(0)*** | 0.881825(0)*** | | -6.50269 | -6.486935 |
| Singapore | 0.000544(0.0003)*** | 0.077022(0)*** | 0.922406(0)*** | | -6.9593 | -6.943542 |
| Thailand | 0.000279(0.0565)** | 0.064058(0)*** | 0.923501(0)*** | | -7.16359 | -7.147838 |
| **Panel-B EGARCH Parameter estimates** | | | | | | |
| | ω | α | β | λ | AIC | SC |
| India | 0.000566(0.0025)*** | 0.104433(0)*** | 0.971584(0)*** | -0.092811(0)*** | -6.70074 | -6.681835 |
| Indonesia | 0.000474(0.0063)*** | 0.143629(0)*** | 0.975646(0)*** | -0.083841(0)*** | -6.7299 | -6.710993 |
| Malaysia | 0.000118(0.2578) | 0.149935(0)*** | 0.976914(0)*** | -0.078105(0)*** | -7.7843 | -7.765397 |
| Phillippines | 0.000387(0.0632)* | 0.132928(0)*** | 0.966412(0)*** | -0.07945(0)*** | -6.51458 | -6.495671 |
| Singapore | 0.000439(0.0035)*** | 0.171011(0)*** | 0.978098(0)*** | -0.079662(0)*** | -6.97679 | -6.957881 |
| Thailand | 0.00015(0.3036) | 0.093419(0)*** | 0.985896(0)*** | -0.069716(0)*** | -7.17902 | -7.160113 |

Source: Author's computation Notes: figures in parentheses are p-values, ***denotes significantat 1%, **denotes significantat 5%, and *denotes significant at 10% . AIC Akaike Information Criterian, SC Schewertz Information Criterian .

a high ARCH parameter implies high short-run volatility and high GARCH parameters indicate high long-run volatility.

The result of univariate GARCH model Table 4, Panel A indicates that the coefficients ARCH (α) and GARCH (β) of ASEAN-5 and India are positive and significant. Results revealed that the sum of the coefficients of α and β is close to unity, which implies that the shocks to the conditional variance are highly persistent in the short run and long run. Hence, these results will help to understand the significance levels of the persistence of shocks in the long and short run. Among all selected indices Malaysia has a high ARCH parameter implies high short-run volatility. However, India has a high GARCH parameter indicates high long-run volatility.

Table 4, Panel-B presents the results of the univariate EGARCH model. EGARCH model indicates the leverage effect. Asymmetry measure λ is negative and significant for all the selected indices which indicates the leverage effect. It conveys that the bad news generates more volatility than good news in these selected indices. GARCH coefficient β is positive and significant for all indices, indicating that the shocks are highly persistent in the long run. The ARCH coefficient α is positive and significant for all indices, conveying that the shocks are persistent in the short term. Among all indices, Singapore has a high ARCH parameter implies high short-run volatility. However, Thailand has high GARCH parameters indicate high long-run volatility. The sum of α and β implies the overall persistence of the series. A statistically significant and close to unity value of (α+β) gives evidence in favour of the persistence of shocks or persistence of volatility. After the univariate estimates, we investigate the Dynamic Conditional Correlations of commodity and equity pairs by using symmetric and asymmetric versions of different DCC .

*Table 5. DCC-GARCH Estimates of ASEAN-5 & India and Inter ASEAN-5 Pairs*

| Panel- A: DCC-GARCH Estimates of ASEAN-India | | | | | | |
|---|---|---|---|---|---|---|
| | $\theta_1$ | $\theta_2$ | DF | AIC | SIC | Log likelihood |
| India-Indonesia | 0.010937(0.15) | 0.955058(0)*** | 7.309913(0)# | -13.5225 | -13.5102 | 11725.91 |
| India-Malaysia | 0.040533(0.13) | 0.269752(0.32) | 7.630519(0)# | -14.4451 | -14.4341 | 12526.03 |
| India-Phillippines | -0.00896(0.57) | 0.611665(0.36) | 9.447471(0)# | -13.2644 | -13.2512 | 11501.62 |
| India-Singapore | 0.024686(0)*** | 0.953407(0)*** | 7.374753(0)# | -13.7983 | -13.7955 | 11973.05 |
| India-Thailand | 0.030527(0)*** | 0.934748(0)*** | 8.153071(0)# | -14.0401 | -14.0418 | 12186.33 |
| **Panel- B: DCC-GARCH Estimates of Inter ASEAN** | | | | | | |
| | $\theta_1$ | $\theta_2$ | DF | AIC | SIC | Log likelihood |
| Indonesia-Malaysia | 0.011262(0.12) | 0.96769(0)*** | 6.58568(0)# | -14.4669 | -14.4562 | 12545.19 |
| Indonesia-Phillippines | 0.017337(0)*** | 0.958095(0)*** | 6.794939(0)# | -13.3911 | -13.3764 | 11610.08 |
| Indonesia-Singapore | 0.015776(0.05)** | 0.954186(0)*** | 6.158478(0)# | -13.8209 | -13.8059 | 11982.07 |
| Indonesia-Thailand | 0.024423(0.00)*** | 0.961213(0)*** | 6.769447(0)# | -14.0416 | -14.0515 | 12194.72 |
| Malaysia-Phillippines | 0.020978(0.38) | 0.347491(0.47) | 8.650803(0)# | -14.2625 | -14.2466 | 12363.69 |
| Malaysia-Singapore | -0.0046(0.72) | 0.811313(0)*** | 6.668518(0)# | -14.7225 | -14.7131 | 12767.68 |
| Malaysia-Thailand | -0.004316(0.71) | 0.873679(0)*** | 7.715303(0)# | -14.932 | -14.9167 | 12943.94 |
| Phillipines-Singapore | 0.008502(0.14) | 0.981127(0)*** | 8.080985(0)# | -13.5432 | -13.5291 | 11742.3 |
| Phillippines-Thailand | 0.00766(0.09)* | 0.977962(0)*** | 9.044403(0)# | -13.7986 | -13.785 | 11963.91 |
| Singapore-Thailand | 0.030229(0)*** | 0.94929(0)*** | 7.292572(0)# | -14.3101 | -14.3108 | 12419.27 |

Source: Author's own computation Notes: figures in parentheses are p-values, ***denotes significantat 1%, **denotes significantat 5%, and *denotes significant at 10% . statistical stability condition ($\theta_1 + \theta_2 < 1$)is proved for all pairs. DF represents degree of freedom, # denotes the significance of DF at 1% significance.

Table 5, Panel A represents the result of the DCC-GARCH parameter estimates of ASEAN-5 & India. Results revealed that the coefficient θ1 denotes the short-term persistence of shocks to conditional correlation. θ1 is positive and significant for India-Singapore, India-Thailand conveys that the shocks to the dynamic conditional correlations are positive and highly persistent in the short run. The coefficient θ2 denotes long-term persistence of shocks. θ2 is positive and significant for India-Singapore, India-Thailand and India-Indonesia conveys that the shocks to the dynamic conditional correlations are positive and highly persistent in the long run. θ1 and θ2 are positive and significant, and (θ1 + θ2)is also found to be less than one. Thus, the overall stability condition of DCCMGARCH method is met. Significance of DCC parameters implies a substantial time-varying co-movement.

Table 6 panel -A presents the results of the ADCC- EGARCH estimates of ASEAN-5 & India. The coefficient of an asymmetric DCC EGARCH, θ3 is positively significant for India-Malaysia. India-Singapore and India-Thailand conveys that good news generates more volatility to the dynamic conditional correlations than bad news. The researcher noticed that the coefficient θ2, long run persistence of shocks to the conditional correlations is positive and highly significant for the all the series except India-Phillippines.

*Table 6. ADCC-EGARCH Estimates of ASEAN-5 & India and Inter ASEAN-5 Pairs*

| Panel A: ADCC EGARCH parameters of ASEAN-5 & India | | | | | | | |
|---|---|---|---|---|---|---|---|
| | $\theta_1$ | $\theta_2$ | $\theta_3$ | DF | AIC | SIC | Log likelihood |
| India-Indonesia | 0.014682(0.27) | 0.941527(0)*** | -0.001372(0.86) | 7.598327(0) # | -13.5465 | -13.5364 | 11754.65 |
| India-Malaysia | 0.000401(0)*** | 0.220608(0)*** | 0.103738(0)*** | 8.060243(0) # | -14.4668 | -14.4889 | 12579.51 |
| India-Phillippines | -0.028985(na) | 0.541043(na) | 0.016047(na) | 10.24007(na) | -13.2935 | -13.2786 | 11531.4 |
| India-Singapore | 0.015667(0.037)** | 0.953332(0)*** | 0.009544(0.031)** | 8.057258(0) # | -13.8317 | -13.8333 | 12011.79 |
| India-Thailand | 0.018298(0.029)** | 0.931959(0)*** | 0.015738(0)*** | 9.425009(0) # | -14.0653 | -14.0696 | 12216.42 |
| **Panel B: ADCC EGARCH parameters of inter ASEAN-5** | | | | | | | |
| | $\theta_1$ | $\theta_2$ | $\theta_3$ | DF | AIC | SIC | Log likelihood |
| Indonesia-Malaysia | -0.009807(0.67) | 0.735997(0.05)** | 0.024357(0.33) | 6.869886(0) # | -14.4837 | -14.473 | 12565.8 |
| Indonesia-Phillippines | 0.008657(0.19) | 0.961094(0)*** | 0.007273(0.08)* | 7.6231(0) # | -13.4093 | -13.3963 | 11633.3 |
| Indonesia-Singapore | 0.012111(0.14) | 0.953357(0)*** | 0.003749(0.49) | 6.636178(0) # | -13.8448 | -13.8345 | 12012.79 |
| Indonesia-Thailand | 0.02127(0)*** | 0.961212(0)*** | 0.004703(0.29) | 7.610571(0) # | -14.0687 | -14.0713 | 12217.9 |
| Malaysia-Phillippines | 0.024328(0.43) | 0.355914(0.51) | -0.006452(0.91) | 9.553589(0) # | -14.2779 | -14.261 | 12382.19 |
| Malaysia-Singapore | -0.008829(0.61) | 0.832616(0)*** | 0.001513(0.93) | 7.150134(0) # | -14.7438 | -14.7332 | 12791.08 |
| Malaysia-Thailand | -0.022743(0)*** | 0.354259(0)*** | 0.037931(0.12) | 8.541937(0) # | -14.9462 | -14.934 | 12964.95 |
| Phillipines-Singapore | 0.006229(0.23) | 0.972801(0)*** | 0.003657(0.42) | 9.292881(0) # | -13.5694 | -13.5566 | 11772.13 |
| Phillippines-Thailand | 0.004877(0.29) | 0.973545(0)*** | 0.006471(0.09)* | 10.68216(0) # | -13.818 | -13.803 | 11985.56 |
| Singapore-Thailand | 0.019589(0)*** | 0.954879(0)*** | 0.010803(0.026)** | 8.508362(0) # | -14.3323 | -14.338 | 12448.83 |

Source: Author's own computation Notes: figures in parentheses are p-values, ***denotes significantat 1%, **denotes significantat 5%, and *denotes significant at 10% . statistical stability condition ($\theta_1 + \theta_2 < 1$) is proved for all pairs. DF represents degree of freedom, # denotes the significance of DF at 1% significance Table 5, Panel-B presents the results of DCC-GARCH parameter estimates of inter ASEAN-5. Results revealed that the coefficient θ1 denotes the short-term persistence of shocks to conditional correlation. θ1 is positive and significant for 5 pairs out of 10 pairs conveys that the shocks to the dynamic conditional correlations are positive and highly persistent in the short run. The coefficient θ2 denotes long-term persistence of shocks. θ2 is positive and significant for all indices except Malaysia-Singapore conveys that the shocks to the dynamic conditional correlations are positive and highly persistent in the long run.

However, the optimization is failed for India-Phillippines pair. The coefficient θ1, short-run persistence of shocks to the conditional correlations is positive and significant for India-Malaysia, India-Singapore and India-Thailand. The coefficients of θ1 + θ2 are less than one which proves the stability condition of all pairs. Significance of DCC parameters implies a substantial time-varying co-movements.

Table 6 panel -B presents θ3 is positively significant only for Indonesia-Phillippines, Phillippines-Singapore and Singapore-Thailand conveys that good news generates more volatility to the dynamic conditional correlations than bad news. The researcher noticed that the coefficient θ2, long run persistence of shocks to the conditional correlations is positive and highly significant for the all the series except Malaysia-Phillippines. The coefficient θ1, short-run persistence of shocks to the conditional correlations is positive and significant for Indonesia-Thailand Singapore-Thailand. however, θ1 is negative and significant for Malaysia-Thailand.

### 5.1. Pairwise Time-Varying Co-Movements of ASEAN-5 and India and Inter ASEAN-5 Pairs

Figure 2 depicts the pairwise time-varying comovements of ASEAN-5 & India pairs. Figure 2. is subdivided into 5 plots. Plot 1 represents India-Indonesia pair comovements, plot 2 depicts India-Malaysia comovements, plot 3 for India-Philippines and plot 5 for India-Thailand. Indian capital market should recognize the opportunities presented by the ASEAN's growth and economic integration. It will be important for Indian Businesses to understand their strength in the ASEAN market and to tailor their strategies accordingly. India need to think of ASEAN as a whole with distinctive regions to get a clear undestanding.

Plots of India-Indonesia, India-Singapore and India-Thailand pairs show the time-varying comovements between the pairs during the sample period. Conditional correlations between the above-mentioned pairs are time-varying, remained positive and relatively high during the time span. Throughout the time frame Correlations of these pairs are found to be more than 0.3 for most of the time. On the other hand, plot-2 and plot-3 depict the time-varying comovements of India-Malaysia and India-Phillippines. Indeed, study denotes evidence of moderate co-movements in these pairs. Throughout the time frame correlations of these pairs are found to be less than 0.2 for most of the time. The researcher found a steep decline in correlations in most of the pairs in 2012 and in 2014 the possible reason may be After annual economic growth of nearly 9% in 2009-10 and 2010-11, the country is likely to grow at 6.5% in 2011-12. The expectations for 2012-13 are not too encouraging. On the other hand, Euro zone crisis impact on ASEAN and India. Steep decline 2014 may be the large rupee depreciation in 2013–14 . It can thus be inferred that Indian asset markets are more vulnerable to internal shocks as well as external shocks.

Figure 3 depicts the pairwise time-varying comovements of inter ASEAN-5 pairs. Figure 3. is subdivided into 10 plots. Plots 2, 3 and 4 of Indonesia-Phillippines, Indonesia-Singapore and Indonesia-Thailand respectively have a similar kind of dynamic correlations behaviour. On the other hand plot, 1 Indonesia-Malaysia pair shows the low comovements throughout the time frame Indeed, the researcher found evidence of moderate co-movements in this pair. Throughout the time frame correlations of these pairs are found to less than 0.5 for most of the time.

Plots 5, 6 and 7 of Malaysia-Phillippines, Malaysia-Singapore and Malaysia-Thailand shows the low comovements throughout the time frame. Throughout the time frame correlations of these pairs are found to less than 0.1 for most of the time. Low time varying co-movements of these pairs may offer diversification benefits to the investors.

Plots 8, 9 Philippines-Singapore and Philippine-Thailand show pairs are time-varying, remained positive and relatively high during the time span. Throughout the time frame Correlations of these pairs are found to be more than 0.5 for most of the time.plot 10

### 5.2 Granger Causality

Granger causality (Granger, 1969) analyses to what extent the change of past values of one variable account for the later variation of other variables. Granger causality test usually analyses two variables together, testing their interaction. All of the possible permutations of the two variables are:

*Figure 2. Dynamic conditional correlations of ASEAN-India pairs*

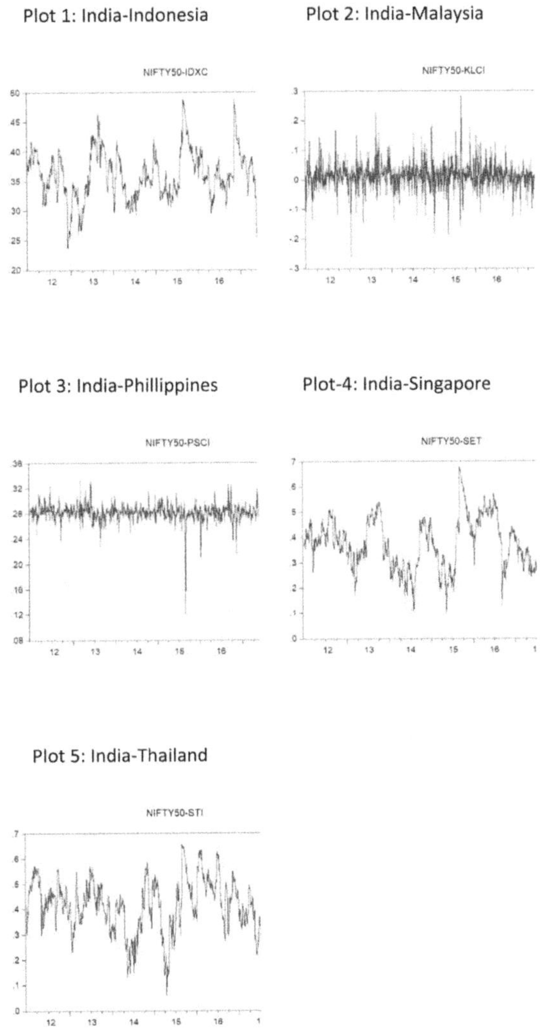


- Unidirectional Granger causality from variables yt to variables xt,
- Unidirectional Granger causality from variables xt to variables yt,
- Bi-directional casualty,
- No causality.

In all possible cases, a common assumption is that the data are stationary. Stationarity in a Random Process implies that its statistical characteristics do not change with time. If not the Granger causality on non-stationary time data can lead to false casual relation (Cheng, 1996). After confirming that all selected series are stationary at level, the study proceeds to perform the pairwise Granger's causality analysis for the ASEAN-5 & India and inter ASEAN.

*Figure 3. Dynamic conditional correlations of inter ASEAN pairs*

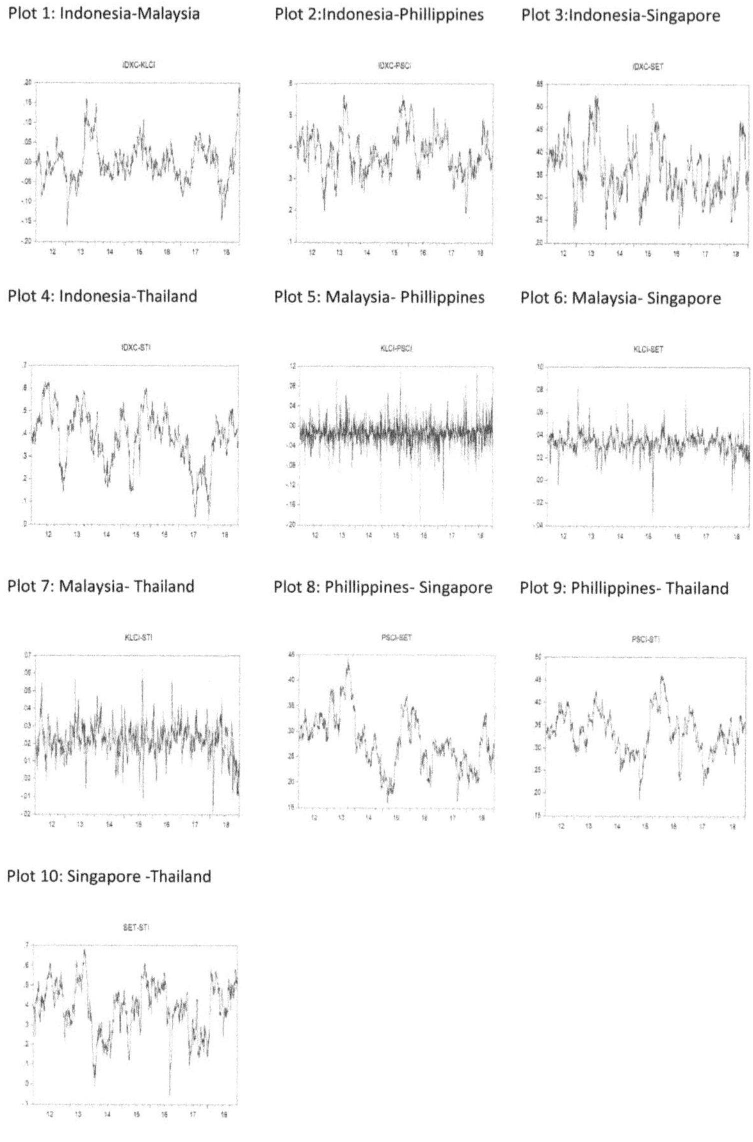


According to the Granger causality test with the usual level of significance of 5%, the study rejects the null hypothesis with the assumption that the change of one country's equity market return does not affect other country's Returns. The results of the Granger causality test are consolidated in Table 7 panel A and Panel B. The Granger causality indicates the short term integration of returns among the markets (Granger 1969). Table 7 panel A presents the causality results of ASEAN-5& India pairs. The causality study shows that the bidirectional causal relation in the following pairs India-Indonesia, and India-Singapore. It indicates that the predictions of the one country's equity returns are based on its past values along with past values of another country's equity market returns, which is better than the predictions based only on the past values of own index returns vice-versa. The researcher found a unidirectional causal relationship between the returns of India-Phillippines and India-Thailand. It indicates that the predictions of the one

*Table 7. Results of pairwise Granger Causality*

| **Panel A: Pairwise Granger Causality of India-ASEAN 5** | | | |
|---|---|---|---|
| Null Hypothesis: | Obs | F-Statistic | Prob. |
| Indonesia returns does not Granger Cause India returns | 1730 | 3.08823 | 0.0458*** |
| India returns does not Granger Cause Indonesia returns | | 8.49144 | 0.0002*** |
| Malaysia returns does not Granger Cause India returns | 1730 | 0.45233 | 0.6362 |
| India returns does not Granger Cause Malaysia returns | | 2.59812 | 0.0747 |
| Phliphines returns does not Granger Cause India returns | 1730 | 2.37826 | 0.093 |
| India returns does not Granger Cause phillippines returns | | 25.2464 | 0.0002*** |
| Singapore returns does not Granger Cause india returns | 1730 | 3.94187 | 0.0196*** |
| India returns does not Granger Cause Singapore returns | | 5.73989 | 0.0033*** |
| Thailand returns does not Granger Cause india returns | 1730 | 1.7694 | 0.1707 |
| India returns does not Granger Cause Thailand returns | | 8.57536 | 0.0002*** |
| **Panel B: Pairwise Granger Causality of Intra ASEAN 5** | | | |
| Null Hypothesis: | Obs | F-Statistic | Prob. |
| Malaysia returns does not Granger Cause Indonesia returns | 1730 | 0.48679 | 0.6147 |
| Indonesia returns does not Granger Cause Malaysia returns | | 1.71566 | 0.1802 |
| phillippines returns does not Granger Cause Indonesia returns | 1730 | 0.58463 | 0.5574 |
| Indonesia returns does not Granger Cause phillippines returns | | 21.9246 | 0.0004*** |
| Singapore returns does not Granger Cause Indonesia returns | 1730 | 4.61381 | 0.01*** |
| Indonesia returns does not Granger Cause Singapore returns | | 4.14978 | 0.0159*** |
| Thailand returns does not Granger Cause Indonesia returns | 1730 | 0.96973 | 0.3794 |
| Indonesia returns does not Granger Cause Singapore returns | | 4.05491 | 0.0175*** |
| Phillippines returns does not Granger Cause Malaysia returns | 1730 | 1.58251 | 0.2058 |
| Malassia returns does not Granger Cause Phillippines returns | | 1.58027 | 0.2062 |
| Singapore returns does not Granger Cause Malaysia returns | 1730 | 4.2123 | 0.015*** |
| Malaysia returns does not Granger Cause Singapore returns | | 2.71474 | 0.0665 |
| Thailand returns does not Granger Cause Malaysia returns | 1730 | 2.05078 | 0.1289 |
| Malaysia returns does not Granger Cause Thailand returns | | 1.96676 | 0.1402 |
| Singapore returns does not Granger Cause Phillippines returns | 1730 | 27.4144 | 0.0002*** |
| Phillippines returns does not Granger Cause Singapore returns | | 0.09901 | 0.9057 |
| Thailand returns does not Granger Cause Phillippines returns | 1730 | 10.0408 | 0.00005*** |
| Phillippines returns does not Granger Cause Thailand returns | | 4.28331 | 0.0139*** |
| Thailand returns does not Granger Cause Singapore returns | 1730 | 0.56527 | 0.5683 |
| Singapore returns does not Granger Cause Thailand returns | | 5.18456 | 0.0057*** |

Source: Author's own computation Notes: *** Denotes rejections of null hypothesis at 5%,

country's equity returns are based on its past values along with past values of another country's equity returns which is better than the predictions based only on the past values of its own index returns. No causal relationship found in between India and Malysia. Finally, from the results researcher concluded that the strong interdependency between the ASEAN-5 & Indian equity market returns.

Table 7 panel B presents the causality results of inter ASEAN-5 pairs. Singapore-Indonesia, Thailand-Phillippines shows the bidirectional causation. However, Indonesia-Philippines, Indonesia-Singapore, Singapore-Malaysia, Singapore-Phillippines and Singapore-Thailand show the unidirectional causation.

## 5.3 Variance Decomposition Method

The study used variance decomposition analysis to assess to what extent shocks to certain markets are explained by other markets in the system. This forecast error can be accounted for by its own innovations and the innovations of other variables in the system.

Table 8 shows the results of variance decomposition of ASEAN 5-India pairwise. The results revealed that Indonesia,Phillippines,Singapore and Thailand equity indices returns have a negligible contribution to the variance of the India's equity index. However, India's equity explains approximately 15%, 12%, 18% and 22% variations in Indonesia,Phillippines,Singapore and Thailand respectively. This result suggests India's equity index volatility contributes more fluctuations in the variance of Indonesia, Phillippines, Singapore and Thailand

On the other hand, Malaysia's equity index has a negligible contribution to the variance of the India's equity index. However, the India's equity index also explains negligible variations in Malaysia . The researcher found the strong endogenous impact on its own index (India) from the short run to the long run. Because own index exhibit strong influence and other variables have negligible impact in predicting the variations in the variance. However, the India's equity index is a week predictor in predicting the variations in the variances of the other equity index of Malaysia. The researcher computed variance decomposition for 10 periods from 1 to 10 but presented only significant estimates in the table 8. Finally, variance decomposition study revealed that India's equity index volatility contributes moderate fluctuations in the variance of Indonesia, Phillippines, Singapore and Thailand.

## 5.4 Markov 2 Regime Switchings of ASEAN-5 and India

The integration relation between India-Asian stock markets is characterized as a two regime process, The first regime is characterized by low stock market integration between India-ASEAN-5. The second regime is characterized by high integration. Investors can use the conditional information of regime switching model to outperform portfolio investment strategies. Keeping in view of place constraint researcher didn't explain methodology in detail. Based on requirement researcher will produce the methodology. The regime-switching model used in the study allows for the identification of dominant market integration regimes.

Table 9 reports the results of the two-regime Markov switching transition probabilities of the ASEAN-India stock markets. The table provides the transition probabilities of switching from each respective regime over the period Jan 2012 to December 2018. The integration relation between ASEAN-5 and Indian stock markets is characterized as a two-regime process, where regimes are stable and generally persistent.

*Table 8. Summary Results of Pairise Variance Decomposition*

| India | | | | Indonesia | | | |
|---|---|---|---|---|---|---|---|
| **Period** | **s.e** | **India** | **Indonesia** | **Period** | **s.e** | **India** | **Indonesia** |
| 1 | 0.009055 | 100 | 0 | 1 | 0.009498 | 14.85933 | 85.14067 |
| 3 | 0.009096 | 99.64616 | 0.353841 | 3 | 0.009569 | 15.93217 | 84.06783 |
| 5 | 0.009096 | 99.64412 | 0.355878 | 5 | 0.009569 | 15.93884 | 84.06116 |
| 10 | 0.009097 | 99.64412 | 0.355883 | 10 | 0.009569 | 15.93886 | 84.06114 |
| **India** | | | | **Malaysia** | | | |
| **Period** | **s.e** | **India** | **Malaysia** | **Period** | **s.e** | **India** | **Malaysia** |
| 1 | 0.009069 | 100 | 0 | 1 | 0.005502 | 0.075635 | 99.92436 |
| 3 | 0.009096 | 99.95007 | 0.049933 | 3 | 0.005528 | 0.415831 | 99.58417 |
| 5 | 0.009096 | 99.94887 | 0.051131 | 5 | 0.005528 | 0.417643 | 99.58236 |
| 10 | 0.009096 | 99.94887 | 0.051133 | 10 | 0.005528 | 0.417645 | 99.58235 |
| **India** | | | | **Philipphines** | | | |
| **Period** | **s.e** | **India** | **Philipphines** | **Period** | **s.e** | **India** | **Philipphines** |
| 1 | 0.009059 | 100 | 0 | 1 | 0.009916 | 8.609374 | 91.39063 |
| 3 | 0.009096 | 99.71274 | 0.28726 | 3 | 0.010095 | 11.66146 | 88.33854 |
| 5 | 0.009096 | 99.7121 | 0.2879 | 5 | 0.010096 | 11.6702 | 88.3298 |
| 10 | 0.009096 | 99.7121 | 0.287905 | 10 | 0.010096 | 11.67024 | 88.32976 |
| **India** | | | | **Singapore** | | | |
| **Period** | **s.e** | **India** | **Singapore** | **Period** | **s.e** | **India** | **Singapore** |
| 1 | 0.009051 | 100 | 0 | 1 | 0.008551 | 17.67596 | 82.32404 |
| 3 | 0.009096 | 99.56637 | 0.433627 | 3 | 0.008588 | 18.36219 | 81.63781 |
| 5 | 0.009096 | 99.55981 | 0.440195 | 5 | 0.008588 | 18.36189 | 81.63811 |
| 10 | 0.009096 | 99.55976 | 0.440241 | 10 | 0.008588 | 18.3619 | 81.6381 |
| **India** | | | | **Thailand** | | | |
| **Period** | **s.e** | **India** | **Thailand** | **Period** | **s.e** | **India** | **Thailand** |
| 1 | 0.009062 | 100 | 0 | 1 | 0.007264 | 20.60594 | 79.39406 |
| 3 | 0.009096 | 99.80817 | 0.191829 | 3 | 0.007301 | 21.23654 | 78.76346 |
| 5 | 0.009096 | 99.80748 | 0.19252 | 5 | 0.007301 | 21.23701 | 78.76299 |
| 10 | 0.009096 | 99.80748 | 0.192523 | 10 | 0.007301 | 21.23701 | 78.76299 |

Source: Author's own computation Notes: s.e denotes Standard Error.

Table 9 reports the results of transition probabilities and expected durations of ASEAN-5& India. *India-Indonesia:* transition probability of p11 = 0.513588 means that there is a 51.35% chance of remaining in Regime 1 and a 48.65% chance of shifting to Regime 2, whereas the p22 = 0.528427 means that there is a 52.85% chance of remaining in Regime 2 and a 47.15% chance of shifting to Regime 1. The low transition probabilities of p11 and p22 indicate a low degree of regime stability.

*Table 9. Markov 2 regime transition probabilities of ASEAN-5 & India*

| Transition probabilities | | | Expected Duration | |
|---|---|---|---|---|
| **India- Indonesia** | | | | |
| | 1 | 2 | 1 | 2 |
| 1 | 0.513588 | 0.486412 | 2.055868 | 2.120877 |
| 2 | 0.471503 | 0.528497 | | |
| **India- Malaysia** | | | | |
| | 1 | 2 | 1 | 2 |
| 1 | 0.984732 | 0.015268 | 65.49646 | 66.79491 |
| 2 | 0.014971 | 0.985029 | | |
| **India-Phillippines** | | | | |
| | 1 | 2 | 1 | 2 |
| 1 | 0.983179 | 0.016821 | 59.44941 | 70.51827 |
| 2 | 0.014181 | 0.985819 | | |
| **India- Singapore** | | | | |
| | 1 | 2 | 1 | 2 |
| 1 | 0.606157 | 0.393843 | 2.53908 | 1.887262 |
| 2 | 0.529868 | 0.470132 | | |
| **India- Thailand** | | | | |
| | 1 | 2 | 1 | 2 |
| 1 | 0.984303 | 0.015697 | 63.70671 | 91.27481 |
| 2 | 0.010956 | 0.989044 | | |

Source: Author's own computation

*India-Malaysia*:The transition probability of p11 = 0.984732 means that there is a 98.47% chance of remaining in Regime 1 and a 1.53% chance of shifting to Regime 2, whereas the p22 = 0.985029 means that there is a 98.50% chance of remaining in Regime 2 and a 1.50% chance of shifting to Regime 1. The high transition probabilities of p11 and p22 indicate a high degree of regime stability.

India-Philippines:The transition probability of p11 = 0.983179 means that there is a 98.31% chance of remaining in Regime 1 and a 1.69% chance of shifting to Regime 2, whereas the p22 = 0.985819 means that there is a 98.58% chance of remaining in Regime 2 and a 1.42% chance of shifting to Regime 1. The high transition probabilities of p11 and p22 indicate a high degree of regime stability.

India-Singapore: The transition probability of p11 = 0.606157 means that there is a 60.61% chance of remaining in Regime 1 and a 30.39% chance of shifting to Regime 2, whereas the p22 = 0.470132 means that there is a 47.01% chance of remaining in Regime 2 and a 52.99% chance of shifting to Regime 1. The low transition probabilities of p11 and p22 indicate a low degree of regime stability.

India-Thailand: The transition probability of p11 = 0.984303 means that there is a 98.43% chance of remaining in Regime 1 and a 1.57% chance of shifting to Regime 2, whereas the p22 = 0.985819 means that there is a 98.90% chance of remaining in Regime 2 and a 1.10% chance of shifting to Regime 1. The high transition probabilities of p11 and p22 indicate a high degree of regime stability.

*Figure 4. India-Indonesia's Markov switching filtered regime probabilities*

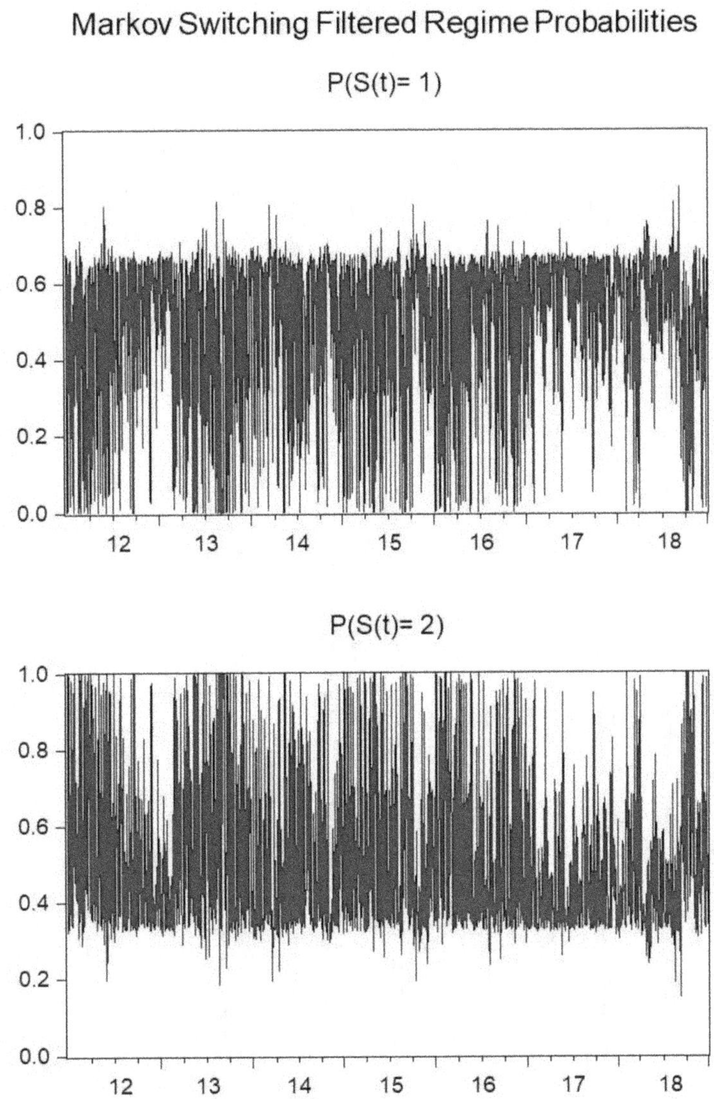


Markov Switching Filtered Regime Probabilities of ASEAN-5 and India

Figure 4 to figure8 show the Markov Switching Filtered Regime Probabilities of ASEAN-5 & India.

Figure 4 shows filtered transition probabilities of India-indonesia stock markets. Figure4 dipicts that about 36 percent of the entire sample belongs to Regime 1 (no or low India-Indonesia equity market integration). Thus, about 64 percent of the remaining sample represents a period of high India-Indonesia equity market integration (Regime 2). The periods of high India-Indonesia equity market integration occur (a) January 2012 to dec 2012; (b) April 2013 to feb 2014(this represents the longest connected period in Regime 2); (c)may 2015 to aug 2015 ; (d) dec 2015 to oct 2015; (e) dec 2015 to June 2016; f) oct 2016 to jan 2017 and finally, (f) july 2018 to Nov 2018.

*Figure 5. India-Malaysia's Markov switching filtered regime probabilities*

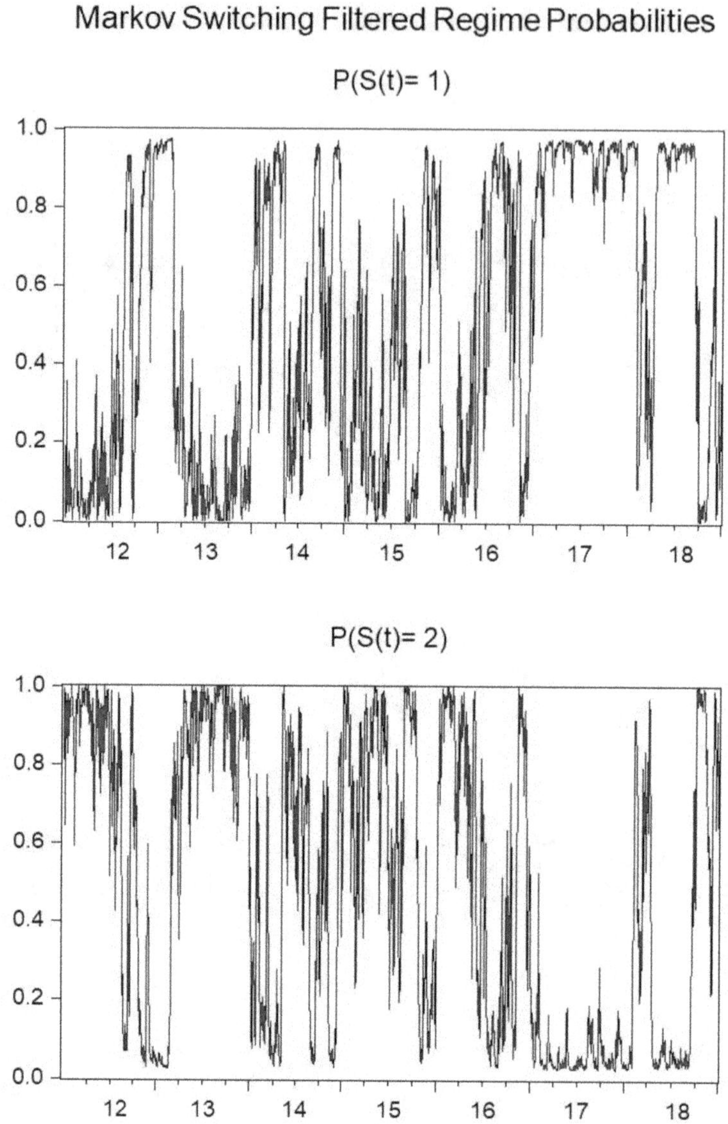


Figure 5 shows filtered transition probabilities of India-Malaysia stock markets. Figure5 shows that about 65 per cent of the entire sample belongs to Regime 1 (no or low India-Malaysia equity market integration). Thus, about 35 per cent of the remaining sample represents a period of high India-Malaysia equity market integration (Regime 2). The periods of high India-Malaysia equity market integration occur (a) January 2012 to August 2012; (b) March 2013 to October 2013 (this represents the longest connected period in Regime 2); (c) May 2014 to July 2014 ; (d) March 2015 to may 2015; (e) January 2016 to June 2016; and finally, (f) September 2018 to Nov 2018.

*Figure 6. India-Philipphines's Markov switching filtered regime probabilities*

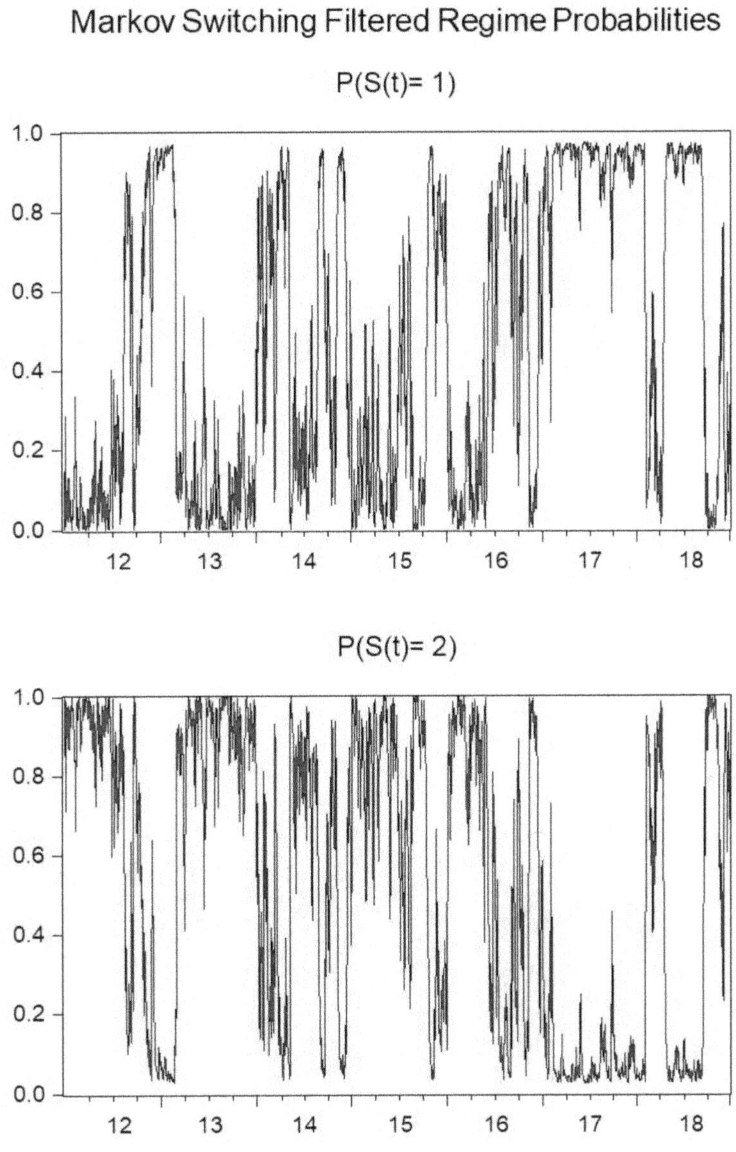


Notes: figure4 to figure8 shows filtered transition probabilities which are obtained by Hamilton's (1989) filter. Filtered transition probability estimates (blue spikes) greater than 50% indicate that this day belongs to the respective regime. The sample period is from Jan 2012 to December 2018. Regime 1 is a period of low integration and Regime 2 is a period of high integration between India and ASEAN-5.

Figure 6 shows filtered transition probabilities of India-Philippines stock markets. Figure6 shows that about 52 per cent of the entire sample belongs to Regime 1 (no or low India-Philippines equity market integration). Thus, about 48 per cent of the remaining sample represents a period of high India-Philippines equity market integration (Regime 2). The periods of high India-Phillippines equity market

*Figure 7. India-Singapore Markov switching filtered regime probabilities*

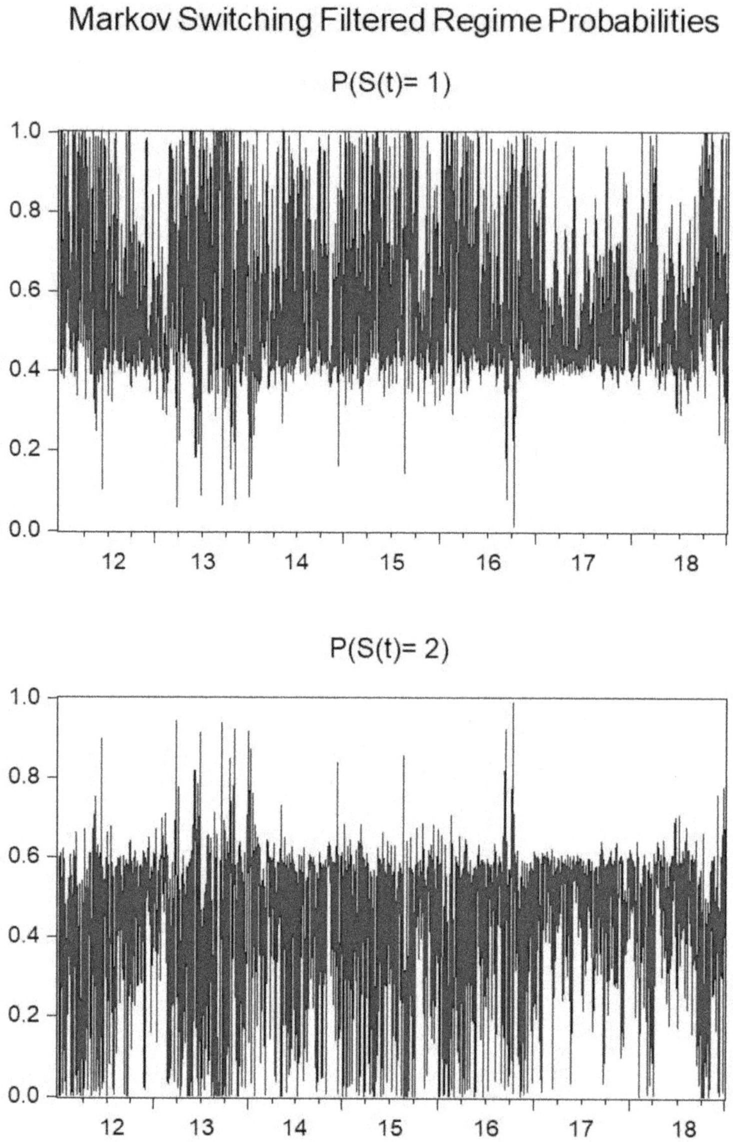


integration occur (a) January 2012 to August 2012; (b) March 2013 to Jan 2014 (c) May 2014 to July 2015(this represents the longest connected period in Regime 2); (d) nov 2015 to June 2016; (e) Oct 2016 to Dec 2016; and finally, (f) March 2018 to April 2018.

Figure 7 shows filtered transition probabilities of India-Singapore stock markets. Figure7 shows that about 60 per cent of the entire sample belongs to Regime 1 (no or low India-Singapore equity market integration). Thus, about 40 per cent of the remaining sample represents a period of high India-Singapore equity market integration (Regime 2). The low transition probabilities of p11 and p22 indicate a low degree of regime stability.

*Figure 8. India-Thailand Markov switching filtered regime probabilities*

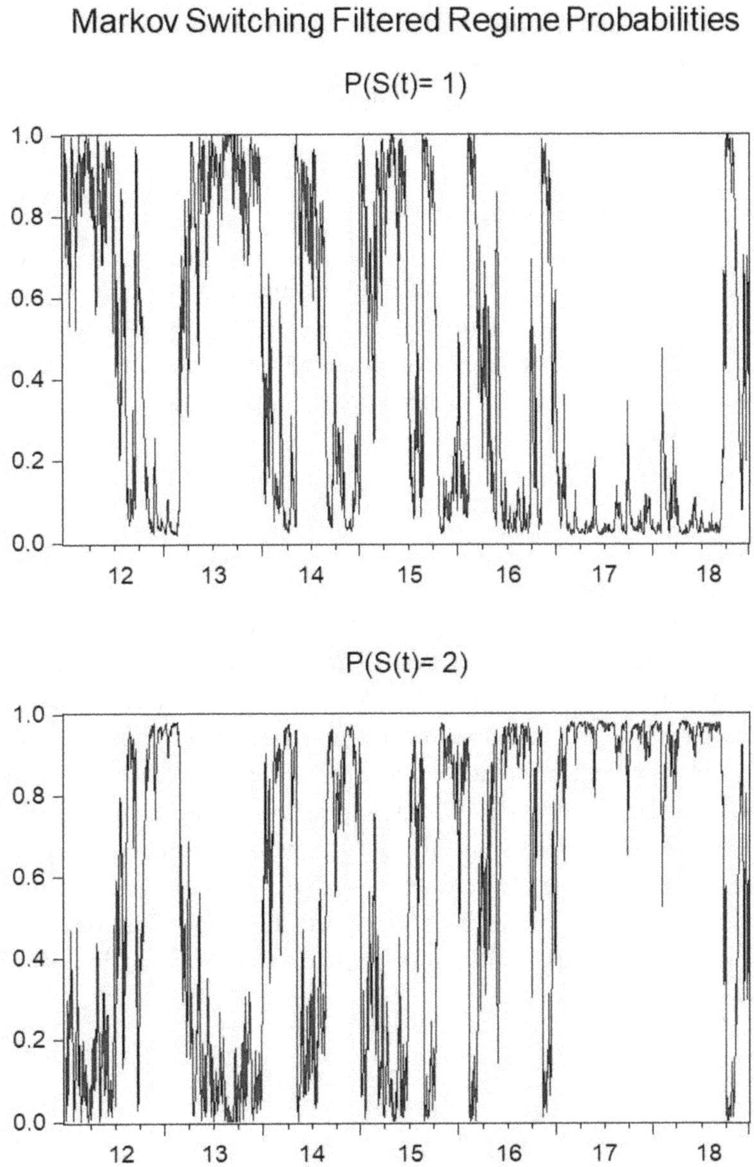


Figure 8 shows filtered transition probabilities of India-Thailand stock markets. Figure8 shows that about 31 per cent of the entire sample belongs to Regime 1 (no India-Thailand equity market integration). Thus, about 69 per cent of the remaining sample represents a period of high India-Thailand equity market integration (Regime 2). The periods of high India-Thailand equity market integration occur (a) January 2012 to March 2013; (b) Jan 2014 to March 2014; (c)July 2014 to feb 2015 ; (d) June 2015 to aug 2015; (e) oct 2015 to March 2016; f) April 2016 to Oct 2016 and finally, (f) Dec 2016 to Sep 2018(this represents the longest connected period in Regime 2).

Authors identified the existence of two stable regimes in India-Malaysia, India-Phillippines and India-Thailand pairs. p11 and p22 indicate a high degree of regime stability over the various stock market samples in this study. The possible reason may be herding behavior of investors and increasing timevarying co-movements of cross country markets due to contagion impact. Impact of the Eurozone crisis of 2010–12, and large rupee depreciation of 2013–14.

## 6. FUTURE RESEARCH DIRECTIONS

In this study, the authors focused on ASEAN-5 & Indian equity market relations and integration hence, other ASEAN countries can also be considered for analysis. In this study, the authors explored the time-varying comovements and integrations of the equity market however, the macroeconomic factors impact on ASEAN equity markets can be studied. Furthermore, commodity and bond markets integrations can be studied.

## 7. CONCLUSION

In sum, this study finds time-varying co-movements and volatility spillovers between ASEAN-5 & Indian Equity markets. Granger causality test results reveal that the strong interdependence between the ASEAN-5 & Indian equity market returns. Variance decomposition results show that India's equity market volatility contributes moderate fluctuations in the variance of Indonesia, Phillippines, Singapore and Thailand. Finally, The integration relation between ASEAN-5 & Indian equity market returns is characterized as a two-regime process. Results show that the high transition probabilities of p11 and p22 for India-Malaysia, India-Philippines and India-Thailand which indicates a high degree of regime stability. Furthermore, the knowledge of the interdependency and integrations of the cross markets is important for domestic and international investors for portfolio diversifications. The study on Financial integration provides important inputs to investors in sharing risk internationally since restrictions on investment are removed.

## REFERENCES

Abdennadher, E., & Hellara, S. (2018). Causality and contagion in emerging stock markets. *Borsa Istanbul Review*, *18*(4), 300–311. doi:10.1016/j.bir.2018.07.001

Alotaibi, A. R., & Mishra, A. V. (2017). Time varying international financial integration for GCC stock markets. *The Quarterly Review of Economics and Finance*, *63*, 66–78. doi:10.1016/j.qref.2016.03.001

Aloui, R., Aïssa, M. S. B., & Nguyen, D. K. (2011). Global financial crisis, extreme interdependences, and contagion effects: The role of economic structure? *Journal of Banking & Finance*, *35*(1), 130–141. doi:10.1016/j.jbankfin.2010.07.021

Asian Development Bank Annual Report. (2018). Asian Development Bank. Retrieved from https://www.adb.org/documents/adb-annual-report-2018

Baele, L., Ferrando, A., Hördahl, P., Krylova, E., & Monnet, C. (2004). *Measuring financial integration in the euro area* (No. 14). ECB occasional paper.

Balli, F., Balli, H. O., & Luu, M. N. (2014). Diversification across ASEAN-wide sectoral and national equity returns. *Economic Modelling*, *41*, 398–407. doi:10.1016/j.econmod.2014.05.035

Batten, J. A., Kinateder, H., Szilagyi, P. G., & Wagner, N. F. (2019). Time-Varying Energy and Stock Market Integration in Asia. *Energy Economics*, *80*, 777–792. doi:10.1016/j.eneco.2019.01.008

Beirne, J., Caporale, G. M., Schulze-Ghattas, M., & Spagnolo, N. (2013). Volatility spillovers and contagion from mature to emerging stock markets. *Review of International Economics*, *21*(5), 1060–1075. doi:10.1111/roie.12091

Bekaert, G., Harvey, C., & Lundblad, C. (2006). Growth volatility and financial liberalization. *Journal of International Money and Finance*, *25*(3), 370–403. doi:10.1016/j.jimonfin.2006.01.003

Bollerslev, T. (1986). Generalized autoregressive conditional heteroskedasticity. *Journal of Econometrics*, *31*(1), 307–327. doi:10.1016/0304-4076(86)90063-1

Boubakri, S., & Guillaumin, C. (2015). Regional integration of the East Asian stock markets: An empirical assessment. *Journal of International Money and Finance*, *57*, 136–160. doi:10.1016/j.jimonfin.2015.07.011

Cappiello, L., Engle, R. F., & Sheppard, K. (2006). Asymmetric dynamics in the correlations of global equity and bond returns. *Journal of Financial Econometrics*, *4*(4), 537–572. doi:10.1093/jjfinec/nbl005

Chevallier, J., Nguyen, D. K., Siverskog, J., & Uddin, G. S. (2018). Market integration and financial linkages among stock markets in Pacific Basin countries. *Journal of Empirical Finance*, *46*, 77–92. doi:10.1016/j.jempfin.2017.12.006

Chowdhury, M. A. F., Haque, M. M., & Islam, M. N. (2017). Contagion effects on stock market of Bangladesh: An empirical study on Dhaka stock exchange Shariah (DSES) index. *International Journal of Asian Business and Information Management*, *8*(2), 114. doi:10.4018/IJABIM.2017040101

Dewandaru, G., Masih, R., & Masih, A. M. M. (2016). Contagion and interdependence across Asia-Pacific equity markets: An analysis based on multi-horizon discrete and continuous wavelet transformations. *International Review of Economics & Finance*, *43*, 363–377. doi:10.1016/j.iref.2016.01.002

Dewandaru, G., Masih, R., & Masih, M. (2018). Unraveling the Financial Contagion in European Stock Markets During Financial Crises: Multi-Timescale Analysis. *Emerging Markets Finance & Trade*, *54*(4), 859–880. doi:10.1080/1540496X.2016.1266614

Dickey, D. A., & Fuller, W. A. (1979). Distribution of the estimates for autoregressive time series with a unit root. *Journal of the American Statistical Association*, *74*, 427–482.

Engle, R. (2002). Dynamic conditional correlation: A simple class of multivariate generalized autoregressive conditional heteroskedasticity models. *Journal of Business & Economic Statistics*, *20*(3), 339–350. doi:10.1198/073500102288618487

Frankel, J. A., & Rose, A. K. (1998). The endogenity of the optimum currency area criteria. *Economic Journal (London)*, *108*(449), 1009–1025. doi:10.1111/1468-0297.00327

Gil-Alana, L., Carcel, H., & Abakah, E. J. A. (2018). On the linkages between Africa's emerging equity markets and global markets: Evidence from fractional integration and cointegration. *Review of development finance, 8*(2), 96-105.

Gourinchas, P.-O., & Jeanne, O. (2006). The elusive gains from international financial integration. *The Review of Economic Studies, 73*(3), 715–741. doi:10.1111/j.1467-937X.2006.00393.x

Granger, C. (1969). Investigating causal relations by econometric models and cross-spectral methods. *Econometrica, 37*(3), 424–438. doi:10.2307/1912791

Hamilton, J. (1989). A new approach to the economic analysis of nonstationary time series and the business cycle. *Econometrica, 57*(2), 357–384. doi:10.2307/1912559

Hamilton, J. D. (2005). *Regime-Switching Models* (2nd ed.). The New Palgrave Dictionary of Economics.

Kenourgios, D., & Samitas, A. (2011). Equity market integration in emerging Balkan markets. *Research in International Business and Finance, 25*(3), 296–307. doi:10.1016/j.ribaf.2011.02.004

Komatsubara, T., Okimoto, T., & Tatsumi, K. I. (2017). Dynamics of integration in East Asian equity markets. *Journal of the Japanese and International Economies, 45*, 37–50. doi:10.1016/j.jjie.2017.07.002

Majdoub, J., & Mansour, W. (2014). Islamic equity market integration and volatility spillover between emerging and US stock markets. *The North American Journal of Economics and Finance, 29*, 452–470. doi:10.1016/j.najef.2014.06.011

Mensah, J. O., & Premaratne, G. (2018). Integration of ASEAN Banking Sector Stocks. *Journal of Asian Economics, 59*, 48–60. doi:10.1016/j.asieco.2018.10.001

Obstfeld, M. (1994). Risk-taking, global diversification, and growth. *The American Economic Review, 84*(5), 1310–1329.

Pentecôte, J. S., & Rondeau, F. (2015). Trade spillovers on output growth during the 2008 financial crisis. *Inter Economics, 143*, 36–47. doi:10.1016/j.inteco.2015.04.003

Perumandla, S., & Kurisetti, P. (2018). Time - Varying Correlations, Causality, and Volatility Linkages of Indian Commodity and Equity Markets: Evidence from DCC - GARCH. *Indian Journal Of Finance, 12*(9), 21–40. doi:10.17010/ijf/2018/v12i9/131558

Perumandla, S., & Kurisetti, P. (in press). An empirical examination of correlation dynamics between commodity and equity derivative indices: Evidence from India using DCC-GARCH models. *Afro Asian Journal of Finance and Accounting*.

Phillips, P. C. B., & Perron, P. (1988). Testing for Unit Roots in Time Series Regression. *Biometrika, 75*(2), 335–346. doi:10.1093/biomet/75.2.335

Teng, K. T., Yen, S. H., Chua, S. Y., & Lean, H. H. (2013). Financial market integration of ASEAN-5 with China and India. In *Emerging markets and the global economy: A handbook* (pp. 825-859). Elsevier.

Vinayak, H. V., Thompson, F., & Tonby, O. (2014). Understanding ASEAN: Seven things you need to know. *growth, 2000*, 13.

Virk, N., & Javed, F. (2017). European equity market integration and joint relationship of conditional volatility and correlations. *Journal of International Money and Finance*, *71*, 53–77. doi:10.1016/j.jimonfin.2016.10.007

Wälti, S. (2011). Stock market synchronization and monetary integration. *Journal of International Money and Finance*, *30*(1), 96–110. doi:10.1016/j.jimonfin.2010.07.004

Wright, M. L. (2005). On the gains from international financial integration. *Economics Letters*, *87*(3), 379–386. doi:10.1016/j.econlet.2005.01.007

# Chapter 3
# Do Team Dynamics Influence the Organizations to Be Innovative?

**Ikramul Hasan**
https://orcid.org/0000-0001-5510-1341
*School of Business, Independent University, Bangladesh*

**MD. Nazmul Islam**
https://orcid.org/0000-0002-6778-508X
*Asia-Europe Institute, University of Malaya, Malaysia*

**Mohammad Ashraful Ferdous Chowdhury**
https://orcid.org/0000-0001-8540-1353
*Department of Business Administration, Shahjalal University of Science and Technology, Bangladesh*

## ABSTRACT

*Innovating products or services are not an emerging issue rather a reality in today's marketplace. Organizations are relentlessly chasing innovation to compete with their rivals. Considering these phenomena, literature from the different management studies suggest that to being mechanistic it would better to become organic in its operation. The organic structure of the firm also highlights teamwork as a priority to become successful. This study illustrates the effects of team dynamics and mediating effects of organizational learning on organizational innovativeness. By proposing the conceptual model, the chapter presents different practical implications to the practitioners, researchers, and academicians connected to the industry and can be a source for future research Bangladesh and other developing economies to develop different insights.*

DOI: 10.4018/978-1-7998-1566-2.ch003

## INTRODUCTION

The world is always evolving, starting from technologies to business and economics. Increasing competition changes the necessities for the organizations to performance better. Innovating product or service is not an emerging issue rather a reality in today's business. To enhance the company's performance, it requires numerous qualities in firm operation (Jiménez-Jiménez et al., 2011; Aragón-Correa et al., 2007). In particular, to comprehend superior quality in operation, innovation can play a vital role for the businesses to adapt change based on the necessities (Garcia et al., 2008; Cho et al., 2007)? An organization is always dynamic with its competitive strategies (Grant, 2005). To deal with the current externalities organizations have to acquire new knowledge and skills to build competitive strategies to improve their performances (Child et al., 2005; Ortenblad, 2001). Kalshoven et al. (2012); Baker et al., (1999) also found in their studies that the new knowledge and skills obtained through organizational learning may enhance a firm's innovative capabilities to confirm competitiveness. Mavondo et al. (2005) argued that the most effective strategy for sustaining and improving a firm's competitive advantage is organizational learning. Team dynamics plays an important role to motivate employees to learn new things and provide new challenges to administer their team efforts. It also helps the employees to understand different organizational contextual ties of organizational trust, interpersonal communication, team expertise, and empowerment (Remedios et al., 2004).

However, very few studies have highlighted the importance of team dynamics and organizational learning for ensuring organizational innovativeness. In order to determine the innovative capabilities for the organizations literature from the different study connects the team dynamics, learning strategies with organizational innovation isolatedly. Especially, as per the knowledge previous studies have not highlighted the effects of team dynamics and organizational learning on organizational innovativeness in the context of business organization in Bangladesh. Hence, model that helps to enhance innovativeness of business organization in Bangladesh would add new knowledge in the contemporary literature of organizational study.

Recently, Bangladesh has shifted from a lower income country to lower-middle income country. The business sector of Bangladesh has played a major role for achieving the current status. In addition, PwC reported that Bangladesh would be placed as a 21st largest economy in the world by 2050 (Hawksworth et al., 2017). However, in order to go further, Bangladesh has to compete with the mature competitions for which business sector in Bangladesh have to bring innovation. On the other hand, research studies have not explored the effects of team dynamics and organizational learning on organizational innovativeness in the context of business organization in Bangladesh. In this circumstance, proper understanding of the mechanism for ensuring organizational innovativeness for business organizations in Bangladesh can enable the organizations to compete in highly competitive market.

Hence, the purpose of this study to develop a conceptual framework to illustrate the effects of team dynamics on organizational learning towards organizational innovativeness. This study fills a gap by proposing a framework for developing country like Bangladesh in an area of organizational innovation, which creates a new scope for the further studies. This study integrated team dynamics as an independent variable and organizational learning as a mediating variable. Different literature witnessed that team dynamics is strongly connected with organizational innovativeness, but this study also unfolds the mediating effects of organizational learning between team dynamics and organizational innovation. Mediating effect explores whether relationship of team dynamics and organizational innovativeness remain same when organizational learning strategy is stronger or weaker. Hence, independent and mediation effect of this proposed model delivers a new knowledge to the researcher, practitioners, policy makers as well as top management of a business organization how to improve organizational innovativeness with the support of team dynamics and organizational learning.

This paper is organized in three steps. First of all, dependent variable (organizational innovativeness), independent variable (team dynamics) and mediating variable (organizational learning) are defined based on the existing literature. Secondly, connection between each variables and mediating effects are explored. Finally, based on the discussion new framework is proposed followed by the implications, future research and conclusion.

## METHODS

In order to develop the literature as many studies as possible were explored and to limit biases caused the study-identification process, three procedures were considered as proposed by Islam et al. (2018) and Rosenbusch et al., (2011). A computerized keyword searches in the databases (e.g., ISI, Web of Knowledge, ABI, Google Scholar) was conducted to explore the relevant studies on the factors that influence team dynamics and organizational innovation. In addition, most relevant journals in organizational innovation (e.g., Journal of Knowledge Management, International Journal of Knowledge Management Studies, and Knowledge Management Research & Practice) and conference proceedings (e.g., European Conference on Organizational Knowledge, Learning, and Capabilities) were searched manually. As a third step, the reference sections of the relevant articles were searched. Three keywords "Team dynamics effects on organizational innovation" were used in the literature search.

## THEORY

### Organizational Innovation (OI)

Organizational Innovation (OI) refers to the reception or acceptance of new ideas or knowledge's, which either can be a new technology, product, process, service or administrative practice to the organization (Wood 1998; Walton 1987). According to Mol and Birkinshaw (2009), organizational innovation is known as "new to the specific organization". Moreover, Innovation is a process of bringing new idea for the organization in order to confirm value addition to the customers as well as for the organizational knowledge by making small or large, radical or incremental changes in the products, services or the process of organization. Similarly, organizational innovation is set up fresh or different process or model to the organization to ensure advancement in the organizational systems and especially, features of the new model will vary significantly from features of the existing model of the organization (Bloch and Bugge, 2013). Mostly, features of the innovation may vary depending on the types of the organization or types of the innovation. These types of features are known as the secondary characteristics of innovation (Downs and Mohr, 1979) and some features of innovation are related closely to the industrial context and do not vary from one organization to another are known as the primary characteristics of innovation (Popa et al., 2010). Notably, types of the organizational innovation are also differing in types such as technological innovation, product innovation, process innovation and administrative innovation. Furthermore, internal factors such as "market orientation, learning and technological policy" have direct impact on organizational innovation; similarly, "industry concentration and barriers to entry" as external factors also have direct impact on organizational innovation (Salavou et al., 2004). Scholars and practitioners in the field of organizational innovation shows great deal of interest in this area since organizational innovation enhances the performance of the organization by minimizing administrative or operational cost as well as increases workers and organizational productivity (OECD, 2005) by implementing updated structural and managerial policy or knowledge (Osterloh et al., 2001) to the organization to gain competitive edge.

## Team Dynamics

Team dynamics are known as the forces which control over the action and behavior of a team's. According to Zachary and Kuzuhara (2005) team dynamics are "the characteristics of the process through which members of a team interact with each other. This includes patterns of communication, conflict resolution, decision-making styles, and the culture of the team". Dynamism in team can be assured by the atmosphere of the organization as well as team environment, team member's traits and relationship among members of the team. Moreover, Team dynamics have significant impact on performance of organization through team performance (Myers, 2013). Also team expertise, team empowerment, trust among team members, team balance and mutual supports are also closely linked with team dynamics and have strong influence over enhancement of organizational performance, productivity as well as organizational learning strategy (Kirkman and Rosen, 1999; Hut and Molleman, 1998; Gondal and Khan, 2008; DeOrtentiis et al., 2013). In the light of the aforementioned discussion, this study also interested to consider trust, interpersonal communication, team empowerment, and team expertise as important components of team dynamics and following section discuss about the components of team dynamics.

### Trust

Trust is known as the faith or confidence about a partner, which results from knowledge, trustworthiness and goodwill of the partner (Cheng et al., 2008; Claro et al. 2006). Basically, trust is related to the positive believes or expectations (Paliszkiewicz, 2011). According to Butler (1991) definitions of trust must have eight following common features: justice, clearness, uniformity in attitudes, directness, maintains confidentiality, practical expertise, common sense and business knowledge. In an organization, trust is created when traits of both parties are coherent with common principles, beliefs and goals (Shockley-Zabalak and Morely, 2000). Moreover, Trust in organization can be two types such as interpersonal and institutional (Isik and Aliyev, 2015). Researchers and practitioners have identified the role of trust in team performance and team dynamics and have revealed that to ensure team dynamism and better team performance in an organization trust is the hygienic factor and will work as a synergy for the organization (Erdem et al., 2003; Costa, 2002; Hakanen and Soudunsaari, 2012; Erdem and Ozen, 2000). Additionally, Sharkie (2005) have found that in high trust environment in an organization supporting attitudes, spontaneous communication and knowledge sharing as well as organizational learning are highly linked. Similarly, Lewicki and Bunker (1996) have also found that trust, better team partnership and knowledge sharing as well as learning among the team members are closely linked with each other.

## Interpersonal Communication

Many scholars describe interpersonal communication in different ways (Dainton & Zelley, 2004). Basically, interpersonal communication is known as the personal communication between two people to share knowledge or experience of something. Canary et al. (2008) define interpersonal communication as a medium of achieving certain goals and objectives through the contact between each other's. Moreover, Singh (2014) highlighted interpersonal communication in the organizational perspective and point out that interpersonal communication is an essential practice for the organization through which one person of the organization can convey knowledge, skills and know-how to the another person of the organization. Also, Wilson (2005) believes that interpersonal communication is not only the way of sharing knowledge and feelings between the employee but also it is a medium of reaching beliefs, values and thinking of others in the organization. Scholars in the field of interpersonal communication realized that interpersonal communication is very essential to ensure organizational effectiveness as well as its paly key role to increase employee involvement, additionally interpersonal communication enhance employee participation in the team and allow them to share their knowledge, thoughts and increase team dynamism and organizational effectiveness. (Smith et al, 1994; Peterson & Hicks, 1996).

## Team Expertise

Team expertise is the collective sum of specialized knowledge and skills of the team members and generally known as a performance predictor (Rasch and Tosi, 1992). Moreover, individual performance differs from the performance of a team (McGrath, 1984). In light of that researchers highlighted the importance of coordination to ensure team effectiveness (Faraj and Sproull, 2000). Generally, team expertise is formed with applied and potential expertise. Previous research studies have failed to distinguished applied and potential expertise. Principally, applied expertise is a type of expertise that team needs to accept in between the project in order to achieve certain objective on the other hand potential expertise is known as the collective expertise which are relevant to the project. Many researchers and practitioners in their studies have found that team expertise is key element to maintain team dynamics as well as to ensure better team performance. Cleland (1999) stated that expert team is needed for successful completion of project. Similarly, numbers of research studies have concluded that knowledge expertise of team members contribute importantly to enhance team performance even in complex task (Yoo and Kanawattanachai, 2001; Faraj and Spruill, 2000). Moreover, in other studies have found strong relationship between team expertise and knowledge sharing (Stasser et al., 2000; Stasser et al., 1995) between team members.

## Team Empowerment

Team empowerment is known as the increased job motivation which results from positive impression of the team members about their tasks in the organization (Kirkman & Rosen, 2000). Moreover, Kirkman & Rosen (1997) have identified that empowerment in a team can be experienced in four dimensions such as potency, meaningfulness, autonomy and impact. These four dimensions jointly create team empowerment in organization. Although, in a team among the four dimensions team may have little autonomy but team can still enjoy empowerment (Spreitzer, 1995). According to Johnson et al. (2000) team empowerment is highly connected with the success of teams and team empowerment exists when team members can have freedom of working, are allowed to take risk, can make decision jointly rather than from one way direction and also when their activities are linked to the strategy of the organization. Bradley and Benson (1990) in their study have found that team empowerment has positive impact on organizational performance and have identified that more empowered team are more productive, high job satisfaction and team commitment exist in between them. Also, Hoegl and Gemuenden (2001) highlighted that success of innovative projects depends on the teamwork of the organization. Similarly, Piral et al. (2005) in their studies have found positive association between team empowerment and team performance as well as attitudes of the organization.

## Organizational Learning Strategy

The concept of organizational learning has not yet established accurately due to its multidisciplinary characteristics (Saadat and Saadat, 2016). According to Crossan et al. (1999) as a multilevel process organizational learning starts with individual learning and at next level it moves on to the team learning and finally transfers to the organizational learning. Additionally, Alvani (2008) defined organizational learning as a system of correcting mistakes by identifying and resolving those mistakes or errors. Similarly, Argyris and Schon (1978) described organizational learning as learning of detecting mistakes and learning of correcting those discovered mistakes or errors. Whereas, many researchers think organizational learning as an experience from the change in the organizational knowledge (Fiol and Lyles, 1985). According to Argote (2011) organizational learning is a "three sub-processes: creating, retaining and transferring knowledge". Similarly, Daft and Weick (1984) divided process of organizational learning in three stages those are gathering, understanding and learning the practice of knowledge or information. For all types of organizations, learning has both impact and advantages on organizations. On the word of Bowen et al. (2006) organizational competitive advantage is ensured by the organizational learning. Also, researchers in different studies have revealed that organizational learning is significantly linked to improve organizational efficiency, productivity (Akhavan and Jafari, 2006; Sharma, 2003) as well as sustainability of the organization (Smith, 2012) by generating or implementing new ideas and knowledge's.

## Proposition Development

### Team Dynamics and Organizational Learning

Researchers and practitioners around the world in different studies have indicated that team dynamics have found better connection with learning (Jha and Kumar, 2016; Prugsamatz, 2010; Lee and Yang, 2015; Brueller and Carmeli, 2011). Bennett (2001) has identified the importance of learning to determine team success. Similarly, Edmondson (2002) has found positive effect of team dynamics in organizational learning. Moreover, team dynamics ensures interpersonal trust between the team members in the organization which allows organization to develop knowledge sharing and learning environment in the organization in order to foster the organizational innovation (Edmondson and Moingeon, 1999; Dodgson, 1993). Because, team learning is not sum of individual learning, it is a synergic process for organization (Pedler et al., 1997). Edmondson (1999) highlighted the importance of trust among the team members in order to become more engage with the organizational learning process. Bakker et al., (2006) define trust as a set of beliefs among team player, which leads one (trustor) to believe that the trustee's actions will have positive consequences for the trustor's self. In organizations trust works as a multidimensional construct which grows the belief, sentiment or expectation about an exchange partner that results from the partner's expertise, reliability and intentionality or from the partner's honesty and benevolence (Islam et al., 2011; Cheng, Yeh and Tu, 2008; Claro, de Oliveira and Hagelaar, 2006). Similarly, Swift and Hwang (2013) have also revealed the strong association between team trust and organizational learning. Hence, this study proposes:

**Proposition 1:** Trust has positive effect on organizational learning strategy

Similarly, team dynamism in an organization also ensures better communication in team members (Garavan et al., 2007; Koster et al., 2007; Liao, 2006; Jacobs and Coghlan, 2005).

Hoegl and Gemuenden (2001) mentioned that collaboration and interpersonal communication between the team members enhance the chances to develop learning skills such as creative, technical, social and project management. Kasl et al. (1992) argued that an unite team and individual learning is key for team learning. Moreover, researchers around the world in their research studies have relealed that better communication support better learning and moreover highlighted the connection between organizational learning and communication (Zeng et al., 2019; Chen et al., 2019, Scharp et al., 2019; Beamer, 1992; Bateson, 1972). Hence, this study proposes:

**Proposition 2:** Interpersonal communication has positive effect on organizational learning strategy

Likewise, trust and interpersonal communication, team expertise is another important factor for expediting team learning and organizational learning (Aggarwal et al., 1995). Team expertise allows organizational team members to share specialized knowledge among each other as well as develop learning culture in an organization. Vygotsky (1978) stated organizational learning as a connection between different expertise. Also, Kayes and Burnett (2006) in their research studies have underlined the importance of different expertise in a team and highlighted team expertise as an important basis of team learning. Hence this study proposes:

**Proposition 3:** Team expertise has positive effect on organizational learning strategy

Also, team empowerment is strongly associated with organizational learning because individual in a team will have autonomy to ask their query, can share their thoughts as well as can participate spontaneously (Johnson et al., 2000) and such culture or atmosphere improve the organizational learning process. Researchers and practitioners around the world have indicated the importance of team empowerment for ensuing positive knowledge sharing and organizational learning culture (Srivastava et al., 2006; Kirkman et al., 2004; Marsick and Watkins, 2003). Hence, this study proposes:

**Proposition 4:** Team empowerment has positive effect on organizational learning strategy.

## Organizational Learning and Organizational Innovation

Many scholars and experts in the field of organizational study have argued that organizational learning and organizational innovation are closely related (García-Morales et al, 2007; Cefis & Marsili, 2005; Ussahawanitchakit, 2008). To support the statement (Sampaio & Perin, 2004) stated that to create innovative environment in an organization, organizational learning plays very vital role. Similarly, Chaveerug (2008) in his study has found that to ensure innovativeness of the firm as well as its performance, generation of new knowledge is very important and organizational learning is connected with new knowledge creation. Additionally, Ruiz Moreno et al. (2005) have found supportive elements in their study which favors the existence of innovation in organization and identified that organizational learning strengthens the innovative capacity of the organization and brings new ideas in organization. Also, Baker & Sinkula (2002) highlighted that organizational learning is a prerequisite to ensure further improvement of the process and practice of the organization as well as important for new product development and bring more organizational innovation. Alike, Teo & Wang (2005) stated that organizational performance can be enhanced by the organizational learning and in addition to that organizational innovation was also determined by the organizational learning. Additionally, Hovland (2003) considered that in organization learning also may be modified by the management practice through employee motivation, employee development and innovative behavior. Likewise, Mansoor and Ratna (2014) found strong relationship between organizational learning and organizational innovation, also said that through organizational learning, organization can update their knowledge level which will help them to become more innovative and more adaptive to the changing environment as well as ensure better position among the rivals and competitors in terms of change responsiveness. Moreover, many studies have found that organizational learning has impact on organizational innovation as well as organizational performance. Their study has found that organizations which emphasize more on learning are more innovative and innovative organization tend to perform better than their competitors (Kitapch et al, 2012; Chen and Cheng, 2010; Jansen et al, 2006; Darroach & McNaughton, 2002).

**Proposition 5:** Organizational learning has positive effect on organizational innovation.

## Mediating Effects of Organizational Learning

Many scholars and experts have found the relationship between organizational learning and organizational innovation (García-Morales et al., 2007; Cefis & Marsili, 2005; Ussahawanitchakit, 2008). Sampaio & Perin, (2004) mentioned in their study that to create innovative environment in an organization, organizational learning plays very vital role. Innovativeness of the firm to produce new product or services the generation of new knowledge is very important and organizational learning is closely connected with new knowledge creation (Chaveerug, 2008). Therefore, to bring new ideas in the organizations learning works as a prerequisite to ensure further improvement of the process and practice of the organization (Baker & Sinkula, 2002). On the contrary Hovland, (2003) found that organization learning may be modified by the management practice through employee motivation, employee development and innovative behavior. In this study team dynamics are considered as influencing factors for the learning to ensure innovativeness for the organizations. Considering the above-mentioned connections among the variables indicate the potential mediating effect of organizational learning between team dynamics and organizational innovativeness. Moreover, previous research studies have also witnessed mediating effect

*Figure 1. A conceptual model of team dynamics*

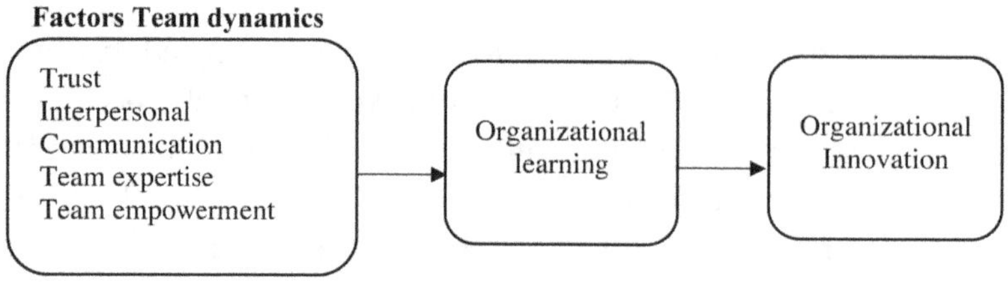


of organizational learning in their study (Mansoor & Ratna, 2014; Kitapch et al., 2012; Chen and Chen, 2010; Jansen et al, 2006). Tippins and Sohi (2003) in their study have found organizational learning plays an effecting mediating role between IT competency and firm performance. Similarly, Real et al. (2014) also have highlighted the mediating effects of organizational learning, their study revealed that organizational learning fully mediates the relationship between learning orientation and performance on the other hand in similar study also revealed partial mediating effect between entrepreneurial orientation and performance. Similarly, organizational learning mediates the relationship between team dynamics and organizational innovation. Because theoretical connection also supports the mediating effects of organizational learning between team dynamics and organizational innovativeness. Also, this study explained the relationship among team dynamics and organizational learning (propositions 1, 2, 3, 4) and organizational learning and organizational innovation (proposition 5) in our justification on the positions mentioned above. Hence, the issue comes arise if the organization learning strategy is better or weaker, whether it will make any improvement or changes on the status of the current relationship between team dynamics and organizational innovation or not. Hence we propose:

**Proposition 6:** organizational learning strategy mediates the relationship between team dynamics and organizational innovation

## DISCUSSION AND IMPLICATIONS

Despite a limited literature that exhibits the relationships between team dynamics, organizational learning and organizational innovation, there is a general agreement among academics and practitioners that organizational learning is influenced by factors (Team dynamics), which eventually lead to organizational innovativeness. The paper proposes an integrative research model to suggest the relationships among the factors, organizational learning and organizational effectiveness. Learning serves as both a dependent variable (to team dynamics) and independent (mediating) variable (to organizational innovations).

The proposed framework will help to link team dynamics associated with organizational learning that seem to enhance its organizational innovation. The present research reveals, in principle, how factors influence organizational learning that enhances their organizational innovation. Several propositions are formulated using various issues that have been emerged through the review of the relevant literature. Figure 1 depicts the conceptual model developed in this study.

This study puts forward several propositions that the team dynamics to be involved in enhancing organizational learning to confirm organizational innovations. The results indicate three factors that may be conceptualized in the form of a diagram, which will help enhance organizational innovation, taking the mediation effect of organizational learning.

The paper contributes to our understanding of knowledge sharing theory and practice in various ways. This research has significant implications for both scholars and practitioners.

## Theoretical Implications

The conceptual model developed will enrich the literature in organizational innovation and knowledge management in several ways. Firstly, it addresses a gap in the literature by connecting team dynamics with organizational learning towards organizational innovativeness. Secondly, this study proposes several factors that influence organizational learning. Therefore, to the knowledge of the authors, this study is probably the first of this kind to establish an integrated view that utilizes a multidimensional constructs (i.e., trust, interpersonal communication, team expertise and team empowerment) of factors, and explains how they influence on the improvement of organizational innovations.

## Practical Implications

From a practical perspective, the conceptual model could be useful for organizations, since it highlights that factors facilitates organizational learning, which is, crucial to achieve organizational innovations. This study promotes guidelines for organizations with respect to the development of organizational innovations policies from the team perspective. The factors (team dynamics) exert an influence on the learning process, and the outcome of this study reveals that the competitive ideas associated with assumptions that managers could gather and would be useful in achieving organizational innovations. Achieving competitive advantage as an organizational innovations goal involves an understanding of trust, interpersonal communication, team expertise and team empowerment, which are considered to be potential determinants of organizational learning. Hence, the factors can be considered unique with its efforts. .

## Implication for Bangladesh

This paper also reveals some implications for the telecom organizations are operating in Bangladesh. Digital Bangladesh is one of the components of social development goal (SDGs) and vision 2021 of Bangladesh. Most of the people currently use mobile phones for their personal communication as well as for their business and trade activities. This is because mobile communication devices are now easily available, more affordable; and is rapidly increasing in numbers to adopt many capabilities with different functions in various services, such as:

- Mobile banking (e.g. bKash)
- File transfer (e.g. Shareit)
- Email (e.g. Gmail)
- Online messaging (e.g. WhatsApp)
- FM radio (e.g. Radio Foorti 88.0)
- Online ticket purchase (e.g. shohoz.com)

- Online shopping (e.g. bikroy.com, pickabo.com)
- Disaster management information network (DMIN) portal
- Mobile remittance
- BBC Janala
- The agricultural management information system (AMIS)

The growing popularity of mobile phone use and related services adoption suggests interesting research questions, since it is a source of income for companies with high user demand. This paper provide evidence on factors influencing the organizational innovativeness to deliver more interesting services to the subscribers. In particular, the team dynamism have received considerable attention from the literature to have effective learning process to enhance innovative activities for the organizations.

## CONCLUSION AND FUTURE RESEARCH

The paper explores the extensive literature and present through the conceptual development. It would be interesting to note that the literature presented in this paper has proposed a worthwhile starting point and demonstrate several new ideas that will urge further investigation and analysis. This paper postulates a number of influential relationships among the constructs which have an impact on innovativeness in organizations, and also put forward the issues for further empirical investigation. Indeed, team dynamics which allows employees learning could increase the employee's efficiency to develop new ideas to generate competitive advantage. Therefore, an empirical research on team dynamics and its influence on learning could possibly offer new insights for better understanding of organizational innovation literature in addressing new product or services development.

From the proposed framework it can be concluded that, for ensuring organizational innovation, team dynamics and organizational learning plays a vital role. Knowledge of this study will provide new scope to the researchers, practitioners and policy makers of business organization in Bangladesh to emphasize more on team dynamics and organizational learning in order to ensure organizational innovativeness. Moreover, empirical justification of this proposed framework will confirm the strength of the model as well as will create space for further study. Also, this study has considered overall business organization in Bangladesh. However, in future study should consider specific business sector of Bangladesh.

## REFERENCES

Abbas, W., & Asghar, I. (2010). The role of leadership in organizatinal change: relating the successful organizational change with visionary and innovative leadership.

Aggarwal, I., Woolley, A. W., Chabris, C. F., & Malone, T. W. (2019). The impact of cognitive style diversity on implicit learning in teams. *Frontiers in Psychology*, *10*, 112. doi:10.3389/fpsyg.2019.00112 PMID:30792672

Akhavan,P., & Jafari, M .(2006). Learning organizations. the necessity of knowledge era. *Journal of Tadbir*, *169*.

Alvani, M. (2008). Public Management (32th ed.). Tehran: Ney publication.

Amar, A. D. (2004). Motivating knowledge workers to innovate: A model integrating motivation dynamics and antecedents. *European Journal of Innovation Management*, *7*(2), 89–101. doi:10.1108/14601060410534366

Aragón-Correa, J. A., García-Morales, V. J., & Cordón-Pozo, E. (2007). Leadership and organizational learning's role on innovation and performance: Lessons from Spain. *Industrial Marketing Management*, *36*(3), 349–359. doi:10.1016/j.indmarman.2005.09.006

Argote, L. (2011). Organizational learning research: Past, present and future. *Management Learning*, *42*(4), 439–446. doi:10.1177/1350507611408217

Argyris, C., & Schon, D. A. (1978). *Organizational Learning: A theory of action perspective*. Reading, MA: Addison Wesley.

Bain, P. G., Mann, L., & Pirola-Merlo, A. (2001). The innovation imperative: The relationships between team climate, innovation, and performance in research and development teams. *Small Group Research*, *32*(1), 55–73. doi:10.1177/104649640103200103

Baker, W. E., & Sinkula, J. M. (1999). The synergistic effect of market orientation and learning orientation on organizational performance. *Journal of the Academy of Marketing Science*, *27*(4), 411–427. doi:10.1177/0092070399274002

Baker, W. E., & Sinkula, J. M. (2002). Market orientation, learning orientation and product innovation: delving into the organization's black box. *Journal of market-focused management, 5*(1), 5-23.

Bakker, M., Leendes, T. A. J., Gabby, S. M., Krazer, J., & Engelen, J. M. L. V. (2006). Is trust really social capital? Knowledge sharing in product development projects. *The Learning Organization*, *13*(6), 594–605. doi:10.1108/09696470610705479

Bateson, G. (1972). The logical categories of learning and communication. In *Steps to an Ecology of Mind* (pp. 279-308). Academic Press.

Beamer, L. (1992). Learning intercultural communication competence. *The Journal of Business Communication, 29*(3), 285-303.

Bloch, C., & Bugge, M. M. (2013). Public sector innovation—From theory to measurement. *Structural Change and Economic Dynamics*, *27*, 133–145. doi:10.1016/j.strueco.2013.06.008

Brueller, D., & Carmeli, A. (2011). Linking capacities of high-quality relationships to team learning and performance in service organizations. *Human Resource Management*, *50*(4), 455–477. doi:10.1002/hrm.20435

Butler, J. K. Jr. (1991). Toward understanding and measuring conditions of trust: Evolution of a conditions of trust inventory. *Journal of Management*, *17*(3), 643–663. doi:10.1177/014920639101700307

Canary, D. J., Cody, M. J., & Manusov, V. L. (2008). Interpersonal communication competence. In *Interpersonal communication: A goals-based approach* (pp. 502-520). Academic Press.

Černe, M., Jaklič, M., & Škerlavaj, M. (2013). Authentic leadership, creativity, and innovation: A multilevel perspective. *Leadership*, *9*(1), 63–85. doi:10.1177/1742715012455130

Chen, P. T., & Cheng, J. Z. (2010). Unlocking the promise of mobile value-added services by applying new collaborative business models. *Technological Forecasting and Social Change*, *77*(4), 678–693. doi:10.1016/j.techfore.2009.11.007

Chen, Z., Duan, L. Y., Wang, S., Lou, Y., Huang, T., Wu, D. O., & Gao, W. (2019). Towards knowledge as a service over networks: A deep learning model communication paradigm. *IEEE Journal on Selected Areas in Communications*, *37*(6), 1349–1363. doi:10.1109/JSAC.2019.2904360

Cheng, J. H., Yeh, C. H., & Tu, C. W. (2008). Trust and knowledge sharing in green supply chains. *Supply Chain Management*, *13*(4), 283–295. doi:10.1108/13598540810882170

Cho, S., Mathiassen, L., & Robey, D. (2007). Dialectics of Resilience: A Multi-level Analysis of a Telehealth Innovation. *Journal of Information Technology*, *22*(1), 24–35. doi:10.1057/palgrave.jit.2000088

Claro, D. P., Claro, P. B. D. O., & Hagelaar, G. (2006). Coordinating collaborative joint efforts with suppliers: The effects of trust, transaction specific investment and information network in the Dutch flower industry. *Supply Chain Management*, *11*(3), 216–224. doi:10.1108/13598540610662112

Cleland, D. I. (1999). *Project management: strategic design and implementation*. McGraw-Hill.

Costa, A. C. (2002). The role of trust for the functioning of teams in organisations. Delft University.

Crossan, M. M., Lane, H. W., & White, R. E. (1999). An organizational learning framework: From intuition to institution. *Academy of Management Review*, *24*(3), 522–537. doi:10.5465/amr.1999.2202135

Daellenbach, U. S., McCarthy, A. M., & Schoenecker, T. S. (1999). Commitment to innovation: The impact of top management team characteristics. *Research Management*, *29*(3), 199–208.

Daft, R. L., & Weick, K. E. (1984). Toward a model of organizations as interpretation systems. *Academy of Management Review*, *9*(2), 284–295. doi:10.5465/amr.1984.4277657

Dainton, M., & Zelley, E. D. (2004). Explaining theories of interpersonal communication.

Darroch, J., & McNaughton, R. (2002). Examining the link between knowledge management practices and types of innovation. *Journal of Intellectual Capital*, *3*(3), 210–222. doi:10.1108/14691930210435570

DeOrtentiis, P. S., Summers, J. K., Ammeter, A. P., Douglas, C., & Ferris, G. R. (2013). Cohesion and satisfaction as mediators of the team trust–team effectiveness relationship: An interdependence theory perspective. *Career Development International*, *18*(5), 521–543. doi:10.1108/CDI-03-2013-0035

Dodgson, M. (1993). Learning, trust, and technological collaboration. *Human Relations*, *46*(1), 77–95. doi:10.1177/001872679304600106

Downs, G. W. Jr, & Mohr, L. B. (1979). Toward a theory of innovation. *Administration & Society*, *10*(4), 379–408. doi:10.1177/009539977901000401

Edmondson, A., & Moingeon, B. (1999). Learning, trust and organizational change. *Organizational learning and the learning organization*, 157-175.

Edmondson, A., & Moingeon, B. (1999). Learning trust and organizational change. In M. Easterby-Smith, J. Burgoyne, & L. Araujo (Eds.), *Organizational Learning and the Learning Organisation* (pp. 157–175). Sage.

Edmondson, A. C., & Mogelof, J. P. (2006). Explaining psychological safety in innovation teams: organizational culture, team dynamics, or personality? In *Creativity and innovation in organizational teams* (pp. 129–156). Psychology Press.

Eisenberger, R., Fasolo, P., & Davis-LaMastro, V. (1990). Perceived organizational support and employee diligence, commitment, and innovation. *The Journal of Applied Psychology*, *75*(1), 51–59. doi:10.1037/0021-9010.75.1.51

Erdem, F., & İşbaşı, J. Ö. (2000). Takım Çalışmalarında Güven ve Güvensizlik: Performans için Koşulsuz Güven mi, Optimum Güven mi? *Erciyes Üniversitesi*, *8*, 25–27.

Erdem, F., Ozen, J., & Atsan, N. (2003). The relationship between trust and team performance. *Work Study*, *52*(7), 337–340. doi:10.1108/00438020310502633

Fairbank, J. F., & Williams, S. D. (2001). Motivating creativity and enhancing innovation through employee suggestion system technology. *Creativity and Innovation Management*, *10*(2), 68–74. doi:10.1111/1467-8691.00204

Faraj, S., & Sproull, L. (2000). Coordinating expertise in software development teams. *Management Science*, *46*(12), 1554–1568. doi:10.1287/mnsc.46.12.1554.12072

Fiol, C. M., & Lyles, M. A. (1985). Organizational learning. *Academy of Management Review*, *10*(4), 803–813. doi:10.5465/amr.1985.4279103

Garcia-Morales, V. J., Lloréns-Montes, F. J., & Verdu-Jover, A. J. (2007). Influence of personal mastery on organizational performance through organizational learning and innovation in large firms and SMEs. *Technovation*, *27*(9), 547–568. doi:10.1016/j.technovation.2007.02.013

Gondal, A. M., & Khan, A. (2008). Impact of team empowerment on team performance: Case of the telecommunications industry in Islamabad. *International Review of Business Research Papers*, *4*(5), 138–146.

Grant, R. (2005). *Contemporary strategy analysis*. Malden, MA: Blackwell Publishers.

Gumusluoğlu, L., & Ilsev, A. (2009). Transformational leadership and organizational innovation: The roles of internal and external support for innovation. *Journal of Product Innovation Management, 26*(3), 264–277. doi:10.1111/j.1540-5885.2009.00657.x

Hakanen, M., & Soudunsaari, A. (2012). Building trust in high-performing teams. *Technology Innovation Management Review, 2*(6).

Hoch, J. E. (2013). Shared leadership and innovation: The role of vertical leadership and employee integrity. *Journal of Business and Psychology, 28*(2), 159–174. doi:10.100710869-012-9273-6

Hoegl, M., & Gemuenden, H. G. (2001). Teamwork quality and the success of innovative projects: A theoretical concept and empirical evidence. *Organization Science, 12*(4), 435–449. doi:10.1287/orsc.12.4.435.10635

Hovland, I. (2003). *Knowledge management and organisational learning: An international development perspective: An anotated bibliography*. London: Overseas Development Institute.

Hut, J., & Molleman, E. (1998). Empowerment and team development. *Team Performance Management: An International Journal, 4*(2), 53–66. doi:10.1108/13527599810214193

Isik, M., Timuroglu, K., & Aliyev, Y. (2015). The relationship between teamwork and organizational trust. *International Journal of Research in Business and Social Science, 4*(1), 113-132.

Islam, M. Z., Ahmed, S. M., Hasan, I., & Ahmed, S. U. (2011). Organizational culture and knowledge sharing: Empirical evidence from service organizations. *African Journal of Business Management, 5*(14), 5900–5909.

Jansen, J. J., Van Den Bosch, F. A., & Volberda, H. W. (2006). Exploratory innovation, exploitative innovation, and performance: Effects of organizational antecedents and environmental moderators. *Management Science, 52*(11), 1661–1674. doi:10.1287/mnsc.1060.0576

Jha, B., & Kumar, A. (2016). Employee engagement: A strategic tool to enhance performance. *DAWN: Journal for Contemporary Research in Management, 3*(2), 21–29.

Johnson, P. R., Heimann, V. L., & O'Neill, K. (2000). The wolf pack: Team dynamics for the 21st century. *Journal of Workplace Learning: Employee Counselling Today, 12*(4), 159–164. doi:10.1108/13665620010332813

Kalshoven, K., Den Hartog, D. N., & de Hoogh, A. H. (2013). Ethical leadership and followers' helping and initiative: The role of demonstrated responsibility and job autonomy. *European Journal of Work and Organizational Psychology, 22*(2), 165–181. doi:10.1080/1359432X.2011.640773

Kayes, D. C., & Burnett, G. (2006). G.: Team learning in organizations A review and integration. Proceedings of the OLKC 2006 Conference. Academic Press.

Kirkman, B. L., & Rosen, B. (1997). A model of work team empowerment. *Research in Organizational Change and Development, 10*, 131–167.

Kirkman, B. L., & Rosen, B. (1999). Beyond self-management: Antecedents and consequences of team empowerment. *Academy of Management Journal, 42*(1), 58–74.

Kirkman, B. L., & Rosen, B. (2000). Powering up teams. *Organizational Dynamics*, *28*(3), 48–66. doi:10.1016/S0090-2616(00)88449-1

Kirkman, B. L., Rosen, B., Tesluk, P. E., & Gibson, C. B. (2004). The impact of team empowerment on virtual team performance: The moderating role of face-to-face interaction. *Academy of Management Journal*, *47*(2), 175–192.

Kitapch, H., Aydin, B., & Celiki, V. (2012). The Effects of Organizational Learning Capacity and Innovativeness on Financial Performance: An Empirical Study. *African Journal of Business Management*, *6*(6), 2332–2341. doi:10.5897/AJBM11.403

Lee, H. H., & Yang, T. T. (2015). Employee goal orientation, work unit goal orientation and employee creativity. *Creativity and Innovation Management*, *24*(4), 659–674. doi:10.1111/caim.12118

Lewicki, R. J., & Bunker, B. B. (1996). Developing and maintaining trust in work relationships. *Trust in organizations: Frontiers of theory and research, 114*, 139.

Mansoor, S. & Ratna, R. (2014). Impact of Organizational Learning on Organizational Innovation.

Marsick, V. J., & Watkins, K. E. (2003). Demonstrating the value of an organization's learning culture: The dimensions of the learning organization questionnaire. *Advances in Developing Human Resources*, *5*(2), 132–151.

Mavondo, F. T., Chimhanzi, J., & Stewart, J. (2005). Learning orientation and market orientation: Relationship with innovation, human resource practices and performance. *European Journal of Marketing*, *39*(11/12), 1235–1263. doi:10.1108/03090560510623244

McGrath, J. E. (1984). *Groups: Interaction and performance* (Vol. 14). Englewood Cliffs, NJ: Prentice-Hall.

Mol, M. J., & Birkinshaw, J. (2009). The sources of management innovation: When firms introduce new management practices. *Journal of Business Research*, *62*(12), 1269–1280. doi:10.1016/j.jbusres.2009.01.001

Myers, S. P. (2013). Definition of Team Dynamics. Team Technology. Retrieved from http://www.teamtechnology.co.uk/team/dynamics/definition/

OECD. (2005). *The measurement of scientific and technological activities Oslo Manual. Guidelines for collecting and interpreting innovation data* (3rd ed.). Paris: OECD EUROSTAT.

Örtenblad, A. (2001). On differences between organizational learning and learning organization. *The Learning Organization*, *8*(3), 125–133. doi:10.1108/09696470110391211

Osterloh, M., Rota, S., & von Wartburg, M. (2001). Open source-New rules in software development. University of Zurich, Institute of Organization and Administrative Science (IOU). Retrieved from http://www.iou.uzh.ch/orga/downloads/OpenSourceAoM.pdf

Paliszkiewicz, J. (2011). Inter-organizational trust: Conceptualization and measurement. *International Journal of Performance Measurement*, *1*(1), 15–28.

Peterson, D. B., & Hicks, M. D. (1996). *Leader as coach*. Minneapolis: Personnel Decisions Inc.

Pieterse, A. N., Van Knippenberg, D., Schippers, M., & Stam, D. (2010). Transformational and transactional leadership and innovative behavior: The moderating role of psychological empowerment. *Journal of Organizational Behavior, 31*(4), 609–623. doi:10.1002/job.650

Popa, I. L., Preda, G., & Boldea, M. (2010). A theoretical approach of the concept of innovation. *Managerial Challenges of the Contemporary Society. Proceedings*. Academic Press.

Prugsamatz, R. (2010). Factors that influence organization learning sustainability in non-profit organizations. *The Learning Organization, 17*(3), 243–267. doi:10.1108/09696471011034937

Rasch, R. H., & Tosi, H. L. (1992). Factors affecting software developers' performance: An integrated approach. *Management Information Systems Quarterly, 16*(3), 395–413. doi:10.2307/249535

Real, J. C., Roldán, J. L., & Leal, A. (2014). From entrepreneurial orientation and learning orientation to business performance: Analysing the mediating role of organizational learning and the moderating effects of organizational size. *British Journal of Management, 25*(2), 186–208. doi:10.1111/j.1467-8551.2012.00848.x

Remedios, R., & Boreham, N. (2004). Organisational learning and employees' intrinsic motivation. *Journal of Education and Work, 17*(2), 219–235. doi:10.1080/13639080410001677419

Rosenbusch, N., Brinckmann, J., & Bausch, A. (2011). Is innovation always beneficial? A meta-analysis of the relationship between innovation and performance in SMEs. *Journal of Business Venturing, 26*(4), 441–457. doi:10.1016/j.jbusvent.2009.12.002

Ruiz Moreno, A., García Morales, V., & Lloréns Montes, F. J. (2005). Learning during the quality management process: Antecedents and effects in service firms. *Industrial Management & Data Systems, 105*(8), 1001–1021. doi:10.1108/02635570510624419

Saadat, V., & Saadat, Z. (2016). Organizational learning as a key role of organizational success. *Procedia: Social and Behavioral Sciences, 230*, 219–225. doi:10.1016/j.sbspro.2016.09.028

Salavou, H., Baltas, G., & Lioukas, S. (2004). Organisational innovation in SMEs: The importance of strategic orientation and competitive structure. *European Journal of Marketing, 38*(9/10), 1091–1112. doi:10.1108/03090560410548889

Sanz-Valle, R., Naranjo-Valencia, J. C., Jiménez-Jiménez, D., & Perez-Caballero, L. (2011). Linking organizational learning with technical innovation and organizational culture. *Journal of Knowledge Management, 15*(6), 997–1015. doi:10.1108/13673271111179334

Scharp, K. M., Seiter, J. S., & Curran, T. (2019). Learning supportive communication through an adaptation of the board game Apples to Apples. *Communication Teacher, 33*(1), 5–10. doi:10.1080/17404622.2018.1530798

Sharkie, R. (2005). Perceptions of trust as a determinant of the level of knowledge sharing. *International Journal of Knowledge, Culture and Change Management, 4*(a), 1795-1801.

Sharma, R. K. (2003). Understanding organizational learning through knowledge management. *Journal of Information & Knowledge Management*, 2(4), 343–352. doi:10.1142/S021964920300053X

Shockley-Zabalak, P., & Morely, D. D. (2000). *Trust in Organizations*. San Francisco, CA: IABC Research Foundation.

Singh, I. A. K. (2014). Role of Interpersonal communication in Organizational effectiveness. *International Journal of Research in Management and Business Studies*, *1*(4), 36–39.

Smith, K. G., Smith, K. A., Olian, J. D., Sims, H. P., O'Bannon, D. P., & Scully, J. A. (1994). Top management team demography and process: The role of social integration and communication. *Administrative Science Quarterly*, *39*(3), 412–438. doi:10.2307/2393297

Smith, P. A. (2012). The importance of organizational learning for organizational sustainability. *The Learning Organization*, *19*(1), 4–10. doi:10.1108/09696471211199285

Soltani, I., Damirchi, Q. V., & Darban, M. Z. (2011). Organizational culture and organizational innovation. *Arabian journal of business and management review, 4*.

Spreitzer, G. M. (1995). Psychological empowerment in the workplace: Dimensions, measurement, and validation. *Academy of Management Journal*, *38*(5), 1442–1465.

Srivastava, A., Bartol, K. M., & Locke, E. A. (2006). Empowering leadership in management teams: Effects on knowledge sharing, efficacy, and performance. *Academy of Management Journal*, *49*(6), 1239–1251. doi:10.5465/amj.2006.23478718

Stasser, G., Stewart, D. D., & Wittenbaum, G. M. (1995). Expert roles and information exchange during discussion: The importance of knowing who knows what. *Journal of Experimental Social Psychology*, *31*(3), 244–265. doi:10.1006/jesp.1995.1012

Stasser, G., Vaughan, S. I., & Stewart, D. D. (2000). Pooling unshared information: The benefits of knowing how access to information is distributed among group members. *Organizational Behavior and Human Decision Processes*, *82*(1), 102–116. doi:10.1006/obhd.2000.2890

Swift, P. E., & Hwang, A. (2013). The impact of affective and cognitive trust on knowledge sharing and organizational learning. *The Learning Organization*, *20*(1), 20–37. doi:10.1108/09696471311288500

Teo, H. H., & Wang, X. (2005). Organizational Learning Capacity and Attitude toward Complex Technological Innovations: An Empirical Study. *International Science*, *57*(2), 264–279.

Tippins, M. J., & Sohi, R. S. (2003). IT competency and firm performance: Is organizational learning a missing link? *Strategic Management Journal*, *24*(8), 745–761. doi:10.1002mj.337

Vygotsky, L. S. (1978). *The mind in society: The development of higher psychological processes*. Cambridge, MA: Harvard University Press.

Walton, R. E., Allen, C., & Gaffney, M. (1987). *Innovating to compete: Lessons for diffusing and managing change in the workplace*. Jossey-Bass.

Wilson, D. (2005). *New perspectives in applied communication*. Ibadan: Sterlin-Hordan Publishers.

Wood, P. (1998). The rise of consultancy and the prospect for regions. *Paper presented at the 38th Congress of the European Regional Science Association*. Academic Press.

WorkforceSkills Preparedness Report. (2016). PayScale.

Yoo, Y., & Kanawattanachai, P. (2001). Developments of transactive memory systems and collective mind in virtual teams. *The International Journal of Organizational Analysis*, *9*(2), 187–208. doi:10.1108/eb028933

Zachary, W. B., Zachary, W. B., & Kuzuhara, L. W. (2005). *Organizational behavior: Integrated models and applications*. South-Western Pub.

Zeng, Y., Zhang, J., He, K., & Cheng, L. (2019). Who cares what parents think or do? Observational learning and experience-based learning through communication in rice farmers' willingness to adopt sustainable agricultural technologies in Hubei Province, China. *Environmental Science and Pollution Research International*, 26(12), 1–15. PMID:30850982

# Chapter 4
# Enhancing Business Performance of Pakistani Manufacturing Firms via Strategic Agility in the Industry 4.0 Era:
## The Role of Entrepreneurial Bricolage as Moderator

**Qaisar Iqbal**
https://orcid.org/0000-0002-6619-7243
*Universiti Sains Malaysia, Malaysia*

**Noor Hazlina Ahmad**
*Universiti Sains Malaysia, Malaysia*

**Heru Kurnianto Tjahjono**
*Universitas Muhammadiyah Yogyakarta, Indonesia*

**Adeel Nasim**
*PricewaterhouseCoopers (PwC), Canada*

**Muhammad Mustafa Muqaddis**
*PriceWaterhouseCoopers (PwC), Canada*

**Majang Palupi**
*Universitas Islam Indonesia, Indonesia*

DOI: 10.4018/978-1-7998-1566-2.ch004

## ABSTRACT

*Manufacturing plays a substantial role in the economic development of any country because of its multiplier impact on the growth of value addition. Currently, industry 4.0 requires manufacturers to deliver highly customized products without compromising on quality at a reduced life cycle. The objective of this study was to find out a solution for the optimum operation of manufacturing firms. By applying resource-based view, dynamic capability, and effectuation theory, this study has proposed an integrated framework of the organizational network, entrepreneurial bricolage, strategic agility and business performance in the context of the industry 4.0. Moreover, the positive effect of the organizational networks on the strategic agility ultimately improves the business performance of manufacturers. Furthermore, strategic agility is also claimed to play its role as mediator between organizational networks and business performance.*

## INTRODUCTION

The characteristics of the industry 4.0 such as advanced manufacturing technologies, big data analytics, and efficient automation has changed the shape of manufacturing on the large scale. The industry 4.0 has actually integrated human, objects, and machines with each other that has enabled dealing with complex manufacturing systems (Bauer, Hämmerle, Schlund, & Vocke, 2015). Such integration has been categorized into three classes i.e. horizontal, vertical, and end-to-end integration (Brettel, Friederichsen, Keller, & Rosenberg, 2014) where manufacturing industry has emerged as highly dynamic (Kagermann, Helbig, Hellinger, & Wahlster, 2013). Such dynamic manufacturing market requires manufacturer to deliver highly customized and high-quality products on an efficient and effective basis (de Sousa Jabbour, Jabbour, Godinho Filho, & Roubaud, 2018) where customary process and workflows are not suitable (Srai et al., 2016).

Enormous challenges such as poor quality products, low productivity, lack of innovation, low value-added creation, high cost of production, weak infrastructure and issues in the regulatory framework/ policies hinder the manufacturers to compete and ultimately make their survival hard in the realm of fourth industrial revolution (Kemal, 2006; Khan & Turowski, 2016). There has been a lack of academic research in order to investigate and claim the enablers and inhibitors in this domain (Khan & Turowski, 2016; Memon & Tahir, 2012). Fourth industrial revolution brings about simultaneously; the opportunities and challenges for manufacturers in parallel (Khan & Turowski, 2016). Challenges of industry 4.0 varies across industries based on the nature of their operations.

Advancement in the technologies of industry 4.0 has shortened the life cycles of products than before, therefore large manufacturers need high level of innovation and swift response to customer's requirements to compete (Khan & Turowski, 2016). Industry 4.0 results in the diverse product requirements of customers so that even with small demand of traditional products such as textiles and food, consumers exhibit diverse preferences for customization and quality products (Yousfi, Saidi, & Dey, 2016). Therefore, the lack of innovation and diversification makes it difficult for large manufacturers to perform in today's dynamic business environment. Given that the contemporary business landscape has changed with emergence of sophisticated and customized products with higher standard of quality and shorter life cycles of various technologies signifying that only manufacturers with innovative characteristics will compete (Kemal, 2006; Memon & Tahir, 2012). An efficient, dynamic, diversified and rapidly growing manufacturing sector ensures maximum value addition (Khan & Turowski, 2016; Yousfi et al., 2016).

The above-mentioned challenges clearly depict that Industry 4.0 necessitates novel ideas and adaptive behaviors to perform at optimum level (Shin, Lee, Kim, & Rhim, 2015). Manufacturing firms are required to quickly and efficiently respond to the emergence of new technologies (Tipu & Fantazy, 2014). Dynamic characteristics of the manufacturing market requires manufacturer to be highly agile enabling them to deliver with high flexibility and customization (Shin et al., 2015). According to Shin et al. (2015), strategic agility that encompasses technology capability, collaborative innovation, organizational learning, and internal alignment is crucial to survive in this type of environment. The flexibility-agility association is similar to the competency-capability affiliation (Shin et al., 2015). Being internally focused competency, flexibility is considered as antecedent of agility (P. M. Swafford, S. Ghosh, & N. Murthy, 2006a). Furthermore, Christopher (2000) has claimed that manufacturing firms can respond to the market requirements on the basis of competencies of their networking that will ultimately make their performance better (Shin et al., 2015). Such agility is also depended upon organizational networks (Nadkarni & Narayanan, 2007). It has been argued that organizational networks are vital to enhance agility of firms because networks allow firms to be more flexible in nature and less vulnerable towards adverse external forces (Yousaf & Majid, 2018).

While the practical gaps seems to be directed towards the need to integrate organizational network and strategic agility ensuring effecient performance of manufacturers, especially given the upsurge in the challenges brought about by Industry 4.0 (Khan & Turowski, 2016), there also seems to be a huge void in the studies for understanding organizational performance. Hence, from a theoretical standpoint, examining the impacts of these constructs on organizational performance could offer better insight into the integration of RBV (organizational networks), dynamic capability and knowledge-based view (represented by strategic agility construct) that connects various resources and organizational processes to create positive outcomes. Also, in view of the limited resources available to manufacturing firms, this study introduces a new construct labeled as Entrepreneurial Bricolage as a moderator that is conjectured to strengthen the relationships among organizational networks and strategic agility based on effectuation theory. Following Sarasvathy, Dew, Read, and Wiltbank (2008) earlier work on the Effectuation Theory, Gundry, Kickul, Griffiths, and Bacq (2011) had elaborated effectuation theory as the acts of combining resources at hand to create new strategic goal which has been labelled as Entrepreneurial Bricolage. Entrepreneurial bricolage may help firms to "make do" with whatever resources they possess in order to enhance their performance especially among manufacturers operating within limited resources.

Extent literature exhibits different point of views of researchers about the factors influencing performance based on theoretical arguments (Green, Covin, & Slevin, 2008; Yousaf & Majid, 2018). Literature claims various factors effecting performance such as response to competitors (Chung, 2011), dynamic capabilities (Gelhard, von Delft, & Gudergan, 2016), strategic alignment (Yousaf & Majid, 2016), organizational learning (Chung, Yang, & Huang, 2015), strategies relevant to finance, customer, marketing, inventory (Kim, Ko, Kim, & Koh, 2008). Additionally, organizational networks along with organizational flexibility is viewed as most promising factor with substantial impact on the performance of the firm (Lans, Blok, & Gulikers, 2015). Organizational networking is a set of nodes and relationships which make the patterns of relations among individuals, groups, or organizations (Rasmussen, Mosey, & Wright, 2015), whereas organizational flexibility is defined as an ability to adapt under dynamic situation (Volberda, 1996). There is lacking of research that how organizational networking improve performance of strategically agile firms (Lin & Lin, 2016). Therefore, there is need to investigate, how firms especially large-scale manufacturing firms perform with strategic agility in the context of industry

4.0. Researchers attempt to fill this gap by considering whether performance increases to the extent firms are strategically agile in presence of strong organizational networks.

Academic researchers are still confused with inconsistencies and contradictions (Zahra, Sapienza, & Davidsson, 2006) because of contradicting and overlapping of definitions, and inconsistent results about the influence of dynamic capabilities on other capabilities and business performances e.g. which capability can be viewed as dynamic capability is still questionable. The research on dynamic capability has been mostly "case and theoretical-based" (Zheng, Zhang, & Du, 2011).

Therefore, this research is unique in the sense that it integrates dynamic capability theory, and effectuation theory in understanding the ability of manufacturing firms to achieve strategic agility and better performance. This research has proposed the substantial influence of organizational networking on strategic agility and subsequently business performance of large-scale manufacturing firms. Entrepreneurial bricolage is also deemed as a moderating variable that strengthens the relationships between organizational networking and strategic agility.

## LITERATURE REVIEW

### Large-Scale Manufacturing Sector of Pakistan

The manufacturing sector is known as backbone of the economic development of any country (Raj, 2011). Lack of availability of adequate information influences badly the research in manufacturing sector of Pakistan. The only major source about Pakistani manufacturing sector is the Census of Manufacturing Industries (CMI) (Ara, 2004), that was lastly published in 2005-06. The total number of large-scale manufacturing companies in Pakistan are 6417 (Pakistan Bureau of Statistics, 2006), Where 50% companies come in the categories of food and beverages, and textile sub-sectors. 52.53% of these large-scale companies are in the single province-Punjab (Pakistan Bureau of Statistics, 2006).

The large-scale manufacturing sector of Pakistan included a handful of industrial units such as textiles, cement, sugar, tea blending and vegetable ghee in 1947 and had only 1.8% share in GDP initially (Kemal, 2006). The latest statistics showed that large-scale manufacturing contributes 10.8% share of GDP by the FY 2017-18 (Ministry of Finance, 2018). It is reported that various structural problems exist in the manufacturing sector of Pakistan, that has resulted in the form of lower investment, exports and productivity in the fourth industrial revolution (Khan & Turowski, 2016; Memon & Tahir, 2012). Poor quality products, low level diversification, low level of research & development activities and high technical inefficiencies have made hard international market for Pakistani Products (Kemal, 2006; Memon & Tahir, 2012; Rehman, Zhang, & Ali, 2014). The traditional industries i.e. textile and food units still contribute 20.915% and 12.370%_of the total large-scale manufacturing in FY 2017-18 respectively (Ministry of Finance, 2018). Even, industries highly linked to modern technologies such as automobile, electrical industries account for just 4.613% and 4.963% for large-scale growth respectively (Ministry of Finance, 2018).

Value added of the large manufacturing at international prices is just a fraction of that in the local market (Khan & Turowski, 2016). The productivity of manufacturing industry has declined where Pakistani exporters feel difficult to compete in the international market (Memon & Tahir, 2012). The performance of Large-scale manufacturing in Pakistan face severe issues such as lacking innovation, low value addition, lower quality products, problems in the regulatory frameworks, weak infrastructure and lower productivity that clearly highlights the need for major reconstructing of the manufacturing sector (Kemal, 2006; Khan & Turowski, 2016; Rehman et al., 2014). Industry 4.0 requires manufacturers to create smart products with high quality standards under greater customization (Kagermann et al., 2013) whereas disruptive technologies of industry 4.0 results in lower value addition with increased flexibility (Balasingham, 2016), which has raised questions over the survival of Pakistani manufacturers in this era.

Rapid product development, flexible production and volatile market are main characteristics of industry 4.0. In this sense, researchers have agreed on the importance of strategic agility in the light of current challenges such as turbulent market, accelerating rate of innovation, and globalization (Weber & Tarba, 2014). Strategic agility is considered as primary determinant of the manufacturing firms in the dynamic market (Weber & Tarba, 2014). Strategic agility is an ability of the firm to remain flexible and adjust the strategic direction with disruptive nature of technologies and develop innovative ways to remain competitive in the market. It involves creation and adoption of new methods of business transformation, development of dynamic capabilities, organizational learning and maintenance of high organizational flexibility, as well as planning of post-acquisition integration techniques (Popadić, Černe, & Milohnić, 2015). It also enable firms to to continuously and effectively adapt its course of action to an ever-changing market in order to remain competitive (Agarwal & Helfat, 2009; Weber & Tarba, 2014).

Given the need to rethink about design of traditional processes to capture the full potential of industry 4.0 (Balasingham, 2016), strategic agility is seen as an important capability that enables firms take to perform in the rapid and uncertain environment (Weber & Tarba, 2014). Agile firms can adapt themselves with reference to the disruptive changes (Franken & Thomsett, 2013) which allow firms to handle the changes that are continuous, and systematic in an organization's product, services, and structures in more effective and efficient manner (Rudtsch, Gausemeier, Gesing, Mittag, & Peter, 2014). Nevertheless, strategic agility is not possible without substantial support of resources in order to maintain high flexibility and necessary speed to respond to last-minute changes (Weber & Tarba, 2014).

In order to be agile, a firm must also embrace organizational flexibility and such flexibility can be enhanced through organizational networking (Yousaf & Majid, 2018). According to Yousaf and Majid (2018), organizational networks make organizations more flexible, dynamic, and independent from hierarchical structures hence play a crucial role in the development of agile firms (Borgatti & Foster, 2003). Hence, organizational network can also enhance strategic agility and ultimately manufacturing firm will perform better. Against this background and mixed interest in entrepreneurial leadership, organizational networking, strategic agility and business performance, this study will examine the effects of: a) organizational networking on strategic agility; b) the mediating role of strategic agility between the relationship of organizational networking and business performance.

## Organizational Networking

Organizational Networking refers to the social skills of a person to leverage the relationship to benefit from them (Ferris et al., 2007; Semrau, Ambos, & Kraus, 2016). According to Coleman (1994), social capitalists consider all the benefits associated with a network structure and all other types of relationships. This study has considered the organizational networking and avoided personal networking (Ferris et al., 2007; Semrau et al., 2016). Organizational networking refers to the different types of direct and indirect relationships of a focal firm (Thornton, Henneberg, & Naudé, 2014). Such relationship provides diverse opportunities and threads to the focal firm. The set of prospective resources is cognizant to the scale of network (Zaheer & Bell, 2005). Similarly, the decision making of the focal firm is highly influenced by the dynamics derived from its web of relationship (Astley, 1984).

There are different definitions of organizational networking. Organizational networking refers to the pattern of relations among individuals, groups or organizations (Aldrich, Reese, & Dubini, 1989). According to (Lechner & Dowling, 2003), organizational networking comprises of the nodes and relationships which connect individuals, groups and organizations. (Kreiner & Schultz, 1993) has defined organizational networking as informal collaboration.

This literature elaborates both positive and negative aspects of the organizational networking. According to Håkansson and Ford (2002), networking reduces the autonomous level of organizations because of shared resources and experience in previous collaboration. Similarly, (Koka & Prescott, 2008) concludes that organizational networking provides non-redundant relationship which services quality information at the cost of efficiency and performance. On the same note, Inkpen and Tsang (2005) claim that effect of organizational networking is cognizant to the nature of context and quality of knowledge. Networking affects the performance negative in case low quality knowledge is shared through it (Inkpen & Tsang, 2005). Though, positive side of networking is more prominent that nits' negative effects. Extent literature significantly relate the positive influence of networking to the opportunity recognition, start-up innovation, quality decision-making, competitiveness, business growth, and firm performance (Huang, Lai, & Lo, 2012; Lechner & Dowling, 2003; Owen-Smith, Cotton-Nessler, & Buhr, 2015; Song, Min, Lee, & Seo, 2017; Zhao, Frese, & Giardini, 2010) (Lechner & Dowling, 2003). Besides all this, business performance is deemed as substantial outcome of the organizational networking (Chung et al., 2015).

## Strategic Agility

Strategic agility has garnered the significant attentions from researchers (Inman, Sale, Green Jr, & Whitten, 2011; Tallon & Pinsonneault, 2011; Vickery, Droge, Setia, & Sambamurthy, 2010). Agility enables the organizations to be flexible, to respond efficiently and effectively, cope with market risk and uncertainty (Sherehiy, Karwowski, & Layer, 2007). According to Kumkale (2016), strategic agility is highly sensitive to the internal and external environments. Strategic agile firms make changes with respect to market fluctuation, learns about it swiftly, exploits such changes, and customize the operations on individual basis (Braunscheidel & Suresh, 2009). By responding to market changes, agile firms convert them into market opportunities (Shin et al., 2015). Yet, there is a lack of research in this domain (Kale, Aknar, & Başar, 2019). According to Kumkale (2016), strategic agility requires firms to constantly monitor the internal and external environment, garner and process the information efficiently and responding to market queries effectively. Strategic agility not only enhances the competitiveness of the firms but also increases its performance (Tallon & Pinsonneault, 2011).

The concept "agility" has been first used in the Iacocca Institute survey in the United States in 1991 where focus was on flexible and agile production to meet the rapidly changes in market (Dove, 1991). Agility is viewed as the transformation of the continuous and uncertain variation in the customer's requirements into opportunity in the dynamic market (Dove, 2002). To better comprehend the agility, an integrative perspective is a requisite. Partnership play vital role in the formation of the agility. According to Kidd (1994), synthesis of the organizations with different basic skills and competencies respond to the changing demands of the customers. Strategic agility of the firms requires the integration of the people, organization, and technology. Gehani (1995) has claimed different functions of the agile operations such as timely introduction of the new products, efficient response to customer's queries, revisiting of the strategic networking effectively. Agile is a source of competitive edge where it facilitates high quality, short delivery time, innovative behaviors, adaptability, low operational cost (Ileri & Soylu, 2010).

There is substantial need to operate with strategic agility across national and international market based on highly changing demands of customers, introduction of new products, and high-quality services (Oyedijo, 2012). Strategic agility refers to the ability of the firm to compete in the dynamic environment (Lengnick-Hall & Beck, 2016). According to Al-Azzam, Irtaimeh, and Khaddam (2017), human performance, processes and technologies of the organization influences the strategic agility. Three meta-skills i.e. strategic sensitivity, leadership unity, and resource fluidity are necessary to develop the strategic agility (Doz & Kosonen, 2010).

The researchers in the production and information technologies have conducted research on the effects of agility and strategic agility with firm performance (Kale et al., 2019). According to P. M. Swafford, S. Ghosh, and N. N. Murthy (2006b), there is direct positive relationship between value chain agility and business performance. According to Ojha (2008), ability to get information about market fluctuation is strong determinant of the strategic agility where it has not linkage to the financial performance but is beneficial in the moderately dynamic environment. Vickery et al. (2010) has concluded with the significant impact of strategic agility on the firm performance. There is empirical evidence about the significant impact of agility on the firm performance (Tallon & Pinsonneault, 2011).

Similarly, Inman et al. (2011) reports that agile manufacturing significantly enhances financial, marketing and operational performance. Roberts et al. (2012) has measured agility based on customer sensing, responding capabilities. Customer sensing significantly influences firm performance but responding capabilities does not. Not only this, Teoh, Lee, and Muthuveloo (2017) has confirmed the mediating role of agility between corporate risk management practices and firm performance.

## Entrepreneurial Bricolage

Lévi-Strauss (1966) has initially introduced the term "bricolage." Generally, bricolage is known as making do with resources at hand. Bricolage is getting increasing attention among the entrepreneurial research and the emerging theories of the entrepreneurship (Fisher, 2012; Phillips & Tracey, 2007). With bricolage, employees are in position to deliver unique services by recombining resources at hand for new objectives that hinders them (Baker & Nelson, 2005). The bricolage helps the employees to deal with market uncertainties, cope with challenges, and even compete in spite of limited resources (Digan, Sahi, Mantok, & Patel, 2019). Bricolage influences the relationship between entrepreneurial efforts and value creation, assists development and exploitation of opportunities (Senyard, Baker, & Davidsson, 2009), and enhance creativity and innovation among employees (Di Domenico, Haugh, & Tracey, 2010; Senyard, Baker, Steffens, & Davidsson, 2014). According to Senyard et al. (2014), bricolage at organizational level significantly influence new product creation under resources at hand. Additionally, creative ways to scarcity enriches the knowledge of entrepreneurs (Cunha, Rego, Oliveira, Rosado, & Habib, 2014).

According to Digan et al. (2019), there is high level of association between empowerment and revenue in the presence of high level of bricolage. self-determination and autonomous are source of high level of bricolage (M. Cunha, 2005); absence of autonomous leads to reduction in its level (Bojica, Istanbouli, & Fuentes-Fuentes, 2014). Hmieleski and Corbett (2008) found that self-efficacy also moderates the relationship between bricolage and firm performance. According to Digan et al. (2019), gains from empowerment is further enhanced through bricolage in the resource-constraint environment.

## Application of Resource-Based View (RBV), Dynamic Capability and Effectuation Theory

Resource-based view (RBV) standalone is not helpful in order to cope with the challenges of dynamic market (Eisenhardt & Martin, 2000). Nevertheless, the integration of the resource-based view (RBV), and dynamic capability works where optimal performance is cognizant of superior resources (Wernerfelt, 1984). In this study, organizational networking is deemed as resource. Dynamic market calls for the continuous and perpetual reinvention, reconfiguration, and recreation of the capabilities to deliver optimum performance (Eisenhardt & Martin, 2000). Strategic agility is considered as dynamic capability that enables firms to operate in the uncertain environment. This study has applied dynamic capability theory to investigate the impact of strategic agility on the business performance in the perspective of industry 4.0 and had postulated that organizational networking (resource) enables strategic agility (dynamic capability) that ultimately results into better business performance.

The effectuation theory is derived from the research work of Sarasvathy et al. (2008), where employees behave entrepreneurially to exploit the opportunities through the reconfiguration of limited resources on hand known as bricolage. The effectuation theory works on the affordable loss criterion where risk to a certain degree is accepted. Because, firms always face resource constraints and other limitations so the application of bricolage under the effectuation theory will strengthen the relationship between organizational networking and strategic agility.

## Organizational Networking and Strategic Agility

Superior performance always relies on the superior resources (Wernerfelt, 1984). Resources enhances the dynamic capability of organization where new resource configuration emerges as market changes, collide, split, and evolve (Eisenhardt & Martin, 2000). Therefore, it is claimed that organizational networking being resources enhances strategic agility which is viewed as dynamic capability. Organizational networks enable organizations to adapt to the changes through coordinated activities with the help of complementary relationship (Borgatti & Foster, 2003).

Organizational networks enhance organizational flexibility by responding positively to change (Walter, Auer, & Ritter, 2006). Organizational networks admire cooperation among units to reformulate and implement strategies to respond to dynamic markets, therefore influences the adaptability, flexibility, and agility (Lans et al., 2015). Organizational networks enable the organization to become more flexible through sharing resources, minimizing additional costs, and updating knowledge (Walter et al., 2006). Organizational networking believes in support, collaboration, and coordination among different stakeholders (Mohr & Spekman, 1994) which is mandatory for the successful collaborative innovation where customer's sensing and responding is conducted (Swafford et al., 2006a). Organizational networks enable firms to own high IT capability based on diverse IT resources, IT infrastructure, and IT-enabled intangibles (Chae, Koh, & Prybutok, 2014; Chan, 2000).

According to Roth (1996), agile firms rely on the organizational learning and require highly skilled employees to gain the information and knowledge about the current opportunities and challenges in the market about their business (Li, Chung, Goldsby, & Holsapple, 2008). Organizational network is a source to gain reliable and valid information about customers, suppliers and market variations. Although, organizational network requires robust internal communication processes (Sivadas & Dwyer, 2000). As, dissemination of the quality information prevents the miscommunication and redundancy if possible, among different stakeholders that ultimately increase the synergetic effect of the networking (Cohen & Levinthal, 2000).Whereas, Iqbal and Nawaz (2019) have claimed that polluted information is increasing exponentially in the presence of linearly rising quality information. Such polluted information is not only negatively impacting the organizations but individual employees as well (Iqbal, Hassan, & Ahmad, 2018).

Organizations need to be highly cautious about information in terms of its representation, accessibility, intrinsic and contextual nature (Iqbal, Yang, Nawaz, & Iqbal, 2019). Based on the strong and effective communication, organizational network enables organizations to extract the quality information that ultimately help them to align the organizational goals, objectives, needs, and structure with each other (Boyer & McDermott, 1999). Strong internal communication creates consensus among leaders and functional managers by enhancing consistency among different units and departments. Based on this discussion, following proposition is developed;

Proposition 1: Organizational networking significantly enhances the strategic agility among manufacturing firms

## Strategic Agility to Business Performance

Although, the association between strategic decision making and business performance has been a customary theme of research (Capon, Farley, & Hoenig, 1990), extent literature has claimed the mixed empirical results between performance and agility variable (Shin et al., 2015). According to Swafford et al. (2006a), there is positive correlation between value chin agility and return on asset. Vickery et al. (2010) has claimed that agility positively affects the financial performance. In an integrated study of IT alignment, agility and performance, Tallon and Pinsonneault (2011) have concluded with positive linkage of agility to the financial performance. Wu, Straub, and Liang (2015) has conducted an empirical study in 131 Taiwanese companies where a significant and impactful linkage was claimed between alignment and organizational performance. Contrary to these findings, there is negative association between manufacturing agility and financial performance (Inman et al., 2011; Jacobs, Droge, Vickery, & Calantone, 2011). On the final note, Iqbal, Ahmad, and Ahmad (2018) have recommended to alter the job characteristics with respect to the required business performance.

Furthermore, Roberts and Grover (2012) has also studied the relationship between agility (customer sensing and responding capabilities) and business performance (financial performance and marketing performance). But, concluded that customer sensing capability and customer responding capability both exhibit contradictory results (Roberts & Grover, 2012). Sustaining collaborative innovation enables the firms to meet customer's requirements (Choi & Krause, 2006). Information and learning substantially contributes to the innovativeness of the firms. Big data is a crucial part of the industry 4.0 where polluted information rises exponentially and equally hazardous for the firms (Iqbal & Nawaz, 2019). Not only this, Iqbal et al. (2019) has recommended that organization learning deal effectively with rising information pollution (Iqbal & Nawaz, 2019) which is blatantly good for the firms operating in the industry 4.0 where quality information contributes significantly to the performance.

Operational perspective emphasizes on the improvised efficient operations that ignites the strategic agility (Braunscheidel & Suresh, 2009; Swafford et al., 2006a). According to Inman et al. (2011) and Green Jr, Whitten, and Inman (2012), agility enhances operational performance significantly. Additionally, Inman et al. (2011) has recommended the significant positive relationship between agile manufacturing and operational performance. According to Shin et al. (2015), strategic agility significantly influences operational and marketing performance. Since strategic agility aims to respond effectively and efficiently to unexpected changes in market and customer requirements. The above discussion leads to the following proposition;

Proposition 2: Strategic agility significantly enhances the business performance among manufacturing firms

## Strategic Agility as Mediator

This literature recommends the significant positive impact of organizational networking on the business performance where efficiency and effectiveness heavily relies on the effort, time, and cost (Yousaf & Majid, 2018), which are prime attributes of the strategic agility. It means, organizational network work as enabler of the strategic agility. According to Yousaf and Majid (2018), networks based on the relationship and coordination enhances the agility that ultimately results into the improved business performance. From the perspective of the fourth industrial revolution, network enables organizations to deliver flexibility and consistently (Lin & Lin, 2016). Furthermore, networks enhance the synergistic efforts of the firms and ensure flexibility (Kozjek & Ferjan, 2015), that is sub-part of organizational agility.

According to Borgatti and Foster (2003), there is strong association between organizational networks and strategic agility. Organizational networks enable organization through strategic agility (collaborative innovation, organizational learning, internal alignment, and IT capability) realize the opportunities with high level of adaptability. According to Mitchell, Schlegelmilch, and Mone (2016); agile firms exploit opportunities through organizational networks based on shared information, resources, costs and access to another's market. Firms align business performance through organizational agility provided through organizational network (Dastmalchian & Blyton, 1998).

Organizational network is useful to anticipate the market opportunities and market-oriented resource deployment (Walter et al., 2006). The effective communication within network is essential for the innovative products based on customer's requirements. Additionally, locus of innovation relies on the networks beside individual employees or individual firms (Håkansson & Ford, 2002). Valued inputs within network have significant importance to the knowledge-based organizations. Such relations provide quality about novel ideas and potential opportunities in parallel to keeping away from the polluted information where it generates upon exponential basis (Iqbal & Nawaz, 2019), ultimately enhancing strategic agility to improve business performance (Majid, Yasir, & Yasir, 2017). Such Shared resources, information and solid communication within network may enhances the organizational learning, IT capability, internal alignment and collaborative innovation that results in the effective delivery of customer's needs. This discussion leads to the proposition of the following statement;

Proposition 3: strategic agility mediates the relationship between organizational networking and business performance in industry 4.0.

*Figure 1. Research framework*

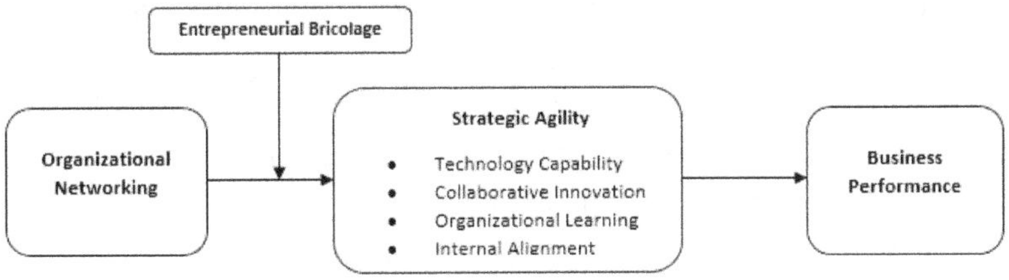


## The Moderating Role of Entrepreneurial Bricolage

According to effectuation theory, entrepreneurs are able to deliver in the presence of limited and scare resources (Sarasvathy et al., 2008). Strong support, collaboration and coordination among internal and external units within organizational networking (Mohr & Spekman, 1994) are useful for collaborative innovation. Organizational networking play significant role in the success of organization thorough shared resources, minimizing additional costs, and updating knowledge (Walter et al. 2006), where these attribute are equally important for IT capability, organizational learning within strategically agile organizations (Chae et al., 2014; Chan, 2000). Availability of current information about all stakeholders prevent miscommunication and influences internal alignment within organization. Based on effectuation theory, entrepreneurial bricolage enables organization to achieve strategic agility with limited organizational networking. Therefore, following proposition is claimed;

Proposition 4: Stronger the entrepreneurial bricolage, the higher the significant relationship between organizational networking and technology capability.

## CONCLUSION

This research has integrated the organizational networking, strategic agility and business performance of the manufacturing firms in the context of fourth industrial revolution, where practitioners can find it worthy. The above propositions based on the critical literature review yielded the integrated framework of the organizational networking, strategic agility and business performance as shown in the below figure 1.

The robust competitive nature of industry 4.0 offers an appropriate field for academic researchers and practitioners to enhance the aspects associated to the high business performance when examined from the perspective for resource-based view (RBV), knowledge-based view (KBV) and dynamic capability. Therefore, with reference to resource-based view, there was a need to investigate the role of organizational networking; when understood and incorporated in the firm, could provide the dynamic capability of manufacturing companies. This, in turn, may bring in positive impact on business performance and could facilitate large-scale manufacturers a competitive edge over their competitors in terms of strategic agility and overall business performance.

Disruptive technologies in industry 4.0 have substantially affected the manufacturing industries such as reduced product life cycles and high-quality products (Rose et al. 2008). This scenario works as a stimulus for large scale manufacturers to look beyond the short-term goals linked with profit margin to long term survival of firms by going strategically agile. Therefore, this research looks reasonable provided it could be useful in achieving competitive edge through strategical aigility. The comprehensive understating of the organizational networking will possibly enhance the overall business performance, could increase sensitivity of decision makers in manufacturing sector towards these resources to strengthen their competitive edge.

Gross domestic product (GDP) is equal to the sum of total consumption, investment, net exports and government spending (McEachern, 2009). A strong manufacturing sector enhances consumption, net exports, and consumption. Therefore, if empirical findings of such propositions would be helpful to boost the economy of the nations. This may prove useful for policy makers and practitioners to formulate strategies to develop large scale manufacturing.

## Implications

This study has extended the domain of organizational networks by revealing strategic agility as a driver to enhance business performance in industry 4.0. This study has also contributed to the literature by suggesting mediating role of the strategic agility between organizational networking and business performance. Therefore, this study enhances the literature by claiming the strategic agility as the major outcome of organizational networking to make business performance better.

The present study opens the new horizons for academic and industry experts while elaborating the moderating role of the entrepreneurial bricolage. The degree to which organizational networking influence strategic agility could be low when entrepreneurial bricolage moderates in terms of internal and external relations of any organization. The significant moderating impact of the entrepreneurial bricolage on the association of organizational networking-strategic agility indicates that entrepreneurial bricolage can preserve dynamic capability of the firm by ensuring high organizational learning, high level of collaborative innovation, strong internal alignment, and technology capability.

This study introduces different practical implications. First, based on the proposed model, it is recommended to the industry experts of large manufacturing firms that there is dire need to strengthen the strategic agility through organizational networking. large manufacturing firms are facing myriad of challenges in the shape of reduced product life cycle, highly customized products, lacking financial resources and uncertain and dynamic market (Brettel et al., 2014), in such cruelty, the organizational networking could provide numerous opportunities to enhance sustainable performance. The present study suggests that practitioners may comprehend the market situation through networking, efficiently respond to the extensive range of information and internal and external stimulus to the firm (Walter et al., 2006).

This study recommends focussing on the relational skills via organizational networking for strengthening the mechanism of strategic agility to improve the business performance (Owen-Smith et al., 2015). Therefore, it is highly recommended to large manufacturing firms in industry 4.0 to optimize their strategic agility by contacting other firms to build strong relationships (Song et al., 2017).

## Limitations

The present study claims interesting propositions however it is not free of certain limitations. Industrialists and practitioners are recommended to first analyze not only their networking mechanism, but also their level of strategic agility before making any decision about investment. As this study has claimed that strategic agility could have vital impact on a wide range of operational measures and hence, should be religiously tackled. Despite the significant variations of large manufacturing firms' operations, it could possess some common processes with SME, such as manufacturing SMEs. The proposed solutions may be applicable for these firms, although an empirical analysis of such propositions could be an exciting topic for future research.

## ACKNOWLEDGMENT

We are thankful to Institute of Postgraduate Studies (IPS), Universiti Sains Malaysia (USM), for providing funding under the Postgraduate Research Attachment Program (PGRA) 2018.

## REFERENCES

Agarwal, R., & Helfat, C. E. (2009). Strategic renewal of organizations. *Organization Science*, *20*(2), 281–293. doi:10.1287/orsc.1090.0423

Al-Azzam, Z. F., Irtaimeh, H. J., & Khaddam, A. A. H. (2017). Examining the Mediating Effect of Strategic Agility in the Relationship between Intellectual Capital and Organizational Excellence in Jordan Service Sector. *The Journal of Business*, *6*(1), 7–15.

Aldrich, H., Reese, P. R., & Dubini, P. (1989). Women on the verge of a breakthrough: Networking among entrepreneurs in the United States and Italy. *Entrepreneurship and Regional Development*, *1*(4), 339–356. doi:10.1080/08985628900000029

Ara, I. (2004). Is Pakistan's manufacturing sector competitive? *Pakistan Development Review*, *43*(4II), 685–706. doi:10.30541/v43i4IIpp.685-706

Astley, W. G. (1984). Toward an appreciation of collective strategy. *Academy of Management Review*, *9*(3), 526–535. doi:10.5465/amr.1984.4279700

Baker, T., & Nelson, R. E. (2005). Creating something from nothing: Resource construction through entrepreneurial bricolage. *Administrative Science Quarterly*, *50*(3), 329–366. doi:10.2189/asqu.2005.50.3.329

Balasingham, K. (2016). *Industry 4.0: Securing the Future for German Manufacturing Companies*. University of Twente.

Bauer, W., Hämmerle, M., Schlund, S., & Vocke, C. (2015). Transforming to a hyper-connected society and economy–towards an "Industry 4.0". *Procedia Manufacturing*, *3*, 417–424.

Bojica, A. M., Istanbouli, A., & Fuentes-Fuentes, M. D. M. (2014). Bricolage and growth strategies: Effects on the performance of Palestinian women-led firms. *Journal of Developmental Entrepreneurship, 19*(4). doi:10.1142/S108494671450023X

Borgatti, S. P., & Foster, P. C. (2003). The network paradigm in organizational research: A review and typology. *Journal of Management, 29*(6), 991–1013. doi:10.1016/S0149-2063(03)00087-4

Boyer, K. K., & McDermott, C. (1999). Strategic consensus in operations strategy. *Journal of Operations Management, 17*(3), 289–305. doi:10.1016/S0272-6963(98)00042-4

Braunscheidel, M. J., & Suresh, N. C. (2009). The organizational antecedents of a firm's supply chain agility for risk mitigation and response. *Journal of Operations Management, 27*(2), 119–140. doi:10.1016/j.jom.2008.09.006

Brettel, M., Friederichsen, N., Keller, M., & Rosenberg, M. (2014). How virtualization, decentralization and network building change the manufacturing landscape: An Industry 4.0 Perspective. *International Journal of Mechanical. Industrial Science and Engineering, 8*(1), 37–44.

Capon, N., Farley, J. U., & Hoenig, S. (1990). Determinants of financial performance: A meta-analysis. *Management Science, 36*(10), 1143–1159. doi:10.1287/mnsc.36.10.1143

Chae, H.-C., Koh, C. E., & Prybutok, V. R. (2014). Information technology capability and firm performance: Contradictory findings and their possible causes. *Management Information Systems Quarterly, 38*(1), 305–326. doi:10.25300/MISQ/2014/38.1.14

Chan, Y. E. (2000). IT value: The great divide between qualitative and quantitative and individual and organizational measures. *Journal of Management Information Systems, 16*(4), 225–261. doi:10.1080/07421222.2000.11518272

Choi, T. Y., & Krause, D. R. (2006). The supply base and its complexity: Implications for transaction costs, risks, responsiveness, and innovation. *Journal of Operations Management, 24*(5), 637–652. doi:10.1016/j.jom.2005.07.002

Christopher, M. (2000). The agile supply chain: Competing in volatile markets. *Industrial Marketing Management, 29*(1), 37–44. doi:10.1016/S0019-8501(99)00110-8

Chung, H. F. (2011). Market orientation, guanxi, and business performance. *Industrial Marketing Management, 40*(4), 522–533. doi:10.1016/j.indmarman.2010.12.008

Chung, H. F., Yang, Z., & Huang, P.-H. (2015). How does organizational learning matter in strategic business performance? The contingency role of guanxi networking. *Journal of Business Research, 68*(6), 1216–1224. doi:10.1016/j.jbusres.2014.11.016

Cohen, W. M., & Levinthal, D. A. (2000). Absorptive capacity: A new perspective on learning and innovation. In *Strategic Learning in a Knowledge economy* (pp. 39–67). Elsevier. doi:10.1016/B978-0-7506-7223-8.50005-8

Coleman, J. S. (1994). *Foundations of social theory*. Harvard university press.

Cunha, M. (2005). Bricolage in organizations. *SSRN Electronic Journal*.

Cunha, M. P., Rego, A., Oliveira, P., Rosado, P., & Habib, N. (2014). Product innovation in resource-poor environments: Three research streams. *Journal of Product Innovation Management*, *31*(2), 202–210. doi:10.1111/jpim.12090

Dastmalchian, A., & Blyton, P. (1998). Organizational flexibility in cross-national perspective: An introduction. *International Journal of Human Resource Management*, *9*(3), 437–444. doi:10.1080/095851998340991

de Sousa Jabbour, A. B. L., Jabbour, C. J. C., Godinho Filho, M., & Roubaud, D. (2018). Industry 4.0 and the circular economy: A proposed research agenda and original roadmap for sustainable operations. *Annals of Operations Research*, 1–14.

Di Domenico, M., Haugh, H., & Tracey, P. (2010). Social bricolage: Theorizing social value creation in social enterprises. *Entrepreneurship Theory and Practice*, *34*(4), 681–703. doi:10.1111/j.1540-6520.2010.00370.x

Digan, S. P., Sahi, G. K., Mantok, S., & Patel, P. C. (2019). Women's Perceived Empowerment in Entrepreneurial Efforts: The Role of Bricolage and Psychological Capital. *Journal of Small Business Management*, *57*(1), 206–229. doi:10.1111/jsbm.12402

Dove, R. (1991). *The 21s Century Manufacturing Enterprise Strategy. Iacocca Institute*, Lehigh University.

Dove, R. (2002). *Response ability: the language, structure, and culture of the agile enterprise*. John Wiley & Sons.

Doz, Y. L., & Kosonen, M. (2010). Embedding strategic agility: A leadership agenda for accelerating business model renewal. *Long Range Planning*, *43*(2-3), 370–382. doi:10.1016/j.lrp.2009.07.006

Eisenhardt, K. M., & Martin, J. A. (2000). Dynamic capabilities: What are they? *Strategic Management Journal*, *21*(10-11), 1105–1121. doi:10.1002/1097-0266(200010/11)21:10/11<1105::AID-SMJ133>3.0.CO;2-E

Ferris, G. R., Treadway, D. C., Perrewé, P. L., Brouer, R. L., Douglas, C., & Lux, S. (2007). Political skill in organizations. *Journal of Management*, *33*(3), 290–320. doi:10.1177/0149206307300813

Fisher, G. (2012). Effectuation, causation, and bricolage: A behavioral comparison of emerging theories in entrepreneurship research. *Entrepreneurship Theory and Practice*, *36*(5), 1019–1051. doi:10.1111/j.1540-6520.2012.00537.x

Franken, A., & Thomsett, H. (2013). When it takes a network: Creating strategy and agility through wargaming. *California Management Review*, *55*(3), 107–133. doi:10.1525/cmr.2013.55.3.107

Gehani, R. R. (1995). Time-based management of technology: A taxonomic integration of tactical and strategic roles. *International Journal of Operations & Production Management*, *15*(2), 19–35. doi:10.1108/01443579510080391

Gelhard, C., von Delft, S., & Gudergan, S. P. (2016). Heterogeneity in dynamic capability configurations: Equifinality and strategic performance. *Journal of Business Research*, *69*(11), 5272–5279. doi:10.1016/j.jbusres.2016.04.124

Green, K. M., Covin, J. G., & Slevin, D. P. (2008). Exploring the relationship between strategic reactiveness and entrepreneurial orientation: The role of structure–style fit. *Journal of Business Venturing, 23*(3), 356–383. doi:10.1016/j.jbusvent.2007.01.002

Green, K. W. Jr, Whitten, D., & Inman, R. A. (2012). Aligning marketing strategies throughout the supply chain to enhance performance. *Industrial Marketing Management, 41*(6), 1008–1018. doi:10.1016/j.indmarman.2012.02.003

Gundry, L. K., Kickul, J. R., Griffiths, M. D., & Bacq, S. C. (2011). Creating social change out of nothing: The role of entrepreneurial bricolage in social entrepreneurs' catalytic innovations. In *Social and sustainable entrepreneurship* (pp. 1–24). Emerald Group Publishing Limited. doi:10.1108/S1074-7540(2011)0000013005

Håkansson, H., & Ford, D. (2002). How should companies interact in business networks? *Journal of Business Research, 55*(2), 133–139. doi:10.1016/S0148-2963(00)00148-X

Hmieleski, K. M., & Corbett, A. C. (2008). The contrasting interaction effects of improvisational behavior with entrepreneurial self-efficacy on new venture performance and entrepreneur work satisfaction. *Journal of Business Venturing, 23*(4), 482–496. doi:10.1016/j.jbusvent.2007.04.002

Huang, H.-C., Lai, M.-C., & Lo, K.-W. (2012). Do founders' own resources matter? The influence of business networks on start-up innovation and performance. *Technovation, 32*(5), 316–327. doi:10.1016/j.technovation.2011.12.004

Ileri, Y. Y., & Soylu, Y. (2010). Bir Rekabet Üstünlüğü Araci Olarak Çeviklik Kavrami Ve Örgüt Yapisina Olasi Etkileri. *Selçuk Üniversitesi Sosyal Bilimler Meslek Yüksekokulu Dergisi, 13*(1-2), 13–28.

Inkpen, A. C., & Tsang, E. W. (2005). Social capital, networks, and knowledge transfer. *Academy of Management Review, 30*(1), 146–165. doi:10.5465/amr.2005.15281445

Inman, R. A., Sale, R. S., Green, K. W. Jr, & Whitten, D. (2011). Agile manufacturing: Relation to JIT, operational performance and firm performance. *Journal of Operations Management, 29*(4), 343–355. doi:10.1016/j.jom.2010.06.001

Iqbal, Q., Ahmad, N. H., & Ahmad, B. (2018). Enhancing sustainable performance through job characteristics via workplace spirituality: A study on SMEs. *Journal of Science and Technology Policy Management*. doi:10.1108/JSTPM-02-2018-0022

Iqbal, Q., Hassan, S. H., & Ahmad, N. H. (2018). The assessment of perceived information pollution in banking sector: A scale development and validation study. *Business Information Review, 35*(2), 68–76. doi:10.1177/0266382118772891

Iqbal, Q., & Nawaz, R. (2019). Rife information pollution (infollution) and virtual organizations in industry 4.0: within reality causes and consequences. In Big Data and Knowledge Sharing in Virtual Organizations (pp. 117–135). Hershey, PA: IGI Global. doi:10.4018/978-1-5225-7519-1.ch005

Iqbal, Q., Yang, S., Nawaz, R., & Iqbal, K. (2019). *A multi-dimensional construct of perceived information pollution in the era of rife infollution. VINE Journal of Information and Knowledge Management Systems*. doi:10.1108/VJIKMS-10-2018-0093

Jacobs, M., Droge, C., Vickery, S. K., & Calantone, R. (2011). Product and process modularity's effects on manufacturing agility and firm growth performance. *Journal of Product Innovation Management, 28*(1), 123–137. doi:10.1111/j.1540-5885.2010.00785.x

Kagermann, H., Helbig, J., Hellinger, A., & Wahlster, W. (2013). *Recommendations for implementing the strategic initiative INDUSTRIE 4.0: Securing the future of German manufacturing industry; final report of the Industrie 4.0 Working Group*. Forschungsunion.

Kale, E., Aknar, A., & Başar, Ö. (2019). Absorptive capacity and firm performance: The mediating role of strategic agility. *International Journal of Hospitality Management, 78*, 276–283. doi:10.1016/j.ijhm.2018.09.010

Kemal, A. R. (2006). Key Issues in Industrial Growth in Pakistan. *Lahore Journal of Economics, 11*.

Khan, A., & Turowski, K. (2016). A survey of current challenges in manufacturing industry and preparation for industry 4.0. *Paper presented at the Proceedings of the First International Scientific Conference "Intelligent Information Technologies for Industry" (IITI'16)*. Academic Press.

Kidd, P. T. (1994). *Agile Manufacturing "Forging new Frontiers." Addison-Wesley* Publication Company Inc.

Kim, E. Y., Ko, E., Kim, H., & Koh, C. E. (2008). Comparison of benefits of radio frequency identification: Implications for business strategic performance in the US and Korean retailers. *Industrial Marketing Management, 37*(7), 797–806. doi:10.1016/j.indmarman.2008.01.007

Koka, B. R., & Prescott, J. E. (2008). Designing alliance networks: The influence of network position, environmental change, and strategy on firm performance. *Strategic Management Journal, 29*(6), 639–661. doi:10.1002mj.679

Kozjek, T., & Ferjan, M. (2015). Organizational flexibility, employee security, and organizational efficiency–a case study of Slovenian public and private sector organizations. *Organizacija, 48*(1), 3–21. doi:10.1515/orga-2015-0001

Kreiner, K., & Schultz, M. (1993). Informal collaboration in R & D. The formation of networks across organizations. *Organization Studies, 14*(2), 189–209. doi:10.1177/017084069301400202

Kumkale, İ. (2016). Organization's Tool for Creating Competitive Advantage: Strategic Agility. *Balkan and Near Eastern Journal of Social Sciences, 2*(3), 118–124.

Lans, T., Blok, V., & Gulikers, J. (2015). Show me your network and I'll tell you who you are: Social competence and social capital of early-stage entrepreneurs. *Entrepreneurship and Regional Development, 27*(7-8), 458–473. doi:10.1080/08985626.2015.1070537

Lechner, C., & Dowling, M. (2003). Firm networks: External relationships as sources for the growth and competitiveness of entrepreneurial firms. *Entrepreneurship and Regional Development, 15*(1), 1–26. doi:10.1080/08985620210159220

Lengnick-Hall, C. A., & Beck, T. E. (2016). Resilience capacity and strategic agility: Prerequisites for thriving in a dynamic environment. In *Resilience Engineering Perspectives* (Vol. 2, pp. 61–92). CRC Press.

Lévi-Strauss, C. (1966). *The savage mind.* Chicago: The University of Chicago Press.

Li, X., Chung, C., Goldsby, T. J., & Holsapple, C. W. (2008). A unified model of supply chain agility: The work-design perspective. *International Journal of Logistics Management, 19*(3), 408–435. doi:10.1108/09574090810919224

Lin, F.-J., & Lin, Y.-H. (2016). The effect of network relationship on the performance of SMEs. *Journal of Business Research, 69*(5), 1780–1784. doi:10.1016/j.jbusres.2015.10.055

Majid, A., Yasir, M., & Yasir, M. (2017). Individual and work dynamics affecting the determinants of functional flexibility in SMEs: Evidence from Pakistan. *Journal of Entrepreneurship in Emerging Economies, 9*(2), 144–160. doi:10.1108/JEEE-03-2016-0008

Memon, M. A., & Tahir, I. M. (2012). Performance analysis of manufacturing companies in Pakistan. *Business Management Dynamics, 1*(7), 12–21.

Ministry of Finance. (2018). Pakistan economic survey 2017-18. Retrieved from http://www.finance.gov.pk/survey_1718.html

Mitchell, V.-W., Schlegelmilch, B. B., & Mone, S.-D. (2016). Why should I attend? The value of business networking events. *Industrial Marketing Management, 52*, 100–108. doi:10.1016/j.indmarman.2015.05.014

Mohr, J., & Spekman, R. (1994). Characteristics of partnership success: Partnership attributes, communication behavior, and conflict resolution techniques. *Strategic Management Journal, 15*(2), 135–152. doi:10.1002mj.4250150205

Nadkarni, S., & Narayanan, V. K. (2007). Strategic schemas, strategic flexibility, and firm performance: The moderating role of industry clockspeed. *Strategic Management Journal, 28*(3), 243–270. doi:10.1002mj.576

Ojha, D. (2008). *Impact of strategic agility on competitive capabilities and financial performance.* Clemson University.

Owen-Smith, J., Cotton-Nessler, N. C., & Buhr, H. (2015). Network effects on organizational decision-making: Blended social mechanisms and IPO withdrawal. *Social Networks, 41*, 1–17.

Oyedijo, A. (2012). Strategic agility and competitive performance in the Nigerian telecommunication industry: An empirical investigation. *American International Journal of Contemporary Research, 2*(3), 227–237.

Pakistan Bureau of Statistics. (2006). Census of Manufacturing Industries (CMI) 2005-06. Retrieved from http://www.pbs.gov.pk/content/census-manufacturing-industries-cmi-2005-06

Phillips, N., & Tracey, P. (2007). Opportunity recognition, entrepreneurial capabilities and bricolage: Connecting institutional theory and entrepreneurship in strategic organization. *Strategic Organization, 5*(3), 313–320. doi:10.1177/1476127007079956

Popadić, M., Černe, M., & Milohnić, I. (2015). Organizational Ambidexterity, Exploration, Exploitation and Firms Innovation Performance. *Organizacija, 48*(2), 112–119. doi:10.1515/orga-2015-0006

Rasmussen, E., Mosey, S., & Wright, M. (2015). The transformation of network ties to develop entrepreneurial competencies for university spin-offs. *Entrepreneurship and Regional Development*, *27*(7-8), 430–457. doi:10.1080/08985626.2015.1070536

Rehman, R. U., Zhang, J., & Ali, R. (2014). Firm Performance And Emerging Economies. *Journal of Applied Business Research*, *30*(3), 701. doi:10.19030/jabr.v30i3.8554

Roberts, N., & Grover, V. (2012). Investigating firm's customer agility and firm performance: The importance of aligning sense and respond capabilities. *Journal of Business Research*, *65*(5), 579–585. doi:10.1016/j.jbusres.2011.02.009

Roth, A. V. (1996). Achieving strategic agility through economies of knowledge. *Planning Review*, *24*(2), 30–36. doi:10.1108/eb054550

Rudtsch, V., Gausemeier, J., Gesing, J., Mittag, T., & Peter, S. (2014). Pattern-based Business Model Development for Cyber-Physical Production Systems. *Procedia CIRP*, *25*, 313–319. doi:10.1016/j.procir.2014.10.044

Sarasvathy, S. D., Dew, N., Read, S., & Wiltbank, R. (2008). Designing organizations that design environments: Lessons from entrepreneurial expertise. *Organization Studies*, *29*(3), 331–350. doi:10.1177/0170840607088017

Semrau, T., Ambos, T., & Kraus, S. (2016). Entrepreneurial orientation and SME performance across societal cultures: An international study. *Journal of Business Research*, *69*(5), 1928–1932. doi:10.1016/j.jbusres.2015.10.082

Senyard, J., Baker, T., & Davidsson, P. (2009). Entrepreneurial bricolage: Towards systematic empirical testing. *Frontiers of Entrepreneurship Research*, *29*(5), 5.

Senyard, J., Baker, T., Steffens, P., & Davidsson, P. (2014). Bricolage as a path to innovativeness for resource-constrained new firms. *Journal of Product Innovation Management*, *31*(2), 211–230. doi:10.1111/jpim.12091

Sherehiy, B., Karwowski, W., & Layer, J. K. (2007). A review of enterprise agility: Concepts, frameworks, and attributes. *International Journal of Industrial Ergonomics*, *37*(5), 445–460. doi:10.1016/j.ergon.2007.01.007

Shin, H., Lee, J.-N., Kim, D., & Rhim, H. (2015). Strategic agility of Korean small and medium enterprises and its influence on operational and firm performance. *International Journal of Production Economics*, *168*, 181–196. doi:10.1016/j.ijpe.2015.06.015

Sivadas, E., & Dwyer, F. R. (2000). An examination of organizational factors influencing new product success in internal and alliance-based processes. *Journal of Marketing*, *64*(1), 31–49. doi:10.1509/jmkg.64.1.31.17985

Song, G., Min, S., Lee, S., & Seo, Y. (2017). The effects of network reliance on opportunity recognition: A moderated mediation model of knowledge acquisition and entrepreneurial orientation. *Technological Forecasting and Social Change*, *117*, 98–107. doi:10.1016/j.techfore.2017.01.004

Srai, J. S., Kumar, M., Graham, G., Phillips, W., Tooze, J., Ford, S., ... Tiwari, M. K. (2016). Distributed manufacturing: Scope, challenges and opportunities. *International Journal of Production Research, 54*(23), 6917–6935. doi:10.1080/00207543.2016.1192302

Swafford, P. M., Ghosh, S., & Murthy, N. (2006a). The antecedents of supply chain agility of a firm: Scale development and model testing. *Journal of Operations Management, 24*(2), 170–188. doi:10.1016/j.jom.2005.05.002

Swafford, P. M., Ghosh, S., & Murthy, N. N. (2006b). A framework for assessing value chain agility. *International Journal of Operations & Production Management, 26*(2), 118–140. doi:10.1108/01443570610641639

Tallon, P. P., & Pinsonneault, A. (2011). Competing perspectives on the link between strategic information technology alignment and organizational agility: Insights from a mediation model. *Management Information Systems Quarterly, 35*(2), 463–486. doi:10.2307/23044052

Teoh, A. P., Lee, K. Y., & Muthuveloo, R. (2017). The impact of enterprise risk management, strategic agility, and quality of internal audit function on firm performance. *International Review of Management and Marketing, 7*(1), 222–229.

Thornton, S. C., Henneberg, S. C., & Naudé, P. (2014). Conceptualizing and validating organizational networking as a second-order formative construct. *Industrial Marketing Management, 43*(6), 951–966. doi:10.1016/j.indmarman.2014.05.001

Tipu, S. A. A., & Fantazy, K. A. (2014). Supply chain strategy, flexibility, and performance: A comparative study of SMEs in Pakistan and Canada. *International Journal of Logistics Management, 25*(2), 399–416. doi:10.1108/IJLM-01-2013-0001

Vickery, S., Droge, C., Setia, P., & Sambamurthy, V. (2010). Supply chain information technologies and organisational initiatives: Complementary versus independent effects on agility and firm performance. *International Journal of Production Research, 48*(23), 7025–7042. doi:10.1080/00207540903348353

Volberda, H. W. (1996). Toward the flexible form: How to remain vital in hypercompetitive environments. *Organization Science, 7*(4), 359–374. doi:10.1287/orsc.7.4.359

Walter, A., Auer, M., & Ritter, T. (2006). The impact of network capabilities and entrepreneurial orientation on university spin-off performance. *Journal of Business Venturing, 21*(4), 541–567. doi:10.1016/j.jbusvent.2005.02.005

Weber, Y., & Tarba, S. Y. (2014). Strategic Agility: A State of the Art Introduction to the Special Section on Strategic Agility. *California Management Review, 56*(3), 5–12. doi:10.1525/cmr.2014.56.3.5

Wernerfelt, B. (1984). A resource-based view of the firm. *Strategic Management Journal, 5*(2), 171–180. doi:10.1002mj.4250050207

Wu, S. P.-J., Straub, D. W., & Liang, T.-P. (2015). How information technology governance mechanisms and strategic alignment influence organizational performance: Insights from a matched survey of business and IT managers. *Management Information Systems Quarterly, 39*(2), 497–518. doi:10.25300/MISQ/2015/39.2.10

Yousaf, Z., & Majid, A. (2016). Strategic performance through inter-firm networks: Strategic alignment and moderating role of environmental dynamism. *World Journal of Entrepreneurship, Management and Sustainable Development, 12*(4), 282–298. doi:10.1108/WJEMSD-03-2016-0015

Yousaf, Z., & Majid, A. (2018). Organizational network and strategic business performance: Does organizational flexibility and entrepreneurial orientation really matter? *Journal of Organizational Change Management, 31*(2), 268–285. doi:10.1108/JOCM-12-2016-0298

Yousfi, A., Saidi, R., & Dey, A. K. (2016). Variability patterns for business processes in BPMN. *Information Systems and e-Business Management, 14*(3), 443–467. doi:10.100710257-015-0290-7

Zaheer, A., & Bell, G. G. (2005). Benefiting from network position: Firm capabilities, structural holes, and performance. *Strategic Management Journal, 26*(9), 809–825. doi:10.1002mj.482

Zahra, S. A., Sapienza, H. J., & Davidsson, P. (2006). Entrepreneurship and dynamic capabilities: A review, model and research agenda. *Journal of Management Studies, 43*(4), 917–955. doi:10.1111/j.1467-6486.2006.00616.x

Zhao, X.-y., Frese, M., & Giardini, A. (2010). Business owners' network size and business growth in China: The role of comprehensive social competency. *Entrepreneurship and Regional Development, 22*(7-8), 675–705. doi:10.1080/08985620903171376

Zheng, S., Zhang, W., & Du, J. (2011). Knowledge-based dynamic capabilities and innovation in networked environments. *Journal of Knowledge Management, 15*(6), 1035–1051. doi:10.1108/13673271111179352

# Chapter 5
# Drivers of Consumer Loyalty in the Wellness Sector

**Shivani Saini**
*Dr. B. R. Ambedkar National Institute of Technology, Jalandhar, India*

**Jagwinder Singh**
*Dr. B. R. Ambedkar National Institute of Technology, Jalandhar, India*

## ABSTRACT

*Facing a severely competitive environment and unique consumer expectations, cultivating consumer loyalty seems crucial for every firm. Rather than to cultivate, marketers find it more difficult to sustain a loyal consumer base. Understanding loyalty drives is of utmost importance now, and it is the main concern of this study. Faced with lack of structural management approaches in reference to relationship practices, this chapter provides a cohesive understanding of loyalty drivers in application of acquisition, retention, and experience strategies. The study involved a survey of 200 consumers of wellness firms. A confirmatory factor analysis has been employed to evaluate consumers' priority among acquisition, retention, and experience-based marketing strategies. The study offers valuable insights to wellness firms to fine-tune their CRM programmes in view of traditional and modern practices. As experience is very complex while acquisition and retention are highly imitable, firms need to embrace it with clarity to select the right elements to gain consumer loyalty.*

## INTRODUCTION

Indian wellness industry witnesses vigorous presence of national and international firms across beauty, therapy, fitness, and slimming services. The industry has registered a strong growth and is expected to reach US$220 billion by 2022 with a twelve percent compound annual growth rate (Ernst and Young wellness industry report 2018). The factors like larger youth population, rising consumer income, changing lifestyle, awareness about new products/services, high capital access, increasing expenditure on personal grooming, digitalization, and brand emergence are responsible to drive the growth of industry.

DOI: 10.4018/978-1-7998-1566-2.ch005

Consumers' growing consciousness about physical and mental well-being potentially plays a bigger role in the service consumption. Additionally, the consumers' increasing interactions on digital social media networks such as Facebook, Instagram, Twitter, Pinterest, YouTube, and MySpace are making them aware about the new trends and practices. Consumers with different age-groups and personality types are not very easy to satisfy however, gaining their loyalty seems crucial. Loyalty states a consumer's commitment to the firm. The commitment indicates a consumer's readiness to make sacrifices in order to maintain mutually benefited long-term relationships (Oliver, 1999; Singh & Saini, 2016; Rather et al., 2018). In today's competitive markets, the business success depends upon its competency to recognize the factors that not just arouses consumers' buying interests but develops a higher level of commitment. Gaining the consumer loyalty helps a firm to maintain long-term, cost-effective interpersonal relationships. The technology has also fostered enhanced interactions between a firm and its stakeholders which in further develops the two-way relationships with valuable consumers. A consumer expresses his or her true loyalty both in attitudinal and behavioural forms (Dick & Basu, 1994). The attitudinal loyalty exhibits a consumer's psychological attachment and positive attitude towards the firm (Dick & Basu, 1994). While a consumer's consistent buy with a single firm ensures his or her behavioural loyalty (Cunningham, 1956). True loyalty benefits a firm in gaining the consumers who are willing to do word-of-mouth publicity (Gremler & Brown, 1996), ready to make additional buy and pay premium prices (Zeithaml et al., 1996), resist competitors' offers (Dick & Basu, 1994) and being adjustable in temporarily breakdowns of services (Narayandas, 1998). In order to achieve these loyalty benefits, a firm needs to focus on delivering the best value proposition. A firm's understanding about consumers' needs/preferences is significant to make right assessment about loyalty drivers. It is not just the focus on a single transaction but contacts across multiple touch-points (from initial to ongoing) need to be perfect. It largely depends on a firm's efficiency to perform with excellence in the customer's journey. Understanding consumer expectations need a deep consumer insight. It is very much necessary that a firm should perform an in-depth market study to have the right consumer inferences. Consumers' attitudinal strength and behavioral patterns display their connection to a brand. To understand a consumer's mindset about loyalty, marketers are always encouraged to explore the factors behind consumers' attitudinal strength and behavioural patterns. The acquisition, retention, and experience are the three strategic activities that can strengthen the firm-consumer relationships (Singh & Saini, 2016). The purpose of this study is to provide a managerial framework to design acquisition, retention, and experience strategies in order to gain the consumer loyalty.

Despite the large body of existing literature that focuses on the importance of consumer loyalty, very fewer attempts have been made to suggest the factors behind consumer loyalty in a specific industry. The purpose of this study is to emphasize the relative importance of various acquisition, retention, and experience-related factors to gain consumer loyalty in wellness industry. The chapter is organized as follows. Following this introduction, the literature review and methodology have been presented. Thereafter, results of empirical analysis have been presented. Finally, the discussion, implications, limitations, and future research directions have been suggested.

## BACKGROUND LITERATURE

### Consumer Loyalty as an Outcome of CRM

The consumer loyalty has been defined in many ways. The broader conceptualization that gained greater acceptance states it as a composition of behavioural and attitudinal aspects (Dick & Basu, 1994; Oliver, 1997). The behavioural aspect signifies a consumer's repurchases with a single seller (Cunningham, 1956) while attitudinal loyalty can be understood as a consumer's deep psychological attachment and advocacy towards a brand (Chaudhari & Holbrook, 2001).

Consumer loyalty has been a strong research area in the field of marketing (Tartaglione et al., 2019). It has been considered as an outcome of successful CRM strategies (Zineldin, 2006; Mithas et al., 2005; Singh & Saini, 2016). CRM is a useful tool to increase consumer retention, loyalty, and lifetime value by maintaining positive consumer relationships (Blattberg & Deighton, 1996; Ahn, Kim, & Han, 2003). In most-recent study, Singh & Saini (2016) argue that acquisition, retention, and experience are the key strategies to gain consumer loyalty in attitudinal or behavioural forms. Other researchers like Bradshaw and Brash (2001) and Massey Montoya-Weiss, and Holcom (2001) also view CRM as a management approach that involves identifying, attracting, developing, and maintaining successful relationships over the time in order to increase retention of profitable consumers. The firms are often encouraged to have strong relational bonds with the present consumers to gain life-long loyalty (Mithas et al., 2005, Tartaglione et al., 2019; Rather et al., 2018). In sum, existing literature of relationship marketing guides the marketers to operate through acquisition, retention, and experience strategies to maintain long-term consumer relationships. Based on the assessment of positive relationship between CRM and consumer loyalty, the factors of acquisition, retention (Buttle, 2004; Blattberg et al., 2001; Thomas, 2001; McKim & Hughes, 2001), and experience (Palmer, 2010; Berry & Carbone, 2007; Greenberg, 2012) are considered to be possible drivers of consumer loyalty.

Acquisition includes the significant marketing strategies that develop and manage profitable consumer relationships (Buttle, 2004; Kaplan & Norton, 1992). It focuses on search of prospective consumers (Buttle, 2004). It is a useful tool to identify the right consumer groups to maintain long-term profitable relationships (Thomas, 2001). In this competitive era, it is no longer enough to attract and acquire new consumers, but it matters what types of consumers are being acquired (Newell, 2000). Thomas (2001) suggests that the right acquisition helps in accurate assessment about the consumer retention, lifetime value, and impact of marketing actions. A firm's successful delivery across all processes largely depends on the right consumer acquisition (Blatterberg et al., 2001; Thomas, 2001). It can be understood that the success of consumer retention and development strategies largely depends on the right operation of acquisition efforts (Buttle, 2004). The acquisition activities include promotional tools; advertising, sales-promotion, public relations, direct-marketing or others to communicate about merchandises information to prospective consumer groups (Buttle, 2004; Saini, 2016). With the use of internet technology an access to larger audience is an easy and informed choice for the marketers. The word-of-mouth and word-of-mouse are helping the companies to connect with new consumers through referrals, brand promotion, and customer engagement campaigns.

Next, the retention strategy includes various sales-promotion tools such as discounts, coupons, free premiums, bundled offers, bonus packs, rebates and cash-back based schemes to increase the consumer interests (Buttle, 2004). Retaining the acquired consumers is crucial for managing successful relationships (Buttle, 2004). The employees' participation is must to make a deep assessment about consumer retention and defection rates. It helps a firm to take appropriate measures in order to operationalize the retention efforts at the right time (Zineldin, 2006). Many firms in retail and hospitality sectors prefer loyalty schemes and reward programmes as a promising retention strategy. Loyalty schemes provide rewards to consumers in return of their cumulative purchases. Firms encourage their consumers to subscribe in loyalty schemes to get rewards and earn more incentives (Lewis, 2006).

The recent IBM study about a brand's growth highlights the importance of experience in consumer management. Today, consumers are demanding unique and pleasant experiences rather than just product or services (Singh & Saini, 2016; Palmer, 2010). This may instigate the firms' to look for meaningful and idiosyncratic measures to understand consumers' expectations (Zolkiewski et al., 2017). Managing consumer journey with deep insights across experience parameters is a key to have successful business and competitive advantage (Følstad & Kvale, 2018). Most recent researches in experiential marketing have communicated consumer experience as a multi-dimensional construct (Gentile et al., 2007). Owning to the lack of a practical, reliable, and valid measurement of consumer experience, Singh & Saini (2016) in their study on a link between CRM and consumer loyalty has developed a multi-dimensional scale of experience. The study conducted by Singh & Saini (2016) suggests five dimensions of consumer experience: functional, affective, sensorial, social, lifestyle, and pragmatic. Functional experience relates a consumer's expectations on functional outputs. Affective experience signifies the consumer's emotional journey with a firm. Sensorial reflects through a consumer's positive/negative responses across various touch points in terms of sight, hear, touch, taste, and smell experiences. Social experience expresses a consumer's interest to attach with a brand that hold good image in the society while lifestyle experience promotes acceptance/promotion of same value system that a consumer follows. Lastly, Pragmatic signifies a consumer's practical convenience to contact a firm for buying. The discussion so far has shown that acquisition, retention and experience are three strategic activities to maintain long-term profitable relationships.

## METHODOLOGY

### Sampling and Data Collection

A field survey using a probability sampling method was used. The data was collected from five major cities (Amritsar, Ludhiana, Jalandhar, Patiala, and Chandigarh) of the northern region of India. The consumers of 16 beauty-salons and 14 fitness centers have been contacted on random basis. Initially, the consumers of these wellness firms have been contacted to invite in for participation in a survey and subsequently a meeting has been scheduled to record their responses in a questionnaire. The researcher's assistance in regard to clarification of questions has been provided. The respondents completed or filled their questionnaires in researcher's presence. Additionally, in some cases, in-depth interviews were arranged for assistance. In total, 365 consumers have initially contacted. After excluding incomplete or unusable responses, finally 200 questionnaires with a response rate of 54.7 percent were processed. The sample of the beauty salons and fitness centers is representative of the overall population as most respondents are young, middle income group, and educated. The sample has balanced gender distribution. This sample characteristic appears to be representative of wellness consumers in India.

## Measures

The measures of all constructs have been newly developed because of incompatibility of validated scale in the proposed framework. The measure of latent constructs; acquisition retention and experience efforts have been drawn from previous studies for the context references. The measurement items have been drafted to depict specific characteristics of wellness sector. An extensive literature review, expert opinions, and pilot study with 50 wellness consumers have been performed in order to assure the reliability and validity of measures. The constructs used a seven point likert-type scale with the descriptive equivalents ranging from 1-strongly disagree to 7-strongly agree. All the measurement items as finalized for the questionnaire have been presented in Appendix A.

The measures of acquisition efforts including fifteen statements (items) stating; Service Quality, Customized Services, Wide-variety, Partnerships with other Firms, Reasonable Price, Nearby Locations, Advertisement's Reliability, Emotional Response to Advertisement, Price Discounts and other Benefits, Well-known Image, Contact by Awareness Camps, Skilled and Expertized Services, Contact by Recommendations, Contact by Emails and SMSs, and Contact by Automated and Manual Calls developed from Blattberg et al. (2001); Oztasyi et al. (2011); Buttle (2004); Verhoef & Donkers, (2005); Lewis (2006); Kaplan & Nortan (1992); Kotler et al. (2011); Smith & Chang, (2010) and Han & Kim (2009).

Nineteen items under the construct consumer retention efforts; Consistent Quality, Additional Product/Service Categories, Uniqueness, Good Quality at Low Price, Competitive Price, Switching Costs, Convenient Location, Ease in Parking Facility, Advertisement as a Reminder, Rational Advertisement, Loyalty Card Programmes, Surety of Promotional Offers, Reminder by Emails & SMSs, Problem Solving, Caring Attitude, Skilled & Experienced Employees, Familiarity with Service Staff, Recognition as a Regular and Special Consumer, and Familiarity with Service Surroundings have been considered for scale measurement. All these content measures have been adopted and customized from Blattberg et al. (2001); Buttle (2004); Oztasyi et al. (2011); Lewis (2006); Kaplan & Nortan (1992); Kotler et al. (2011); and Smith & Chang, (2010).

For the measurement of consumer experience efforts, twenty-six items (statements) have been developed under six distinct dimensions; functional, affective, sensorial, relational, pragmatic, and lifestyle of experimental marketing. Specifically, four functional, five affective, six sensorial, two social, two lifestyle, and seven pragmatic items have been designed to measure the consumer experiences. The measurement items have been designed in reference to previous scales of Schmitt (1999); Fournerino, Helme-Guizon, & Gaudemaris (2006); Gentile et al. (2007); Batterbee & Koskinen, (2005); Singh & Saini (2016) and Nam, Ekinci, & Whyatt (2011). The related statements of items under acquisition, retention, and experience efforts have been presented in appendix table.

## Data Analysis

### Reliability, Validity, and Fit Indices

Using AMOS17.0, the confirmatory factor analysis (CFA) was employed to verify the factors of three latent constructs. Subsequently, the classification has been done by factor loadings. It divides the drivers into four groups; Most Important, Very Important, Important, and Requisites. Four groups identify the priority of factors in reference to acquisition, retention, and experience constructs. The standardized factor loadings above 0.4 have been considered relevant to ensure the reliability and validity of proposed scale

*Table 1. Discriminant validity of proposed constructs*

| Constructs | CA | CR | CE |
|---|---|---|---|
| CA | **0.50** | | |
| CR | 0.002 | **0.55** | |
| CE | 0.016 | 0.057 | **0.58** |

Notes: *AVE values in the diagonal and squared correlations off-diagonal.

(Bagozzi & Baumgartner, 1994). The reliability indexes; composite reliability and regression weights of latent constructs have been calculated. The composite reliability index has been used to check an overall reliability of varied indicators and their relationship with same construct. The composite reliability of all three constructs: consumer acquisition efforts (0.87), consumer retention efforts (0.91) and consumer experience efforts (0.95) has been found greater than 0.6 as recommended by Bagozzi and Yi, 1988.

The content validity of the scale has been assessed through review of literature, expert opinions, and subjects' opinions in a pilot study. Next the convergent validity suggests that all indicators of one construct have high proportion of variance in common (Hair et al., 2010). As evidence of convergent validity, average variance extracted (AVE) of all three constructs: consumer acquisition efforts (0.50), consumer retention efforts (0.55) and consumer experience efforts (0.58) have been accepted by minimum of 0.5 recommended by Fornell & Lacker, 1981. The discriminant validity has been calculated to prove the differentiation among different constructs. All AVE values have been reported greater than square of inter-construct correlations of respective constructs and it validates the discriminant validity of constructs. As shown in table 1, discriminant validity of all constructs has been accepted by recommended limits (Fornell & Larcker, 1981). The fit indices of measurement model (CFA) reported above recommended thresholds: CIMIN/DF = 1.821, RMSEA = 0.36, CFI = 0.9, PCFI = .677.

CA- Consumer Acquisition Efforts, CR-Consumer Retention Efforts, CE- Consumer Experience Efforts

## Results of CFA Model

The drivers of consumer loyalty (refer table 2 and 3) have been classified into four groups; 'most important' (factor loadings [β] more or equal to 0.98), 'very important' (β more or equal to 0.87 less than 0.98), 'important' (β more or equal to 0.77 less than 0.87), and 'requisites' (β more or equal to 0.5 less than 0.77).

## Most Important Drivers

As shown in table 2, an acquisition effort; an offer to customize as per consumers' preferences (CA2: $\beta = 0.998$, $p < 0.05$) has been identified 'most important' driver of consumer loyalty in wellness services.

## Very Important Drivers

Table 2 presents the grouping of 'very important' drivers drawn from acquisition, retention, and experience efforts. The service quality (CA1: $\beta = 0.894$, $p < 0.05$) has been considered relevant to drive consumer loyalty.

*Table 2. Categorization of drivers under CRM efforts*

| Most Important | | Very Important | | Important | | Requisites | |
|---|---|---|---|---|---|---|---|
| **Item Code** | **SRW or β** | **Item Code** | **SRW or β** | **Item Code** | **SRW or β** | **Item Code** | **SRW or β** |
| CA2 | 0.998 | CA1 | 0.894 | CE15 | 0.85 | CE16 | 0.766 |
| | | CR17 | 0.891 | CA10 | 0.848 | CA13 | 0.756 |
| | | CR18 | 0.89 | CE6 | 0.825 | CE23 | 0.742 |
| | | CR14 | 0.889 | CE10 | 0.824 | CR4 | 0.739 |
| | | CR15 | 0.879 | CA3 | 0.81 | CE21 | 0.731 |
| | | CR19 | 0.878 | CR1 | 0.809 | CR12 | 0.686 |
| | | CE9 | 0.873 | CE13 | 0.805 | CE4 | 0.685 |
| | | CE12 | 0.873 | CE14 | 0.802 | CR16 | 0.671 |
| | | CE7 | 0.872 | CE11 | 0.796 | CE26 | 0.66 |
| | | CE8 | 0.87 | CA6 | 0.795 | CE1 | 0.659 |
| | | CE3 | 0.868 | CE2 | 0.791 | CR2 | 0.648 |
| | | | | CR11 | 0.786 | CR13 | 0.619 |
| | | | | CA5 | 0.785 | CE25 | 0.611 |
| | | | | | | CR5 | 0.608 |
| | | | | | | CR6 | 0.607 |
| | | | | | | CE24 | 0.607 |
| | | | | | | CE5 | 0.589 |
| | | | | | | CE22 | 0.587 |
| | | | | | | CA8 | 0.575 |
| | | | | | | CA14 | 0.563 |
| | | | | | | CA4 | 0.535 |
| | | | | | | CR7 | 0.535 |
| | | | | | | CE20 | 0.531 |
| | | | | | | CA12 | 0.529 |
| | | | | | | CR8 | 0.529 |
| | | | | | | CA9 | 0.517 |

Notes: 1 SRW (Standardized Regression Weight) or β (Beta Coefficient) values have been extracted from CFA output in AMOS17.0; 2 All values are significant at $p < 0.05$.

Five of consumer retention efforts such as problem solving (CR14: $\beta = 0.889$, $p < 0.05$), caring attitude (CR15: $\beta = 0.879$, $p < 0.05$), familiarity with service staff (CR17: $\beta = 0.891$, $p < 0.05$), recognition as regular and special consumer (CR18: $\beta = 0.89$, $p < 0.05$), and familiarity with service surroundings (CR19: $\beta = 0.878$, $p < 0.05$) have been found in 'very important' category of drivers.

Among consumer experience factors; functional experience by wide-variety (CE3: $\beta = 0.868$, $p < 0.05$), affective experiences by feel-good factor (CE7: $\beta = 0.872$, $p < 0.05$), by pleasant surprises (CE8: $\beta = 0.87$, $p < 0.05$), and by familiarity (CE9: $\beta = 0.873$, $p < 0.05$), sensorial experience by interior (CE12: $\beta = 0.873$, $p < 0.05$) have been found 'very important' to create positive consumer experiences which-in-turn creates true consumer loyalty.

*Table 3. Distribution of drivers under CRM efforts*

| | Categories | | | |
|---|---|---|---|---|
| **CRM Efforts** | **Most Important** | **Very Important** | **Important** | **Requisites** |
| CA | CA2 | CA1, CA10 | CA3, CA5, CA6, CA10 | CA4, CA8, CA9, CA12, CA13, CA14 |
| CR | | CR14, CR15, CR17, CR18, CR19 | CR1, CR11 | CR2, CR4, CR5, CR6, CR7, CR8, CR12, CR13, CR16 |
| CE | | CE3, CE7, CE8, CE9, CE12 | CE2, CE3, CE6, CE10, CE11, CE13, CE14, CE15 | CE1, CE4, CE5, CE16, CE20, CE21, CE22, CE23, CE24, CE25, CE26 |

Note: Table 3 shows the factors grouping as per standardized regression weights.

## Important Drivers

As shown in table 2 and table 3, The 'important' category of drivers includes consumer experience factors such as functional experience by consistent quality (CE2: $\beta = 0.791$, $p < 0.05$), affective experience by personalized dealings (CE6: $\beta = 0.825$, $p < 0.05$), and sensorial experiences by welcome (CE10: $\beta = 0.824$, $p < 0.05$), by cleanliness (CE11: $\beta = 0.796$, $p < 0.05$), by dress-code (CE13: $\beta = 0.805$, $p < 0.05$), by music (CE14: $\beta = 0.802$, $p < 0.05$), and by fragrance (CE15, $\beta = 0.85$, $p < 0.05$).

Additionally among consumer acquisition efforts; wide-variety (CA3: $\beta = 0.81$, $p < 0.05$), reasonable price (CA5: $\beta = 0.785$, $p < 0.05$), nearby locations (CA6: $\beta = 0.795$, $p < 0.05$) and well-known image (CA10: $\beta = 0.848$, $p < 0.05$) while in consumer retention efforts; consistent quality (CR1: $\beta = 0.809$, $p < 0.05$) and loyalty card programmes (CR11: $\beta = 0.786$, $p < 0.05$) have been accounted as 'important' to drive consumer loyalty.

## Requisites Drivers

The requite category includes acquisition factors such as partnership with other firms (CA4: $\beta = 0.535$, $p < 0.05$), emotional response to advertisement (CA8: $\beta = 0.575$, $p < 0.05$), price discounts & other benefits (CA9: $\beta = 0.517$, $p < 0.05$), skilled and expertized services (CA12: $\beta = 0.529$, $p < 0.05$), contact by recommendations (CA13: $\beta = 0.756$, $p < 0.05$), and contact by emails and SMSs (CA14: $\beta = 0.563$, $p < 0.05$). In addition, retention efforts such as wide-variety (CR2: $\beta = 0.648$, $p < 0.05$), good quality at low price (CR4: $\beta = 0.739$, $p < 0.05$), less price in comparison to competitors (CR5: $\beta = 0.608$, $p < 0.05$), switching costs (CR6: $\beta = 0.607$, $p < 0.05$), convenient location (CR7: $\beta = 0.535$, $p < 0.05$), ease in parking facility (CR8, $\beta = 0.529$, $p < 0.05$) price discounts and other benefits (CR12: $\beta = 0.686$, $p < 0.05$), reminder by emails & SMSs (CR13: $\beta = 0.619$, $p < 0.05$), and skilled and experienced employees (CR16: $\beta = 0.671$, $p < 0.05$) have been considered for requisites category. Under consumer experience efforts, functional experience by purpose (CE1: $\beta = 0.659$, $p < 0.05$), and value (CE4: $\beta = 0.685$, $p < 0.05$), affective experience by skilled services (CE5: $\beta = 0.589$, $p < 0.05$), social experience by good impression on others (CE16: $\beta = 0.766$, $p < 0.05$), pragmatic experiences by understanding of offers (CE20: $\beta = 0.531$, $p < 0.05$), by good place (CE21: $\beta = 0.731$, $p < 0.05$), by handling operations (CE22: $\beta = 0.587$, $p < 0.05$), by complaint management processes (CE24: $\beta = 0.607$, $p < 0.05$), by dealings of staff (CE23: $\beta = 0.742$, $p < 0.05$), by firm's involvement (CE25: $\beta = 0.611$, $p < 0.05$), and by feedback processes (CE26, $\beta = 0.66$, $p < 0.05$) have been understood as 'requisites' to drive consumer loyalty.

## DISCUSSION AND IMPLICATIONS

To gain a loyal consumer base, the study provides valuable insights to wellness firms. First, the findings indicate major differences in consumers' priority among loyalty drivers across acquisition, retention, and experience factors. The experience plays an important role in maintaining consumer satisfaction and loyalty (Roy, 2018; Singh & Saini, 2016). Managers of wellness are suggested to design their offerings on experience factors on priority over and above the acquisition and retention strategies.

Second, the results emphasize prominence of certain factors as crucial; most important, very important, important and others as mediocre; requisites. In wellness services, customization has been identified as a 'most important' driver to build consumer loyalty. Customization promises personalized solutions to consumers' unique needs and expectations (Pine et al., 1995). A consumer attracts to a firm that promises to deliver desired service quality standards. Moreover, a consumer's perception to get the desired benefits encourages him or her to buy with a single seller (Parra-Lopez et al., 2018). Thus, it may further draw consumer satisfaction that builds true consumer loyalty (Singh & Saini, 2016). Recognizing the importance of customization, managers of fitness centers/beauty salons are advised to tailor the services by consumers' preferences based on specific brands, location, staff, timings, loyalty programmes, or other service benefits.

Next, 'very important' group of drivers includes five of consumer experience efforts. First functional experience by wide-variety indicates consumers' expectations to have one-shop access to similar nature services. From the managerial perspective, beauty-salon firms need to be efficient in delivering certified solutions to all grooming options; skin, hair, make-up, apparel, etc. Next, affective experience by familiarity, feel-good factor, pleasant surprises, and sensorial experience by interiors have been found positive to generate pleasant experiences. Good interiors increase interactions between consumers and serving staff. In wellness services, for example, the good layout of a gym encourages enhanced interactions among trainers and consumers. Moreover, familiarity with service staff and positive interactions make it comfortable to contact on regular basis. The wellness firms need to couple people-centric skills and sensorial elements in their service experience.

Next, an acquisition factor; service quality and five consumer retention factors; problem solving, caring attitude, familiarity with service staff, familiarity with service surroundings and recognition as a regular and special consumer have been found 'very important' to drive consumer loyalty. Service quality serves the consumer's major purpose to contact a firm. In the wellness sector, every consumer has a unique set of expectations. It varies by their concern to contact a firm for some specialized services in styling, haircuts, makeup (in beauty salons), and physical fitness (in gym centers). It can be related with outputs, professional staff, timings, ambience, infrastructure, equipment, operational processes, safety assurances, and staff behavior, etc. It indicates that marketers should be aware of a consumer's expectations in detail so that the right quality can be offered at right time. Subsequently, the retention factors such as familiarity with service staff, familiarity with service surroundings, and recognition as a regular and special consumer signifies that the consumer seeks convenience in their contact to a wellness firm. The problem solving and caring attitude of employees minimize consumers' emotional worries related to cancellation or reschedule of appointments, long-waits, lack of options, and product or service deficiencies.

Subsequently, seven consumer experience efforts, three consumer acquisition efforts, and two of retention efforts have been grouped under 'important' category of drivers. The 'Important' list includes sensorial, affective, and functional experience factors. First, the functional experience by consistent quality has been found positive to generate utilitarian value in order to drive consumer loyalty. Moreover, consistency in quality provides surety of getting same experiences that they had enjoyed earlier. Second, affective experience by personalization might help in more positive interpersonal interactions between employees and consumers. A firm's effort to personalize service experience creates consumers' benevolence trust, which in further develops consumer satisfaction and loyalty (Ball, Coelho, & Vilares, 2006). In wellness sector, personalization provides a better platform to design, communicate, and deliver right consumer experience (Price & Arnould, 1999). Similarly, consumers' interface with sensorial elements such as cleanliness, music, dress, fragrance, and welcome notes ensures visibility and careful attitude. It uplifts consumers' moods and senses that further creates emotional bond with a brand (Kalra, 2012). The good fragrances provides comfort in stay (Mitchell, Kahn, &. Knasko, 1995; Spangenberg, Sprott, Grohmann, & Tracy, 2006). Additionally, it comprises two consumer retention efforts; loyalty card programmes, consistent quality, and four consumer acquisition factors; well-known image, wide-variety, nearby location and reasonable price. The brand image, wide-variety, reasonable price, and nearby locations can help a firm to build a new consumer base. This is primarily because consumers may seek these factors as basic elements for initial contacts. In addition, beauty salons/fitness centers must maintain their consistency in service quality and should offer attractive loyalty schemes or reward programmes to retain their once acquired consumers.

In the 'Requisites' category of drivers, eight of experience factors, seven of retention and one of acquisition factors have been considered. The experience factors include social, pragmatic and functional factors. To this end, service managers should draft their experience strategies focusing on practical convenience, functional outputs, and brand identity measures to maintain long-term relationships. Likewise, a firm's successful efforts in promoting the brand on social media networks initiate the two-way communications with consumers which in further build the strong brand identity.

The retention factors include various price, product, communication and recommendation strategies to attract and retain new consumers. Thus, the present study provides a guide to wellness firms in order to choose and implement right acquisition, retention and experience strategies to gain the consumer loyalty.

## LIMITATIONS AND FUTURE RESEARCH DIRECTIONS

The present study has few limitations which do not undermine the value of findings but provide scope for future research. The present study was restricted to wellness service only. So, future researches may consider other services to find out differential effects. Second, the study took place within specific region in one country and it would be interesting to explore the effects of other cultures in consideration to know their loyalty parameters. Future studies should investigate the individual effects of these CRM practices on attitudinal or behavioural dimensions of loyalty.

## REFERENCES


Ahn, J. Y., Kim, S. K., & Han, K. S. (2003). On the design concepts for CRM system. *Industrial Management & Data Systems, 103*(5), 324–331. doi:10.1108/02635570310477370

Bagozzi, R. P., & Baumgartner, H. (1994). The evaluation of structural equation models and hypothesis testing In R. P. Bagozzi (Ed.), Principles of Marketing Research (pp. 386–422). Cambridge, MA: Blackwell.

Bagozzi, R. P., & Yi, Y. (1988). On the evaluation of structural equation models. *Journal of the Academy of Marketing Science, 16*(1), 74–94. doi:10.1007/BF02723327

Berry, L. L., & Carbone, L. (2007). Building loyalty through experience management. *Quality Progress, 40*(9), 26–32.

Blattberg, R. C., Getz, G., & Thomas, J. S. (2001). *Customer Equity: Building and Managing Relationships as Valuable Assets*. Boston, MA: Harvard Business School Press.

Bradshaw, D., & Brash, C. (2001). Management consumer relationships in the e-business world: How to personalise computer relationships for increased profitability. *International Journal of Retail & Distribution Management, 29*(12), 520–530. doi:10.1108/09590550110696969

Chaudhuri, A., & Holbrook, M. B. (2001). The chain of effects from brand trust and brand affect to brand performance: The role of brand loyalty. *Journal of Marketing, 65*(2), 81–93. doi:10.1509/jmkg.65.2.81.18255

Cunningham, R. (1956). Brand loyalty- what where how much? *Harvard Business Review, 34*(1), 116–128.

Dick, A. S., & Basu, K. (1994). Customer loyalty: Toward an integrated conceptual framework. *Journal of the Academy of Marketing Science, 22*(2), 99–113. doi:10.1177/0092070394222001

Dwayne, B., Pedro, C. S., & Manuel, V. J. (2006). Service personalization and loyalty. *Journal of Services Marketing, 20*(6), 391–403. doi:10.1108/08876040610691284

Ernst and Young. (2018). Wellness Industry Report. Retrieved from https://www.indiaservices.in/medical

Følstad, A., & Kvale, K. (2018). Customer journeys: A systematic literature review. *Journal of Service Theory and Practice, 28*(2), 196–227. doi:10.1108/JSTP-11-2014-0261

Fornell, C., & Larcker, D. F. (1981). Evaluating structural equation models with unobservable variables and measurement error. *JMR, Journal of Marketing Research, 18*(1), 39–50. doi:10.1177/002224378101800104

Fournerino, M., Helme-Guizon, A., & de Gaudemaris, C. (2006). Mesurer L'immersion dans une experience de consommation: Premiers developpements. *Proceedings of the XXIIth Congress de l'AFM*. Academic Press.

Gentile, C., Spiller, N., & Noci, G. (2007). How to sustain the customer experience: an overview of experience components that co-create value with the customer. *European Management Journal, 25*(5), 395-410. Retrieved from http://www.sciencedirect.com/science/article/pii/S0263237307000886

Greenberg, P. (2012). Customer Experience Finally.... and once again. *Customer Relationship Management*, (October), 38-39. Retrieved from www.destinationCRM.com

Gremler, D. D., & Brown, S. (1996). Service Loyalty: its nature, importance and implications In B. Edvardsson, S.W. Brown, R. Johnston, & E.E. Schewing (Eds), Advancing Service Quality: A Global Perspective (pp. 171-180). International Service Quality Association.

Hair, J. F., Black, W. C., Babin, B. J., & Anderson, R. E. (2010). *Multivariate Data Analysis* (7th ed.). Upper Saddle River, NJ: Prentice Hall.

Han, H., & Kim, W. (2009). Outcomes of relational benefits: Restaurant customers' perspective. *Journal of Travel & Tourism Marketing*, *26*(8), 820–835. doi:10.1080/10548400903356236

Kalra, R. (2012, March 5). Something in the Air. Retrieved from http://www.adweek.com/news/advertising-branding/something-air-138683

Kaplan, R. S., & Nortan, D. P. (1992). The balance scorecard-measures that drive performance. *Harvard Business Review*, *70*(1), 71–79. PMID:10119714

Katja, B., & Ilpo, K. (2005). Co-experience: User experience as interaction. *CoDesign*, *1*, 1, 5–18. doi :10.1080/15710880412331289917

Kotler, P., Armstrong, G., Agnihotri, P. Y., & Haque, E. (2011). *Principles of Marketing-A south Asian Perspective*. New Delhi: Pearson Education.

Lewis, M. (2006). Customer acquisition promotions and customer asset value. *JMR, Journal of Marketing Research*, *43*(2), 195–203. doi:10.1509/jmkr.43.2.195

Massey, A. P., Montoya-Weiss, M. M., & Holcom, K. (2001). Re-engineering the customer relationship: leveraging knowledge assets at IBM. *Decision Support Systems, 32*(2), 155-170. Retrieved from http://www.sciencedirect.com/science/article/pii/S0167923601001087

McKim, B., & Hughes, A. M. (2001). How to measure consumer relationship management success. *Journal of Database Marketing*, *8*(3), 224–231.

Mitchell, D. J., Kahn, B. E., & Knasko, S. C. (1995, September). There's something in the air: effects of congruent or incongruent ambient odor on consumer decision making. *Journal of Consumer Research, 22*(2), 229-238. Retrieved from http://www.jstor.org/stable/2489814

Mithas, S., Krishnan, M. S., & Fornell, C. (2005). Why do customer relationship management applications affect customer satisfaction? *Journal of Marketing*, *69*(4), 201–209. doi:10.1509/jmkg.2005.69.4.201

Nam, J., Ekinci, Y., & Whyatt, G. (2011). Brand equity, brand loyalty and consumer satisfaction. *Annals of Tourism Research*, *38*(3), 1009–1030. doi:10.1016/j.annals.2011.01.015

Narayandas, D. (1998). Measuring and managing the benefits of customer retention: An empirical investigation. *Journal of Service Research*, *1*(2), 108–128. doi:10.1177/109467059800100202

Newell, F. (2000). *Loyalty.com: Customer Relationship Management in the New Era of Internet Marketing*. New York, NY: McGraw-Hill.

Oliver, R. L. (1997). Loyalty and profit: long term effects of satisfaction. In *Satisfaction: A Behavioural Perspective on the consumer*. New York, NY: McGraw-Hill.

Oliver, R. L. (Special Issue 1999). Whence consumer loyalty? *Journal of Marketing, 63*, 33-44.

Oztaysi, B., Sezgin, S., & Ozak, A. F. (2011). A measurement tool for consumer relationship management processes. *Industrial Management & Data Systems, 111*(6), 943–960. doi:10.1108/02635571111144982

Palmer, A. J. (2010). Customer experience management: A critical review of an emerging idea. *Journal of Services Marketing, 24*(3), 196–208. doi:10.1108/08876041011040604

Parra-Lopez, E., Martínez-González, J.A., & Chinea-Martin, A. (2018). Drivers of the formation of e-loyalty towards tourism destinations. *European Journal of Management and Business Economics, 27*(1), 66 82. doi:. doi:10.1108/EJMBE-11-2017-0050

Peters, L., Raddats, C., & Robinson, W. (2017). Strategic B2B Customer Experience Management: The Importance of Outcomes-Based Measures. *Journal of Services Marketing, 31*(2). doi:10.1108/JSM-10-2016-0350

Pine, B. J. II, Peppers, D., & Rogers, M. (1995). Do you want to keep your customers forever? *Harvard Business Review, 73*(2), 103–114.

Rather, A. R., Tehseen, S., & Parrey, H. S (2018). Promoting customer brand engagement and brand loyalty through customer brand identification and value congruity. *Spanish Journal of Marketing - ESIC, 22*(3), 319-337. doi:10.1108/ SJME-06-2018-0030

Saini, S. (2016). *Role of Consumer Experience Management versus Consumer Relationship Management in Consumer Loyalty.* Unpublished Doctoral dissertation, Dr. B R Ambedkar National Institute of Technology.

Schmitt, B. H. (1999). *Experiential Marketing: How to Get Consumers to SENSE FEEL THINK ACT and RELATE to Your Company and Brands*. New York, NY: The Free Press.

Singh, J., & Saini, S. (2016). Managing consumer loyalty through acquisition, retention and experience efforts: an empirical study on service consumers in India. *Vision- The Journal of Business Perspective, 20*(3), 121-134.

Smith, M., & Chang, C. (2010). Improving customer outcomes through the implementation of customer relationship mangement evidence from Taiwan. *Asian Review of Accounting, 18*(3), 260–285. doi:10.1108/13217341011089658

Spangenberg, E. R., Sprott, D. E., Grohmann, B., & Tracy, D. L. (2006). Gender-congruent ambient scent influences on approach and avoidance behaviors in a retail store. *Journal of Business Research, 59*(November), 1281–1287. doi:10.1016/j.jbusres.2006.08.006

Tartaglione, A. M., Cavacece, Y., Russo, G., & Granata, G. (2019). A Systematic Mapping Study on Customer Loyalty and Brand Management, *Administrative Sciences, MDPI. Open Access Journal, 9*(1), 1–21.

Thomas, J. S. (2001). A methodology for linking customer acquisition to customer retention, *Journal of Marketing Research, 38*(2), 262-268. Retrieved from http://www.jstor.org/stable/1558629

Verhoef, P. C., & Donkers, B. (2001). Predicting customer potential value: An application in the insurance industry. *Decision Support Systems*, *32*(2), 189–199. doi:10.1016/S0167-9236(01)00110-5

Zeithaml, V. A., Berry, L. L., & Parasuraman, A. (1996). The behavioral consequences of service quality. *Journal of Marketing, 60*(2), 31-46. Retrieved from http://www.jstor.org/stable/1251929

Zineldin, M. (2006). The royalty of loyalty: CRM quality and retention. *Journal of Consumer Marketing*, *23*(7), 430–437. doi:10.1108/07363760610712975

Zolkiewski, J., Story, V., Burton, J., Chan, P., Gomes, A., Hunter-Jones, P., ... Robinson, W. (2017). Strategic B2B customer experience management: The importance of outcomes-based measures. *Journal of Services Marketing*, *31*(2), 172–184.

# APPENDIX

*Table 4. Questionnaire for field survey*

| This survey deals with your opinions about your dealings with a beauty salon or gym. Kindly understand 'a firm'/'this firm'/'here' terms in the statements as your current or preferred Beauty Salon.<br>If you strongly agree that your service firm possess that effort, then tick in box against the 7-strongly agree. If you strongly disagree that your service firm possess that effort, then tick in the box against the 1-strongly disagree. If your feelings are not strong then chose appropriately in the middle ones. There is no wrong and right answer. | 1- Strongly Disagree | 2- Somewhat Disagree | 3- Disagree | 4- Neutral | 5- Agree | 6- Somewhat Agree | 7- Strongly Agree |
|---|---|---|---|---|---|---|---|
| 1. I am willing to make special efforts to get very best quality services. | | | | | | | |
| 2. I like to contact a firm where I can design my service offerings as I want. | | | | | | | |
| 3. I expect wide variety of products/ brands/services from a firm. | | | | | | | |
| 4. I prefer buying from a firm that does partnership with other firms. | | | | | | | |
| 5. A firm's reasonable prices make me attentive for a new deal. | | | | | | | |
| 6. I prefer to visit nearby locations. | | | | | | | |
| 7. Advertisement helps me to make the right purchases. | | | | | | | |
| 8. Advertisement makes me feel good. | | | | | | | |
| 9. The price discounts and other special benefits always take my interest. | | | | | | | |
| 10. I like to attach with a well-known firm. | | | | | | | |
| 11. I find that the participation in consumer awareness camps is always useless. | | | | | | | |
| 12. Assured skilled and expertise service attracts me for a new deal. | | | | | | | |
| 13. I am less interested in a firm that contacts me by someone's recommendations. | | | | | | | |
| 14. E-mails and SMSs provide good source of information to contact a firm. | | | | | | | |
| 15. I like to receive automated/ manual calls for making new service deals. | | | | | | | |
| 16. I like to revisit a firm where I get same quality services as I had experienced earlier. | | | | | | | |
| 17. Adding new lines of products/brands/services motivates me to revisit. | | | | | | | |
| 18. I visit here to get exclusive service offerings. | | | | | | | |
| 19. Getting good quality service at low price is the reason I visit here. | | | | | | | |
| 20. I often find this firm's prices less to other firms. | | | | | | | |
| 21. It is costly for me to shift to a new firm for a trail. | | | | | | | |
| 22. I revisit a firm only if it is conveniently located. | | | | | | | |
| 23. The easy parking facility of a firm is one of the reasons to revisit. | | | | | | | |
| 24. Advertisements remind me to plan my next visit. | | | | | | | |
| 25. An advertisement makes me easy to find good deals from this firm. | | | | | | | |
| 26. I am more likely to visit here for loyalty card programmes. | | | | | | | |
| 27. The price discounts and other special benefits are the main reasons I visit here. | | | | | | | |
| 28. The regular messages through emails/SMSs encourage me to revisit. | | | | | | | |

continued on following page

Table 4. Continued

| | | | | | | | |
|---|---|---|---|---|---|---|---|
| 29. This firm attends and solves my problem as I expect. | | | | | | | |
| 30. I am more likely to revisit a firm if they really care me. | | | | | | | |
| 31. I often visit here to get skilled and expertise service. | | | | | | | |
| 32. My familiarity with service staff comforts me to plan my next visit. | | | | | | | |
| 33. My recognition as regular and special consumer encourages me to being attached. | | | | | | | |
| 34. I am habitual to visit here as of my familiarity with service surroundings. | | | | | | | |
| 35. This firm fulfills my main purpose of visit. | | | | | | | |
| 36. This firm's consistent quality efforts enrich my experiences. | | | | | | | |
| 37. I feel delighted to get wide variety of services here. | | | | | | | |
| 38. Good quality service at reasonable price is enough for me to get a good deal. | | | | | | | |
| 39. Getting skilled and expertise service is meaningful to get a pleasant experience. | | | | | | | |
| 40. I feel honored to get personal attention from service staff. | | | | | | | |
| 41. My contact with this firm makes me feel good about myself. | | | | | | | |
| 42. I feel happy to get pleasant surprises here. | | | | | | | |
| 43. My familiarity with the staff makes me feel relaxed. | | | | | | | |
| 44. I feel good to have warm welcome from the service staff. | | | | | | | |
| 45. The cleanness at firm's premise is very important to me. | | | | | | | |
| 46. The clumsy interior of a firm makes me upset. | | | | | | | |
| 47. The well-dressed employees of a firm leave good impression on me. | | | | | | | |
| 48. The listening of soothing music at firm premise makes me feel relaxed. | | | | | | | |
| 49. The fragrance at a firm premise appeals my senses. | | | | | | | |
| 50. My connection with this firm helps me to make a good impression on other people. | | | | | | | |
| 51. I visit here to make new friends. | | | | | | | |
| 52. This firm promotes the same values and beliefs as I believe. | | | | | | | |
| 53. This firm fits in my lifestyle. | | | | | | | |
| 54. It is easy for me to understand the promotional offers of this firm. | | | | | | | |
| 55. I feel relaxed to visit a firm that is conveniently located. | | | | | | | |
| 56. Sometimes I really feel tensed with the mismanagement of this firm. | | | | | | | |
| 57. I feel amazed with the quick and responsible dealings of service staff. | | | | | | | |
| 58. I feel impressed with a firm's positive attitude to handle the problems if arise. | | | | | | | |
| 59. I feel good with a firm whenever they plan my next visit. | | | | | | | |
| 60. I feel acknowledged to give advices for making improvements in services. | | | | | | | |

# Chapter 6
# Inter-Relational Dynamics of Various HR Aspects in High Altitude Illness Attrition

**Ajeya Jha**
https://orcid.org/0000-0003-0491-5008
*Sikkim Manipal Institute of Technology, Sikkim Manipal University, India*

**Ajay Dheer**
*Sikkim Manipal Institute of Medical Sciences, India*

**Vijay Kumar Mehta**
*Sikkim Manipal Institute of Medical Sciences, India*

**Saibal Kumar Saha**
*Sikkim Manipal Institute of Technology, Sikkim Manipal University, India*

## ABSTRACT

*The adverse health effects of high altitude are of considerable importance since they may seriously interfere with working efficiency of an organization that is actively involved with inescapable duties. The objective of the current study is to explore inter-relational dynamics of various HR aspects in HAIA. The HR aspects included are job delay, poor team, motivation, less leave, high working hours, poor decision making, personal stress, family stress, personal discomfort, uncertainty, poor relations, health, accidents, quality and performance. A decision-making trial and an evaluation laboratory have been used to explore the inter-relation dynamics of various factors of HR. The results indicate that personal stress has the highest impact priority which is followed by poor performance, poor team and motivation. Uncertainty, less leave, and high working hours has the least impact priority. It is also found that high working hours, less leave and poor health are the major causes whereas decrease in motivation and poor quality of work are the major results.*

DOI: 10.4018/978-1-7998-1566-2.ch006

## INTRODUCTION

High altitude (HA), with its attendant hypobaric hypoxia and cold, results in certain stresses to the human body. To overcome these adverse effects of the environment, certain physiological changes occur which, if successful, leads to acclimatization of the individual. Failure, on the other hand leads to the various maladies of high altitude. Apart from hypoxia and cold, other factors at HA which could affect performance and lead or contribute to illness include low humidity, solar and ultraviolet radiation. However, hypoxia remains the single most important challenge to the system and is responsible for the various life-threatening problems have become compounded and new problems have arisen.

Mountains cover one-fifth of the earth's surface. 38 million people live permanently at altitudes ≥2400 m, and 100 million people travel to high-altitude locations each year. Skiers in Aspen, religious pilgrims to Lhasa, trekkers and climbers to Kilimanjaro or Everest, and personnel employed at high-altitude locales are all at risk of developing illnesses. High Altitude illness is likely to occur above 2500 m but has been documented even at 1500–2500 m.

The adverse health effects of high altitude are of considerable importance since they may seriously interfere with working efficiency of an organization that is actively involved with inescapable duties. Organisations operating in high altitude, on induction from the plains, the manifestation of various forms of altitude related ailments leads to considerable attrition of manpower i.e. high altitude illness attrition (HAIA). The significance is amplified further because of the fact that human efficiency also reduces considerably at higher altitudes. These occur with increase in the altitude because of decrease in atmospheric pressure resulting in decreased availability of oxygen.

Workplace attendance and its converse absenteeism are complex multi-dimensional issues involving the interaction and subtle interplay between worker, employer, workplace, social, societal and economic factors.

Disease and illness are rarely discussed in most models of workplace absenteeism, yet they form the bulk of the cause and provide a legitimate explanation for much of its occurrence. Many of these factors are prevalent in society and are open to interventions in the public arena as well as the workplace. For example, the workplace can be a vehicle for the amplification of public health messages dealing with cigarette smoking, drug and alcohol use, diabetes and influenza for mutual benefit. Thus, illness related leave might be unfairly considered unavoidable from a workplace or medical perspective.

High altitude deployment in large numbers is common for the armed forces and organizations building roads, rail lines, bridges, tunnels and power projects and other infrastructures. Mountains along with rivers, ocean and desert have served as borders of two countries. Pyrenees separate France and spain, Jura mountains separate France and Switzerland, Himalayas separate India and China, Tian Shan Mountain separate China and Kirgizstan, Altai mountains separate Mongolia and China, Hkakabo Razi mountains separate China and Myanmar, Mount Fitz Roy separates Chile and Argentina and so on. Warring neighbors is an age-old phenomenon and therefore mountains around the world have been the key regions where armed forces have been deployed. Armed forces need infrastructures for quick movements and hence roads, bridges and tunnels have to be built on large scale. This necessitates the deployment of a large personnel on mountains where weather is harsh and living conditions tough.

A prime example is Siachen conflict. This conflict embroiling India and Pakistan is commonly stated to as the coldest war, or the endless war atop the roof of the world. The hostile conditions nurtured by high altitude and extreme climate result in more deaths than the war itself. The soldiers are stationed at heights of up to 6700 msl. (Tahir-Kheli, & Biringer, 2001; Baghel & Nüsser, 2015).

## HR Aspects

The HR aspects which have been identified to study the effect of attrition due to de-induction of professionally competent and trained personnel on account of high-altitude related illnesses. These factors are deemed to impact the effectiveness, functional ability, the quality of output and various other aspects of a non-profit organization. Such organization that employs highly trained manpower for specific roles in sensitive areas depends solely on the performance of its manpower, hence performance levels may get influenced due to attrition of manpower.

This is an overlooked aspect of Human Resource (HR) management in organizations working in high altitude areas. Adverse health effects of high altitude are of considerable importance since they may seriously interfere with working efficiency of an organization that is actively involved with inescapable duties. On induction into high altitude from the plains, manifestation of various forms of altitude related ailments leads to considerable attrition of manpower. The significance is amplified further because of the fact that human efficiency also reduces considerably at higher altitudes.

It is estimated that fatalities in industrial countries some 5–7% is attributed to work-related illnesses and occupational injuries (Päivi Hämäläinen, 2017). This percentage is somewhat smaller in developing countries where non-occupational health problems have a bigger share. Globally 2.3 million deaths take place due to occupational injuries (318,000 deaths) and work-related diseases (2,022,000 deaths) annually. The effects of high-altitude adversities are of considerable importance since the attrition will interfere with the working efficiency of an organization because of various altitude related ailments. The significance is amplified further because of the fact that human efficiency also reduces considerably at higher altitudes.

The study is significant because adverse health effects of high altitude are of considerable importance since they may seriously interfere with working efficiency of an organization due to loss of manpower. On induction into High Altitude from the plains, manifestation of various forms of altitude related ailments leads to considerable attrition of manpower.

Thus, an organization will lose trained manpower on account of de-induction which does impact the organisational functioning responsible for a sensitive role. There are no studies available wherein the rate of attrition due to de-induction in High Altitude Area has been assessed. The present research study fills this gap by analyzing data of all inductees in high altitude over a period of two years and the data of attrition due to morbidity and mortality in high altitude area. The human resource perspective of this attrition in the organisation is analyzed.

Since we are studying a macro-phenomenon and trying to understand the interrelationship of various factors and broadly understand cause and effect relationships in a broad sense, the tool that appear appropriate is Decision making trial and evaluation laboratory (DEMATEL). It is an efficient tool that helps to identify cause-effect chain components of a large and intricate and multifaceted system. It helps in evaluating interdependent relationships across factors and in pin-pointing the critical ones. Important feature of DEMATEL is a visual structural model. Very many studies have been undertaken on the application of DEMATEL and quite a few variants have been put forward in the literature.

## LITERATURE REVIEW

The available studies pertaining to absenteeism/attrition and its impact on productivity/profitability in an organisation have been reviewed.

**Absenteeism**: The effect of system restructuring on absenteeism due to sickness in the workplace (Indulski et al., 1997). The study dwells on the issue of sickness absenteeism and restructuring of the organization. The study showed that changes in the size and the structure of employment resulting from the plant restructuring influence the level of sickness absenteeism. These issues do not complement or assist our study.

**Attrition Revisited**: Identifying the Problem and Its Solutions (Laurence, 1996). The attrition due to terrain specific medical problem has not been dealt with. Job Stress, Depression, Work Performance, and Perceptions of Supervisors in Military Personnel (Pflanz. 2006). The study as the title suggest aims at identifying the job stressors and other factors and has not identified attrition due to illness causing extra burden on others as one of the possible causes.

**Aspects in HR**: The HR aspects which have been identified to study the effect of attrition due to de-induction of professionally competent and trained personnel on account of high-altitude related illnesses. The available literature on some of these aspects were reviewed, attrition/absenteeism and ease of substitution / timely completion of job. The negative impact of attrition on these aspects in diverse industries have been established (Katou, 2015). Less absenteeism increases performance (Thiruchelvan, 2017). It is seen from the studies above that attrition negatively impacts the ease of substitution and timely completion of job in an organization thereby affecting the productivity/outcome. Attrition in organisations functioning in high altitude have not been studied.

Attrition/Absenteeism on Individual Performance and Team Performance (Kalkavan, 2014; Coffeng, 2014; Fong, 2015). It is seen from the studies above that Attrition negatively impacts the employee and team performance organization thereby affecting the productivity/outcome. Attrition in organisations functioning in high altitude have not been studied.

**Attrition/Absenteeism and Motivation and Morale:** There are various studies that relate motivation levels to attrition/absenteeism. There is a negative impact of attrition on motivation (Dhingra, 2017; Malhotra, 2016; Kempegowda; 2016, Liang; 1983).

Attrition and Absenteeism Affects Teamwork: Weiss (1985) conludes that low rates of absenteeism are highly valued when production involves teamwork. According to him if absenteeism falls below a critical number the output would be zero. Other scholars drawing similar conclusions include Kalisch (2010), Heywood (2008), Kinjerski, & Skrypnek (2008), and Demerouti (2008)

Attrition and absenteeism lead to high work pressures and which in turn results in poor health (Siegrist, 1996; Hayashi, 1996; Höge, 2009). As work pressure affects work-life balance and which in turn affects health (Frone, 2009). Many more scholars have studied work-pressure and health issues (Cooper & Payne 1988, 1991; French, Caplan, & Harrison, 1982; Hall, 1990; Johnson & Johannson, 1991; Karasek, Steptoe, & Appels, 1989; Warr, 1987; Assunção & Abreu, 2019; Kleiner & Wallace, 2017; Adjei, Jonsson & Brand, 2018).

The relationship between work pressure and accidents is also well understood and documented. Work pressure is also a result of attrition (Lusa et al., 2002; Ghasemi et al., 2018; van der Klauw et al., 2016; Barlas & Izci, 2018).

Attrition/Absenteeism on Personal relations and Boss relations. Citations below mention the impact of attrition/absenteeism on personal relations. The impact is not positive in diverse industries (Diestel, 2014; Banerjee, 2016; Gottschling, 2016).

There is not enough literature to understand the impact of attrition of manpower due to high altitude illness. No literature is available to assess the impact of attrition on present human resource aspects and their interdimensional relationships.

## METHODOLOGY

With the help of Decision-Making Trial and Evaluation Laboratory (DEMATEL) an attempt has been made to understand the inter-relational dynamics of major HR Aspects that impact the organisation due to HAIA. The HR aspects are: job delay, poor team, motivation, less leave, high working hours, poor decision making, personal stress, family stress, personal discomfort, uncertainty, poor relations, health, accidents, quality and performance. This was done based on experts' opinion. The level of dependence of these variables on each other was identified using paired comparison. The number of experts providing information stood at 50. Each respondent was asked to estimate the dependence of factors by a numeral score ranging between 0 to 10 with 0 implying no influence and 10 implying highest influence.

DEMATEL was introduced about 4 decades back. It found wide acceptance as a robust tool to determine the cause and effect relationship while evaluating multi-dimensional decision situation (Chiu et al., 2006, Liou et al., 2007, Tzeng et al., 2007, Wu & Lee, 2007, Lin & Wu, 2008). It is a useful method to examine and develop cause and effect association while exploring a complex and a less known situation (Yang et al., 2008). It is also applicable to develop connectivity amongst competing and correlating issues in an intricate setup (Lin & Tzeng, 2009). Yet another benefit of this tool is its ability to display the associations between various criteria and prioritize them as per the nature of relationships and their interdependencies (Govindan, Khodaverdi, & Vafadarnikjoo, 2015; Fontela & Gabus, 1976; Yu & Tseng, 2006; Liou et al., 2007; Tzeng, et al., 2007; Yang, et al., 2008; Wu & Lee, 2007; Shieh et al., 2010).

## RESULT AND DISCUSSION

With the help of DEMATEL, an attempt has been made to understand the inter-relational dynamics of major HR Aspects that impact the organisation due to HAIA. The HR aspects are: job delay, poor team, motivation, less leave, high working hours, poor decision making, personal stress, family stress, personal discomfort, uncertainty, poor relations, health, accidents, quality and performance. This was done based on experts' opinion. The level of dependence of these variables on each other was identified using paired comparison. The number of experts providing information stood at 50. Each respondent was asked to estimate the dependence of factors by a numeral score ranging between 0 to 10 with 0 implying no influence and 10 implying highest influence.

*Table 1. Inter-relational dynamics of the HR aspects*

| Aspects | Job delay | Poor Team | Motivation | Less leave | High working hours | Poor decision making | Personal stress | Family stress | Personal discomfort | Uncertainty | Poor relations | Poor Health | Accidents | Poor quality | Poor performance | Σaij |
|---|---|---|---|---|---|---|---|---|---|---|---|---|---|---|---|---|
| Job delay | 0 | 3 | 3 | 0 | 4 | 2 | 4 | 6 | 6 | 2 | 5 | 2 | 3 | 2 | 4 | 46 |
| Poor Team | 6 | 0 | 6 | 0 | 3 | 4 | 5 | 2 | 4 | 2 | 6 | 3 | 5 | 4 | 4 | 54 |
| Motivation | 0 | 4 | 0 | 0 | 0 | 3 | 4 | 0 | 1 | 0 | 5 | 2 | 4 | 6 | 8 | **37** |
| Less leave | 0 | 4 | 3 | 0 | 7 | 4 | 8 | 8 | 6 | 0 | 4 | 6 | 5 | 3 | 4 | **62** |
| High working hours | 2 | 5 | 7 | 0 | 0 | 4 | 8 | 8 | 7 | 2 | 4 | 6 | 6 | 4 | 5 | **68** |
| Poor decision making | 2 | 4 | 6 | 0 | 0 | 0 | 3 | 2 | 2 | 1 | 6 | 2 | 3 | 8 | 8 | 47 |
| Personal stress | 3 | 5 | 5 | 0 | 0 | 2 | 0 | 6 | 7 | 2 | 4 | 4 | 6 | 5 | 6 | 55 |
| Family stress | 2 | 4 | 4 | 0 | 0 | 4 | | 0 | 3 | 1 | 2 | 3 | 3 | 4 | 4 | **34** |
| Personal discomfort | 3 | 3 | 3 | 0 | 0 | 4 | 4 | 2 | 0 | 1 | 4 | 4 | 4 | 5 | 6 | 43 |
| Uncertainty | 4 | 4 | 4 | 0 | 0 | 3 | 2 | 2 | 3 | 0 | 3 | 1 | 1 | 3 | 4 | **34** |
| Poor relations | 2 | 8 | 6 | 0 | 0 | 6 | 3 | 4 | 2 | 2 | 0 | 1 | 2 | 5 | 6 | 47 |
| Health | 3 | 2 | 6 | 0 | 0 | 3 | 5 | 4 | 4 | 2 | 2 | 0 | 4 | 6 | 7 | 48 |
| Accidents | 6 | 3 | 6 | 0 | 0 | 6 | 5 | 3 | 4 | 5 | 5 | 2 | 0 | 5 | 6 | **56** |
| Quality | 2 | 4 | 6 | 0 | 0 | 2 | 3 | 4 | 4 | 1 | 2 | 1 | 1 | 0 | 5 | **35** |
| Performance | 4 | 4 | 5 | 0 | 0 | 1 | 5 | 2 | 4 | 3 | 4 | 2 | 1 | 1 | 0 | **36** |
| Σaij | 39 | 57 | 70 | **0** | **14** | 48 | **59** | 53 | 57 | **24** | 56 | 39 | 48 | **61** | **77** | |

The steps of DEMATEL method calculation are described as follows:

Step 1: Generation of Average Matrix / Original Impact Matrix (A) to analyse cause and effect.

The calculations of the Original Impact Matrix are conducted by pairwise comparisons of different dimensions to evaluate the perceived level of impact of each respondent regarding the dimensions.

The respondents were experts from the field and had more than 10 years' experience in the field.

Table (i): Average Matrix / Original Impact Matrix (A)

## Observations of the Inter-Relational Dynamics of the HR Aspects

From the original impact matrix it is seen that:

- Job delay has a high degree of impact on family stress and personal discomfort (6) and is influenced by poor team (6) and accidents (6)
- Poor Team has a high degree of impact on job delay, motivation and poor relations (6) and is influenced by poor relations (8), high working hours (5) and personal stress (5)
- Motivation has a high degree of impact on poor performance (8), poor quality (6), and poor relations (5) and is influenced by poor team (6), poor relation (6), poor health (6), accidents (6) and quality (6)
- Less leave has a high degree of impact on personal stress (8) and family stress (8) and is influenced by nil aspects
- High working hours has a high degree of impact on personal stress (8) and family stress (8) and is influenced by mainly less leave (7)
- Poor decision making has a high degree of impact on poor quality (8) and poor performance (8) and is influenced by poor relations (6) and accidents (6)

- Personal stress has a high degree of impact on personal discomfort (7), accidents (6), poor performance (6) and family stress (6) and is influenced by less leave (8) and high working hours (8)
- Family stress has a high degree of impact on poor team (4), motivation (4), poor decision making (4), poor quality (4) and poor performance (4) and is influenced by less leave (8) and high working hours (8)
- Personal discomfort has a high degree of impact on poor performance (6) and poor quality (5) and is influenced by high working hours (7) and personal stress (7)
- Uncertainty has a high degree of impact on job delay, poor team, motivation and poor performance (4) and is influenced by accidents (5)
- Poor relations have a high degree of impact on poor team (8), motivation (6), poor decision making (6) and poor performance (6) and is influenced by poor team (6) and poor decision making (6)
- Poor Health has a high degree of impact on poor performance (7), motivation (6) and accidents (6) and is influenced by less leave (6) and high working hours (6)
- Accidents has a high degree of impact on Job delay (6), Motivation (6), Poor decision making (6) and Poor performance (6) and is influenced by high working hours (6) and personal stress (6)
- Poor quality has a high degree of impact on Motivation (6), poor performance (5) and is influenced by poor decision making (8), motivation (6) and health (6)
- Poor performance has a high degree of impact on motivation (5) and personal stress (5) and is influenced by motivation (8) and poor decision making (8)

## Analysis of Inter-relational Dynamics of HR Aspects: Cause and Effect

The table facilitates us to identify / differentiate between the Cause and Effect of various aspects with their relative degree of importance and their correlation. The last column gives the horizontal total score on various aspects and reflect the Cause and the last row gives the vertical total score of various aspects and reflect the Effect. The inter-relation of one aspect in relation to the other aspects are compared. Following is seen from the table above:

- The top three Causes as shown in bold are High working hours (68), Less Leave (62) and Accidents (56) and the top three Effects as shown in bold are Poor performance (77), Poor quality (61) and Personal stress (59) and lowest effect on less Leave (0), High working hours (14) and Uncertainty (24).

The impact of attrition of manpower in an organisation therefore causes persons working overtime, the leave gets affected and accidents happen which effects the person to perform poorly, the quality of work also deteriorates and the personal stress also increases. However, the HR aspects that have lowest effect are Less leave, High working hours and Uncertainty

The characteristic of the organization studied where the peculiarity of work culture discipline and high training is reflected in the low level causes of Family stress(34), Uncertainty(34), Quality of work(35) and low level effects of less Leave(0), High working hours(14) and Uncertainty(24).

Step 2: Calculation of maximum values of all rows and columns of the impact matrix (***A***).

Max of Σaij (column and rows) = 77

Step 3: Calculation of k

*Table 2.*

| M=K*A | Job delay | Poor Team | Motivation | Less leave | High working hours | Poor decision making | Personal stress | Family stress | Personal discomfort | Uncertainty | Poor relations | Poor Health | Accidents | Poor quality | Poor performance | $\Sigma a_{ij}$ |
|---|---|---|---|---|---|---|---|---|---|---|---|---|---|---|---|---|
| Job delay | 0.0000 | 0.0390 | 0.0390 | 0.0000 | 0.0519 | 0.0260 | 0.0519 | 0.0779 | 0.0779 | 0.0260 | 0.0649 | 0.0260 | 0.0390 | 0.0260 | 0.0519 | 0.5974 |
| Poor Team | 0.0779 | 0.0000 | 0.0779 | 0.0000 | 0.0390 | 0.0519 | 0.0649 | 0.0260 | 0.0519 | 0.0260 | 0.0779 | 0.0390 | 0.0649 | 0.0519 | 0.0519 | 0.7013 |
| Motivation | 0.0000 | 0.0519 | 0.0000 | 0.0000 | 0.0000 | 0.0390 | 0.0519 | 0.0000 | 0.0130 | 0.0000 | 0.0649 | 0.0260 | 0.0519 | 0.0779 | 0.1039 | 0.4805 |
| Less leave | 0.0000 | 0.0519 | 0.0390 | 0.0000 | 0.0909 | 0.0519 | 0.1039 | 0.1039 | 0.0779 | 0.0000 | 0.0519 | 0.0779 | 0.0649 | 0.0390 | 0.0519 | 0.8052 |
| High working hours | 0.0260 | 0.0649 | 0.0909 | 0.0000 | 0.0000 | 0.0519 | 0.1039 | 0.1039 | 0.0909 | 0.0260 | 0.0519 | 0.0779 | 0.0779 | 0.0519 | 0.0649 | 0.8831 |
| Poor decision making | 0.0260 | 0.0519 | 0.0779 | 0.0000 | 0.0000 | 0.0000 | 0.0390 | 0.0260 | 0.0260 | 0.0130 | 0.0779 | 0.0260 | 0.0390 | 0.1039 | 0.1039 | 0.6104 |
| Personal stress | 0.0390 | 0.0649 | 0.0649 | 0.0000 | 0.0000 | 0.0260 | 0.0000 | 0.0779 | 0.0909 | 0.0260 | 0.0519 | 0.0519 | 0.0779 | 0.0649 | 0.0779 | 0.7143 |
| Family stress | 0.0260 | 0.0519 | 0.0519 | 0.0000 | 0.0000 | 0.0519 | 0.0000 | 0.0000 | 0.0390 | 0.0130 | 0.0260 | 0.0390 | 0.0390 | 0.0519 | 0.0519 | 0.4416 |
| Personal discomfort | 0.0390 | 0.0390 | 0.0390 | 0.0000 | 0.0000 | 0.0519 | 0.0519 | 0.0260 | 0.0000 | 0.0130 | 0.0519 | 0.0519 | 0.0519 | 0.0649 | 0.0779 | 0.5584 |
| Uncertainty | 0.0519 | 0.0519 | 0.0519 | 0.0000 | 0.0000 | 0.0390 | 0.0260 | 0.0260 | 0.0390 | 0.0000 | 0.0390 | 0.0130 | 0.0130 | 0.0390 | 0.0519 | 0.4416 |
| Poor relations | 0.0260 | 0.1039 | 0.0779 | 0.0000 | 0.0000 | 0.0779 | 0.0390 | 0.0519 | 0.0260 | 0.0260 | 0.0000 | 0.0130 | 0.0260 | 0.0649 | 0.0779 | 0.6104 |
| Poor Health | 0.0390 | 0.0260 | 0.0779 | 0.0000 | 0.0000 | 0.0390 | 0.0649 | 0.0519 | 0.0519 | 0.0260 | 0.0260 | 0.0000 | 0.0519 | 0.0779 | 0.0909 | 0.6234 |
| Accidents | 0.0779 | 0.0390 | 0.0779 | 0.0000 | 0.0000 | 0.0779 | 0.0649 | 0.0390 | 0.0519 | 0.0649 | 0.0649 | 0.0260 | 0.0000 | 0.0649 | 0.0779 | 0.7273 |
| Poor quality | 0.0260 | 0.0519 | 0.0779 | 0.0000 | 0.0000 | 0.0260 | 0.0390 | 0.0519 | 0.0519 | 0.0130 | 0.0260 | 0.0130 | 0.0130 | 0.0000 | 0.0649 | 0.4545 |
| Poor performance | 0.0519 | 0.0519 | 0.0649 | 0.0000 | 0.0000 | 0.0130 | 0.0649 | 0.0260 | 0.0519 | 0.0390 | 0.0519 | 0.0260 | 0.0130 | 0.0130 | 0.0000 | 0.4675 |
| $\Sigma a_{ij}$ | 0.5065 | 0.7403 | 0.9091 | 0.0000 | 0.1818 | 0.6234 | 0.7662 | 0.6883 | 0.7403 | 0.3117 | 0.7273 | 0.5065 | 0.6234 | 0.7922 | 1.0000 | |

$$k = Min\left(\frac{1}{\max\limits_{1\le i\le n}\sum_{j=1}^{n}|a_{ij}|}, \frac{1}{\max\limits_{1\le i\le n}\sum_{i=1}^{n}|a_{ij}|}\right), \qquad i, j = 1,2,3,\ldots,n$$

Putting the value of Max of Σaij (column and rows) = 77 in the above equation, we get

k = 0.01299

Step 4: Calculation of the Direct Impact Matrix (M)
Table (ii): Direct Impact Matrix (M)

## Priority Table of HR Aspects

Step 5: Creation of priority ranking of the Direct Impact Matrix (M). It is done by taking the summation of row sum and column sum.
From the Direct Impact Matrix, the priority table is formed.

**Analysis**: The impact priority matrix helps us to identify priority of impact of various aspects. It is seen that personal stress has the highest impact priority which is followed by poor performance, poor team and motivation. Uncertainty, less leave and high working hours has the least impact priority.

Step 6: Calculation of Total Impact Matrix (T). It is done by using the following formula:

$T = M\,(I-M)^{-1}$

*Table 3. Priority table of HR aspects*

| Σaij column | Σaij row | Row sum + column sum | Priority | |
|---|---|---|---|---|
| 0.5974 | 0.5065 | 1.1039 | 12 | Job delay |
| 0.7013 | 0.7403 | 1.4416 | 3 | Poor Team |
| 0.4805 | 0.9091 | 1.3896 | 4 | Motivation |
| 0.8052 | 0.0000 | 0.8052 | 14 | Less leave |
| 0.8831 | 0.1818 | 1.0649 | 13 | High working hours |
| 0.6104 | 0.6234 | 1.2338 | 9 | Poor decision making |
| 0.7143 | 0.7662 | 1.4805 | 1 | Personal stress |
| 0.4416 | 0.6883 | 1.1299 | 10 | Family stress |
| 0.5584 | 0.7403 | 1.2987 | 7 | Personal discomfort |
| 0.4416 | 0.3117 | 0.7532 | 15 | Uncertainty |
| 0.6104 | 0.7273 | 1.3377 | 6 | Poor relations |
| 0.6234 | 0.5065 | 1.1299 | 11 | Poor Health |
| 0.7273 | 0.6234 | 1.3506 | 5 | Accidents |
| 0.4545 | 0.7922 | 1.2468 | 8 | Poor quality |
| 0.4675 | 1.0000 | 1.4675 | 2 | Poor performance |

*Table 4. (I – M) matrix*

| I - M | Job delay | Poor Team | Motivation | Less leave | High working hours | Poor decision making | Personal stress | Family stress | Personal discomfort | Uncertainty | Poor relations | Poor Health | Accidents | Poor quality | Poor performance |
|---|---|---|---|---|---|---|---|---|---|---|---|---|---|---|---|
| Job delay | 1.0000 | -0.0390 | -0.0390 | 0.0000 | -0.0519 | -0.0260 | -0.0519 | -0.0779 | -0.0779 | -0.0260 | -0.0649 | -0.0260 | -0.0390 | -0.0260 | -0.0519 |
| Poor Team | -0.0779 | 1.0000 | -0.0779 | 0.0000 | -0.0390 | -0.0519 | -0.0649 | -0.0260 | -0.0519 | -0.0260 | -0.0779 | -0.0390 | -0.0649 | -0.0519 | -0.0519 |
| Motivation | 0.0000 | -0.0519 | 1.0000 | 0.0000 | 0.0000 | -0.0390 | -0.0519 | 0.0000 | -0.0130 | 0.0000 | -0.0649 | -0.0260 | -0.0519 | -0.0779 | -0.1039 |
| Less leave | 0.0000 | -0.0519 | -0.0390 | 1.0000 | -0.0909 | -0.0519 | -0.1039 | -0.1039 | -0.0779 | 0.0000 | -0.0519 | -0.0779 | -0.0649 | -0.0390 | -0.0519 |
| High working hours | -0.0260 | -0.0649 | -0.0909 | 0.0000 | 1.0000 | -0.0519 | -0.1039 | -0.1039 | -0.0909 | -0.0260 | -0.0519 | -0.0779 | -0.0779 | -0.0519 | -0.0649 |
| Poor decision making | -0.0260 | -0.0519 | -0.0779 | 0.0000 | 0.0000 | 1.0000 | -0.0390 | -0.0260 | -0.0260 | -0.0130 | -0.0779 | -0.0260 | -0.0390 | -0.1039 | -0.1039 |
| Personal stress | -0.0390 | -0.0649 | -0.0649 | 0.0000 | 0.0000 | -0.0260 | 1.0000 | -0.0779 | -0.0909 | -0.0260 | -0.0519 | -0.0519 | -0.0779 | -0.0649 | -0.0779 |
| Family stress | -0.0260 | -0.0519 | -0.0519 | 0.0000 | 0.0000 | -0.0519 | 0.0000 | 1.0000 | -0.0390 | -0.0130 | -0.0260 | -0.0390 | -0.0390 | -0.0519 | -0.0519 |
| Personal discomfort | -0.0390 | -0.0390 | -0.0390 | 0.0000 | 0.0000 | -0.0519 | -0.0519 | -0.0260 | 1.0000 | -0.0130 | -0.0519 | -0.0519 | -0.0519 | -0.0649 | -0.0779 |
| Uncertainty | -0.0519 | -0.0519 | -0.0519 | 0.0000 | 0.0000 | -0.0390 | -0.0260 | -0.0260 | -0.0390 | 1.0000 | -0.0390 | -0.0130 | -0.0130 | -0.0390 | -0.0519 |
| Poor relations | -0.0260 | -0.1039 | -0.0779 | 0.0000 | 0.0000 | -0.0779 | -0.0390 | -0.0519 | -0.0260 | -0.0260 | 1.0000 | -0.0130 | -0.0260 | -0.0649 | -0.0779 |
| Poor Health | -0.0390 | -0.0260 | -0.0779 | 0.0000 | 0.0000 | -0.0390 | -0.0649 | -0.0519 | -0.0519 | -0.0260 | -0.0260 | 1.0000 | -0.0519 | -0.0779 | -0.0909 |
| Accidents | -0.0779 | -0.0390 | -0.0779 | 0.0000 | 0.0000 | -0.0779 | -0.0649 | -0.0390 | -0.0519 | -0.0649 | -0.0649 | -0.0260 | 1.0000 | -0.0649 | -0.0779 |
| Poor quality | -0.0260 | -0.0519 | -0.0779 | 0.0000 | 0.0000 | -0.0260 | -0.0390 | -0.0519 | -0.0519 | -0.0130 | -0.0260 | -0.0130 | -0.0130 | 1.0000 | -0.0649 |
| Poor performance | -0.0519 | -0.0519 | -0.0649 | 0.0000 | 0.0000 | -0.0130 | -0.0649 | -0.0260 | -0.0519 | -0.0390 | -0.0519 | -0.0260 | -0.0130 | -0.0130 | 1.0000 |

Where I is the Identity matrix.

$T = M\,(I\text{-}M)^{-1}$

Cause - Effect Relationship of HR Aspects

*Table 5. (I – M) matrix*

| (I - M) inverse | Job delay | Poor Team | Motivation | Less leave | High working hours | Poor decision making | Personal stress | Family stress | Personal discomfort | Uncertainty | Poor relations | Poor Health | Accidents | Poor quality | Poor performance |
|---|---|---|---|---|---|---|---|---|---|---|---|---|---|---|---|
| Job delay | 1.0526 | 0.1107 | 0.1251 | 0.0000 | 0.0590 | 0.0854 | 0.1160 | 0.1308 | 0.1398 | 0.0579 | 0.1320 | 0.0702 | 0.0951 | 0.1026 | 0.1466 |
| Poor Team | 0.1348 | 1.0856 | 0.1766 | 0.0000 | 0.0493 | 0.1184 | 0.1410 | 0.0922 | 0.1269 | 0.0640 | 0.1582 | 0.0871 | 0.1280 | 0.1407 | 0.1656 |
| Motivation | 0.0451 | 0.1088 | 1.0731 | 0.0000 | 0.0066 | 0.0829 | 0.1042 | 0.0440 | 0.0651 | 0.0285 | 0.1190 | 0.0571 | 0.0920 | 0.1342 | 0.1758 |
| Less leave | 0.0729 | 0.1492 | 0.1609 | 1.0000 | 0.1005 | 0.1324 | 0.1933 | 0.1809 | 0.1683 | 0.0457 | 0.1461 | 0.1399 | 0.1456 | 0.1470 | 0.1854 |
| High working hours | 0.1022 | 0.1651 | 0.2130 | 0.0000 | 1.0117 | 0.1350 | 0.1936 | 0.1776 | 0.1807 | 0.0726 | 0.1521 | 0.1378 | 0.1581 | 0.1643 | 0.2053 |
| Poor decision making | 0.0766 | 0.1221 | 0.1610 | 0.0000 | 0.0087 | 1.0556 | 0.1028 | 0.0771 | 0.0873 | 0.0448 | 0.1432 | 0.0639 | 0.0890 | 0.1717 | 0.1935 |
| Personal stress | 0.1003 | 0.1433 | 0.1621 | 0.0000 | 0.0108 | 0.0937 | 1.0759 | 0.1346 | 0.1594 | 0.0636 | 0.1312 | 0.0980 | 0.1372 | 0.1501 | 0.1860 |
| Family stress | 0.0640 | 0.1004 | 0.1137 | 0.0000 | 0.0072 | 0.0915 | 0.0496 | 1.0368 | 0.0823 | 0.0363 | 0.0778 | 0.0667 | 0.0765 | 0.1063 | 0.1213 |
| Personal discomfort | 0.0869 | 0.1037 | 0.1194 | 0.0000 | 0.0086 | 0.1020 | 0.1110 | 0.0759 | 1.0600 | 0.0440 | 0.1151 | 0.0871 | 0.0996 | 0.1326 | 0.1642 |
| Uncertainty | 0.0878 | 0.1019 | 0.1121 | 0.0000 | 0.0085 | 0.0785 | 0.0730 | 0.0637 | 0.0837 | 1.0227 | 0.0903 | 0.0426 | 0.0527 | 0.0924 | 0.1199 |
| Poor relations | 0.0785 | 0.1692 | 0.1615 | 0.0000 | 0.0107 | 0.1301 | 0.1026 | 0.1000 | 0.0877 | 0.0564 | 1.0732 | 0.0538 | 0.0802 | 0.1385 | 0.1706 |
| Poor Health | 0.0899 | 0.0967 | 0.1605 | 0.0000 | 0.0084 | 0.0933 | 0.1271 | 0.1029 | 0.1144 | 0.0578 | 0.0962 | 1.0413 | 0.1040 | 0.1498 | 0.1838 |
| Accidents | 0.1346 | 0.1226 | 0.1748 | 0.0000 | 0.0118 | 0.1398 | 0.1373 | 0.1005 | 0.1245 | 0.0993 | 0.1458 | 0.0725 | 1.0630 | 0.1513 | 0.1882 |
| Poor quality | 0.0636 | 0.1026 | 0.1370 | 0.0000 | 0.0073 | 0.0672 | 0.0858 | 0.0871 | 0.0961 | 0.0357 | 0.0788 | 0.0445 | 0.0548 | 1.0566 | 0.1339 |
| Poor performance | 0.0907 | 0.1063 | 0.1281 | 0.0000 | 0.0089 | 0.0584 | 0.1127 | 0.0678 | 0.1004 | 0.0622 | 0.1059 | 0.0584 | 0.0582 | 0.0734 | 1.0764 |

*Table 6. (I – M) inverse matrix*

| T | Job delay | Poor Team | Motivation | Less leave | High working hours | Poor decision making | Personal stress | Family stress | Personal discomfort | Uncertainty | Poor relations | Poor Health | Accidents | Poor quality | Poor performance |
|---|---|---|---|---|---|---|---|---|---|---|---|---|---|---|---|
| Job delay | 0.0000 | 0.0043 | 0.0049 | 0.0000 | 0.0031 | 0.0022 | 0.0060 | 0.0102 | 0.0109 | 0.0015 | 0.0086 | 0.0018 | 0.0037 | 0.0027 | 0.0076 |
| Poor Team | 0.0105 | 0.0000 | 0.0138 | 0.0000 | 0.0019 | 0.0061 | 0.0092 | 0.0024 | 0.0066 | 0.0017 | 0.0123 | 0.0034 | 0.0083 | 0.0073 | 0.0086 |
| Motivation | 0.0000 | 0.0057 | 0.0000 | 0.0000 | 0.0000 | 0.0032 | 0.0054 | 0.0000 | 0.0008 | 0.0000 | 0.0077 | 0.0015 | 0.0048 | 0.0105 | 0.0183 |
| Less leave | 0.0000 | 0.0078 | 0.0063 | 0.0000 | 0.0091 | 0.0069 | 0.0201 | 0.0188 | 0.0131 | 0.0000 | 0.0076 | 0.0109 | 0.0095 | 0.0057 | 0.0096 |
| High working hours | 0.0027 | 0.0107 | 0.0194 | 0.0000 | 0.0000 | 0.0070 | 0.0201 | 0.0185 | 0.0164 | 0.0019 | 0.0079 | 0.0107 | 0.0123 | 0.0085 | 0.0133 |
| Poor decision making | 0.0020 | 0.0063 | 0.0125 | 0.0000 | 0.0000 | 0.0000 | 0.0040 | 0.0020 | 0.0023 | 0.0006 | 0.0112 | 0.0017 | 0.0035 | 0.0178 | 0.0201 |
| Personal stress | 0.0039 | 0.0093 | 0.0105 | 0.0000 | 0.0000 | 0.0024 | 0.0000 | 0.0105 | 0.0145 | 0.0017 | 0.0068 | 0.0051 | 0.0107 | 0.0097 | 0.0145 |
| Family stress | 0.0017 | 0.0052 | 0.0059 | 0.0000 | 0.0000 | 0.0048 | 0.0000 | 0.0000 | 0.0032 | 0.0005 | 0.0020 | 0.0026 | 0.0030 | 0.0055 | 0.0063 |
| Personal discomfort | 0.0034 | 0.0040 | 0.0047 | 0.0000 | 0.0000 | 0.0053 | 0.0058 | 0.0020 | 0.0000 | 0.0006 | 0.0060 | 0.0045 | 0.0052 | 0.0086 | 0.0128 |
| Uncertainty | 0.0046 | 0.0053 | 0.0058 | 0.0000 | 0.0000 | 0.0031 | 0.0019 | 0.0017 | 0.0033 | 0.0000 | 0.0035 | 0.0006 | 0.0007 | 0.0036 | 0.0062 |
| Poor relations | 0.0020 | 0.0176 | 0.0126 | 0.0000 | 0.0000 | 0.0101 | 0.0040 | 0.0052 | 0.0023 | 0.0015 | 0.0000 | 0.0007 | 0.0021 | 0.0090 | 0.0133 |
| Poor Health | 0.0035 | 0.0025 | 0.0125 | 0.0000 | 0.0000 | 0.0036 | 0.0083 | 0.0053 | 0.0059 | 0.0015 | 0.0025 | 0.0000 | 0.0054 | 0.0117 | 0.0167 |
| Accidents | 0.0105 | 0.0048 | 0.0136 | 0.0000 | 0.0000 | 0.0109 | 0.0089 | 0.0039 | 0.0065 | 0.0064 | 0.0095 | 0.0019 | 0.0000 | 0.0098 | 0.0147 |
| Poor quality | 0.0017 | 0.0053 | 0.0107 | 0.0000 | 0.0000 | 0.0017 | 0.0033 | 0.0045 | 0.0050 | 0.0005 | 0.0020 | 0.0006 | 0.0007 | 0.0000 | 0.0087 |
| Poor performance | 0.0047 | 0.0055 | 0.0083 | 0.0000 | 0.0000 | 0.0008 | 0.0073 | 0.0018 | 0.0052 | 0.0024 | 0.0055 | 0.0015 | 0.0008 | 0.0010 | 0.0000 |

Step 7: Calculation of vector elements.
C = column sum vector
R = row sum vector
R - C = row and column difference vector
R + C = row and column sum vector

*Table 7. Cause - Effect relationship of HR Aspects*

| C | R | R+C | C-R | Factors | |
|---|---|---|---|---|---|
| 0.0675 | 0.0511 | 0.1185 | 0.0164 | Job delay | Cause |
| 0.0921 | 0.0944 | 0.1864 | -0.0023 | Poor Team | Effect |
| 0.0579 | 0.1414 | 0.1993 | **-0.0836** | Motivation | Effect |
| 0.1253 | 0.0000 | 0.1253 | **0.1253** | Less leave | Cause |
| 0.1495 | 0.0141 | 0.1636 | **0.1353** | High working hours | Cause |
| 0.0840 | 0.0682 | 0.1522 | 0.0158 | Poor decision making | Cause |
| 0.0996 | 0.1043 | 0.2039 | -0.0046 | Personal stress | Effect |
| 0.0406 | 0.0867 | 0.1273 | -0.0461 | Family stress | Effect |
| 0.0628 | 0.0960 | 0.1588 | -0.0332 | Personal discomfort | Effect |
| 0.0401 | 0.0206 | 0.0608 | 0.0195 | Uncertainty | Cause |
| 0.0803 | 0.0931 | 0.1735 | -0.0128 | Poor relations | Effect |
| 0.0795 | 0.0474 | 0.1269 | **0.0320** | Poor Health | Cause |
| 0.1014 | 0.0705 | 0.1719 | 0.0308 | Accidents | Cause |
| 0.0448 | 0.1115 | 0.1562 | **-0.0667** | Poor quality | Effect |
| 0.0448 | 0.1707 | 0.2155 | **-0.1260** | Poor performance | Effect |

*Figure 1. Cause - effect relationship of HR aspects*

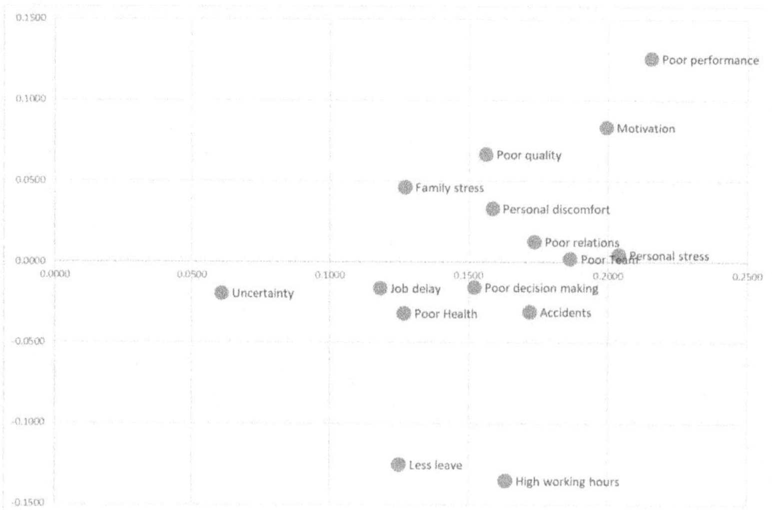


The difference vector (C - R) gives the net impact of the Total Impact Matrix. If *(C - R)* is > 0, it means that the factor has greater impact on other factors. Hence, it is called the Cause. If *(C -R)* is < 0, the factor has a smaller impact on other factors. Hence, it is referred to as the Effect.

## ANALYSIS

From the above tables we have been able to group the HR Aspects as Cause and as Effects. The high-level cause is identified as high working hours, less leave and poor health. It is inferred that the attrition of manpower would lead to increase in the working hours of the available persons. The leave may also be curtailed due to paucity of manpower and in turn health gets affected due to overwork.

The high-level effect as poor performance, motivation, and poor quality.

It is inferred that the Causes result in the decrease in performance levels, lowering of motivation and poor quality of work. Thus, in an organization attrition of manpower mainly causes high working hours, less leave, poor health and effect the poor performance, decrease in motivation, poor quality of work of the individuals. Other aspects studied fall in the middle of the spectrum.

### Summary of DEMATEL Analysis

1. Job delay has a high degree of impact on family stress and personal discomfort (6) and is influenced by poor team (6) and accidents (6)
2. Poor Team has a high degree of impact on job delay, motivation and poor relations (6) and is influenced by poor relations (8), high working hours (5) and personal stress (5)
3. Motivation has a high degree of impact on poor performance (8), poor quality (6), and poor relations (5) and is influenced by poor team (6), poor relation (6), poor health (6), accidents (6)and quality (6)
4. Less leave has a high degree of impact on personal stress (8) and family stress (8) and is influenced by nil aspects
5. High working hours has a high degree of impact on personal stress (8) and family stress (8) and is influenced by mainly less leave (7)
6. Poor decision making has a high degree of impact on poor quality (8) and poor performance (8) and is influenced by poor relations (6) and accidents (6)
7. Personal stress has a high degree of impact on personal discomfort (7), accidents (6), poor performance (6) and family stress (6) and is influenced by less leave (8) and high working hours (8)
8. Family stress has a high degree of impact on poor team (4), motivation (4), poor decision making (4), poor quality (4) and poor performance (4) and is influenced by less leave (8) and high working hours (8)
9. Personal discomfort has a high degree of impact on poor performance (6) and poor quality (5) and is influenced by high working hours (7) and personal stress (7)
10. Uncertainty has a high degree of impact on job delay, poor team, motivation and poor performance (4) and is influenced by accidents (5)
11. Poor relations have a high degree of impact on poor team (8), motivation (6), poor decision making (6) and poor performance (6) and is influenced by poor team (6) and poor decision making (6)

12. Poor Health has a high degree of impact on poor performance (7), motivation (6) and accidents (6) and is influenced by less leave (6) and high working hours (6)
13. Accidents have a high degree of impact on Job delay (6), Motivation (6), Poor decision making (6) and Poor performance (6) and is influenced by high working hours (6) and personal stress (6)
14. Poor quality has a high degree of impact on Motivation (6), poor performance (5) and is influenced by poor decision making (8), motivation (6) and health (6)
15. Poor performance has a high degree of impact on motivation (5) and personal stress (5) and is influenced by motivation (8) and poor decision making (8)

## Summary of the Analysis of Impact of HAIA on HR Aspects in the Organisation:

The HR Aspects can be qualified in the order of importance from lowest to highest and their Inter-relational dynamics as Cause & Effect is as under.

In the Table above the HR Aspects have been prioritised from lowest to highest in the order of importance i.e. Accidents at number 18 is the least important and Health at number 1 is the topmost priority. The Inter dimensional relationship analysis included 15 HR Aspects since 03 Aspects were found to be partly similar in characteristic e.g. morale & motivation, poor relations and boss relations and difficult substitution & team performance. The HR Aspects identified as Cause (1 to 7) and Effect (1 to 8) in the order of their importance in the organisation is indicated by the serial number. It is obvious that high attrition due to HAIA directly results in high working hours and less available leaves. This results

*Table 8. Summary of the analysis of impact of HAIA on HR aspects in the organisation*

| Priority | HR Aspects | Cause / Effect |
|---|---|---|
| 18 | Boss relations | Effect |
| 17 | Accidents | 4 Cause |
| 16 | Team performance | 1 Effect |
| 15 | Poor relations | 6 Effect |
| 14 | Difficult substitution | Cause |
| 13 | Uncertainty | 5 Cause |
| 12 | Quality | 3 Effect |
| 11 | Motivation | 2 Effect |
| 10 | Poor decision making | 7 Cause |
| 9 | Personal discomfort | 5 Effect |
| 8 | Failure (Poor Team) | 8 Effect |
| 7 | High working hours | 1 Cause |
| 6 | Less leave | 2 Cause |
| 5 | Personal stress | 7 Effect |
| 4 | Job delay | 6 Cause |
| 3 | Family stress | 4 Effect |
| 2 | Morale | Effect |
| 1 | Health | 3 Cause |

in high work pressure, and which in turn affects many other attributes. Adverse health and accidents are an outcome and which also contribute in creating effects such as low morale, poor teamwork, low quality and poor performance. When operations are of critical importance, such a condition cannot be accepted. The solution lies in either hiring critically more personnel in the beginning to compensate for attritions later. The other solution could be to empower people with resources and skills so that efficiency is enhanced. A strong communication network for example, can reduce the need for physical movement.

The impact of HAIA on the HR Aspects identified as Cause has implications for the Employer and Effect has implications for the Employees.

The implication of the HR Aspects on the employer are as under:

1. The implication of High working hours in any organisation is well known that it leads to job burn out, accidents, fatigue, stress and health. All these will have a compounding effect in high altitude area where it is a known fact that the working efficiency is also compromised due to harsh terrain. It has a high degree of impact on personal stress and family stress and is influenced by mainly less leave.
2. **Less Leave**: Absenteeism or attrition of manpower disrupts the leave planning in any organisation, keeping in view the shortfall of manpower and execution of the tasks assigned in the required framework of time. In the organisation under study the employees working in the monotonous harsh and difficult terrain look forward very eagerly to go on leave, therefore managing leave becomes a challenge to the managers in the organisation. Less leave has a high degree of impact on personal stress and family stress and is seen to be influenced by nil aspects.
3. It is important for the employer to ensure good health of the employees in high altitude area where health is already compromised due to terrain peculiarities. Poor health has a high degree of impact on poor performance, motivation and accidents and is influenced by less leave and high working hours.
4. **Accident:** The implication of the employer to address this cause is also important because accidents have a high degree of impact on job delay, motivation, poor decision making and poor performance and is influenced by high working hours and personal stress.
5. In the organisation Uncertainty is a highly avoidable HR aspect since it has a high degree of impact on job delay, poor team, motivation and poor performance and is influenced by accidents.
6. **Job Delay**: Job delay impacts family stress and personal discomfort and is influenced by poor team and accidents. The attrition of manpower in an organisation, the job delay may be a result of poor team and the accidents. It is likely to put pressure on the hierarchy which gets translated into family stress and personal discomfort.
7. In the organisation the attrition of manpower in high altitude area, it is observed that Poor decision making has a high degree of impact on poor quality and poor performance and is influenced by poor relations and accidents which is required to be addressed.

Implication for the Employees

1. The team performance in the organisation deteriorates because of impact of HAIA. It is seen that **Poor team** has a high degree of impact on job delay, motivation and poor relations and since it is influenced by poor relations, high working hours and personal stress the employers responsibility is to address these influences adequately.

2. **Motivation** has a high degree of impact on poor performance, poor quality and poor relations and is influenced by poor team, poor relation, poor health, accidents and quality so to improve the motivation levels of the employees the team should be good, their relations and health must be good besides ensuring high quality and reduction in accidents.
3. **Poor quality** has a high degree of impact on motivation, poor performance and it is important for the employer in high altitude area to avoid poor decision making, do the needful to keep motivation levels high and good health.
4. It is seen that the **Family stress** impacts four HR aspects i.e. on poor team, motivation, poor decision making, poor quality and is poor performance and since this aspect is influenced by less leave and high working hours the managers require to plan the leave and working hours keeping in view the attrition of manpower in high altitude.
5. **Personal discomfort** has a high degree of impact on poor performance and poor quality and is influenced by high working hours and personal stress.
6. The implication of **Poor relations** on the employees is that it has large impact on poor team, motivation, poor decision making and poor performance and since it is influenced by poor team and also poor decision making both these aspects need to be addressed by the employer.
7. **Personal stress** has a high degree of impact on personal discomfort, accidents, poor performance, family stress and is influenced by less leave and high working hours.
8. **Poor Team** or Failure has a high degree of impact on job delay, motivation and poor relations and it is seen not be influenced by poor relations, high working hours and personal stress which needs to be addressed to avoid failures due to HAIA.

## CONCLUSION AND IMPLICATIONS FOR THE EMPLOYER

From the discussion it may be concluded that high altitude sickness if not handled properly can result in a severe human resource crunch and which in turn leads to high productivity losses due to high stress and demotivation of team members, accidents, low quality of work performance and enhanced interpersonal issues. It is therefore, recommended that under such circumstances recruitment and acclimatization plays an important role. At the selection stage itself the individuals may be tested for medical fitness. Also, approximately 20% more personnel should be selected so that later attrition does not affect the ideal strength. Finally, the team leaders must understand the effects of high attrition and develop appropriate strategies to counter them.

## REFERENCES

Adjei, N. K., Jonsson, K. R., & Brand, T. (2018). Time spent on work-related activities, social activities and time pressure as intermediary determinants of health disparities among elderly women and men in 5 European countries: A structural equation model. *International Journal for Equity in Health*, *17*(1), 121. doi:10.118612939-018-0840-y PMID:30115073

Assunção, A. A., & Abreu, M. N. S. (2019). Pressure to work, health status, and work conditions of schoolteachers in Basic Education in Brazil. *Cadernos de Saude Publica*, 35. PMID:31141023

Baghel, R., & Nüsser, M. (2015). Securing the heights: The vertical dimension of the Siachen conflict between India and Pakistan in the Eastern Karakoram. *Political Geography*, *48*, 24–36. doi:10.1016/j.polgeo.2015.05.001

Banerjee, S., & Mehta, P. (2016). Determining the antecedents of job stress and their impact on job performance: A study among faculty members. *IUP Journal of Organizational Behavior*, *15*(2), 7.

Barlas, B., & Izci, F. B. (2018). Individual and workplace factors related to fatal occupational accidents among shipyard workers in Turkey. *Safety Science*, *101*, 173–179. doi:10.1016/j.ssci.2017.09.012

Chiu, Y. J., Chen, H. C., Tzeng, G. H., & Shyu, J. Z. (2006). Marketing strategy based on customer behaviour for the LCD-TV. *International Journal and Decision Making*, *7*(2/3), 143–165. doi:10.1504/IJMDM.2006.009140

Coffeng, J. K., Hendriksen, I. J., Duijts, S. F., Twisk, J. W., van Mechelen, W., & Boot, C. R. (2014). Effectiveness of a combined social and physical environmental intervention on presenteeism, absenteeism, work performance, and work engagement in office employees. *Journal of Occupational and Environmental Medicine*, *56*(3), 258–265. doi:10.1097/JOM.0000000000000116 PMID:24603201

Demerouti, E., Bakker, A. B., & Bulters, A. J. (2004). The loss spiral of work pressure, work–home interference and exhaustion: Reciprocal relations in a three-wave study. *Journal of Vocational Behavior*, *64*(1), 131–149. doi:10.1016/S0001-8791(03)00030-7

Dhingra, V., & Dhingra, M. (2017). Industry Framework of Indian Art-Metalware Handicraft Industry: A Case Study. *IUP Journal of Business Strategy*, *14*(1), 52.

Diestel, S., Wegge, J., & Schmidt, K. H. (2014). The impact of social context on the relationship between individual job satisfaction and absenteeism: The roles of different foci of job satisfaction and work-unit absenteeism. *Academy of Management Journal*, *57*(2), 353–382. doi:10.5465/amj.2010.1087

Fong, K. H., & Snape, E. (2015). Empowering leadership, psychological empowerment and employee Outcomes: Testing a multi-level mediating model. *British Journal of Management*, *26*(1), 126–138. doi:10.1111/1467-8551.12048

Fontela, E., & Gabus, A. (Eds.). (1976). *The DEMATEL Observe*. Battelle Institute, Geneva Research Center.

Frone, M. R., Russell, M., & Cooper, M. L. (1997). Relation of work–family conflict to health outcomes: A four-year longitudinal study of employed parents. *Journal of Occupational and Organizational Psychology*, *70*(4), 325–335. doi:10.1111/j.2044-8325.1997.tb00652.x

Ghasemi, F., Kalatpour, O., Moghimbeigi, A., & Mohhamadfam, I. (2018). A path analysis model for explaining unsafe behavior in workplaces: The effect of perceived work pressure. *International Journal of Occupational Safety and Ergonomics*, *24*(2), 303–310. doi:10.1080/10803548.2017.1313494 PMID:28399726

Gottschling, J., Hahn, E., Maas, H., & Spinath, F. M. (2016). Explaining the relationship between personality and coping with professional demands: Where and why do optimism, self-regulation, and self-efficacy matter. *Personality and Individual Differences*, *100*, 49–55. doi:10.1016/j.paid.2016.03.085

Govindan, K., Khodaverdi, R., & Vafadarnikjoo, A. (2015). Intuitionistic fuzzy based DEMATEL method for developing green practices and performances in a green supply chain. *Expert Systems with Applications*, *42*(20), 7207–7220. doi:10.1016/j.eswa.2015.04.030

Hayashi, T., Kobayashi, Y., Yamaoka, K., & Yano, E. (1996). Effect of overtime work on 24-hour ambulatory blood pressure. *Journal of Occupational and Environmental Medicine*, *38*(10), 1007–1011. doi:10.1097/00043764-199610000-00010 PMID:8899576

Heywood, J. S., Jirjahn, U., & Wei, X. (2008). Teamwork, monitoring and absence. *Journal of Economic Behavior & Organization*, *68*(3-4), 676–690. doi:10.1016/j.jebo.2008.09.004

Höge, T. (2009). When work strain transcends psychological boundaries: An inquiry into the relationship between time pressure, irritation, work–family conflict and psychosomatic complaints. *Stress and Health: Journal of the International Society for the Investigation of Stress*, *25*(1), 41–51. doi:10.1002mi.1226

Indulski, J.A., & Szubert, Z. (1997). System for analysing sickness absenteeism in Poland. *International Journal of Occupational Medicine and Environmental Health*, *10*, 159–166. PMID:9278128

Kalisch, B. J., & Lee, K. H. (2010). The impact of teamwork on missed nursing care. *Nursing Outlook*, *58*(5), 233–241. doi:10.1016/j.outlook.2010.06.004 PMID:20934078

Kalkavan, S., & Katrinli, A. (2014). The effects of managerial coaching behaviors on the employees' perception of job satisfaction, organisational commitment, and job performance: Case study on insurance industry in Turkey. *Procedia: Social and Behavioral Sciences*, *150*, 1137–1147. doi:10.1016/j.sbspro.2014.09.129

Katou, A. A., & Budhwar, P. (2015). Human resource management and organisational productivity: A systems approach based empirical analysis. *Journal of Organizational Effectiveness: People and Performance*, *2*(3), 244–266. doi:10.1108/JOEPP-06-2015-0021

Kempegowda, M. (2016). A study on employee retention measures and its impact on employee motivation. *Clear international journal of research in commerce & management*, *7*(4).

Kinjerski, V., & Skrypnek, B. J. (2008). The Promise of Spirit at WorkIncreasing Job Satisfaction and Organizational Commitment and Reducing Turnover and Absenteeism in Long-Term Care. *Journal of Gerontological Nursing*, *34*(10), 17–25. doi:10.3928/00989134-20081001-03 PMID:18942536

Kleiner, S., & Wallace, J. E. (2017). Oncologist burnout and compassion fatigue: Investigating time pressure at work as a predictor and the mediating role of work-family conflict. *BMC Health Services Research*, *17*(1), 639. doi:10.118612913-017-2581-9 PMID:28893255

Laurence, J. H., Naughton, J., & Harris, D. A. (1996). *Attrition Revisited: Identifying the Problem and Its Solutions*. Human resources research organization. doi:10.21236/ADA309172

Liang, t. T. (1983). *Civilian manpower planning model for scientific and engineering personnel in the navy R&D centers*. Navy personnel research and development center.

Lin, C.-J., & Wu, W.-W. (2008). A causal analytical method for group decision-making under fuzzy environment. *Expert Systems with Applications*, *34*(1), 205–213. doi:10.1016/j.eswa.2006.08.012

Lin, C. L., & Tzeng, G. H. (2009). A value-created system of science (technology) park by using DEMETEL. *Expert Systems with Applications*, *36*(6), 9683–9697. doi:10.1016/j.eswa.2008.11.040

Liou, J. J. H., Tzeng, G.-H., & Chang, H.-C. (2007). Airline safety measurement using a hybrid model. *Air Transport Management*, *13*(4), 243–249. doi:10.1016/j.jairtraman.2007.04.008

Lusa, S., Häkkänen, M., Luukkonen, R., & Viikari-Juntura, E. (2002). Perceived physical work capacity, stress, sleep disturbance and occupational accidents among firefighters working during a strike. *Work and Stress*, *16*(3), 264–274. doi:10.1080/02678370210163301

Malhotra, A. (2016). Impact of HRM Practices on Employee Attrition A Study of Selected BPOs in National Capital Region.

Mirzaei Aliabadi, M., Aghaei, H., Kalatpour, O., Soltanian, A. R., & Nikravesh, A. (2018). Analysis of human and organizational factors that influence mining accidents based on Bayesian network. *International Journal of Occupational Safety and Ergonomics*, 1–8. doi:10.1080/10803548.2018.1455411 PMID:29560801

Pflanz, S. E., & Ogle, A. D. (2006). Job stress, depression, work performance, and perceptions of supervisors in military personnel. *Military Medicine*, *171*(9), 861–865. doi:10.7205/MILMED.171.9.861 PMID:17036607

Shieh, J. I., Wu, H. H., & Huang, K. K. (2010). A DEMATEL method in identifying key success factors of hospital service quality. *Knowledge-Based Systems*, *23*(3), 277–282. doi:10.1016/j.knosys.2010.01.013

Siegrist, J. (1996). Adverse health effects of high-effort/low-reward conditions. *Journal of Occupational Health Psychology*, *1*(1), 27–41. doi:10.1037/1076-8998.1.1.27 PMID:9547031

Tahir-Kheli, S., & Biringer, K. L. (2001). *Reducing risk in South Asia: Managing India-Pakistan tensions*. Cooperative Monitoring Center, Sandia National Laboratories.

Tharion, W. J., Lieberman, H. R., Montain, S. J., Young, A. J., Baker-Fulco, C. J., DeLany, J. P., & Hoyt, R. W. (2005). Energy requirements of military personnel. *Appetite*, *44*(1), 47–65. doi:10.1016/j.appet.2003.11.010 PMID:15604033

Thiruchelvan, S. (2017). *Working Environment and Its Influence on Employees' Performance: A Case Study of an Oils and Gas Vendor Company in Malaysia* [Doctoral dissertation]. UTAR.

Tzeng, G. H., Chiang, C. H., & Li, C. W. (2007). Evaluating intertwined effects in e-learning programs: A novel hybrid MCDM model based on factor analysis and DEMATEL. *Expert Systems with Applications*, *32*(4), 1028–1044. doi:10.1016/j.eswa.2006.02.004

van der Klauw, M., Hengel, K. O., Roozeboom, M. B., Koppes, L. L., & Venema, A. (2016). Occupational accidents in the Netherlands: Incidence, mental harm, and their relationship with psychosocial factors at work. *International Journal of Injury Control and Safety Promotion*, *23*(1), 79–84. doi:10.1080/17457300.2014.966119 PMID:25531558

Weiss, A. M. (1985). Absenteeism and Wages. *Economics Letters*, *19*(3), 277–279. doi:10.1016/0165-1765(85)90036-9

Wu, W.-W., & Lee, Y.-T. (2007). Developing global managers' competencies using the fuzzy DEMATEL method. *Expert Systems with Applications*, *32*(2), 499–507. doi:10.1016/j.eswa.2005.12.005

Yang, Y. P., Shieh, H. M., Leu, J. D., & Tzeng, G. H. (2008). A novel hybrid MCDM model combined with DEMATEL and ANP with applications. *International Journal Operational Research*, *5*(3), 160–168.

Yang, Y. P. O., Shieh, H. M., Leu, J. D., & Tzeng, G. H. (2008). A novel hybrid MCDM model combined with DEMATEL and ANP with applications. *International journal of operations research*, *5*(3), 160-168.

Yu, R., & Tseng, G. H. (2006). A soft computing method for multi-criteria decision making with dependence and feedback. *Applied Mathematics and Computation*, *180*(1), 63–75. doi:10.1016/j.amc.2005.11.163

# Chapter 7
# Green Architecture of Malay Traditional House Exhibitions:
## Mini Malaysia and ASEAN Cultural Park (MMACP), Melaka

**Chew Boon Cheong**
*Universiti Teknikal Malaysia Melaka, Malaysia*

**Mohd Syaiful Rizal Abd Hamid**
*Universiti Teknikal Malaysia Melaka, Malaysia*

**Saifuddin Isa**
*Universiti Teknikal Malaysia Melaka, Malaysia*

## ABSTRACT

*Mini Malaysia and ASEAN Cultural Park (MMACP) is one of the tourism spots in Melaka which showcases the Malay traditional houses from thirteen states in Malaysia. Most of the visitors viewing the houses depicting the culture of each states. Contrarily we revisited the park with a curiosity and we would like to know, "How far these Malay traditional houses portray the green architecture?" We designed this research in an exploratory, looking the Malay traditional houses found in MMACP and trace for their green architectural criteria. Through observation, we analysed the data from our field notes, photos, description found on the park's storyboards and explanations from the tour guides. We can conclude that the Malay traditional houses are green buildings that portray green architecture. From this research, we appreciate our forefathers' skills and wisdom, to build the houses that harmonise with the environment.*

DOI: 10.4018/978-1-7998-1566-2.ch007

## ORGANISATION BACKGROUND

Opened on 17 July 1986, Mini Malaysia and ASEAN Cultural Park (MMACP) is one of the tourism hotspots in Melaka. Different from other theme parks which emphasis on fun and excitement; MMACP is an educational theme park, which showcases the Malay traditional houses from thirteen states in Malaysia and also every country in the Association of Southeast Asia Nations (ASEAN).

## INTRODUCTION

Malay traditional houses are constructed by the indigenous ethnic Malay in Malaysia. Nowadays, because of modernisation, most of the Malay houses in Malaysia are influenced by the contemporary design which gradually losing their traditional character.

Therefore, the original traditional Malay houses are hard to be found and could be traced in some deep rural areas of Malaysia. With the effort of Mini Malaysia and ASEAN Cultural Park (MMACP) preserving, conserving and exhibiting these traditional houses, the visitors able to witness how the old-days houses look like.

Houses shown in MMACP conserve the most characteristic Malay traditions. Explained by the Tour Guide 1, these houses are built in the actual sizes that accommodate the actual living. Visitors can walk into these houses to see and experience the setting. Inside the houses, activities such as traditional games, costume rental, cooking demo and product selling are to reflect the unique cultural representation.

Guerrero-Baca and Soria-Lopez (2018) agree, built heritage such as these Malay traditional houses has been one of the most precious assets for the society because these traditional houses bring together a series of values that make them essential for the development of a society. These traditional houses' unique aesthetic value is linked to a historical deed and includes vernacular/traditional architecture which can be inherited to future generations.

Visiting the park, most of the visitors viewing these traditional houses which contain fixtures and decorative work of art, depicting the culture of each state or country. We have visited the park a few times in the past and we learned a lot about the detail aspects of cultural and artistic values of the Malay traditional houses here.

Contrarily, two years ago, we revisited the park with a curiosity that differed from the ordinary quest of traditional value search. We were interested to know, "How far these Malay traditional houses portray the green architecture?" We believe that these Malay traditional houses have a series of socioenvironmental values that justify their existence. A reflection was made around these Malay traditional houses available at the MMACP within the concept of environmental sustainability.

Considering these traditional houses are the actual building which reflect the actual accommodation of the real living in the past; it is interesting to know how our forefathers constructed their houses with the minimalist lifestyle nearly a century ago. Without the most advanced technology, while under the similar tropical climate of Malaysia, can we consider these Malay traditional houses as green buildings? These research questions guided us to have a new insight when we revisited MMACP and we started our study one year ago.

## LITERATURE REVIEW

### Green Building

The U.S. Environmental Protection Agency (2016) provides a comprehensive definition of green building. Green building is the practice of creating structures and using processes that are environmentally responsible and resource efficient.

The processes take in the building's lifecycle: from site selection, building design, construction, operations, maintenance, renovation and deconstruction/removal. Since then, there are many definitions of green building appeared. These definitions are discussing the similar context of green building.

In Malaysia, a green building rating tool (namely Green Building Index-GBI) is established under the Malaysian Institute of Architects and Association of Consulting Engineers Malaysia. The main purpose of this establishment is to administrate GBI accreditation and training of GBI facilitators and certifiers (GBI, 2019)

Under GBI index, the common assessment criteria are (i) energy efficiency, (ii) indoor environmental quality, (iii) sustainable site planning and management, (iv) materials and resources, (v) water efficiency and (vi) innovation. These criteria are set with the marks capped, while the final total of 100 could be determined after the evaluation process are carried out.

### Green Architecture

According to Kubba (2016), the architecture is now at the crossroads since designing for a sustainable future is recognised as a milestone in the history of green building movement.

The National Architectural Accreditation Board (2009) defines sustainability as an ability to design projects that optimises, conserves, or reuses natural and builds resources, provides healthful environments for occupants/users, and reduces the environmental impacts of building construction and operations on future generations through means such as carbon-neutral design, bioclimatic design, and energy efficiency.

Some scholars consider the green building movement is a reaction to the energy crises, to make buildings more efficient while revamp the way energy, water and materials are used. We should note that "green/sustainable architecture" are relatively new; as they represent a whole systems approach incorporating a building's siting, design, construction, and operations in a manner that enhances the well-being of a building's occupants and preserves the environment for future generations by conserving natural resources and safeguarding air and water quality (Kubba, 2016). The key message is, the buildings we build will not only last longer, but also more efficient to operate, and contribute to a healthier living of the occupants and the surrounding environments.

Ragheb et al. (2016) emphasis that green architecture is a theory, science and style of buildings designed and constructed under environmentally friendly principles. Green architecture strives to minimise the number of resources consumed in the building's construction, use and operation, and curtailing the harm done to the environment through the emission, pollution and waste of its components.

Subsequently, Thomas (2009) justified green architecture produces environmental, social and economic benefits. Environmentally, green architecture helps reduce pollution, conserve natural resources and prevent environmental degradation. Economically, it reduces the cost of the building's operations and the cost of using resources (water and energy). Socially, green architecture constructs green buildings that cause minimal strain on the local infrastructure.

Murray-White (2019) explains that green architecture is a sustainable method of green building design: it is designed and constructed with the environment in mind. Green architects working with the key concepts of creating an energy-efficient and environmentally friendly house. Meanwhile Craven (2019) advocates that green architecture is an approach to construct a building that minimises the harmful effects of construction projects on human health and the environment. The green architect or designer attempts to safeguard air, water, and earth by choosing eco-friendly building materials and construction practices. Wines (2019) adds green architecture is a philosophy of architecture that advocates sustainable energy sources, the conservation of energy, the reuse and safety of building materials and the siting of a building with a consideration of its impact on the environment.

## Connecting Green Architecture to Create Green Building

Rinkesh (2018) through his blog provides another insight. He recognises that green architecture is a design technique with the environment in mind, focusing on the larger construction implications of environmental sustainability. It gives careful consideration to energy-efficient design and eco-friendly houses/buildings, while natural ecology is a key concept in green architectural providing the basis for the construction.

The main aim of green architecture is to create buildings that safeguard the natural environment and integrate with the existing environmental setting about space, energy, water and resource consumption. The designs have to be sustainable from the initial consultations, site survey and design, plan modification, material use, harmony with the existing ecology, and choices made to ensure the environmental-friendly criteria are included (Rinkesh, 2018).

Therefore, we summarise that the principle of green architecture is to create environmentally friendly buildings, efficient use of resources with little/no impact on the environment or the occupants' health.

There are three aspects of a green architecture which we can conclude from our literature analysis:

1. Ensuring occupant health: A building is not just a structure that protects the occupants from the external environment, it is also a space where people spend most of their time living in it.
2. Efficient use of resources: From planning, sourcing materials, site selection, designing, construction, operations, maintenance and renovation; all these phases need to consider the efficient use of resources such as energy, water and materials.
3. Preventing pollution and environmental degradation: Production of the building materials, the construction process, the energy used during occupancy, and disposal of construction waste at the end of a building's life should not pollute or create hazardous waste.

From the literature review, there is a gap between the existing academic focus on architecture and vernacular construction. While most of the up-to-date green architecture focusing on the current and future building that aim for transformation, these scholars have overlooked the socioenvironmental value of such traditional building that quietly resided. This knowledge gap drove us to have a new insight when we focus on the traditional Malay houses.

Our motivation is aligned with Guerrero-Baca and Soria-Lopez (2018) latest concept which in recent years, there is a growing concern for the environment and the effects of climate change in the exploration of constructive alternatives that focus on vernacular buildings. Furthermore, Salgin et al. (2017) emphasis, many sustainable architectural designs depend on references to vernacular architecture, and there are many examples found in different parts of the world to which architects can refer. When the world seeks for more sustainable/greener buildings, it is good to revisit the past to understand sustainable features of vernacular architecture since vernacular architecture has knowledge from a sustainability point of view.

## RESEARCH METHODS

We designed this study in an exploratory since green architecture is relatively new and not been widely adopted in Malaysia. Besides, we also take into the consideration of looking the Malay traditional houses found in MMACP and trace for their green architectural criteria, while most of the current and up-to-date research assessing the new/existing buildings (not traditional buildings) to assess its green credential and continue with the categorisation.

We chose Mini Malaysia and ASEAN Cultural Park (MMACP) as the key location of the study since this is an educational theme park, which converses, preserves and exhibits all Malay traditional houses from thirteen states of Malaysia. The complete exhibition of the traditional houses helped easy comparison and to find the similarities/common criteria which we can hypothesize and relate to the green architecture that we intend to investigate.

We visited the MMACP for three times, spent more than 6 hours for each visit, twice, which was guided by the tour guide 1 and tour guide 2 for a short group visit, while the last visit was to affirm the common characteristics of houses from thirteen states in Malaysia that we have synthesized. Through observation, we analysed the data from our visiting field notes recorded, photos taken and also description found on the park's storyboards to understand these houses further. We recorded some explanations from the tour guides minimally since we did not stick to the group visit session as most of them completed the visit shorter than what our time spent.

## DISCUSSION AND ANALYSIS

### Site Selection

Before constructing, a Malay traditional house, the first step starts with the choice of the most favourable site. These traditional houses take on the sensitive eco-balance of the earth and surroundings to prolong the living of the occupants.

The Malay ethics build their houses near to a river (for a water source), next to a seaside (for fishery), or on farmlands (for agriculture). Besides, the availability of underground water is one of the main reasons for houses built around the sources.

Therefore, it is a norm to see, most of the traditional houses are raised on stilts high above the ground and requires stairs to reach the elevated interior. This is to ensure the house is well-protected from damp lands, wild animals' invasion, floods, thieves and to enhance natural ventilation. The space in between the stilts can a parking garage (for a car, motorcycles, bicycles) storage, and to keep calf/pets, a smarter way to create storage space for family members.

*Figure 1. An example of a Malay house*
*Source: Photo taken at Mini Malaysia and ASEAN cultural park (MMACP)*

## Site Design for House Construction

The Malay traditional houses emphasis on maximum natural lighting receiving. Most of the houses are built facing north or south, to tap the maximum daylight under the sunrise-sunset trajectory with minimal heat capture. No artificial lighting is required and no additional energy is required (electric fans) to keep the indoor temperature comfortable.

"With such a setting, Muslims in the Malay house can easily identify Kiblat when praying," explained the Tour Guide 1

Besides, the lands around the houses are planted with flowers, herbs and trees (particularly fruit trees). Not only flowers decorate the landscape, but the trees also keep the house cold under the shading, supply fruits/herbs and oxygen during the daytime. These flora/trees create a natural habitat for some insects and small animals like birds.

## Materials Use for House Construction

Guerrero-Baca and Soria-Lopez (2018) believe that, the historical development of the built environment has originated from continuous processes of trial and error. With the passing of the decades, a certain point of balance was achieved where the cultural development was translated into better systems, adapted to local conditions, which meant an increased and more logical use of available locally sourced materials, and therefore remained as design resources for extended periods of time.

The Malay traditional houses use natural materials such as timber or bamboo, which are locally sourced. Some well-known woods like Cengal-a termite resistant timber and meranti, which promises good workmanship are used to build the main structures of the Malay traditional houses. The main benefit of these natural materials is that they have a lower thermal capacity which does not capture heat for long.

*Figure 2. Climatic design of a Malay house*
*Source: Yuan, 1987*

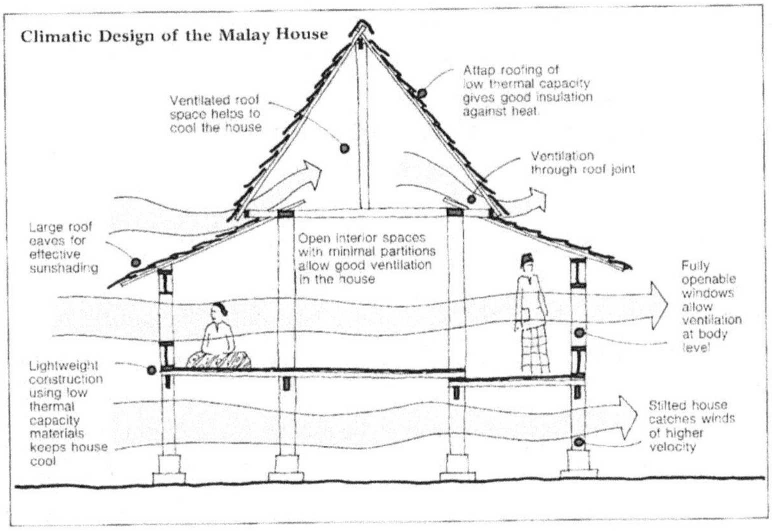


Most of the main structures of the Malay houses are built without nails. Instead, pre-cut holes and grooves are used to fit the joints seamlessly. This is because nails easily got rusted when exposing to the tropical weather and its strengths will gradually decline. Besides, the main structure could be dismantled and reconstructed easily when the repair is required in particular section which damage is found.

Informed by the Tour Guide 2, in the past, the entire house can be carried from one place to another by a group of people when required. The flexibility of reconstructing-ability of the main structure helps the house moving.

## Design of the Tropically Suited Large Roof

Malaysia has an equatorial climate (high temperatures and humidity, heavy rainfall and a climatic year, patterned around the northeast and southwest monsoons). Therefore, the Malay traditional house has a tropically suited large roof, which is wider than the building boundary for effective sun-shading, and the vernacular roof (steep-slopping roof) eases the rain flow.

"Many Malay houses would have a few big jars that collect the rainwater. Collected rainwater could be used for plant watering, washing vehicles (cars, bicycles), washing floors and also cleaning feet before entering the house." With this explanation, we impressed with the concept and use of rain harvesting, which exists nearly a century ago.

Most of the traditional houses' rooftops are supported by wood panels and covered with zinc (a metal which transmit heat quickly). On top of the zinc pieces, dry vegetation (straw, palm branches) or wooden panels are placed. This dry vegetation is lightweight, acting as a thermal insulator. The ceilings of the house are built high and many are uncovered which expose the roof-beams to ensure good ventilation for effective cooling under the hot weather.

## Design of the Doors and Windows

The Malay traditional houses are fitted with many windows and some doors. However, all of them can be closed with wooden panels. This is because, the occupants aim to have good ventilation during the hot day, and also want to keep the room temperature stable at cold/rainy nights.

Most of the windows and doors are fitted with curtains, not only for decoration purpose, but also to limit the daylight (if required), as well as to protect the privacy.

## Design of the Interior Areas

Explained by the Tour Guide 2, traditional houses have two entrances and the guests need to take off their shoes before entering the house. This is a way to show respect, and to ensure the house cleanliness is well taken care of.

"First, the main entrance at the front for male visitors who will gather in the veranda/living room with the male host of the family. Second, the entrance at the back/side of the house which connects to the dinning/kitchen areas are for women and children, told by the Tour Guide 2.

"This back entrance allows the ladies with kids come directly to the kitchen, either helping for some house chores such as cooking; or getting food to children who are visiting," shared by the Tour Guide 1.

Malay houses have large open spaces with minimal partitions. This puts another added value on the ventilation and natural light receives. Kamal et al. (2004) describe, a house is divided into different areas, rather than rooms for various social and household activities.

Most of the traditional houses do not have walls separating the areas. Three main areas-the veranda, main house and kitchen are formed by slight floor level changes (Kamal et al., 2004).

From our observation, all the kitchens of the traditional houses are situated at the back of the house with a few windows that help to channel the smoke out of the house during cooking. Besides, these kitchens have the lowest floor level, to avoid water from getting to the dining areas and helps water flowing out from the kitchens when washing activities are performed.

Next, the veranda is slightly lower than the main house. This is to separate between these two areas, and to avoid dusk/dirt carried by wind and get into the house.

## Maintenance of the House

Shellac, a natural and non-toxic product is used to paint/repaint the wooden panels of the traditional houses. Since the walls or the floors are constructed from individual wood panels, any damaged parts can be replaced specifically without the need to replace the entire section.

## Transformation of Malay Traditional Houses to Modern Malay Houses

Our study echoed Guerrero-Baca and Soria-Lopez (2018) work which, vernacular architecture (in this case the traditional Malay houses) has its richness of knowledge, still alive in the exhibition and practise in some deep rural areas of Malaysia. They are related to the optimisation of the environment and resources in connection to social needs. Its forms and topological features that are the product of accumulated social practices that have been preserved and brought to date for new experiences. It is an understanding that can be taken to face the new social and environmental challenge.

Inevitably, the original Malay traditional houses showed in MMACP and those found in some rural areas have the weakness of durability which requires periodic maintenance compared to modern constructions. However, the modern houses still follow through the similar process of planning, site selection, designing which remain to harmonize with the environment; yet the material sourcing and method of construction have transformed significantly.

Nowadays, to prolong the lifespan of the house, most of the Malay houses found in the rural areas are built/renovated the walls with bricks while the key structure and the foundations are concrete made. Besides, the rooftops are covered by roof tiles. This is to ensure minimal maintenance and the ease of operations, while not looking forward to the deconstruction of a house, because a house is an asset for a family which can be a inherit from one generation to another.

The modern Malay houses have omitted the stilts because of the strength of the concrete foundation which allows double story houses can be directly constructed from the ground. More partitions are erected to separate the areas while windows with glasses are installed. Because of the drastic changes, the natural ventilation of the houses become poorer and rendered fan/air-conditioning needs to be used to generate artificial ventilation which increases the energy consumption and the cost of operations.

## CONCLUSION

Visiting the MMACP theme park which preserves, conserves and exhibits these Malay traditional houses, the educational journey was fruitful, which led us to understand more about the historic, cultural and affirmed our new finding of the green architecture of these traditional houses.

Although the Malay traditional houses have significant weaknesses on the durability (compare with the modern houses which are concrete made), there are many aspects of these houses show the green architecture which we can learn from.

Supported by Kamal et al. (2004), the Malay traditional houses are best reflected by the tropical climatic condition which is hot and humid. Hence, we conclude that the natural ventilation and maximise natural light-receiving are the key concern for Malay traditional house design to reduce energy usage. Next, the natural resources sourcing (wood, bamboo, dry vegetation) ensure the houses take on the sensitive eco-balance of the earth and surroundings to prolong the living of the occupants. The houses harmonise with the environment, and they have practised rain harvesting with no environmentalists to campaign for this purpose.

We relate these traditional houses with the GBI available. Admittedly, the houses ticked many boxes of (i) energy efficiency, (ii) indoor environmental quality, (iii) sustainable site planning and management, (iv) materials and resources, (v) water efficiency and (vi) innovation. Although we could not professionally put a score on these assessment criteria, we can conclude that the Malay traditional houses are green buildings that portray green architecture.

In today's environmental crisis, we learned to use more energy-saving technologies (such as LED, air-conditioning and others) at home as our obligation for being environmentally friendly. Yet, the energy saving paradox takes place because people use more on these energy-efficient technologies that directly causing higher energy consumption. These energy-saving technologies (such as LED) comprise the rare earth materials which are highly polluting of their production even though it promises the energy saving benefits at its utilisation later. This is an example of how a solution to an existing problem causes more problems, or nullify the solution.

The new challenge that we are facing is, these artificial-energy-saving technologies have foreclosed other alternative solutions. Through advertisements and promotions, we forgot that we have the ultimate option to reduce the usage of these technologies instead, to be eco-friendly.

These traditional houses in many academic realms, used to be associated with the sociocultural of the inhabitants. However, this study, which is angled from the socioenvironmental lenses, opened up a fresh insight of green architecture-not only the inhabitants who have made their choice of the houses, but also those who conceptualise and built these traditional houses. From this study, we appreciate our forefathers' wisdom, to build the houses that harmonise with the environment. Besides, we also affirm, being minimalist and being simple in life, are better ways towards being green and sustainable.

## REFERENCES

Craven, J. (2019). *A Primer on Green Architecture and Green Design.* ThoughtCo. Retrieved from https://www.thoughtco.com/what-is-green-architecture-and-green-design-177955

GBI. (2019). *What and Why Green Buildings?* Retrieved from https://new.greenbuildingindex.org/whatandwhy

Guerrero Baca, L., & Soria López, F. (2018). Traditional Architecture and Sustainable Conservation. *Journal of Cultural Heritage Management and Sustainable Development*, *8*(2), 194–206. doi:10.1108/JCHMSD-06-2017-0036

Kamal, K. S., Wahab, L. A., & Ahmad, A. C. (2004). Climate Design of the Traditional Malay House to Meet the Requirement of Modern Living. *Proceedings of the 38th Conference of Architectural Science Association (ANZAScA).* Context of Architecture. Academic Press.

Kubba, S. (2016). *Handbook of Green Building Design and Construction. LEEDS, BREEAM and Green Globes.* Butterworth-Heinemann.

Murray-White, J. (2019). *What is Green Architecture? Retrieved from: Updated on 15 April 2019.* Sustainable Build.

National Architectural Accreditation Board. (2009). *Conditions for Accreditation.* Washington, DC: National Architectural Accrediting Board.

Ragheb, A., El-Shimy, H., & Ragheb, G. (2016). Green Architecture: A Concept of Sustainability. *Procedia: Social and Behavioral Sciences, 216*, 778–787.

Rinkesh. (2018). *What is Green Architecture?* Retrieved from https://www.conserve-energy-future.com/materials-used-and-principles-green-architecture.php

Salgın, B., Bayram, Ö. F., Akgün, A., & Agyekum, K. (2017). Sustainable Features of Vernacular Architecture: Housing of Eastern Black Sea Region as a Case Study. *Arts*, *6*(11), 1–14.

Thomas, R. (2009). The Principles of Green Building Design,

U.S. Environmental Protection Agency. (2016). Green Building. Retrieved from https://archive.epa.gov/greenbuilding/web/html/

Wines, J. (2019). Green Architecture. Encyclopedia Britannia. Retrieved from https://www.britannica.com/art/green-architecture

Yuan, L. J. (1987). *The Malay House: Rediscovering Malaysia's Indigenous Shelter System*. Penang: Institute Masyarakat.

# Chapter 8
# Open Innovation Challenges and Coopetition-Based Open-Innovation Empirical Evidence From Malaysia

**Jawad Iqbal**
*The Islamia University of Bahawalpur, Pakistan*

**Waseem Ul Hameed**
*The Islamia University of Bahawalpur, Pakistan*

## ABSTRACT

*Innovation trends are a highly competitive environment, and have been changed and companies are moving towards the open innovation model rather than to follow close or in a traditional innovation model. Therefore, this chapter demonstrated various determinants of open innovation. For this purpose, a survey was carried out among Malaysian small and medium-sized enterprises (SMEs). The outcomes of the survey highlighted that, the success of open innovation model is based on five major elements, namely, 1) motivating spillovers, 2) incorporation of external knowledge, 3) intellectual property management, 4) maximization of internal innovation, and 5) financial constraints. These five elements are the major challenges for companies while adopting open innovation model. More specifically, the phenomenon of coopetition-based open-innovation is emerging rapidly among the companies. Nowadays, by following the open innovation activities, competitors are collaborating with each other rather than to compete.*

## INTRODUCTION

The innovation trend has been changed and companies are adopting open innovation model rather than to stick with close innovation model (Hameed & Altaf, 2019). Traditionally, the companies were following close innovation model (Hameed, Basheer, Iqbal, Anwar, & Ahmad, 2018). A traditional approach which based on ownership as well as control, this approach also takes a linear approach in which organization

DOI: 10.4018/978-1-7998-1566-2.ch008

*Figure 1. Closed innovation Model*
*Source: Chesbrough (2012)*

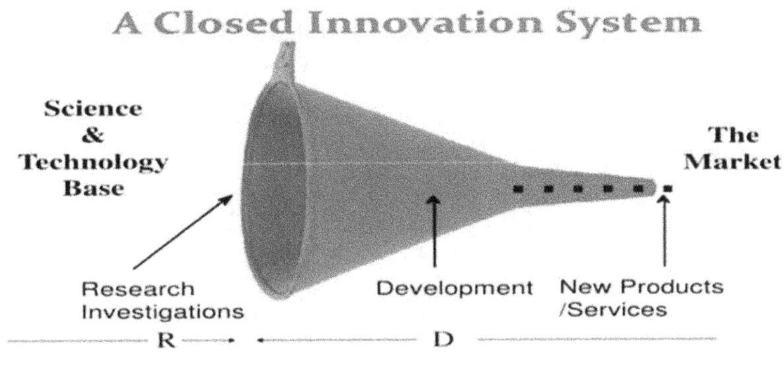


just depends on its internal competences. For example, for the sake of idea generation and value creation, all activities like research and development activities and all practices are according to the innovation project (Lichtenthaler, 2008). More traditionally, in an organization, scientists bring new ideas; develop new products according to the demand of customers remaining within the boundaries of the firm but they never look outside the boundaries of the firm for new idea generation (Conboy & Morgan, 2011). Therefore, it can be further described as, close innovation is one of the isolated process in which a small group of people and certain individuals involve in an idea generation and value creation process which only depends on internal ability of this group or individuals and they do not look outside the firm. Idea generation, production as well as commercialization of innovation, firms only use own channels of distribution like a close innovation model (Bogers, Burcharth, & Chesbrough, 2019). H. Chesbrough (2012) challenging close innovation model explains that all the projects of research depend on the internal knowledge of the firm due to which some projects could not reached to the maturity whereas few projects which seems to be value creating would be selected for future investigation. According to the Figure 1, entry and exit of any project explains the description of close or open model. When a project enters or exit in one way it will be the example of close innovation model, furthermore, when a project introduces in market by just one way, it needs to be defended against competition with the help of intellectual property management and it is the example of close innovation model (See Figure 1).

It is proved from the research that innovation tendency has been replaced; now companies are working differently on commercialization considering that the horizons of companies are turning into permeable (Trott & Hartmann, 2009) and it is noticed that a lot of independent bodies are attending innovation process like client, retailer, suppliers, analysis-based associations, challengers, etc. (Nandonde, 2019; Wallin & Krogh, 2010). As very large ton of knowledge is binding to deliver creativity which can bring fresh ideas and close innovation model which is one of the authoritative as well as individualistic process of innovation, transfer towards a new model of innovation which highlights collaboration with external partners and commercialization of new idea, here the point of departure between two models is collaboration outsider like client, retailer, suppliers, associations based on analysis, challengers etc.

In contrast to the close model of innovation, knowledge as well as ideas cannot flow (inter or exit) through a single way but it flows through different ways like external players, partnership, internal research investigation etc. Transfer of technology can be achieved with the help of missing knowledge from outside of the firm and the knowledge of the employees within the firm like expertise of the employees

*Figure 2. The Open-Innovation Model*
*Source: Chesbrough (2012)*

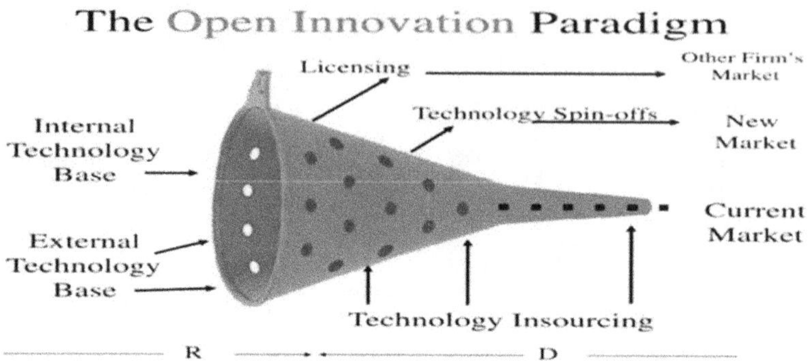


within firm and it can approach to the market with different ways: venturing, outsourcing, with the help of company own channels, joint ventures etc. (Chesbrough, 2012). Open innovation is different from close innovation in which organizations generate their own ideas and then build, develop, market, distribute, finance and support with the help of own internal applications. As described by the experts that the traditional innovation model or closed innovation model is shifted to open innovation model with the help of internal research and development (Chesbrough, 2003; Enkel, Gassmann, & Chesbrough, 2009). Therefore, it can be further explained as, open innovation comes into existence when the close model shifted to open innovation model. When a project enters or exit by two ways it will be the example of open innovation model (Chesbrough, 2012) as shown in Figure 2.

From the model of Chesbrough, it can be described as open innovation is one of the orientations to commercialize internal ideas as well as external ideas. Therefore, it is shift of "not invented here" syndrome to "proud to found everywhere." Use of internal knowledge of the firm to outside the boundaries of the firm is known as outbound process and it is needed to protect this internal knowledge by the help of intellectual property rights because external bodies can use this knowledge for their own interest, even the employees of the firm can use this knowledge for their own purpose (Bogers, Chesbrough, & Moedas, 2018; Brunswicker & Chesbrough, 2018; Chesbrough, 2003). Therefore, it can be further elaborated as, it is better for companies to selling of this knowledge to other firms which is one of the better ways to utilize internal knowledge and it leads to open model as compared to the close model.

Open innovation is the "use of purposive inflows and outflows of knowledge to accelerate innovation inside the firm, and expand the markets for external use of innovation, respectively" (H. Chesbrough, 2003). From the above discussion it can be further defined as, close innovation is one of the process in which firms develop new ideas remaining within the boundaries of the firm, implement these ideas into the product and bring a cheaper product in the market and generate profit. On the other hand, open innovation is one of the process in which firms generate new ideas by the help of internal capabilities as well as external knowledge and then commercialize these ideas by licensing it properly.

However, open innovation adoption is one of the challenges for small and medium size enterprises (SEMs). Particularly, the Malaysian SMEs are facing various issues to adopt open innovation (Hameed, Nadeem, Azeem, Aljumah, & Adeyemi, 2018). The major challenges facing by the companies include; motivating spillovers, external knowledge incorporation, internal innovation, intellectual property management and financial constraint (Hameed & Altaf, 2019). Additionally, companies are involved in coopetition activities in which trust and dependency is imported to carry out open innovation activities (Hameed & Naveed, 2019). Therefore, this chapter demonstrated various determinants of open innovation. Hence, objective of the study is to investigate the of determinants of open innovation.

## LITERATURE REVIEW

Open innovation is the use of targeted objectives and discharge of expertise to advance internal innovation and grow the advertisement market for extraneous use of modernization, commonly. Open innovation is the both outside-in and inside-out deportation of automation technologies and conceptual ideas (Hameed et al., 2018). It is the exemplar which companies can and should use out-side and in-side ideas, use both external and internal method to the advertisement market (Hameed, Altaf, & Ahmed, 2019). Open innovation is the way in which subjective knowledge as well extraneous knowledge get together to design something new. However, successful open innovation activities require certain important sections such as motivating spillovers, external knowledge incorporation, internal innovation, intellectual property management and financial constraint. All the sections are discussed in detail.

### Motivation Spillovers

Motivation is the state of being incited to action (Armstrong, 2019). It is one of the psychological process resulting from reciprocal action among two things, one is individual other is environment, this relation effects individual's choices, efforts of individual and persistence of individual (Latham & Ernst, 2006). In other words, if the goals of an individual achieved, he/she will be motivated but if the goals are not achieved, he/she will not de motivates. So, it is clear that people do something better if they believe that they can achieve their desired results. Desire results mean the personal goals of an individual like choice, promotions, rewards, appreciations etc. Individuals who are well motivated always take actions and believe that desired results will achieve. However, in Malaysian SMEs, it is one of the critical issues in the way of open innovation.

There is a direct relationship between motivation and performance. Therefore, it can be described as there is a direct relationship between motivation and innovation performance. If the people are motivated, they will innovate new things, they will bring new ideas and create value but if they are not motivated, they seem to me lazy and their performance relating to open innovation will be low.

Open innovation study assumes that there should be increase in external source of innovation, external sources include persons and any entity, and it is necessary to motivate them for the further innovation (West & Gallagher, 2006). From this, a question rises, how we can motivate them for better innovation? Incentive problem is related to the creation and value capturing of ideas and knowledge of inventors (López-Nicolás & Meroño-Cerdán, 2011). Above statement explains that innovation is related to the individuals who bring new ideas and there is a relation between motivation and innovation. Therefore, if we make better motivation system then we can improve the performance of people who generate new ideas, and it is one of the challenges of open innovation. According to the author, incentive problem includes, problem of left ideas inside the firm and team production problem.

Cooperation or collaboration among firms in innovation projects can face team production problem (Wiltshire, Steffensen, & Fiore, 2019). Author argues that it is difficult to measure the contribution of each individual in overall output and it is costly, therefore, different participants do not contribute enough to the overall task, because these participants are engage in some other interesting projects, thus, this situation decreases the value of project but it can be overcome with the help of motivation. Firms which are engage in innovation projects should have an effective motivation system to overcome free riding problems.

According to Chesbrough (2003), many ideas do not reach to the market and left within the firm because other ideas gain more intention for commercialization, therefore, this situation create incentives for the employees to pay more intention and develop these ideas. Due to which parent company cannot gain advantage from their own ideas but other companies can do it. Franco and Filson (2006) investigate about spin-out process and say that spin-out process is a source of technology diffusion which is increasing in technological companies. Hence, we can say that when employees will not be motivated, they will not pay intention to the ideas then ideas will not go to the market for commercialization and other companies will use these ideas to complete with original firm. Therefore, it can be described, motivation or incentive problem is one of the important challenges of open innovation and motivation system will also increase the overall cost.

There are many theories are present on motivation which shows the relationship of motivation and performance, and many researchers start their work on motivation, use different known theories as well as new models which support the argument that motivation increases the performance of individuals. Different theories in his book related to rewards of employees. According to the Armstrong and Taylor (2014), Taylor's theory which is related to the rewards and penalties of employees described that there is a relationship between motivation and performance of employees. Taylòr said that rewards motivate people and increase their performance. Maslow's also introduce the hierarchy of need in which he explains some basic needs of people related to motivation. When these needs are fulfilling, people become motivated otherwise they will not be motivated, and their performance will decrease. Armstrong explain some modern theories, for example, Vroom's expectancy theory says that motivation is only achieved when there will be a relationship between performance and outcome. When the employees realize that if they perform, they will achieve their outcome, they will be motivated, otherwise, they will be de motivated. Outcome means the expectations of the employee's like promotion, rewards, benefits, increments etc., like all personal desires. Therefore, it is clear from these theories that motivation is directly related to the performance. So, if we motivate the factors that enhance the innovation, we can make better innovation practices and can overcome this challenge of open innovation.

When people feel that, their efforts will be given intention and will be rewarded then they give their maximum. According to the author, human behaviour is the reflection of their goals. People who are motivated found to be more autonomy-oriented and also more self-driven as compared to the people who are less motivated. Therefore, motivation increases the performance. Finally, it can be described that to promote open innovation; it is needed to motivate the factors which maximize open innovation. Therefore, there is a direct relation between motivation and open innovation.

## External Knowledge Incorporation

Unwillingness of employees to utilize the extra-organizational knowledge is one of the important challenges of open innovation (Chesbrough, 2006). Employees of an organization have not positive attitude towards the utilization of knowledge which is outside the boundaries of an organization like not invented here syndrome (NIH) and employees have also negative attitude towards commercialization of ideas. According to the author, employees do not want to bring external knowledge inside the boundaries of the firm which is one of the main challenges to bring new ideas for promotion of open innovation. According to the definition of open innovation, it is the process in which both internal and external knowledge combine to bring something new. Therefore, it is clear that external knowledge incorporation is positively correlated with open innovation (Pollok, Lüttgens, & Piller, 2018) but from the above discussion, negative attitude toward external knowledge shows that it will not reach inside the boundaries of the firm and open innovation process will not complete. So, external knowledge incorporation is one of the main challenges of open innovation. The better utilization of external knowledge many lead to the higher open innovation performance among Malaysian SMEs.

External knowledge cannot provide any benefit to the firm if the firm cannot sort out the relevant knowledge and cannot incorporate in the innovation activities of the firm (West & Gallagher, 2006). Coordination problem is a problem in which ideas and knowledge lie outside the boundaries of the firm (López-Nicolás & Meroño-Cerdán, 2011). Coordination or incorporation of external knowledge is not limited to the mechanism among activities of organization, but it also includes the searching of ideas, selecting of ideas as well as knowledge to carry out innovation process, and it includes; problem of searching of valuable ideas, problem of networking, problem of divergence. External source of idea searching and knowledge to make better technological development is very first problem of open innovation. On the other hand, coordination requires cost because it requires corporation of multiple suppliers. Therefore, the better incorporation of external knowledge is a solution if different open innovation problems and improves the performance of open innovation system by resolving different problems of open innovation system.

Technological knowledge is not sufficient to gain competitive advantage; however, it comes from coordination/external knowledge incorporation. In this sense, coordination/external knowledge incorporation is one of the important elements of open innovation. Almirall and Casadesus-Masanell (2010) said that divergence is another problem related to coordination. Author argues that some choices which could have been made by the original designer are now undertaken by the different independent firms for their own interest due to the openness of system for the supplier and this loses the control of original designer but restraining of this freedom is much more costly. In this way supplier and complement or maximize their own interest as compared to the interest of original designer. Hence, we can further explain that better incorporation of external knowledge can resolve divergence problem which is related to the open innovation system. That is why; incorporation of external knowledge is positively related to open innovation system and improves the open innovation system.

This challenge of external knowledge incorporation also creates another problem which is related to the top management. Negative attitude of employees create misunderstanding among top management and employees of the firm. According to the author, not all the firms have the capability to absorb the external knowledge because every company has its own culture. Therefore, it is true that external knowledge leads open innovation if it manages properly.

Finally, better incorporation of external knowledge is much more important for open innovation system. Failure in incorporation of external knowledge can affect badly on firm as well as can delay the overall project and may causes financial losses. Therefore, we can say that during open innovation launching, management of external knowledge incorporation is one of the important challenges.

## Intellectual Property (IP) Management

Intellectual property (IP) has significant role in organizations (Amoroso & Link, 2019; Uzuegbunam, Ofem, & Nambisan, 2019; Wiwchar, Walker, & Marsh, 2020; Yadav, Shukla, & Gupta, 2019). The possession of knowledge, experience, organizational technology, relationships with customer as well as professional skills that provide a competitive edge in the market is known as intellectual property (IP). Intellectual property (IP) is one of the assets which secures the commercial success of an innovation (Von Zedtwitz, 2004). On the other hand, management of intellectual property is the problem of open innovation which faces by companies. Imitation and devaluation of idea are the challenges which rise due to the open innovation. In this sense, if intellectual property manages properly, it will become the asset but if does not manage it properly, it will become the weakness that is why intellectual property (IP) management is a challenge of open innovation. Particularly Malaysian SMEs are not focusing to manage their intellectual property which is leading to the low open innovation performance.

Intellectual property (IP) has a vital role in open innovation process due to which inflows and outflows of knowledge. It is an opportunity for the enterprises to out license their intellectual property and obtain more benefits from it but as it is discussed above, management of intellectual property is though. If information managed properly, firms cannot take benefit from it, consequently, management of intellectual property (IP) is important as well as it is a challenge of open innovation.

Inside-out is a process in which companies earn additional profit by selling their intellectual property to outside environment; they transfer their ideas to other companies outside the boundaries of firm. This process also studied by the Lichtenthaler (2008); Lichtenthaler and Lichtenthaler (2009) at operational as well as strategic level and said that inside out process (to sell IP outside the boundaries of the firm) is associated with outbound technology transfer. Therefore, further it can be explained that proper management of intellectual property is important because by managing properly companies can earn profit by selling it to the other companies by the help of licensing. Thus, it is clear that management of intellectual leads open innovation, but it is needed to manage properly. Intellectual property rights and open innovation has special connection with each other, and intellectual property rights has important role in open innovation practices which is studied by the help of licensing. There is a close relationship among licensing and innovation process, also in-licensing as well as out-licensing are connected with each other because companies cannot fulfil all needs by remaining inside the boundaries of the firm, they need to look outside the boundaries of the firm, they need to get technologies from outside (known as in-licensing process) to generate something new and sell this new idea to other firms outside the boundaries of the firm (known as out-licensing process) and generate profit. These processes are interconnected and promote open innovation. It can be further explained that proper management of intellectual property is one of the assets which can bring more value to the firm by out-licensing it and it is also one of the ways to enter in international market.

"Openness" is the nature of open innovation, but firms believe that Intellectual property rights are enough to protect their innovation capabilities (Hagedoorn & Ridder, 2012). On the other hand, intellectual property right may threaten the open innovation as wide range knowledge associability as well as technology is critical part of open innovation. Due to the open nature of open innovation, wide spread of knowledge, information and technology, it is much tough to protect them but can be manageable with the help of licensing and when it managed properly, it improves the open innovation system.

Proper management of intellectual property promote open innovation and it is manageable by the help of licensing which is one of the formal ways to protect intellectual property. Intellectual property decreases the ability of competitor to innovate and gain competitive advantage. Strong intellectual property right protection enhances trade in technology related market (Chesbrough, 2006) and it is one of the bases of "commodification" as well as transfer of technology. Strong intellectual property protection keeps safe companies from imitation and there is a direct relation among intellectual property and product innovation, more intellectual property more will be the product innovation and vice versa. Strong intellectual property protection encourages companies to invest in identification of new ideas, development of new ideas and commercialization of new ideas. Finally, intellectual property management promotes open innovation. Finally, if researcher says that management of intellectual property (IP) is one of the important challenges of open innovation, this will not be wrong but if companies properly manage intellectual property (IP) then they can take much benefit from it in the form of open innovation system.

## Maximizing Internal Innovations

How to best use internal research and development capabilities to gain maximum advantage is the central concern to open innovation (Kapetaniou & Lee, 2019; West & Gallagher, 2006). To feed product pipeline line of company is not enough to gain maximum advantage from internal innovation but its need's wide range approaches. Therefore, if we focus on quantity of product and just try to meet the demand of market by producing more products then this will not provide us maximum advantage from internal innovation, therefore, we need to introduce new methods to promote internal innovation and to enhance the open innovation system which is a challenge. In past, number of industries focused on internal innovation, but other industries focused on co-innovation process.

Venturing in small and medium size enterprises can enhance the internal innovation. Author said that venturing is a process of starting up new organization on the basis of internal knowledge, including spin-out and spin-off process as well the support from parent organization in the form of finance, human capital, valuable advices and administration. Hence, it is clear that maximization of open innovation improves the open innovation system in form of venturing. Potential of venturing activities covered wide range, for instance, total market value of eleven projects which changed into new ventures is more than that of their parent company. Therefore, maximization of internal innovation can enhance the open innovation system. If the internal innovation increases, open innovation system also improves and vice versa.

Communication is one of the key factors to increase employee innovation, both outside as well as inside communication is important for employee innovation. Furthermore, ideas only generate when employee's communication with each other and share their ideas with each other, that's why communication among employees of firm is much beneficial for innovation. Leaders can enhance internal communication among employees by using different strategies as well as different channels. Strategies which increases communication among employees and generate creative thinking are "cross-department meeting" in which different department employees meet with each other and share their ideas, "informal lunch sessions hosted by a named" in which employees gather on lunch and share their views with complete freedom, "chief creative officer," in which chief creative officer of the company announces a meeting in which all employees take part and formally discuss all ideas one by one, and best idea chose to implement, other strategy include unstructured meetings which encourage employees. These strategies like arrangement of meetings increase the overall cost. Therefore, communication is one of the expensive process.

## Financial Constraint

What is the role of financial constraint in the management of open innovation challenge? Open innovation has widespread application, but companies face tough time in implementation due to the financial constraints. Lack of finance is one of the constraints of open innovation. Therefore, we can say that finance has an important role in implementation as well as management of open innovation challenges. Moreover, research and development (R & D) in innovation is considered key to productivity and growth performance but it is expensive process.

Essentially, R&D is the core as well as crucial part of openness of concepts as well as also one of the assets as an extraneous character (Chesbrough, 2003). Since very beyond research and development also described to be essential, by following the Cohen and Levinthal (1989) research and development acts binal aspect in the companies evolution, first of all, it establish the association internally by developing new concepts within the limitations of the firm and then it develops the capability to bibulous size to set as well as to inquiry evolution extraneous from the limitations of the association. It is also noticed that the companies which spent large amount in R & D in relationship to other companies gain more profit.

As discussed above, motivating spillovers; it needs incentives to motivate, intellectual property management; need licensing, maximization of internal innovation; need communication, and incorporation of external knowledge; needs coordination. Again, financial constraint rises in the form of incentives to motivate, to license the intellectual property (IP), communication to maximize internal innovation and coordination for external knowledge incorporation and R & D is necessary condition for open innovation which also need enough finance. External knowledge comes into the boundaries of firm when there is collaboration between firm and external partners as well as suppliers which is one of the expensive process. Collaboration with partners is one of the expensive process. Intellectual property management is one of the costly as well as complex process. Intellectual property rights promote open innovation, but it is not possible for all firms to license its all new ideas because it is one of the expensive process. Increase in intellectual property management, increase the overall cost. Therefore, it be described as, intellectual property significantly improves open innovation, but financial constraint disturbs this relation.

## Coopetition-Based Open-Innovation

Coopetition has drawn an extraordinary attention as it joins two various or antithesis ideas: combination and competition. Coopetition is best considered as—sleeping with the enemy. It is a momentous plan which promises major asset in the range of open innovation. The obligation in different tasks of so-called coopetition climate, essentially depends upon telecommunications in which participation as well as competition occur altogether (Bengtsson & Kock, 2000), normally, can outcome in different antique knowledge-related proceedings among associations. These knowledge-related actions show essential point in the method of open innovation. Because open innovation is binal agenda where extraneous awareness outside the association get in inside the organization and move outside in frame of closing idea.

None the less, coopetition demands confidence on competitor and dependency on each and every one. Trust has an urgent aspect in coopetition anomaly. Different participants taking part in coopetition compete danger and deficiency of trust that cut-down performance and different innovation actions. Dependency is fairly urgent with trust and important act in trust (Bouncken & Fredrich, 2012). At many occasions, capable or big companies design the structure of dependency when they have their owned assets and turn more capable. Accordingly, trust and dependency show a arbitrate role between coopetition and company achievement.

## Trust

Trust is the more important element of each coopetition action (Kraus, Klimas, Gast, & Stephan, 2019). It is described as developing chunk of relationship actions. The factor of trust between challengers is real achievement of coopetition actions. Because coopetition is not only designed on various kinds of challenges, but also on different actions of affiliation. Accordingly, trust has important relationship with coopetition which response the achievement of companies.

Communal force like trust ascribe with different fiscal burdens by effecting importantly on the achievement by a company. Like the open innovation is depending upon both internal and extraneous belongings and coopetition is also one of the methods to employ extraneous assets. Thus, trust as one of the aspects of coopetition has an important act in an open innovation. A study on Malaysian SMEs by W. Hameed and Naveed (2019) also revealed the fact that trust is significant element of coopetition-based open innovation.

## Dependency

Dependency is also equitably significant with trust. Dependency has an important play in coopetition accomplishment. It is illustrated by resource dependency theory, company exchange their resources like knowledge and depend other 's (van den Broek, Boselie, & Paauwe, 2018). Matched with resource dependence theory, powerful companies design dependency conditions when they get the authority of concerned assets and have control over other companies which result on open innovation movements. Accordingly, dependency in coopetition is most an important part of open innovation activities.

In addition, dependency disturbs the contact between the affiliated actions of business and coopetition (Bouncken & Fredrich, 2012). In the practice of coopetition, different challenging companies depending on one and other 's to enhance the action of open innovation. It is one of the important elements in coopetition achievement. So, in alignment with trust, dependency has also essential significance for coopetition achievement. Research demonstrated that dependency fundamental part of an open innovation which is depends on coopetition.

## Hypotheses Development

It is evident from various studies that firm's open innovation performance (FOIP) has many determinants. According to West and Gallagher (2006), motivating spillovers has significant role in open innovation activities. Hameed et al. (2019) found that motivating spillovers has positive effect on FOIP. More the motivation, more will be the open innovation success. Hameed et al. (2018) investigated that, external knowledge and internal innovation are the major determinants of FOIP. Both internal innovation and

*Figure 3. Framework of the chapter showing the relationship between open innovation challenges, coopetition based open innovation and firm's open innovation performance*

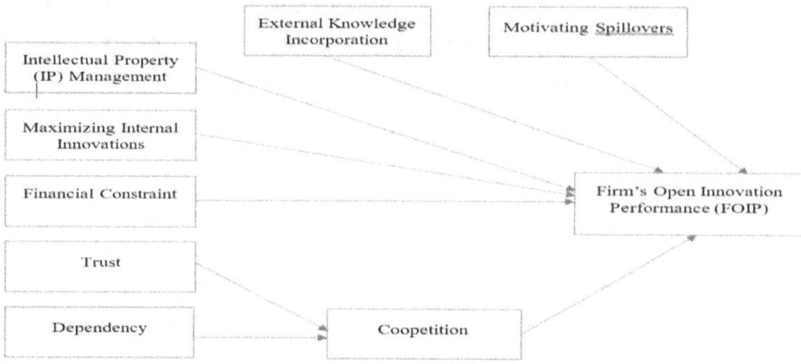


external knowledge has positive relationship with FOIP (Hameed et al., 2018). Increase in external knowledge and internal innovation has positive effect on FOIP (Brunswicker & Vanhaverbeke, 2015; Del Vecchio, Secundo, Rubino, Garzoni, & Vrontis, 2019). Therefore, incorporation of knowledge outside the boundaries of the firm and increase in internal innovation influence positively on FOIP. Moreover, while open innovation activities, intellectual property management has major role. Patents and copyrights are required to protect the ideas which shows positive role in FOIP. As previous studies highlighted the intellectual property has positive effect on FOIP (Davoudi et al., 2018; Schönhals, Hepp, & Gipp, 2018; Tang, Tietze, & Molloy, 2019; Toma, Secundo, & Passiante, 2018). Additionally, while adopting open innovation, companies need R & D department, which requires more finance. Various studies mentioned that financial constraint has significant relationship with open innovation (W. Hameed & Altaf, 2019). Essentially, R&D is the core and crucial part of openness of concepts as well as also one of the assets as an extraneous character (Chesbrough, 2003) which shows positive effect on FOIP. In addition, coopetition has positive role to enhance FOIP which is only based on trust and dependency (Hameed & Naveed, 2019). Better the level of trust and dependency, better will be the performance of coopetition and open innovation. Therefore, from the above discussion, Figure 3 and hypotheses are proposed.

**Hypothesis 1.** Motivating spillovers has positive effect on FOIP.
**Hypothesis 2.** External knowledge incorporation has positive effect on FOIP.
**Hypothesis 3.** Intellectual property management has positive effect on FOIP.
**Hypothesis 4.** Maximizing internal innovation has positive effect on FOIP.
**Hypothesis 5.** Financial constraint has negative effect on FOIP.
**Hypothesis 6.** Trust has positive effect on coopetition.
**Hypothesis 7.** Dependency has positive effect on coopetition.
**Hypothesis 8.** Coopetition has positive effect on FOIP.

*Figure 4. Confirmatory factor analysis (CFA)*

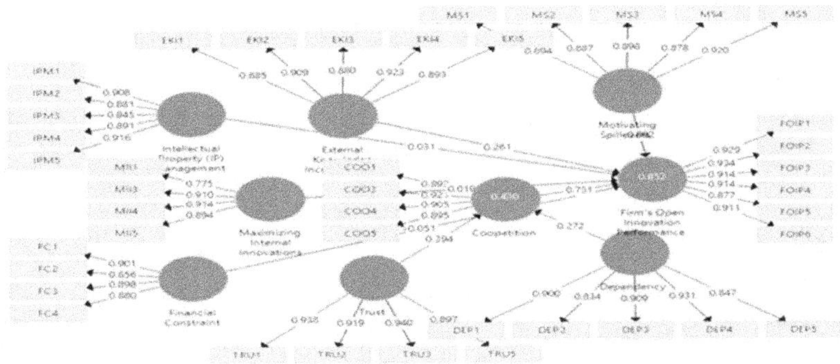

## RESEARCH METHOD

This portion of the study deals with method to attain end results based on empirical evidences. Based on above discussion on open innovation challenges and coopetition based open innovation, a framework was development. It is shown in Figure 3. Based on the framework, eight hypotheses were proposed. Manufacturing SMEs were selected to collect the data.

This study carried out a research survey to collected primary data. A survey questionnaire was developed by using the previous studies. Scale items for external knowledge, internal innovation and FOIP were adopted from Hameed et al. (2018). Scale items for motivating spillovers, intellectual property management and financial constraint were adapted from Hameed and Altaf (2019). Finally, scale items for coopetition, trust and dependency were adopted from Hameed and Naveed (2019). Various close ended questions related to the above discussion on open innovation challenges and coopetition based open innovation were asked on a seven-point Likert scale.

Population of the study were based on the manufacturing SMEs of Malaysia. Managerial employees of these SMEs were selected to collect the data. Therefore, respondents of the study were managerial employees of Malaysian manufacturing SMEs. Thus, unit of analysis was individual. Comrey and Lee (1992) provided sample in a series of inferential statistics. "Sample having less than 50 participants will observed to be a weaker sample; sample of 100 size will be weak; 200 will be adequate; sample of 300 will be considered as good; 500 very good whereas 1000 will be excellent." Therefore, four hundred (400) sample size was selected. Four hundred (400) questionnaires were distributed among the Malaysian SMEs.

## DATA ANALYSIS

Findings of the study is based on the primary data as discussed in above (section 3). Data screening is shown in appendix (Table 1) in which missing value, outlier and normality was examined. Primary data were analysed by using statistical software, namely; Partial Least Square (PLS) version 3. Structural Equation Modeling (SEM) was used to obtain results. PLS-SEM is based on 2 major steps, 1) measurement model, and 2) structural model. Figure 4 shows confirmatory factor analysis (CFA) or measurement model which highlighted that all the items have factor loadings above 0.7. Factor loading is also highlighted in appendix (Table 2). Composite reliability (CR), Cronbach alpha and average variance

*Table 1. Construct reliability and validity*

| | Alpha | rho_A | CR | AVE |
|---|---|---|---|---|
| Coopetition | 0.925 | 0.927 | 0.947 | 0.816 |
| Dependency | 0.93 | 0.933 | 0.947 | 0.783 |
| External Knowledge Incorporation | 0.94 | 0.942 | 0.954 | 0.807 |
| Financial Constraint | 0.907 | 0.909 | 0.935 | 0.781 |
| Firm's Open Innovation Performance | 0.96 | 0.961 | 0.968 | 0.834 |
| Intellectual Property (IP) Management | 0.933 | 0.934 | 0.949 | 0.789 |
| Maximizing Internal Innovations | 0.896 | 0.898 | 0.929 | 0.766 |
| Motivating Spillovers | 0.938 | 0.939 | 0.953 | 0.802 |
| Trust | 0.942 | 0.944 | 0.959 | 0.853 |

Note: MS = Motivating Spillovers, EKI = External Knowledge Incorporation, IPM = Intellectual Property Management, MII = Maximization of Internal Innovation, FC = Financial Constraints, TRU = Trust, DEP = Dependency, COO = Coopetition, FOIP = Firm's Open Innovation Performance

*Table 2. Discriminant validity*

| | COO | DEP | EKI | FC | FOIP | IPM | MII | MS | TRU |
|---|---|---|---|---|---|---|---|---|---|
| Coopetition | 0.904 | | | | | | | | |
| Dependency | 0.641 | 0.885 | | | | | | | |
| External Knowledge Incorporation | 0.618 | 0.695 | 0.898 | | | | | | |
| Financial Constraint | 0.634 | 0.792 | 0.671 | 0.884 | | | | | |
| Firm's Open Innovation Performance | 0.702 | 0.706 | 0.707 | 0.666 | 0.913 | | | | |
| Intellectual Property (IP) Management | 0.654 | 0.743 | 0.728 | 0.723 | 0.72 | 0.888 | | | |
| Maximizing Internal Innovations | 0.669 | 0.778 | 0.768 | 0.708 | 0.707 | 0.791 | 0.875 | | |
| Motivating Spillovers | 0.61 | 0.685 | 0.707 | 0.688 | 0.668 | 0.703 | 0.787 | 0.896 | |
| Trust | 0.649 | 0.737 | 0.721 | 0.714 | 0.694 | 0.769 | 0.701 | 0.7 | 0.924 |

Note: MS = Motivating Spillovers, EKI = External Knowledge Incorporation, IPM = Intellectual Property Management, MII = Maximization of Internal Innovation, FC = Financial Constraints, TRU = Trust, DEP = Dependency, COO = Coopetition, FOIP = Firm's Open Innovation Performance

extracted (AVE) is also above 0.7, 0.7 and 0.5 respectively. It is highlighted in Table 1. Additionally, discriminant validity is presented in Table 2. Figure 4 shows the PLS measurement model.

Figure 5 demonstrated the PLS-structural model. Hypotheses which were development in section 3 were tested by examining the t-value and beta value. The hypotheses with t-value above 1.96 were supported, however, the hypotheses having t-value below 1.96 were not supported. Table 3 highlighted the results of hypotheses which shows that all the hypotheses have t-value above 1.96. Thus, all the hypotheses are supported. Additionally, seven hypotheses show positive relationship. However, the relationship between financial constraint and FOIP is negative.

*Figure 5. Structural model*

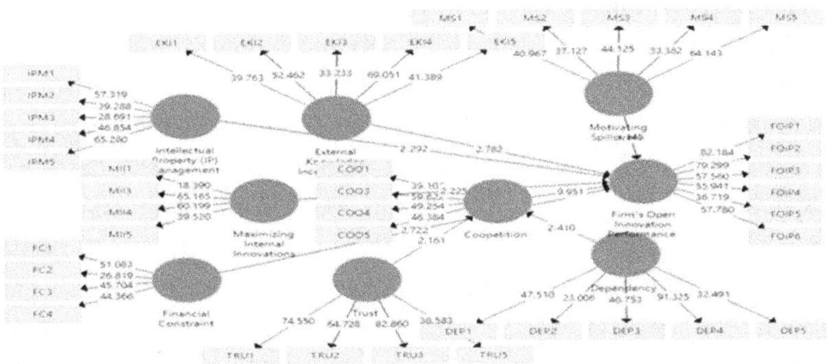

*Table 3. Results of the hypotheses*

| | β | M | SD | T Statistics | P Values |
|---|---|---|---|---|---|
| Coopetition -> Firm's Open Innovation Performance | 0.731 | 0.731 | 0.074 | 9.951 | 0 |
| Dependency -> Coopetition | 0.272 | 0.266 | 0.113 | 2.41 | 0.011 |
| External Knowledge Incorporation -> Firm's Open Innovation Performance | 0.261 | 0.257 | 0.094 | 2.782 | 0.006 |
| Financial Constraint -> Firm's Open Innovation Performance | -0.051 | -0.052 | 0.019 | 2.722 | 0.007 |
| Intellectual Property (IP) Management -> Firm's Open Innovation Performance | 0.031 | 0.04 | 0.017 | 2.292 | 0.026 |
| Maximizing Internal Innovations -> Firm's Open Innovation Performance | 0.019 | 0.017 | 0.019 | 2.225 | 0.028 |
| Motivating Spillovers -> Firm's Open Innovation Performance | 0.092 | 0.098 | 0.042 | 2.145 | 0.032 |
| Trust -> Coopetition | 0.394 | 0.402 | 0.182 | 2.161 | 0.031 |

Note: MS = Motivating Spillovers, EKI = External Knowledge Incorporation, IPM = Intellectual Property Management, MII = Maximization of Internal Innovation, FC = Financial Constraints, TRU = Trust, DEP = Dependency, COO = Coopetition, FOIP = Firm's Open Innovation Performance

## FINDINGS AND DISCUSSION

The current study investigated the major determinants of FOIP. For this purpose, data were collected with the help of survey questionnaire. Questionnaires were distributed among the employee of Malaysian SMEs. Moreover, PLS-SEM was used to analyse the collected and attain the objective of the study.

It is found that motivating spillovers has positive effect on FOIP. Increase in motivating spillovers increases the open innovation activities. These results are consistent with West and Gallagher (2006). West and Gallagher (2006) investigated that motivating spillovers has significant role in open innovation activities. Hameed et al. (2019) also found that motivating spillovers has positive effect on FOIP. In case of external knowledge, it is found that external knowledge also has significant role to enhance open innovation. Increase in external knowledge has the ability to boost FOIP. Same results were found that in case of internal innovation. These results found that both internal innovation and external knowledge are the major determinants of FOIP. Hameed et al. (2018) also found that these are two major determinants of FOIP. Both internal innovation and external knowledge has positive relationship with FOIP (W.-U. Hameed et al., 2018). Increase in external knowledge and internal innovation has positive effect on FOIP (Brunswicker & Vanhaverbeke, 2015; Del Vecchio et al., 2019). This study also found that intellectual property has positive relationship with FOIP. Better management of intellectual property lead to the increase in FOIP. As previous studies highlighted that intellectual property has positive effect on FOIP (Davoudi et al., 2018; Schönhals et al., 2018; Tang et al., 2019; Toma et al., 2018). In addition, it is found that financial constraints show negative effect on FOIP. Increase in financial constraint decreases the FOIP. Various studies mentioned that financial constraint has significant relationship with open innovation (Hameed et al., 2018; Hameed & Altaf, 2019). Finally, this study found that trust and dependency have positive effect on coopetition and coopetition has positive effect on FOIP. These results are consistent with W. Hameed and Naveed (2019); (Wemmer, Emrich, & Koenigstorfer, 2016). Increase in the level of trust and dependency, better will be the performance of coopetition and open innovation.

## CONCLUSION

It is found that success of open innovation in SMEs is based on five major elements, namely, 1) motivating spillovers, 2) incorporation of external knowledge, 3) intellectual property management, 4) maximization of internal innovation, and 5) financial constraints. These elements have significant role in the open innovation activities. First four elements contribute positively to open innovation performance. However, last element, namely; financial constraint effect negatively on open innovation activities. Increase in financial constraint among Malaysian SMEs discourages the open innovation activities. These five elements are the major challenges for companies while adopting open innovation model. Additionally, the phenomenon of coopetition-based open-innovation is emerging rapidly among the companies. It is new form of open innovation in which competitors collaborate with each other's rather than to compete. Now a day, by following the open innovation activities, competitors are adopting coopetition practices which is based on trust and dependency. To get success in coopetition-based open-innovation activities, the competitors should have higher level of trust and dependency. Existence of low trust and dependency among competitors lead to the failure of coopetition-based open-innovation activities. Therefore, trust and dependency promote coopetition which foster the open innovation activity.

The current study has significant contribution to the literature by investigating the determinants of FOIP. This study contributed by exploring that; motivating spillovers, external knowledge incorporation, internal innovation, intellectual property management and financial constraint are the key determinants of FOIP. Moreover, this study contributed by examining the role of coopetition in FOIP. This is one of the pioneer studies which investigated various determinants of FOIP along with coopetition mechanism.

## LIMITATIONS AND FUTURE RESEARCH

Although this study provided important insights to the current phenomenon, however, still the study has few limitations which could be the basis of future directions. The current study is focused on SMEs, small and large. Actually, open innovation is not easy to adopt, particularly it is quite tough for small SMEs. Therefore, future research should be on large SMEs, rather than small SMEs. Moreover, this study only examines the effect of various factors on open innovation, on the other hand, future study should focus on the comparison of SMEs between those who already adopted open innovation and those who are working on close innovation model. It will better clear the benefits of open innovation for companies. Longitudinal research design may also beneficial to get more accurate results as compared to cross-sectional research design. Finally, the current framework of the study is mixed with the SMEs those working with coopetition activities and those do not work with coopetition activities. Results may different if one type of SMEs will be selected. Therefore, future research should be on coopetition-based SMEs or without coopetition-based SMEs.

## ACKNOWLEDGMENT

We would like to acknowledge Muhammad Farhan Basheer and Shazma Razzaq for giving us the valuable instructions and make us capable to produce good research.

## REFERENCES

Almirall, E., & Casadesus-Masanell, R. (2010). Open versus closed innovation: A model of discovery and divergence. *Academy of Management Review*, *35*(1), 27–47.

Amoroso, S., & Link, A. N. (2019). *Intellectual Property Protection Mechanisms and the Characteristics of Founding Teams*. Joint Research Centre.

Armstrong, M., & Taylor, S. (2014). *Armstrong's handbook of human resource management practice*. Kogan Page Publishers.

Armstrong, K. (2019). Mastering Motivation. *APS Observer*, *32*(7).

Bengtsson, M., & Kock, S. (2000). "Coopetition" in business Networks—To cooperate and compete simultaneously. *Industrial Marketing Management*, *29*(5), 411–426. doi:10.1016/S0019-8501(99)00067-X

Bogers, M., Burcharth, A., & Chesbrough, H. (2019). Open innovation in Brazil: Exploring. *Int. J. Innov*, *7*(2), 178–191. doi:10.5585/iji.v7i2.417.2318-9975

Bogers, M., Chesbrough, H., & Moedas, C. (2018). Open innovation: Research, practices, and policies. *California Management Review*, *60*(2), 5–16. doi:10.1177/0008125617745086

Bouncken, R. B., & Fredrich, V. (2012). Coopetition: Performance implications and management antecedents. *International Journal of Innovation Management*, *16*(05), 1–28. doi:10.1142/S1363919612500284

Brunswicker, S., & Chesbrough, H. (2018). The Adoption of Open Innovation in Large Firms: Practices, Measures, and Risks A survey of large firms examines how firms approach open innovation strategically and manage knowledge flows at the project level. *Research Technology Management, 61*(1), 35–45. doi:10.1080/08956308.2018.1399022

Brunswicker, S., & Vanhaverbeke, W. (2015). Open innovation in small and medium-sized enterprises (SMEs): External knowledge sourcing strategies and internal organizational facilitators. *Journal of Small Business Management, 53*(4), 1241–1263. doi:10.1111/jsbm.12120

Chesbrough, H. (2003). *Open innovation*. Boston: Harvard Business School Press.

Chesbrough, H. (2012). Open innovation: Where we've been and where we're going. *Research Technology Management, 55*(4), 20–27. doi:10.5437/08956308X5504085

Chesbrough, H. W. (2006). *Open innovation: The new imperative for creating and profiting from technology*. Harvard Business Press.

Cohen, W. M., & Levinthal, D. A. (1989). Innovation and learning: The two faces of R & D. *Economic Journal (London), 99*(397), 569–596. doi:10.2307/2233763

Comrey, A., & Lee, H. (1992). A First Course in Factor Analysis (2nd ed.). Hillsdale, NJ: Lawrence Earlbaum Associates.

Conboy, K., & Morgan, L. (2011). Beyond the customer: Opening the agile systems development process. *Information and Software Technology, 53*(5), 535–542. doi:10.1016/j.infsof.2010.10.007

Davoudi, S. M. M., Fartash, K., Venera, G. Z., Asiya, M. B., Rashad, A. K., Anna, V. B., & Zhanna, M. S. (2018). Testing the Mediating Role of Open Innovation on the Relationship between Intellectual Property Rights and Organizational Performance: A Case of Science and Technology Park. *EURASIA Journal of Mathematics Science and Technology Education, 14*(4), 1359–1369.

Del Vecchio, P., Secundo, G., Rubino, M., Garzoni, A., & Vrontis, D. (2019). Open innovation in family firms: Empirical evidence about internal and external knowledge flows. *Business Process Management Journal*.

Enkel, E., Gassmann, O., & Chesbrough, H. (2009). Open R&D and open innovation: Exploring the phenomenon. *Research Management, 39*(4), 311–316.

Franco, A. M., & Filson, D. (2006). Spin-outs: Knowledge diffusion through employee mobility. *The Rand Journal of Economics, 37*(4), 841–860. doi:10.1111/j.1756-2171.2006.tb00060.x

Hagedoorn, J., & Ridder, A.-K. (2012). Open innovation, contracts, and intellectual property rights: an exploratory empirical study.

Hameed, W., & Altaf, M. (2019). Open Innovation Challenges: Empirical Evidence from Malaysian Small And Medium-Sized Enterprises (SME's). *Journal of Management Research, 5*(2).

Hameed, W., & Naveed, F. (2019). Coopetition-Based Open-Innovation and Innovation Performance: Role of Trust and Dependency Evidence from Malaysian High-Tech SMEs. *Pakistan Journal of Commerce and Social Sciences, 13*(1), 209–230.

Hameed, W. U., Altaf, M., & Ahmed, A. (2019). Open Innovation Challenges: Empirical Evidence from Malaysian Small and Medium-Sized Enterprises (SME's). *Journal of Management Research, 6*(1), 28–53.

Hameed, W. U., Basheer, M. F., Iqbal, J., Anwar, A., & Ahmad, H. K. (2018). Determinants of Firm's open innovation performance and the role of R & D department: An empirical evidence from Malaysian SME's. *Journal of Global Entrepreneurship Research, 8*(1), 1–20. doi:10.118640497-018-0112-8

Hameed, W.-U., Nadeem, S., Azeem, M., Aljumah, A. I., & Adeyemi, R. A. (2018). Determinants of E-Logistic Customer Satisfaction: A Mediating Role of Information and Communication Technology (ICT). *International Journal of Supply Chain Management, 7*(1), 105–111.

Kapetaniou, C., & Lee, S. H. (2019). Geographical proximity and open innovation of SMEs in Cyprus. *Small Business Economics, 52*(1), 261–276. doi:10.100711187-018-0023-7

Kraus, S., Klimas, P., Gast, J., & Stephan, T. (2019). Sleeping with competitors: Forms, antecedents and outcomes of coopetition of small and medium-sized craft beer breweries. *International Journal of Entrepreneurial Behaviour & Research, 25*(1), 50–66. doi:10.1108/IJEBR-09-2017-0356

Latham, G. P., & Ernst, C. T. (2006). Keys to motivating tomorrow's workforce. *Human Resource Management Review, 16*(2), 181–198. doi:10.1016/j.hrmr.2006.03.014

Lichtenthaler, U. (2008). Open innovation in practice: An analysis of strategic approaches to technology transactions. *IEEE Transactions on Engineering Management, 55*(1), 148–157. doi:10.1109/TEM.2007.912932

Lichtenthaler, U., & Lichtenthaler, E. (2009). A capability-based framework for open innovation: Complementing absorptive capacity. *Journal of Management Studies, 46*(8), 1315–1338. doi:10.1111/j.1467-6486.2009.00854.x

López-Nicolás, C., & Meroño-Cerdán, Á. L. (2011). Strategic knowledge management, innovation and performance. *International Journal of Information Management, 31*(6), 502–509. doi:10.1016/j.ijinfomgt.2011.02.003

Nandonde, F. A. (2019). Building Commitment in Supplier–Retailer Relationship in Developing Economies: The Case of Tanzania. *FIIB Business Review, 8*(1), 39–50. doi:10.1177/2319714518821208

Pollok, P., Lüttgens, D., & Piller, F. T. (2018). How firms develop capabilities for crowdsourcing to increase open innovation performance: The interplay between organizational roles and knowledge processes. *Journal of Product Innovation Management.*

Schönhals, A., Hepp, T., & Gipp, B. (2018). Design Thinking using the Blockchain: Enable Traceability of Intellectual Property in Problem-Solving Processes for Open Innovation. *Paper presented at the 1st Workshop on Cryptocurrencies and Blockchains for Distributed Systems.* Academic Press. 10.1145/3211933.3211952

Tang, A., Tietze, F., & Molloy, J. (2019). Openness in Intellectual Property Strategies of Synthetic Biology Start-ups. *Paper presented at the 2019 IEEE Technology & Engineering Management Conference (TEMSCON).* IEEE Press. 10.1109/TEMSCON.2019.8813740

Toma, A., Secundo, G., & Passiante, G. (2018). Open innovation and intellectual property strategies: Empirical evidence from a bio-pharmaceutical case study. *Business Process Management Journal, 24*(2), 501–516. doi:10.1108/BPMJ-11-2016-0230

Trott, P., & Hartmann, D. (2009). Why 'open innovation' is old wine in new bottles. *International Journal of Innovation Management, 13*(04), 715–736. doi:10.1142/S1363919609002509

Uzuegbunam, I., Ofem, B., & Nambisan, S. (2019). Do Corporate Investors Affect Entrepreneurs' IP Portfolio? Entrepreneurial Finance and Intellectual Property in New Firms. *Entrepreneurship Theory and Practice, 43*(4), 673–696. doi:10.1177/1042258717738247

van den Broek, J., Boselie, P., & Paauwe, J. (2018). Cooperative innovation through a talent management pool: A qualitative study on coopetition in healthcare. *European Management Journal, 36*(1), 135–144. doi:10.1016/j.emj.2017.03.012

Von Zedtwitz, M. (2004). Managing foreign R&D laboratories in China. *Research Management, 34*(4), 439–452.

Wallin, M. W., & Krogh, G. (2010). Organizing for open innovation: Focus on the integration of knowledge. *Organizational Dynamics, 39*(2), 145–154. doi:10.1016/j.orgdyn.2010.01.010

Wemmer, F., Emrich, E., & Koenigstorfer, J. (2016). The impact of coopetition-based open innovation on performance in nonprofit sports clubs. *European Sport Management Quarterly, 16*(3), 341–363. doi:10.1080/16184742.2016.1164735

West, J., & Gallagher, S. (2006). Challenges of open innovation: The paradox of firm investment in open-source software. *Research Management, 36*(3), 319–331.

Wiltshire, T. J., Steffensen, S. V., & Fiore, S. M. (2019). Multiscale movement coordination dynamics in collaborative team problem solving. *Applied Ergonomics, 79*, 143–151. doi:10.1016/j.apergo.2018.07.007 PMID:30103905

Wiwchar, M., Walker, D., & Marsh, R. (2020). *Protecting Your Digital Health Intellectual Property: Fundamentals of Intellectual Property and How It Applies to Software, Hardware and Business Processes. In Digital Health Entrepreneurship* (pp. 103–118). Springer.

Yadav, R., Shukla, N. K., & Gupta, R. (2019). All-Digital Phase Locked Loop (ADPLL) as an Intellectual Property (IP) Core for an Application-specified Integrated Circuit (ASIC) Product: A Survey. *Journal of VLSI Design Tools & Technology, 5*(3), 12–17.

## ADDITIONAL READING

Chesbrough, H. (2003). *Open innovation*. Boston: Harvard Business School Press.

Chesbrough, H., Vanhaverbeke, W., & West, J. (Eds.). (2006). *Open innovation: Researching a new paradigm*. Oxford University Press on Demand.

Gnyawali, D. R., & Park, B. J. (2009). Co-opetition and technological innovation in small and medium-sized enterprises: A multilevel conceptual model. *Journal of Small Business Management, 47*(3), 308–330. doi:10.1111/j.1540-627X.2009.00273.x

Hameed, W., & Naveed, F. (2019). Coopetition-Based Open-Innovation and Innovation Performance: Role of Trust and Dependency Evidence from Malaysian High-Tech SMEs. *Pakistan Journal of Commerce and Social Sciences, 13*(1), 209–230.

Hameed, W. U., Altaf, M., & Ahmed, A. (2019). Open Innovation Challenges: Empirical Evidence from Malaysian Small and Medium-Sized Enterprises (SME's). *Journal of Management Research, 6*(1), 28–53.

Hameed, W. U., Basheer, M. F., Iqbal, J., Anwar, A., & Ahmad, H. K. (2018). Determinants of Firm's open innovation performance and the role of R & D department: An empirical evidence from Malaysian SME's. *Journal of Global Entrepreneurship Research, 8*(1), 1–20. doi:10.118640497-018-0112-8

Wemmer, F., Emrich, E., & Koenigstorfer, J. (2016). The impact of coopetition-based open innovation on performance in nonprofit sports clubs. *European Sport Management Quarterly, 16*(3), 341–363. doi:10.1080/16184742.2016.1164735

West, J., & Gallagher, S. (2006). Challenges of open innovation: The paradox of firm investment in open-source software. *Research Management, 36*(3), 319–331.

## KEY TERMS AND DEFINITIONS

**Coopetition:** A form of business collaboration in which firms collaborate with each other's rather than to compete.

**External Knowledge:** Valuable information of business from external stakeholders.

**Financial Constraint:** Financial problem raised while adopting any innovative process.

**Intellectual Property:** Category of property that comprises intangible creations such as new ideas.

**Internal Innovation:** Generation of latest ideas and products by using the internal capabilities of the firm.

**Motivating Spillovers:** Factor that motivate innovation process.

**Open Innovation:** It is a two-way process in which knowledge inter inside the boundaries of the firm and goes outside the boundaries of the firm for commercialization.

## APPENDIX

*Table 4. Data screening*

| | No. | Missing | Mean | Median | Min | Max | SD | Kurtosis | Skewness |
|---|---|---|---|---|---|---|---|---|---|
| MS1 | 1 | 0 | 4.965 | 5 | 1 | 7 | 1.846 | -0.502 | -0.714 |
| MS2 | 2 | 0 | 5.338 | 6 | 1 | 7 | 1.738 | -0.342 | -0.84 |
| MS3 | 3 | 0 | 5.259 | 6 | 1 | 7 | 1.818 | 0.06 | -1.01 |
| MS4 | 4 | 0 | 5.03 | 6 | 1 | 7 | 1.809 | -0.638 | -0.659 |
| MS5 | 5 | 0 | 5.015 | 5 | 1 | 7 | 1.875 | -0.6 | -0.683 |
| EKI1 | 6 | 0 | 4.945 | 5 | 1 | 7 | 1.745 | -0.713 | -0.538 |
| EKI2 | 7 | 0 | 4.756 | 5 | 1 | 7 | 1.907 | -0.977 | -0.436 |
| EKI3 | 8 | 0 | 4.925 | 5 | 1 | 7 | 1.765 | -0.572 | -0.647 |
| EKI4 | 9 | 0 | 4.791 | 5 | 1 | 7 | 1.797 | -0.832 | -0.352 |
| EKI5 | 10 | 0 | 5.015 | 5 | 1 | 7 | 1.867 | -0.78 | -0.576 |
| IPM1 | 11 | 0 | 4.751 | 5 | 1 | 7 | 1.822 | -0.852 | -0.357 |
| IPM2 | 12 | 0 | 5.139 | 6 | 1 | 7 | 1.796 | -0.209 | -0.87 |
| IPM3 | 13 | 0 | 5.1 | 6 | 1 | 7 | 1.796 | -0.7 | -0.634 |
| IPM4 | 14 | 0 | 4.99 | 5 | 1 | 7 | 1.795 | -0.412 | -0.733 |
| IPM5 | 15 | 0 | 4.915 | 5 | 1 | 7 | 1.803 | -0.563 | -0.652 |
| MII1 | 16 | 0 | 4.891 | 5 | 1 | 7 | 1.741 | -0.417 | -0.685 |
| MII2 | 17 | 0 | 3.925 | 4 | 1 | 7 | 1.898 | -1.16 | 0.183 |
| MII3 | 18 | 0 | 5.194 | 6 | 1 | 7 | 1.744 | 0.039 | -0.879 |
| MII4 | 19 | 0 | 5.08 | 5 | 1 | 7 | 1.752 | -0.538 | -0.637 |
| MII5 | 20 | 0 | 5.159 | 6 | 1 | 7 | 1.8 | -0.092 | -0.886 |
| FC1 | 21 | 0 | 5.169 | 6 | 1 | 7 | 1.762 | -0.549 | -0.727 |
| FC2 | 22 | 0 | 5.184 | 6 | 1 | 7 | 1.807 | -0.169 | -0.869 |
| FC3 | 23 | 0 | 5.1 | 5 | 1 | 7 | 1.745 | -0.305 | -0.737 |
| FC4 | 24 | 0 | 4.935 | 5 | 1 | 7 | 1.872 | -0.606 | -0.681 |
| FC5 | 25 | 0 | 3.985 | 4 | 1 | 7 | 1.881 | -1.103 | 0.117 |
| TRU1 | 26 | 0 | 4.861 | 5 | 1 | 7 | 1.864 | -0.735 | -0.423 |
| TRU2 | 27 | 0 | 5.04 | 5 | 1 | 7 | 1.82 | -0.64 | -0.628 |
| TRU3 | 28 | 0 | 4.801 | 5 | 1 | 7 | 1.742 | -0.765 | -0.374 |
| TRU4 | 29 | 0 | 4.055 | 4 | 1 | 7 | 1.804 | -1.07 | 0.071 |
| TRU5 | 30 | 0 | 4.995 | 5 | 1 | 7 | 1.7 | -0.288 | -0.635 |
| DEP1 | 31 | 0 | 5.035 | 5 | 1 | 7 | 1.706 | -0.657 | -0.515 |
| DEP2 | 32 | 0 | 4.985 | 5 | 1 | 7 | 1.797 | -0.64 | -0.584 |
| DEP3 | 33 | 0 | 4.955 | 5 | 1 | 7 | 1.788 | -0.568 | -0.543 |
| DEP4 | 34 | 0 | 5.134 | 5 | 1 | 7 | 1.727 | -0.384 | -0.671 |
| DEP5 | 35 | 0 | 4.846 | 5 | 1 | 7 | 1.906 | -0.941 | -0.503 |
| COO1 | 36 | 0 | 4.816 | 5 | 1 | 7 | 1.98 | -0.683 | -0.687 |
| COO2 | 37 | 0 | 3.945 | 4 | 1 | 7 | 1.953 | -1.243 | 0.199 |

continued on following page

Table 4. Continued

| | No. | Missing | Mean | Median | Min | Max | SD | Kurtosis | Skewness |
|---|---|---|---|---|---|---|---|---|---|
| COO3 | 38 | 0 | 4.925 | 5 | 1 | 7 | 1.952 | -0.654 | -0.691 |
| COO4 | 39 | 0 | 4.94 | 5 | 1 | 7 | 1.892 | -0.732 | -0.584 |
| COO5 | 40 | 0 | 4.846 | 5 | 1 | 7 | 2.032 | -0.868 | -0.613 |
| FOIP1 | 41 | 0 | 4.955 | 5 | 1 | 7 | 1.948 | -0.782 | -0.616 |
| FOIP2 | 42 | 0 | 4.692 | 5 | 1 | 7 | 2.026 | -0.974 | -0.427 |
| FOIP3 | 43 | 0 | 4.851 | 5 | 1 | 7 | 1.959 | -0.975 | -0.477 |
| FOIP4 | 44 | 0 | 4.731 | 5 | 1 | 7 | 2.012 | -0.961 | -0.473 |
| FOIP5 | 45 | 0 | 4.672 | 5 | 1 | 7 | 1.851 | -0.959 | -0.398 |
| FOIP6 | 46 | 0 | 4.692 | 5 | 1 | 7 | 1.981 | -0.903 | -0.498 |

Note: MS = Motivating Spillovers, EKI = External Knowledge Incorporation, IPM = Intellectual Property Management, MII = Maximization of Internal Innovation, FC = Financial Constraints, TRU = Trust, DEP = Dependency, COO = Coopetition, FOIP = Firm's Open Innovation Performance

*Table 5. Factor loadings*

| | COO | DEP | EKI | FC | FOIP | IPM | MII | MS | TRU |
|---|---|---|---|---|---|---|---|---|---|
| COO1 | 0.893 | | | | | | | | |
| COO3 | 0.921 | | | | | | | | |
| COO4 | 0.905 | | | | | | | | |
| COO5 | 0.895 | | | | | | | | |
| DEP1 | | 0.9 | | | | | | | |
| DEP2 | | 0.834 | | | | | | | |
| DEP3 | | 0.909 | | | | | | | |
| DEP4 | | 0.931 | | | | | | | |
| DEP5 | | 0.847 | | | | | | | |
| EKI1 | | | 0.885 | | | | | | |
| EKI2 | | | 0.909 | | | | | | |
| EKI3 | | | 0.88 | | | | | | |
| EKI4 | | | 0.923 | | | | | | |
| EKI5 | | | 0.893 | | | | | | |
| FC1 | | | | 0.901 | | | | | |
| FC2 | | | | 0.856 | | | | | |
| FC3 | | | | 0.898 | | | | | |
| FC4 | | | | 0.88 | | | | | |
| FOIP1 | | | | | 0.929 | | | | |
| FOIP2 | | | | | 0.934 | | | | |
| FOIP3 | | | | | 0.914 | | | | |
| FOIP4 | | | | | 0.914 | | | | |
| FOIP5 | | | | | 0.877 | | | | |
| FOIP6 | | | | | 0.911 | | | | |
| IPM1 | | | | | | 0.908 | | | |
| IPM2 | | | | | | 0.881 | | | |
| IPM3 | | | | | | 0.845 | | | |
| IPM4 | | | | | | 0.891 | | | |
| IPM5 | | | | | | 0.916 | | | |
| MII1 | | | | | | | 0.775 | | |
| MII3 | | | | | | | 0.91 | | |
| MII4 | | | | | | | 0.914 | | |
| MII5 | | | | | | | 0.894 | | |
| MS1 | | | | | | | | 0.894 | |
| MS2 | | | | | | | | 0.887 | |
| MS3 | | | | | | | | 0.898 | |
| MS4 | | | | | | | | 0.878 | |
| MS5 | | | | | | | | 0.92 | |
| TRU1 | | | | | | | | | 0.938 |
| TRU2 | | | | | | | | | 0.919 |
| TRU3 | | | | | | | | | 0.94 |
| TRU5 | | | | | | | | | 0.897 |

Note: MS = Motivating Spillovers, EKI = External Knowledge Incorporation, IPM = Intellectual Property Management, MII = Maximization of Internal Innovation, FC = Financial Constraints, TRU = Trust, DEP = Dependency, COO = Coopetition, FOIP = Firm's Open Innovation Performance

# Chapter 9
# How Is It Different From Conventional Learning?
## The Growing Trend of Corporate Universities in Indonesia

**Hary Febriansyah**
https://orcid.org/0000-0002-6578-7735
*School of Business and Management, Institut Teknologi Bandung, Indonesia*

**Haifa Labdhagati**
*School of Business and Management, Institut Teknologi Bandung, Indonesia*

**Widi Galih Anggara**
*Ministry of Finance, Indonesia*

## ABSTRACT

*In the current Indonesian business scene, corporate universities have become fashionable. Since their first emergence in the 2010s, Indonesian corporate universities have shown that they are more than just dressed-up learning centers operating under a new name. As prominent organizations in Indonesia start to build their own learning initiatives adopting the model of the corporate university, success stories of the new approach radiate. Bringing an example of a developing country of Indonesia, this chapter attempts to discuss the trends, challenges, and practices of corporate university. This chapter also examines best practices from Indonesian corporate universities and how they differ from the conventional university and their learning techniques differ from traditional learning methods.*

DOI: 10.4018/978-1-7998-1566-2.ch009

## INTRODUCTION

"For training to be effective, it has to maintain a reliable, consistent presence. Employees should be able to count on something systematic, not a rescue effort summoned to solve the problem of the moment. In other words, training should be a continuous process rather than a one-time event." –Andrew S. Grove, former chairman and CEO of Intel Corporation

It is not an exaggeration to say that corporate learning today should all have the perspective that was brought by Andrew S. Grove, the former chairperson of Intel Corporation. The conventional training and development function in companies was typically aimed to develop specific sets of information, skills, or knowledge needed by the employees. The early training and development programs in companies tended to be reactive, decentralized, tactical, focused on the individual job skills, and often operated as a function within the human resource department (Meister, 1998). However, the conventional approach is not robust or versatile enough to survive in the current competition.

The world is changing at a pace never seen before in the history of humankind. This kind of rapid change has happened because knowledge is now produced faster than ever before. In some cases, knowledge is valued more than tangible assets, as it is essential to survive under these volatile conditions. Knowledge workers are now considered to be important assets for a corporation (Drucker, 1999). Consequently, training and education have come to be regarded as valid long-term investments rather than a simple operating cost. As time has passed, the conventional training and education approach evolved into a new approach called the corporate university. This new approach is intended to engage the employees in a continuous learning process, where the employees are regarded as learners (El-Tannir, 2002).

The term corporate university refers to "an educational entity that is a strategic tool designed to assist parent organizations in realizing its mission by engaging in activities that foster individual, organizational learning, knowledge and wisdom" (Allen, 2002). Although "corporate university" is often considered as simply a new term to replace the corporate training and education department, it is actually a major transformation from the old approach. Unlike the conventional training and education department, the corporate university possesses a strategic dimension (Abel, 2008; Rademakers, 2005). The learning process in a corporate university is driven by the company's strategy and goals and aligned to its business needs (Andresen & Lichtenberger, 2007; Ben-Hur, Jaworski, & Gray, 2015). Some corporate universities are not limited to supporting learning within the corporation but also attempt to educate other components within the company's value chain.

Although the concept originated in the Western world, the corporate university has been warmly welcomed in other parts of the world too. Over the past decade, there has been a blooming trend of corporate universities in Indonesia. A large number of companies, ranging across state-owned, private-owned, or even government institutions in Indonesia, have established their own corporate university. The concept of the corporate university is so popular in the country that it is considered trendy enough for a notable business press agency to hold a yearly competition for corporate universities. Questions that this trend raises include Why are all these companies so eager to establish their own corporate university?, Why is it called a 'university'?, and How is it different from the conventional university or the traditional corporate learning department we all knew?

Questions that this trend raises include Why are all these companies so eager to build their own corporate university?, Why does it called a 'university'?, How is it different from the conventional university or the traditional corporate learning department we all knew? Some of the corporate universities in the West have transformed into a 'real' university, is this case possible in Indonesia?, What will the future hold for Indonesian corporate university?, and Will Indonesian corporate universities compete with conventional universities in the future?

This chapter attempts to examine the common practice of corporate universities in a developing nation, which is Indonesia, as well as the matters which is needed to understand the phenomenon of corporate university. The limitations and challenges that lie ahead of Indonesian corporate universities will also be discussed.

## THE BIRTH AND EXPANSION OF CORPORATE UNIVERSITY

Long before the terms "university," "academy," or "institute" came to be used as metaphors for continuous learning, companies have held in-house training to improve their employees' competencies. However, the question of which company started the first corporate university has been much debated. Nixon and Helms (2002) suggested that the first ever corporate university was the GM Institute, established in the United States by General Motors in 1927. Moore (2002) stated that Northrop University, which was established in 1940 is one of the earliest corporate universities. Andresen and Lichtenberger (2007) have a different opinion, believing that Disney University, which was established in 1955, is the first representative of a true corporate university. The recent research from Ewer and Russ-Eft (2017), however, has revealed a somewhat different origin of the corporate university. Their study found that the concept of a corporate university was rooted back in the 1910s when Goodyear Tire and Rubber Company in the US developed an internal workplace training institution and labeled it as Goodyear's Industrial University in 1913. In spite of these different historical findings, it can be concluded that the concept of the corporate university originated in the US as a result of the primary requirement of a skill-based workforce in an information economy that emphasizes the importance of knowledge (Allen, 2002; Meister, 1998; Morin & Renaud, 2004; Andresen & Lichtenberger, 2007; Patrucco et al., 2017).

One of the earliest corporate universities noted by most scholars was the GM Institute (Lytovchenko, 2016; Morin & Renaud, 2009; Nixon & Helms, 2002; Oh & Park, 2011; Ryan, Prince, & Turner, 2015; Thompson, 2000). Initially, the GM Institute focused on engineering and management skills. Later in 1982, it became a private and independent college, and in 1997, it changed its name to Kettering University (Morin & Renaud, 2004). Nowadays, the university also provides an MBA program plus other studies in production management, engineering, lean production, and other areas. A few decades after the first emergence of corporate universities, around the 1950s and 1960s, large corporations in the US built their own learning initiatives. These firms included General Electric, Disney, McDonald's, and Arthur D. Little (Lytovchenko, 2016; Morin & Renaud, 2009; Ryan et al., 2015). According to Lytovchenko (2016), a crucial milestone in the development of the corporate university was the birth of McDonald's Hamburger University in 1961. Hamburger University claims to be the nation's number one training facility, even larger than the US Army in terms of size and scope.

During the mid-to-late 1990s, the concept of corporate university hits the UK and other European countries (Plompen, 2005). Huge companies like STMicroelectronics, Siemens, ABN AMRO, Alcatel, Heineken, and Allianz were transforming their regular learning centers into corporate universities. While corporate universities in the US are mainly lean towards operational aspects such as efficiency and productivity, European corporate universities are generally focused on the development at the strategic level (Hilse & Nicolai, 2004). Most of these corporate universities are partnering with higher education institutions to deliver university-level programs in their corporate education programs (Andresen

& Lichtenberger, 2007). The main drivers of the corporate university expansion in Europe are (Prince & Stewart, 2002): (1) growing awareness among corporations of the management education's role; (2) the need to link education and organization strategy; (3) to change the managers' behavior to promote dialogue rationally and logically; and (4) to retain good employees through incentives.

The concept attracts Asian companies as well. China corporations are the early adopters of corporate university in Asia. The first corporate university in China is Motorola University China, which was founded in 1993 (Qiao, 2009). Many foreign companies established corporate universities in China as a form of investment. However, local Chinese companies, such as Huawei and Shangri-la, also build their own learning facilities. Foreign companies that build corporate universities in China are mostly driven by the urgent need of skilled workforce to operate in the market located outside the parent company (Alagaraja & Li, 2015). Whereas for the local Chinese companies, the main rationales are (Qiao, 2009): (1) to improve talent for the company; (2) to coordinate extensive training resources in the company; and (3) to advocate the company's brand image.

Currently, the concept of corporate university became a trend in corporate learning worldwide. The main driver is globalization and the rise of the knowledge era (Alagaraja & Li, 2015; Lytovchenko, 2016). The rise of the knowledge era has created a new economic opportunity for growth. It also has created a global war for talent, which pushed companies to create a strategic learning approach through the establishment of the corporate university (Ryan et al., 2015). Other than that, the survival needs for productivity and performance results in the best practices in organizational learning so that the corporate university concept has been quickly adopted.

*Table 1. Paradigm shift in corporate learning*

| Component | Conventional Training Department | Corporate University |
|---|---|---|
| Place | Classroom-based learning | On-demand learning anywhere, enabling the use of technology |
| Content | Upgrade technical skills related to job skills | Build core workplace competencies, aligned to organizational goals |
| Methodology | Learn by listening, focus on the instructor | Action learning, focus on the employee |
| Audience | Individual internal employees, limited depth | Intact team of employees, customers, and product suppliers |
| Faculty | External university professors/consultants | International senior manager and a consortium of university professors/consultants |
| Frequency | One-time single event | Continuous learning process |
| Goal | Building individual's inventory of skills on the job | Solve real business issues and improve performance on the job, aligned to organizational goals |
| Focus | Reactive to workplace challenges | Proactive about upcoming changes |
| Organization | Fragmented and decentralized | Cohesive and centralized |
| Scope | Focused on the tactics of training | Deliberate about learning strategy |
| Delivery | Instructor-led | Experience with various delivery technologies |
| Outcome | Increase in job skills; support for specific business unit | Increase in performance on the job, broader impact across entire organization |
| Image | Go get trained | University as a metaphor for learning |
| Evaluation | Little or no evaluation practice | A robust system of measurement and accountability |

Source: adapted from Meister (1998); Abel (2008); Barley (2002).

Furthermore, the gap between what conventional universities offer and the requirements of the industry has forced companies to create their own learning initiatives (Lytovchenko, 2016). Although some might argue that conventional universities are better at shaping individuals to a paradigm and training them to see the bigger picture, sometimes the materials delivered in conventional universities are unable to fulfill the needs of corporations. Textbooks that are used in conventional universities are often lacking in some areas, such as an elevated level of thinking that is now a vital qualification to possess (Upadhyay & Paul, 2019). Meanwhile, corporate universities are able to offer a more effective training that is tailored to the company needs, in a faster way and at considerably lower cost compared to the conventional universities (Nixon & Helms, 2002). As a result, companies prefer to build their own in-house learning initiatives to educate their employees, rather than sending them to study in conventional universities.

## The Paradigm Shift

The first thing to do in establishing a corporate university is to shift from the old learning paradigms to the new one. Several academics have reported this paradigm shift and evolution in corporate learning (Abel, 2008; Barley, 2002; Meister, 1998) as summarized in Table 1.

Although the corporate university is often considered as a mere new term for the old approach, there is an underlying difference in both programs offered by conventional training and the corporate university. Conventional training aims to close the gap between the employees' current competencies and the competencies needed for them to do the work. For example, marketing division employees will undergo training programs that familiarize them with the products and services offered by the company as well as how to market them. This kind of training will help them to do their work in the future. The problem with conventional training is that the provided training is sometimes delivered without first conducting a proper needs analysis. Other than that, the conventional training was also mainly seen as a one-time event and considered as a cost by the company rather than an investment.

Alternatively, corporate university programs mainly act as a business solution provider that aims to solve the business problems faced by the company (Satrijono, Djawahir, & Sugiarsono, 2017). Corporate university programs are aligned with the corporate strategy and aim for several primary goals (Hearn, 2002), including:

1. Organize training.
2. Start and support change in the organization.
3. Get the most out of the investment in education.
4. Bring a common culture, loyalty, and sense of belonging to a company.
5. Remain competitive in today's economy.
6. Retain employees.

The corporate university is part of a change management program that specifically aims to foster the employees' competence. Through recruitment, the company can gather and pick the best talent. However, these potential talents still need to be improved. The corporate university plays a key role in nurturing these talents by providing them with essential knowledge. The curriculum in a corporate university is also curated to integrate the learning process into the employee's job. Through a learning and mentorship program that fosters the development of employee competence, the corporate university gives birth to the company's future leaders.

*Figure 1. Four stages of corporate university evolution (Allen, 2002)*

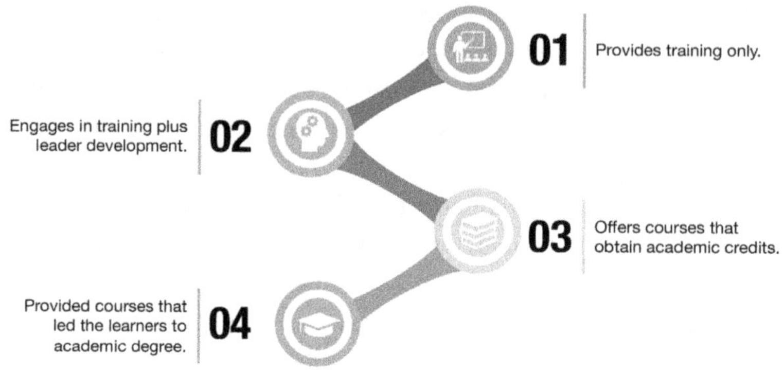


Sometimes the problem in the conventional learning process is that the knowledge acquired from the training programs is not implemented on the job and the employees returning to the job after training have not changed their behaviors (Phillips & Phillips, 2007). Corporate universities make sure that the knowledge they provide causes an impact on the company's overall performance. The success of corporate university programs is usually evaluated by measuring their return on training investment (ROTI).

Corporate university brings the concept of the university to the corporation. Employing the terms "university," "institute," or "academy," the corporate university engages the employees in a continuous learning process. This way, employees are able to adapt with the changing environment and answer the business challenges. The paradigms have now shifted. Corporate learning that was regarded as a cost, is now considered as an investment. In addition, corporate universities also act as the "center of excellence" and "center of certification" for the company (Satrijono et al., 2017). In the current global business scene, certification of competence is important to verify the employees' skills and ability. The corporate university can set a competence standard based on the company's need, evaluating the human capital and providing them with the knowledge to achieve the desired standard. In sum, the corporate university is meant to solve the problems that usually occur with the conventional approach.

The existence of a corporate university in a company does indeed add extra work and effort to learning, but it does not guarantee an improvement. Changes are only possible when every component of the company shares the same vision toward learning. It is important for corporate universities to utilize the symbols and language of higher education. The use of the revolutionary term "university" is meant to promote an atmosphere of learning and begin to influence organizational learning through the expansion of scope and practice.

## CORPORATE UNIVERSITY EVOLUTION

Although corporate universities are considered to be revolutionary, they do not develop and are not implemented as a fully-grown strategic learning program overnight. When the corporate university is first established, most likely it has yet to become the sophisticated corporate university that is fully functional. The corporate university matures and grows alongside the organizational changes and developing experimental approaches over time.

The previous literature has classified the types and stages of corporate university (CU) evolution. Walton (2005) defined a CU as maturing through three generations:

1. The first generation CU focuses on a relatively narrow realm by renaming conventional and developmental activities requiring classroom attendance and heavily emphasizing the promotion and acquisition of corporate values.
2. The second generation CU reflects a broader-based strategy toward organizational learning but still leans toward the usual practices.
3. The third-generation CU possesses a virtual element and demonstrated sophistication in the learning philosophy and the process of human capital development.

On the other hand, Allen (2002) described the corporate university's evolutionary growth as occurring in four stages, as depicted in Figure 1.

Rademakers (2014) proposed a scheme that divides corporate university evolution into three development phases as follows:

1. Operational phase, consisting simply of advanced training departments.
2. Tactical phase, focusing on reproducing and spreading knowledge and transforming corporate universities as the knowledge backbone of the organization.
3. Strategic phase, highlights the transformation of corporate universities into a knowledge factory and the development of a new strategically relevant knowledge in order to gain competitive advantage.

Based on the explanations above, it can be concluded that corporate universities undergo several stages of transformation that eventually differentiate them from the conventional training and development departments. At the beginning, the change of title to corporate university might seem like the mere introduction of a new terminology for the same practice as the conventional training department. As time goes on, the corporate university transforms, becoming more like a 'real' university. In several cases, corporate universities have successfully transformed themselves into a university that provides academic degrees (e.g., Kettering University, formerly the GM Institute).

However, not all corporate universities can be transformed into institutions that issue academic degrees. It might not be possible in several countries, for example, where educational degrees are strictly regulated (Baporikar, 2014). This is not an absolute drawback for corporate universities that are located in such countries, however. Collaboration between corporate universities and conventional universities could break this limitation. Through such a partnership, the learners in the corporate university could not only obtain academic credits from the courses, but they could also be granted a degree. This form of partnership has also happened to corporate universities in Indonesia, where a corporate university bloom is currently happening.

## THE GROWING TRENDS IN INDONESIA

The corporate university trend hit Indonesia in the mid-2000s, almost a century after the concept first emerged in the United States. Anna Maria—a senior consultant in the knowledge management and the corporate university field—as cited in Satrijono, Djawahir, and Sugiarsono (2017), stated that the concept became popular after it was adopted by banking and state-owned corporations. The early adopters of the corporate university in Indonesia were PT Telkom Indonesia, a telecommunications company; PLN, an electricity company; and PT Pertamina, an oil and natural gas company. All three of these corporate university early adopters in Indonesia are state-owned corporations.

In Indonesia, the existence of the corporate university still confuses the general public. PT Telkom Indonesia alone has both Telkom University—which is a private, conventional university—and also Telkom Corporate University—which is a corporate university dedicated to managing knowledge within the corporation. Both universities are also located in the same city. Moreover, not all professionals in Indonesia have understood the benefits of establishing a corporate university, because not all corporations have established one.

Now that the pace of knowledge development has brought many challenges to corporations, the need for knowledge workers has become the priority of companies all around the world. Indonesian corporations started to realize that conventional training did not go far enough to provide their employees with the essential knowledge. Sending employees to study in a conventional university is still an option that is considered by companies. However, it is not possible to send hundreds, or even thousands of employees to study at once. Companies need to find a way that integrates learning with the employees' jobs. Thus, the materials delivered need to be adaptive and flexible at the same time. This phenomenon has caused the concept of the corporate university to be adopted and become fashionable among Indonesian corporations.

Corporate universities in Indonesia are mostly a form of alteration from the previous corporate learning activities. They usually started as a specific department that handled learning activities, was transformed into a learning center, and eventually reached its final form, which is the corporate university. It is an actualization of a paradigm shift in corporate learning. The early concept of workplace learning in an Indonesian company would focus on upgrading the skills by providing training programs. The corporate university shifts the focus from the previous irregular effort into a continuous learning process. They also highlight the importance of aligning the learning to the company's strategy and focusing on solving real business issues.

The corporate university approach is still considered new in Indonesia. A lot of people still believe the programs offered by corporate universities are no different from the ones offered by conventional training departments. What they do not know is that the corporate university is not just a unit that conducts training, but it specifically develops the overall competence and capability of the company's human capital. Aside from providing knowledge, the corporate university also develops the employees' skills, attitudes, and wisdom. In short, the corporate university aims to increase the intelligence level of the employees.

Another aspect that may be confusing in the Indonesian context is that there are corporations that have established both a conventional university and a corporate university (e.g., PT Telkom Indonesia and PT Pertamina). The difference might only rest on the word 'corporate,' but the two institutions serve different purposes. While the conventional university provides knowledge through learning for the students to work toward a formal degree, the corporate university's main purpose is to support the corporate business strategy through learning. The 'students' of corporate universities are the company's human capital and future leaders, while in the conventional university, anyone who wishes to get a higher education degree can apply to become a student.

Conventional universities develop and give birth to state-of-the-art knowledge, which has yet to be done by an Indonesian corporate university. On the other hand, the learning process in the corporate university requires the most recent knowledge so that the company can adapt to the changing environment. Therefore, in order to support their learning process, Indonesian corporate universities build partnerships with conventional universities. For example, Adira Corporate University—parented by a finance company, Adira Finance—partnered with Binus University for the financing education. Telkom Corporate University also holds an annual knowledge sharing event, collaborating with Telkom University, ITB, and Universitas Padjajaran. The event is not just a platform to share knowledge but also an event to share corporate university success stories that deserve to be celebrated.

Success stories are one of the factors that have promoted the blooming of corporate universities in Indonesia. Companies such as PT PLN and PT Telkom Indonesia, which are the pioneers of the corporate university in Indonesia, have showcased an improvement in their business performance thanks to the educational activities in their respective corporate universities. PT Telkom Indonesia is recorded as achieving a double-digit business growth index as a result of their learning process. Pertamina Corporate University claimed that more than 1.250 PT Pertamina employees are certified in corporate safety and this has resulted in a decrease in the workplace accident rate at PT Pertamina.

So far, the rationale of the corporate university movement in Indonesia has been described. Although it may seem like the problems can be solved by improving the conventional training programs the companies already own, the existence of a separate business unit that focuses on developing the company's human resources such as the corporate university shows that the company is committed to their learning initiatives.

However, establishing corporate universities in Indonesia is not without any limitation. In the previous chapter, the stages of corporate university evolution have been discussed. A lot of corporate universities in the US and Europe are able to issue degrees to the learners through partnership with conventional universities or transforming themselves into university that is able to issue degree. In Indonesia, higher education is strictly regulated. Only the accredited institutions can issue a higher education degree. Despite the fact that corporate universities exist as an answer to the gap between what conventional universities offer, courses offered in Indonesian corporate universities are not eligible to be transferred into academic credits. The partnership between Indonesian corporate universities and conventional universities is, so far, unable to bypass the strict regulation. This limitation may reduce the perceived value of corporate university courses, as it cannot be measured with academic credits that will lead the learners to a formal degree. Although the knowledge acquired from the courses is valuable, sometimes it is not enough to motivate the employee and engage them in the learning process.

## CHALLENGES IN ESTABLISHING CORPORATE UNIVERSITY

Establishing a corporate university in a company is not as simple as upgrading the terminology that describes the company's training and development department. It is a long process with many obstacles ahead. The beginning is always the hardest part. Changing what had already become a rooted system and transforming it into an upgraded version sounds hard. Even if the change is perceived as a better

Figure 2. Learning value chain

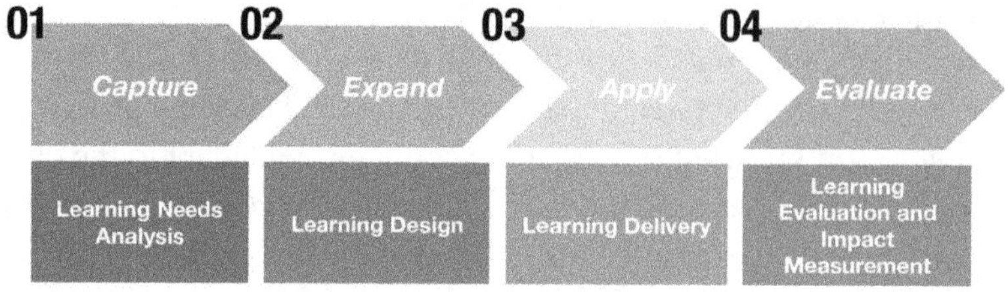


option, sometimes it is not enough to motivate the change. Change needs much more than motivation; it also requires determination.

The main challenge of establishing a corporate university is not related to the infrastructure or technology. Instead, the challenge is how to transform the mindset of people within the company. Changing the terminology is the first step to show the company's determination to improve their learning initiatives. Whether it employs the term "university," "academy," or "institute," these learning initiatives bring the same spirit. This spirit needs to be constantly communicated to all employees within the company to create a sense of transformation, so that they will actively participate in the new learning system brought by the corporate university.

When discussing the transformation of the mindset, one of the most crucial aspects is transforming the paradigm of the company's management leaders. The top management's commitment to managing the new learning initiatives will determine whether the corporate university will have a positive impact on the company. If the top management believes that the corporate university needs to be prioritized and exhibits a commitment toward that end, all the subordinates and employees will hopefully follow and become enthusiastic about the new approach.

Another challenge to starting a corporate university is related to the corporate university strategy. Rademakers (2005, 2014) suggested three corporate university strategies that a company can refer to, namely:

1. School strategy, in which the corporate university acts as a disseminator of knowledge by providing training to individuals in reaction to organization strategy and aimed at enhancing the efficiency of individual training.
2. College strategy, in which the corporate university refers to a redistributor of knowledge within the organization by deriving its training programs from organizational strategy and aimed at aligning organizational goals with individuals' competence development programs.
3. Academy strategy, in which the corporate university acts as a driver of knowledge innovation and helps the organization to gain competitive advantage and achieve its goals.

These three strategies can be put into consideration when choosing a path for the corporate university, with best practices being taken from each. However, there is no one best way to manage a corporate university. Every company has its own unique situation that sometimes cannot be comparable with the others. The challenge is how to choose the best strategy that fits the company, depending on the company's direction. Choosing the best strategy can start by clarifying the answers to questions like these: What kind of company do you want to become through this corporate university? What kind of people are important to achieving this vision? Afterward, the corporate university needs to determine its scope, identify stakeholders and their needs, and develop its operational strategy.

After the establishment of the corporate university, other challenges await. The newly established corporate university will still encounter many challenges, especially those related to its operations. For example, regarding curriculum development. In order to develop the curriculum that fits well with the company, hiring an expert who understands the core business of the company and which learning approach matches well with the company will be very useful. However, hiring an expert can be costly. Another challenge related to the corporate university's operation is how can it build the infrastructure and facilities that support the corporate university learning value chain, from the learning needs analysis to the learning evaluation? Corporate university infrastructure is a sign of the commitment of the company towards learning activities and is more inclusive to all stakeholders than the usual training center.

Global corporate universities are now moving towards the self-funded models, in which corporate universities serve as a business unit and finance their activities, or even provide revenue for the company. However, in the context of Indonesia, most of the corporate universities are funded by the corporate fund through the human capital function. Since Indonesian corporate universities are considered relatively new, they have yet to produce state-of-the-art knowledge that can be copyrighted and generate profits. Nevertheless, Indonesian corporate universities are promising enough and still have much room for improvement in the future.

## CORPORATE UNIVERSITY LEARNING VALUE CHAIN

The learning value chain encompasses the full range of activities or functions required to bring a learning solution for the stakeholders. It consists of the full lifecycle of a learning solution service. According to Satrijono et al. (2017), the learning value chain consists of a learning needs analysis, learning design, learning delivery, and learning evaluation or impact measurement. This section will discuss the Indonesian corporate university learning value chain's best practices.

### Step 1: Capture – Learning Needs Analysis

As stated before, one of the drawbacks of conventional training is the lack of a prior needs analysis, a process that is important for ensuring that the learning is in accordance with the company's goals. A learning needs analysis in a corporate university is the process of identifying factors that deserve more attention to improve the employees' performance. The learning needs analysis is usually conducted by following these steps:

1. Analyze the gap between the existing performance and the expected performance.
2. Identify the challenges and obstacles that might arise preventing performance improvement.
3. Designing a plan to reduce the gap between the existing performance and expected performance.

Surveys, performance appraisal data, and employees' individual career plans can be sources for analyzing the gap. Gaps that usually occur in the company are competency gaps, performance gaps, and generational gaps (Satrijono et al., 2017). Technically, a learning needs analysis should result in information that is needed to design the learning strategy.

*Figure 3. Five-level learning evaluation model*

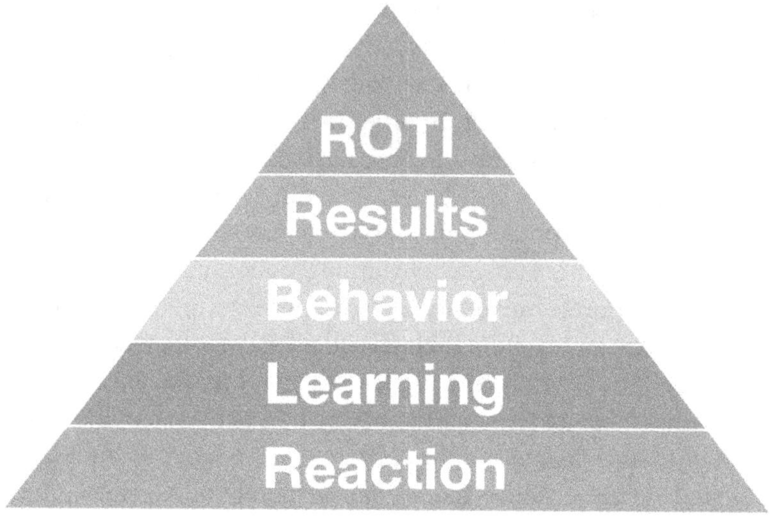


## Step 2: Expand – Learning Design

The next step after the learning needs analysis is designing the learning so that it becomes effective. Learning design is based on the learning needs analysis and targets not just the company's current needs but also its predicted future needs. Several factors are to be emphasized in the learning design step, such as these:

1. Learning curriculum and materials: these consist of the required qualifications and competences as well as the combination of technical skills and the soft skills required in the job description.
2. Learning methods: these ensure that the designed curriculum and materials are delivered effectively to the employees. The common methods employed in Indonesian corporate universities is the blended learning method, in which several learning techniques (e.g., classroom learning, e-learning, assignments, mentorship, action learning, etc.) are combined.
3. Competent instructors: delivery of the learning should be in the hands of instructors that possess the needed competence related to the materials, methods, and curriculum.
4. Learning tools and technology: learning can be assisted by the use of virtual or digital learning, a learning management system (LMS), and the supporting infrastructure. Several Indonesian corporate universities, such as PLN Corporate University, Telkom Corporate University, and Bank Mandiri Corporate University, have integrated an advanced LMS that facilitates tracking, managing, and delivering the learning programs to the learners or end-users.

Alongside the technological development that is getting more advanced for business purposes on a daily basis, new technology adoption has been taken up on a large scale among Indonesian corporate universities. As new technology—such as learning machines, artificial intelligence, and big data analysis—are continuously developing, it is expected that the new technology will lead and transform corporate university practices as well.

## Step 3: Apply – Learning Delivery

Learning delivery is the actualization of the designed learning. This process is highly important because it is the means by which knowledge is transferred to the learners. The key components of learning delivery are these:

1. The instructors that possess the key knowledge and competences must be able to deliver the materials effectively. In the PLN Corporate University, for instance, the instructors have to be qualified in order to ensure the quality of learning.
2. The course materials must be up to date and relevant to improve the employees' competence and performance. For each course delivery, before and after assessments (pre-test and post-test) are usually conducted to judge the extent of improvement resulting from the learning process.
3. The method of delivery can involve the use of conventional, digital, or blended learning techniques. Blended learning, combining face-to-face traditional learning, online or digital learning, and one-on-one hands-on mentoring or coaching, is the most common method applied by Indonesian corporate universities.

## Step 4: Evaluate – Learning Evaluation and Impact Measurement

Learning evaluation and impact measurement is the final cycle of the learning value chain. Of the many evaluation models to evaluate the result of a training program, the most common model used by Indonesian corporate universities is the Kirkpatrick four-level model (1976; in Kirkpatrick & Kirkpatrick, 2006), combined with Phillips's Return on Training Investment (ROTI) model (2011) as the fifth level.

This model evaluates the result of training based on five levels:

1. **Reaction**: This level measures how the employees react after a learning program. It is as simple as asking their thoughts about the overall learning, how they feel about it, and whether it is useful for them, as the employees are finished with the learning.
2. **Learning**: This level measures whether the employees have developed the desired knowledge, skills, and attitude through the learning program.
3. **Behavior**: This level measures whether there is a change in behavior after a learning intervention and whether the knowledge acquired is being applied. Behavioral level evaluation is usually conducted using the 360-degree evaluation (Tyson & Ward, 2004).
4. **Results**: This level evaluates the results of the training as measured against the primary goal. Results evaluation is usually measured by key performance indicators (KPIs).
5. **Return on Training Investment (ROTI)**: This level measures the impact of training investment on the return (benefit) after the learning. This level of measurement can be a bit tricky since it requires an in-depth scientific analysis.

Kirkpatrick's model alone has been widely used by Indonesian corporations' conventional training departments. The corporate university, however, tries to add the fifth level, namely the ROTI measurement level. This level provides a way to quantify the return from the investment spent on learning initiatives. However, the implementation requires rigorous practice to ensure its accuracy. Therefore, not many corporate universities have adopted the fifth level as part of their learning evaluation.

## CONCLUSION

Blooming in Indonesia, local corporate universities have shown that they have become more than a dressed-up learning center with a new name. Instead, they represent a fresh effort to develop human resources competencies in order to be capable of adapting to the changing environment. In order to have such a huge impact on the company, the corporate university is placed at a strategic level and strictly linked to the corporate strategic goals.

Although specific practices vary from one company to the next as they are adapted to each company's specific goals, almost all Indonesian corporate universities are modeled similarly, adopting the previous models that have been documented in the corporate university literature. Most of the corporate universities followed the "best practices" examples and slightly modified them to fit their company. As Indonesian corporate universities keep growing in numbers, more corporations are predicted to build their own corporate universities, and the already established ones are expected to evolve to a more mature one.

In many cases, global corporate universities conduct courses that can be transferred into academic credit or award the learners with formal degrees. Arguably, these corporate universities are "competing" with conventional universities. In the current Indonesian scene, however, corporate universities are unlikely to stand head to head with the conventional university. Although corporate universities existed in Indonesia is partly because of the gap of what conventional university graduates offer and the job requirements of the company, instead of competing, many corporate universities and conventional universities are working hand in hand to deliver knowledge to the learners. Indonesian conventional universities still hold the role as a knowledge creator and disseminator, while the partnering corporate universities benefit from that knowledge.

Since formal degrees are considered essential to verify that the learners have gained specific knowledge, skills and competencies, engaging in corporate university courses that do not worth academic credits might be perceived as less valuable by the learners. Nevertheless, if Indonesian corporate universities were able to issue academic degrees to the learners, they might disrupt the conventional universities and higher education institutions in general. This disruption would likely to happen because the courses offered would not just fit perfectly to the need of the industry, but the learners would also get the prestigious formal degree. Meanwhile, the conventional universities would still lack in meeting the needs of the industry. This possible scenario raised several questions to the current state of Indonesian higher education institutions. What makes them unable to meet the industry's needs? These gaps should be evaluated and addressed by the higher education faculty members so that they would be able to redesign the conventional university programs and narrow the gaps.

Although the concept of corporate university originated in a developed country, this approach starts to get more common in the developing country as well. This chapter sheds light on the corporate university trends, practices, and challenges in a developing country, Indonesia. Most of corporate universities literature came from developed countries, which have different context compared to the ones in developing countries. Therefore, it is essential to understand the example of corporate university practices in a developing country such as Indonesia, since it will likely to have more similarities with other developing countries. For this reason, this chapter is expected to be useful for corporate universities practitioners in developing countries. The practices of Indonesian corporate university explained in this chapter perhaps can be employed by corporate universities in the context of other countries. The challenges that are faced by Indonesian corporate universities can also serve as a remark that should be taken into consideration by corporate university executives. Understanding the case of developing country may assist other corporate universities in developing countries effort to ensure that the main goals and objectives are within reach.

The new institutions are given the name "university," and yet they are not 'real' universities. The term "university' is employed to bring the spirit of continuous learning to the company. Similar with conventional universities, however, corporate universities in Indonesia have established their roles as the center of knowledge and center of excellence in the scope of corporations.

## REFERENCES

Abel, A. T. L. (2008). *The Development of A Conceptual Framework and Taxonomy for Defining and Classifying Corporate Universities*. New York University. doi:10.1017/CBO9781107415324.004

Alagaraja, M., & Li, J. (2015). Utilizing institutional perspectives to investigate the emergence, rise, and (relative) decline of corporate universities. *Human Resource Development International*, *18*(1), 4–23. doi:10.1080/13678868.2014.979003

Allen, M. (Ed.). (2002). *The Corporate University Handbook: designing, managing and growing a successful program*. New York: Amacom.

Andresen, M., & Lichtenberger, B. (2007). The corporate university landscape in Germany. *Journal of Workplace Learning*, *19*(2), 109–123. doi:10.1108/13665620710728484

Baporikar, N. (2014). Corporate university edification in knowledge society. *International Journal of Strategic Change Management*, *5*(2), 125–139. doi:10.1504/IJSCM.2014.060258

Barley, K. (2002). Corporate University Structures that Reflect Organizational Structures. In M. Allen (Ed.), *The Corporate University Handbook: designing, managing and growing a successful program*. New York: Amacom.

Barley, K. (2002). Corporate University Structures that Reflect Organizational Structures. In M. Allen (Ed.), *The Corporate University Handbook: designing, managing and growing a successful program*. New York: Amacom.

Ben-Hur, S., Jaworski, B., & Gray, D. (2015). Aligning Corporate Learning With Strategy. *MIT Sloan Management Review*, *57*(1), 1–12.

Drucker, P. F. (1999). Knowledge-Worker Productivity: The Biggest Challenge. *California Management Review*, *41*(2), 79–94. doi:10.2307/41165987

El-Tannir, A. A. (2002). The corporate university model for continuous learning, training and development. *Education + Training*, *44*(2), 76–81. doi:10.1108/00400910210419973

Ewer, G., & Russ-Eft, D. (2017). 35 Corporate University Theory and Practice: the Case of Platt University, USA. *International Journal of Human Resource Development: Practice. Policy & Research*, *2*(1), 35–49. doi:10.22324/ijhrdppr.2.104

Hearn, D. R. (2002) Education in the Workplace: An Examination of Corporate University Models. New Foundations. Retrieved from http://www.newfoundations.com/OrgTheory

Hilse, H., & Nicolai, A. T. (2004). Strategic learning in Germany's largest companies: Empirical evidence on the role of corporate universities within strategy processes. *Journal of Management Development*, *23*(4), 372–398. doi:10.1108/02621710410529811

Kirkpatrick, D. L., & Kirkpatrick, J. D. (2006). *Evaluating Training Programs*. Tata McGraw-Hill Education.

Lytovchenko, I. (2016). Corporate University As a Form of Employee Training and Development in American Companies. *Advanced Education*, 35. doi:10.20535/2410-8286.62280

Meister, J. C. (1998). *Corporate Universities: Lessons in Building a World-Class Workforce*. New York: McGraw-Hill.

Moore, J. (2002). Running a Corporate University like a Business: a financial model. In M. Allen (Ed.), *The Corporate University Handbook: designing, managing and growing a successful program*. New York: Amacom.

Nixon, J. C., & Helms, M. M. (2002). Corporate universities vs higher education institutions. *Industrial and Commercial Training*, *34*(4), 144–150. doi:10.1108/00197850210429129

Oh, J. R., & Park, C. H. (2011). Corporate Universities and Adult Education: Implications for Theory and Practice. *Adult Education Research Conference*. Retrieved from http://www.adulterc.org/Proceedings/2011/papers/rok_park.pdf

Patrucco, A. S., Pellizzoni, E., & Buganza, T. (2017). The design process of corporate universities: A stakeholder approach. *Journal of Workplace Learning*, *29*(4), 304–318. doi:10.1108/JWL-04-2016-0024

Phillips, J. J. (2011). *Return on Investment in Training and Performance Improvement Programs* (2nd ed.). Routledge. doi:10.1017/CBO9781107415324.004

Phillips, J. J., & Phillips, P. P. (2007). Next-Generation Evaluation: Searching for Value. In M. Allen (Ed.), *Next Generation of Corporate Universities*. San Francisco, CA: Pfeiffer.

Plompen, M. (2005). *Innovative Corporate Learning: Excellent Management Development Practice in Europe*. New York: Palgrave Macmillan. doi:10.1057/9780230288799

Prince, C., & Stewart, J. (2002). Corporate universities – an analytical framework. *Journal of Management Development*, *21*(10), 794–811. doi:10.1108/02621710210448057

Qiao, J. X. (2009). Corporate universities in China: Processes, issues and challenges. *Journal of Workplace Learning*, *21*(2), 166–174. doi:10.1108/13665620910934852

Rademakers, M. (2005). Corporate universities: Driving force of knowledge innovation. *Journal of Workplace Learning*, *17*(1/2), 130–136. doi:10.1108/13665620510574513

Rademakers, M. (2014). *Corporate Universities: Drivers of the Learning Organization*. New York: Routledge. doi:10.4324/9780203073988

Ryan, L., Prince, C., & Turner, P. (2015). The Changing and Developing Role of the Corporate University Post-Millennium. *Industry and Higher Education*, *29*(3), 167–174. doi:10.5367/ihe.2015.0256

Satrijono, W., Djawahir, K. M., & Sugiarsono, J. (Eds.). (2017). *Indonesia's Best Practices of Corporate University*. Jakarta: PLN Corporate University.

Thompson, G. (2000). Unfulfilled Prophecy: The Evolution of Corporate Colleges. *The Journal of Higher Education*, *71*(3), 322–341. Retrieved from http://www.jstor.org/stable/2943781

Tyson, S., & Ward, P. (2004). The Use of 360 Degree Feedback Technique in the Evaluation of Management Development. *Management Learning*, *35*(2), 205–223. doi:10.1177/1350507604043025

Upadhyay, P., & Paul, M. (2019). The linkage between knowledge management practices and organization based projects for better learning outcome. *International Journal of Educational Management*, *33*(1), 166–178. doi:10.1108/IJEM-04-2018-0133

Walton, J. (2005). Would the real corporate university please stand up? *Journal of European Industrial Training*, *29*(1), 7–20. doi:10.1108/03090590510576181

# Chapter 10
# Health Sustainability and Socialization Agents Roles in Organ Donation:
## A Malaysian Youth Case

**Pushpavalli Maniam**
*Tunku Abdul Rahman University College, Malaysia*

**Izzal Asnira Zolkepli**
*Universiti Sains Malaysia, Malaysia*

## ABSTRACT

*This article explores how sustainable health can be encouraged through the role of socialization amongst the youth in Malaysia in the context of the decision-making process towards becoming an organ donor. This is imperative for the nation especially the policymaker and health business sector in designing strategies pertaining to health issues. The concept of the 'duality of structure' is used as a starting point to link between health sustainability and health communication on organ donation in congregating intergenerational equity by uncovering the structural properties or conditions which either enable or constrain the future of health initiatives. Therefore, this article is aimed to focus on the values and norms commonly transferred by the socialization agents regarding the behavioral development of the potential donor. Data was gathered using self-administered questionnaires from 162 youth. The findings illustrate that the supportive influence from agents of socialization affected towards certain degree on the behavioral formation on becoming an organ donor.*

DOI: 10.4018/978-1-7998-1566-2.ch010

## INTRODUCTION

Organ donation is a medical procedure which involves a process of an individual freely giving consent to donate organs or tissue for the purpose of transplantation for the very ill or dying living recipients (Transplant Unit, 2015). In Malaysia, organ donation is facilitated by the Human Tissue Act (1974) and the National Transplantation Program (1975). It is propelled by the National Transplant Resource Centre with the support and cooperation of Ministry of Health and Medical Development. The objective is to enhance public awareness about organ donation in Malaysia. Hence, the relevant authorities have took many initiatives to keep the public aware and informed, subsequently register as an organ donor. However, the number of donors in Malaysia are amongst the lowest in the world (Foong, Sheng, Ong, Oo, Hossain, Baskaran, Haron, Valappil and Varadarajan, 2019). According to the statistics given by the National Transplant Resource Centre, from 1976 until July 2015, only 537 organ donors (deceased) have been registered in the country, and up to September 2019, only 481,833 Malaysians have pledged to donate their organs i.e. 1.3% of total population. The World Health Organization noted that among the 68 countries committed in organ donation, Malaysia has the lowest number of donors. Moreover, 21,018 individuals are still waiting for organ transplants (BH Online, 2019). For instance, in the year 2017, mere 86 cases of organ transplantation were carried out and only 35 deceased organ donations were recorded in Malaysia (Baskaran, Haron & Valappil, 2019).

Organ transplantation is the best form of treatment for patients whom is at the end stage of organ disease. Transplantation will improve the survival rates and quality of life of the patient. Despite organ transplantation being available in Malaysia since 1975, there were only 86 transplants carried out and a mere 35 deceased organ donations made in 2017. The number of patients requiring transplant keep on increasing with over 21,000 currently on the waiting list (Foong et al., 2019). Several factors contribute to the organ donation and transplantation rates in Malaysia that includes the misconceptions of the possible barriers to gain consent from potential organ donors, the surgical procedure and concerns about the mutilation of the body, lack of awareness and knowledge, public and spiritual belief and attitudes, cultural differences and myth, consent process and insufficient investment in the area of transplant infrastructure and workforce (Kopfman, Smith, Yun, & Hodges, 1998; Loch, Hilmi, Mazam, Pillay, & Choon, 2010; Wong, 2010; Mostafa, 2010). Other negative attitudes include fear of organs being used for research and fear of getting less active treatment if the patient known to be a donor (Edmund, Shazlyiana, Nursyahirah, Fam, & Chua, 2018). Also, the reluctance to pledge as an organ donor is also contributed by the organ donation programs itself that do not highlight on the tangible benefit (Lwin, Williams, & Lan, 2002), if the program can be associated with quality of life and spiritual positive outcome of death, the perception probably can e latered (Makmor, Abdillah, Raja Noriza, Nurulhuda, Soo-Kun, & Kok-Peng, 2014). Carducci (1984) have found that one of possible explanations for organ shortage is the perception of the non-donors who believe that the cost of donating organ is greater than its reward (including tangible and intangible rewards). Some other views also emphasize that characteristics of the collectivistic culture, emotional support, and family opinions works also as a strong contributor that can explained the rejecting act of donating organ (Irving, Jan, Tong, Wong, Craig, Chadban, Rose, Cass, Allen, & Howard, 2014). For example, BH Online (2019) reported that 80% potential individuals did not get consent from their family members. In a nutshell, the gesture of volunteering as an organ donor reflect the values, beliefs, attitude and factual knowledge of the person themselves.

Despite such a strong unwillingness to donate organ among Malaysians, very few empirical studies were conducted to find the solution of this problem. Some notable ones are Wong (2010), Normawati, Faridah, Normah & Chang (2011), Nazni, Zaherawati, Adnan, Mahazril, Mohd Zool Hilmie, Mohd Shamsul, Azrul Shahimy, & Kamarudin (2012) and Foong et al. (2019). Their studies basically focused on the reasons for reluctance and not on examining possible areas of communicating health and sustainability issues with regards to becoming organ donor to the society at large, so that organ transplant can be used as one of the solutions to increase health sustainability in Malaysian.

To merge this with the context of generational studies, it is alarming to see the increasing number of youths in Malaysia aged between 15 to 40 that counts at 45% that have a minimum awareness of becoming an organ donor (Yunus & Landau, 2019, Foong 2019). With the progressive increment from this cohort, the need to change conception indeed important to ensure that the above-mentioned hindrances are handled. Therefore, this study is attempted to address the problems regarding the insufficiency of donors by looking precisely into the construct of influence from their socialization agents that is the parents, peers, school and media (old and new media). Most Malaysian youth relied very much on information and guidance from their socialization agents to achieve life goals. With this as a backdrop this study investigates on two issues: (1) the relationship between the informational influence from each agent on youth's attitude and behavior towards the act of donating organ and (2) the relationship between the supportive influence from each agent on youth's attitude and behavior towards organ donation. The outcome of this study will hell the social marketer and policy maker to design a more effective health communication program, that could ensure the sustainability of the society is met.

## LITERATURE REVIEW

This study focuses on the concept of consumer socialization by Moschis and Churchill (1978) which referred to the people and groups that influence an individual's self-concept, emotion, attitude, and behavior. Consumer socialization has been defined as a process by which young people develop skills, knowledge, attitudes, and behavior relevant to their role in society which includes performance as consumers (Alvarez, 2019). Grusec (2002, cited in Arnett, 2006) added that socialization explains how individuals are assisted to acquire skills necessary to function as members of their social group. The process of socialization usually considered to have significant impacts on how individuals learn, develop and change over time (Hota & Bartsch, 2019). Although this concept is majorly used with regards to marketing activities of products and services, research shows that this notion can be transferred to specific areas to understand the patterns of thinking and behaving towards public issues and social causes. The key concept of socialization is the influence of agents on any individuals. Particularly, this paper discusses the informative and supportive influence of socialization agents upon youth attitude towards becoming an organ donor. In understanding this, all socialization agents are assessed based on the Theory of Reasoned Action as proposed by Ajzen and Fishbein (1980). This approach suggested that an individual's intention is shaped by attitude and subjective norm. The main insight of this study referred to an individual's beliefs on whether the significant others think that he or she should engage in a given behavior. Explanations of significant others explicitly narrate the functions of the socialization agents in the life of an individual. These are the people whose opinions are considered important to the individual.

## Socialization Agents

The term "socialization" refers to the process of teaching naïve individuals the skills, behaviors, values, and motivations needed for them to function effectively in a social setting (Maccoby, 2006). Consumption related skills, knowledge and attitude are transferred between generations (Aleti, Brennan, & Parker, 2015). Many agents are involved in human life-learning process, namely parents, peers, educators, and media. The agents serve as teachers and young individuals as learners. For many years, the concept of socialization has received substantial attention in the academic literature because it has presented new directions and opportunities in understanding attitudes and behaviors. Attitude is largely known as "a learned predisposition to respond in a consistently favorable or unfavorable manner with respect to a given object" (Fishbein & Ajzen, 1980). In order to articulate the influence of agents towards the attitude and behavior of youth, past studies have delved into the role and importance of socialization agents.

Parental influence is defined as parent-child interaction which stresses on family values and the need of conforming to social norms (Wiman, 1983). Parents play a strategic role in positively influencing attitude and behavior of their children. Ramaseshan, Wong, and Turner (2011) emphasized that family values are positively associated with willingness to donate organ. Parents may convey values through discussion and encouragement. Family opinions work as a factor in determining community attitudes (Irving et al., 2014). From one generation to the other, parents are changing and altering the actions of the children (Grusec & Davidov, 2006). Parents influence their children directly through communication and discussions to help children's journey through life (Moore, Raymond, Mittelstaedt & Tanner, 2002; Pinto & Mansfield, 2011). Parents communicate to children what are acceptable and unacceptable behaviors in the culture. Family has been reinforced as the major influencer for consumer socialization (Ghouse, Chaudhary & Durrah, 2019). According to Mascarenhas and Higby (1993), parental information exceeds all other socialization agents.

In looking at the peer influence, Bukowski, Brendgen & Vitaro (2006) indicated that peers are influencing self-concept and pro-social behaviors in terms of adopting ideas that communicate an opinion about what could or should be done (Sussman & Siegal, 2003). For example, peers have been found to be preferred when decisions primarily relate to the issues of social acceptance (Moore & Bowman, 2006). As an individual gets older, the peer group becomes the primary source of social values, replacing the influence of parents, which provides additional pathway for socialization (Hota & Bartsch, 2019). In the process of achieving personal autonomy, young individuals turn to peers for stimulation (Bukowski, Brendgen & Vitaro, 2006) and they can influence each other in positive traits (Lee, 2011). Due to the fact that they share the common experience, the informational and supportive influence could be considered as one of the dominant factors in assimilating social values.

Past studies have also given considerable attention to influence from educational institutions and educators. Wentzel (1991) looked at the social relationships and identified that educational institutions and educators have been promoting social behavior in the forms of moral character and conformity to social norms and values. School-based educational programs demonstrated effective results in consumer knowledge (Dündar, 2017). Schools provide youth with skills, attitudes and cognitive bases which are designed to foster critical thinking and self-confidence (Shim, 1996). Subsequently, Pinto & Mansfield (2011) believes that schools help transform children into consumers by providing them with information about consumption-related activities. In relation to public affairs, adolescents who are exposed to organ donation information by their school curriculum are seen to be more open-minded about it. In addition, Trenholm and Rose (1981) documented that teachers communicate social values and expectations to their students. Higher-level educators focus on the development of pro-social behavior and encourage prosocial interactions (Wentzel & Looney, 2006). Social values and prosocial interactions can be developed through informational and supportive influence.

One of the most recognized agents found in the earlier studies is the media. Normawati et al. (2011) have indeed reported that newspapers, televisions, posters, and pamphlets are the main information sources of mass media for Malaysians to seek knowledge about organ and tissue donation. Rady, McGregor & Verheijde (2012) described media as an important tool to communicate accurate and reliable information for the public to make informed decisions on organ donation. According to Dubow, Huesmann, and Greenwood (2006), today's youths are exposed to media-saturated environment which serves as reference for social norms. An individual acquire models, norms and ideas on how the world works through media (Šramová, 2017). The informational and supportive influence of traditional mass media comes mainly from programming and advertising. For instance, the news coverage emphasized stories of children saved by organ transplants and the drama series such as CSI, NCIS, and Bones have portrayed the issue of organ donation which affects viewers' attitude towards organ donation (Harrison, Morgan & Chewning, 2008). In this context, the greater the mass media influence on daily life, the more suggestive is its information in terms of providing credible evidence along with serving as reference for social norms (Uzniene, 2010).

Similarly, La Ferle, Edwards and Lee (2000) referred the Internet as an attractive source of information as it is interactive and easily accessible to simultaneously act as a new social agent. The diversity of new media and the increasing usage by youth makes it a potentially solid agent of socialization which can enhance an individual's learning process (McQuail, 2010). Repeated exposure to online media may enable youth to receive information and support pertaining to various social issues including organ donation. Particularly, social media has been transforming passive individuals into active producers who care share knowledge and opinions (Zolkepli & Kamarulzaman, 2015). The growing popularity of social media has provoked the interest for consumer socialization (Delafrooz, Rahmati & Abdi, 2019). Increased connectivity enables youth to experience a social space far more diverse than their family, peers, and school (Chilton, 2000; Barber, 2013). Urbain, Gonzalez & Roberts (2013) showed that young people is subjected to family, peer and media influences. McMillan and Morrison (2006), for example, supported the fact that this cohort prefers to receive information through social networks. They are using mobile web to socialize and outpaces the earlier generations. Luczak & Younkin (2012) also reported that this cohort consists of group-oriented individuals who are becoming sensitive to societal and environmental problems. As shortage of organ supply is considered as a societal problem, youth may pay substantial attention to messages of organ donation.

## Informative and Supportive Influence

The informative influence may be understood with reference to Bearden, Netemeyer, and Teel (1989) findings which denote the tendency to accept information from knowledgeable others and be guided in the decision-making process. As in taking the right decision, an individual believes that group members' information would be able to help in conforming to social values (Arpita, Ceeba, & Ankita, 2012) and take the position advocated by the acquired information (Liu & Zhang, 2010). Sussman and Siegal (2003) have studied how people treat received advice and found that the information embedded in the advice was useful in solving one's problem.

An individual who receives information from the socialization agents would be expected to attain self-regulation; accept standards, attitudes, and values; and learn appropriate ways of behaving (Arnett, 2006). He further added that informative influence operates at the family level when young adults become notably more adept at role-taking and accept their parents' perspectives. Informative influences help individuals to seek conformance (Khare, 2014). Informative influence is said to occur when the parents teach role-taking skills and communicate the acceptable and unacceptable behavior in the culture. As an individual gets older, they tend to rely more on peers' information. Informative influence may be accomplished in educational institutions through either formal or informal education. Educators focus on the development of prosocial behaviors by communicating acceptable attitudes and values (Wentzel & Looney, 2006). Thus, a student would identify by taking on the behaviours and opinions he or she obtained from the educational institutions. Dubow, Huesmann, and Greenwood (2006) point out the power of media consumption which provides learning opportunities. Adolescents observe through the mass media's "window of the world" and alter their beliefs, attitudes, and behaviors. Similarly, new media may also appear to develop knowledge and beliefs about the appropriate social attitudes and behaviors.

Research has shown that supportive influences of socialization agents affect the psychosocial functioning of adolescents (Baumrind, 1989). Usually, the social support of parents has been most frequently documented in the socialization context, but little is known about parental supportive influence within a specific context (Swenson & Prelow, 2005). Across peer relationship, supportive influence is positively related to the internalization of values and enhancement of the well-being and negatively related to alienation or truancy (Ryan & Powelson, 1991). For example, friends provide suggestions and options which change adolescents' motivation and engagement (Ryan, 2000).

According to supportive influence literature, people develop cultural values of the society through the hidden curriculum found in educational institutions (Barkan, 2012). To be more specific, educational institutions frequently attempt to achieve a range of social goals and positive forms of behavior (Wentzel & Looney, 2006). Thus, educational institutions which solicit opinions and feelings, make social goals more salient to students. Huesmann (1995) states that mass media gained influence in socializing children while other traditional agents steadily losing influence. It is inclusive of the support system provided by both traditional and new media. A person's thoughts, feelings, and behavior are susceptible to repeated exposure to specific supportive media content. As socialization agents found to be a dominant influence towards attitude and behavior, hence the following hypotheses have been formulated.

H1a: Parental informative influence effect youths' attitude towards organ donation
H1b: Parental supportive influence effect youths' attitude towards organ donation
H2a: Peer informative influence effect youths' attitude towards organ donation
H2b: Peer supportive influence effect youths' attitude towards organ donation
H3a: The informative influence from educational institutions and educators effect youths' attitude towards organ donation
H3b: The supportive influence from educational institutions and educators effect youths' attitude towards organ donation

H4a: The informative influence from traditional mass media effect youths' attitude towards organ donation
H4a: The supportive influence from traditional mass media effect youths' attitude towards organ donation
H5a: The informative influence from new media effect youths' attitude towards organ donation
H5b: The supportive influence from new media effect youths' attitude towards organ donation
H6: Attitude towards organ donation has a significant effect on the intention to donate the organ among Malaysian youth.

# METHODOLOGY

## Quantitative Research

Quantitative approach was used to analyze the relationship between the informational and supportive influence from each agent towards the act of donating organ. The amount of information and level of support from each agent served as the foundation for the research. The study used self-administered questionnaires which were distributed to the targeted respondents via an online survey. The target population for this research was youth in Malaysia aged between 15 to 40. According to the Department of Statistics Malaysia, the estimated population of youth in Malaysia is about 14.5 million.

The questionnaire was pilot tested on 50 respondents to test the consistency and stability of the items measuring the concept of the study. The reliabilities, measured with Cronbach's alpha, ranged from .453 to .950. The accepted Cronbach value for this study is above .5 which was based on the study conducted by Moschis and Churchill (1978) on consumer socialization. As such, the item scored less than .5 was eliminated from the attitude scale to produce acceptable reliability. The remaining items were factor-analyzed for validity. For the peer influence scale, one item on the supportive influence falls on a separate component and following the recommendations by Kothari & Garg (2014), the said item was dropped in order to classify variables accordingly. Hence, seven primary factors emerged explaining 69% of the variance. The factor loadings ranged from .593 to .880. As a result, 38 item questions were retained with 162 usable questionnaires.

## Measures

All constructs in the study involved multiple items and they are measured using a 5-point Likert Scale. The data of this study were analyzed using the SPSS program. Multiple regression was used to explore the relationship between the dependent variable and the independent variables. Table 1 summarizes the sources used to operationalize the model constructs.

## Analyses

Data was keyed-in into the Statistical Package for Social Science (SPSS) for descriptive statistical analysis, bivariate correlation and multiple regression analysis. Pearson correlation was obtained to examine the relationships among the variables. Multiple regression was used to explore the relationship between the agents and the attitude and behavior of youth. A p-value less than 0.05 was considered as significant. Standardized coefficients BETA determined which socialization agents are more important in explaining the attitude and behavior of Malaysian youth towards becoming an organ donor.

*Table 1. Measurements of variables and sources*

| Construct | Variable | Item Questions | Sources |
|---|---|---|---|
| Parental influence | Informational influence | 1. My parents discuss organ donation with me.<br>2. My parents give me detailed information about organ donation.<br>3. My parents share updated information on organ donation with me. | Moschis & Churchill (1978), Soh, Charlton & Chew (2014) |
| | Supportive influence | 1. My parents support organ donation.<br>2. My parents give me advice about organ donation.<br>3. My parents encouraged me to sign up as an organ donor. | Lacanche & Legault (2007), Soh, Charlton & Chew (2014) |
| Peer influence | Informational influence | 1. My friends discuss organ donation with me.<br>2. My friends give me detailed information about organ donation.<br>3. My friends share updated information on organ donation with me. | Moschis & Churchill (1978), Soh, Charlton & Chew (2014) |
| | Supportive influence | 1. My friends support organ donation.<br>2. My friends give me advice about organ donation. | Lacanche & Legault (2007), Barber (2013), Soh, Charlton & Chew (2014) |
| Influence of Education | Informational influence | 1. My educational institution provides me useful information about organ donation.<br>2. My educational institution distributes materials about organ donation.<br>3. My teachers/lecturers talk about organ donation to me.<br>4. My teacher/lecturer communicates the values of organ donation. | Moschis & Churchill (1978), Kamaruddin & Mokhlis (2003), Lacanche & Legault (2007) |
| | Supportive influence | 1. My educational institution supports organ donation.<br>2. My educational institution encourages me to sign up as an organ donor.<br>3. My teacher/lecturer supports organ donation.<br>4. My teacher/lecturer encourages me to sign up as an organ donor. | Moschis & Churchill (1978) |
| Influence of traditional media | Informational influence | 1. I always get information about organ donation from mass media.<br>2. I consider the mass media a good source of organ donation information.<br>3. I read, hear and watch stories of people who need organs. | Wong (2010), Barber (2013) |
| | Supportive influence | 1. The mass media shows testimonials from the family members of the deceased organ donors and organ recipients.<br>2. There is good advice about organ donation on mass media.<br>3. The mass media provides me with convenient options for becoming an organ donor. | Wong (2010), Barber (2013) |
| Influence of new media | Informational influence | 1. I pay attention to organ donation information on the Internet.<br>2. I consider the Internet a good source of organ donation information.<br>3. I read and watch stories of people who need organs. | Wong (2010), Barber (2013) |
| | Supportive influence | 1. The Internet shows testimonials from the family members of the deceased organ donors and organ recipients.<br>2. There is good advice about organ donation on the Internet.<br>3. The Internet provides me with convenient options for becoming an organ donor. | Wong (2010), Barber (2013) |
| Attitude towards organ donation | | 1. I believe in the importance of organ donation.<br>2. I believe that organ donation is morally justified.<br>3. I view organ donation as a benefit to humanity. | Mostafa (2010) |

# RESULTS

## Profile of Respondents

Out of the 162 respondents, 62 (38.3%) were male and 100 (61.7%) were female. The Chinese constituted the largest ethnic group, accounting for 51.9% of the respondents, followed by Malays (25.9%), Indians (18.5%) and others (3.7%). In terms of age, the majority group is 20 years old (39.5%) with a mean of 21.17 and standard deviation of 1.64. This usable sample consisted of 12 respondents (7.4%) who have completed secondary school, 91 respondents (56.2%) completed Diploma and 59 respondents (36.4%) at the Degree level.

## Agents' Influence and Hypotheses Testing

### Parental Influence

There was no significant relationship between informational influence from parents and attitude towards organ donation. It was negatively associated with the items *My parents discuss about organ donation with* me (p = .375; r = .070); My parents give me detailed information about organ donation (p = .523; r = .051) and My parents share updated information about organ donation (p = .367; r = .071). H1a, which states that the parental informative influence has a significant effect on the youth's attitude, was not supported. On the other hand, there was a significant positive relationship between the supportive influence of parents and attitude towards the act of donating organ (p = .000; r = .329). The item My parents support organ donation was positively associated with mean attitude (p = .000; r = .412) and item My parents encourage me to sign up as an organ donor (p = .000; r = .368). These findings suggest that parents are encouraging their children to become an organ donor but there is no sufficient information given to them. This is in line with studies by Wang, Zhang & Wang (2008) and Ramaseshan, Wong & Turner (2011) who found that youth's parents are supportive of organ donation. Therefore, H1b which tested for parental influence effects on youth's attitude was supported.

### Peer Influence

Interestingly, the opposite result was found with peers. It was associated negatively for informational influence (p = .912; r = -.009). Therefore, H2a is not supported. However, supportive influence from peers showed positive correlation (p = .030; r = .170). Thus, the results of the test showed that H2a garnered support. In other words, youth who are supporting organ donation seems to be encouraging their peers to become an organ donor too although sufficient information is not being exchanged.

### The Influence of Educational Institutions and Educators

Similar to peer influence, the informational influence of education (p = .054; r = .152) was statistically insignificant but the supportive influence (p = .004; r = .224) showed positive relationship. These results rejected H3a and supported H3b. It can be concluded that educational institutions and educators are supporting youth towards organ donation.

## The Influence of Traditional Mass Media

The correlation between the informational influence and youth's attitude was statistically significant ($p = .004$; $r = .225$) and this suggests that youth is getting sufficient information from the traditional mass media. Therefore, the hypotheses (H4a) regarding the informative influence of traditional mass media were supported. Additionally, the correlation between the supportive influence of mass media and youth's attitude was also significant ($p = .007$; $r = .210$) and hence the data provided support for H4b. These data suggested that although the strength of the relationship for both informational influence and supportive influence is little the mass media is considered an influential socialization agent for youth.

## The Influence of New Media

Similarly, the two correlations for the influence of new media were significant ($p = .002$; $r = .241$ for informational influence; and $p = .004$; $r = .224$ for supportive influence). Therefore, H5a and H5b were fully accepted. As hypothesized, the statistical results with new media functioning as a predictor variable indicated that new media is a good source of organ donation information ($p = .000$; $r = .281$) and it provides good advice about organ donation ($p = .003$; $r = .230$). This finding suggests that new media is an influential socialization agent in the formation of youth's attitude towards organ donation.

## Relationship between Attitude and Behaviour

The test for correlation was done for attitude with current behavior and intention to sign up as an organ donor. The analysis demonstrated that attitude is significantly important for current donors ($p = .042$; $r = -.160$). Anyhow, the r value of -.160 shows negative relationship between attitude and the current donor behavior which means although they have negative attitude towards organ donation, they still sign up as an organ donor. Furthermore, the results also indicated that there is significant relationship between attitude and intention to sign up as an organ donor ($p = .000$; $r = -.283$). Based on the r value (-.283), the strength of the relationship showed that these two variables were having low relationship and a negative r value indicated a negative relationship between the two variables which means as the attitude increases, the intention to sign up decreases. This explains the situation that the decision to sign up as an organ donor is not supported by the positive attitude towards organ donation. Therefore, H6 is supported with negative correlation between attitude and behavior. Attitude is considered significant in predicting behavior but as indicated by this study, the negative attitude explains youth's intention of not signing the donation card.

## Regression Analysis

To understand the amount of influence of each agent upon youth's attitude and behavior, multiple regression test was run (Table 2). Parental supportive influence emerged as the main predictor ($\beta = .410$; $p = .000$) of youth's attitude towards organ donation with a significant relationship. The other predictors for the attitude of youth were peer supportive influence, new media informational influence, education supportive influence and both influence from traditional mass media. The beta value shows that the greater the amount of influence from these agents, the better the youth's attitude towards organ donation. However, all these agents were not significant predictors for attitude. Alternatively, the coefficient

*Table 2. Regression coefficient ff the influence of socialization agent]*

| Dependent Variable | Beta (β) coefficients | | | | | | | | | |
|---|---|---|---|---|---|---|---|---|---|---|
| | Parental Influence | | Peer Influence | | Influence of Education | | Influence of Mass Media | | Influence of New Media | |
| | Info | Support | Info | Support | Info | Support | Info | Support | Info | Support |
| Attitude | -.189 | .410 | -.282 | .226 | -.178 | .200 | .099 | .063 | .232 | -.157 |
| Current Behaviour | -.038 | -.143 | .179 | -.243 | .144 | -.163 | .226 | -.198 | -.182 | .065 |
| Intention to sign up | -.021 | -.181 | .085 | -.250 | .185 | -.183 | .084 | -.022 | -.327 | .122 |

*Figure 1. Framework and findings*

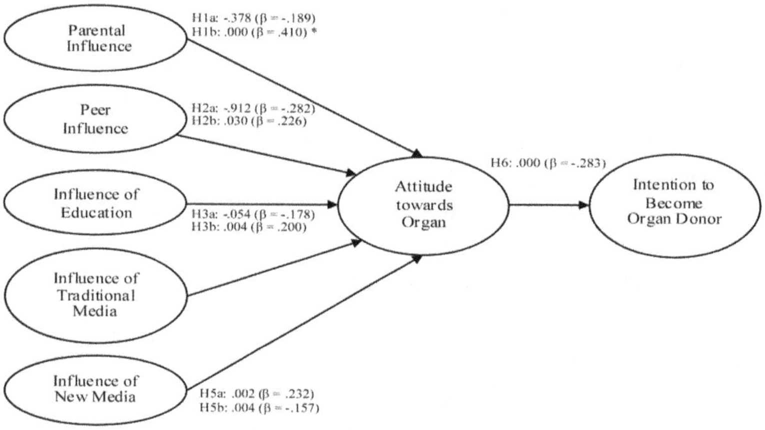


for parental informative influence, peer informative influence, educational informative influence, and new media supportive influence showed negative relationship, which indicated the greater the influence they received, the poorer their attitude towards organ donation. These results are noteworthy for the discussion later.

Next, current behavior refers to the number of donors registered at the material time of this research. Donors are more likely to be influenced by information given by the mass media ($\beta = .226$; $p = .084$), peers ($\beta = .179$; $p = .174$) and education ($\beta = .144$; $p = .420$) but none of the predictors shown significant correlation. The positive direction of coefficient indicated that the greater the information received from the mass media, peers and education, the better their behavior towards organ donation. The supportive influence from new media ($\beta = .065$; $p = .690$) was also a predictor for the current donors but not having significant relationship. On the other hand, the remaining variables showed negative coefficients suggesting that the low level of influence received from these agents does not inspire the youth to be an organ donor. The results of the final theoretical framework with the coefficient paths are presented in Figure 1. Analysis reveals that informational and supportive influence affect attitude and behavior of Malaysian youth.

## DISCUSSION AND IMPLICATIONS

The goal of the present study was to investigate the relationship between informational influence and supportive influence from each agent on Malaysian youth's attitude and behavior towards organ donation. Overall, the final framework as indicated in Figure 1 indicated that most youth have a negative attitude towards organ donation. Consistent with previous researches, for instance Sirois et al. (2005), Makmor et al. (2014) and Riyanti et al. (2014), youth does not trust the process of organ donation and due to the insufficiency of information, they may create a perception that organ donation is a terrifying activity which is dangerous to their life. Fear was also a major reason for not pledging. All these suggest that information and material directed at youth should be redeveloped.

The findings suggested that among all the socialization agents examined in this study, parents had the most effect on the attitude of youth with regards to organ donation. One explanation for this finding is consistent with past research on that attributed it to the role of parents in influencing their children's attitude (Wang, Zhang & Wang, 2008). The impact of parents appears to be strong in terms of supportive influence but in terms of strength of their support, it has lessened effects on youth's attitude. Hence, it appeared that parents are not considering much about getting their children to pledge as an organ donor. They may not perceive organ donation as a highly important issue. The study also revealed that the informative influence of parents is not significant at all. It should be noted that parental influence is associated with less information on organ donation. This finding suggested that parents are not getting sufficient information about organ donation and because of that, they could not communicate it to their children. Another possibility is parents are not paying attention to organ donation information and thus they are not equipped with sufficient knowledge to share them with their children. This is unfavourable for the relevant authority.

Peer influence is associated with an erosion of attitude towards organ donation. As opposed to the literature, results demonstrated that youth is getting little support from their peers on becoming a donor. Peer informational influence was totally not significant with their attitude and this may reflect youth's status as a generation who is unknown of their surrounding's phenomenon and they do not share information on social causes. In addition, there is a general assumption that the young generation has less concern on health issues, hence there is no impact on their awareness and action towards organ donation. Unlike parents, peers are considered not favorable in shaping potential donors. It is important and necessary to change their mindsets about organ donation.

It appeared that educational institutions and educators had mixed results. Institutions and educators appeared to influence only in terms of supportive and are associated negatively in terms of information sent to youth. Similar to peer influence, the support given by the institutions and educators are associated with a positive attitude of youth. Information influence from these entities had no significant effects on youth. This suggests that educational institutions and educators can serve as important mechanism to support and enhance the notion of organ donation and inform the members of youth about the values and facts of organ donation in a way that peers may be unable to do so.

The results of this study supported the previous research done by Pinto & Mansfield (2011) that attitude is influenced by the exposure of the media. It showed that both informative and supportive influenced by the traditional mass media improve the attitude of youth. These desirable effects of traditional mass media appear to increase the likelihood that this is one of the best sources to transmit the information and messages about organ donation directed at youth. In such situations, communication conveyed through the mass media may improve the knowledge and attitude of youth.

Besides, it is interesting to note that the influence of new media is found to be in the same direction as the traditional mass media. For both informational and supportive influence, it is observed that the influence of new media is greater and having a stronger effect on the formation of youth's attitude towards organ donation. Consistent with the literature and the findings on time spent, the degree to which youth is influenced by the new media appears to be significantly higher. This is likely to occur because among youth, the new media is available to them everywhere and they are exposed to it all the time. Consequently, the influence of new media is relatively powerful in developing the attitude of youth.

Clearly, the results of the study bring into light a few important things that social marketers need to be cognizant of. First, this study found that the youth care about organ donation as a majority of them agreed on the importance of organ donation. Thus, health promotional strategy should be developed to promote more persuasive slogan, image or emotional feeling (rewards and benefit) about the value of organ donation will likely be noticed. Authorities should use strategies to disperse the negative perception that holds youth back from signing in. Particularly, the health communication strategy should not be directed with fear but to focus on the charitable feelings of the donor and his/her family coupled with the benefits that the organ receiver will get with the appreciative values shown by the family members. As attitude determines behaviour, the authority needs to consider how to create appropriate communication programs that appeal to this cohort by emphasizing features such as "one sight, one sound, one concept" as proposed by Schultz & Shultz (2004) being spanned across various channels to change their attitude. This concept ensures that the communication remains consistent and achieves maximum clarity.

Second, since parental influence is the most powerful agent facilitating youth's attitude, social marketers should encourage the parents of youth to frequently share organ donation information with their children so that youth will be equipped with sufficient knowledge. One suggestion would be to cultivate a charitable feeling at the point of communication. Advertising message should promote the knowledge and values of organ donation. Another way to stimulate active involvement of parents who are currently not supportive of organ donation is to use affective or emotional appeals in their promotional messages. Parents should be educated first about the low rate of organ donation and the consequences faced by the awaiting patients. Subsequently, they would be able to transmit the message effectively to fit their child's needs.

Another option would be to provide information for youth in order for them to make knowledgeable decisions as well as clarifying the misconceptions about organ donation. Efforts at a ministerial-level can be done to insert the ideas of organ donation in any part of the curriculum and co-curriculum in order to maximize students' awareness and understanding. Social marketers should consider lobbying through the mass media, specifically the key editors, to include more programs and materials that motivate organ donation as well as creating Public Service Announcement (PSA) on organ donation. At the same time, as indicated by Quick, Morgan, Lavoie, and Bosch (2014), it is important to ensure that there's no negative content being portrayed in any program or materials that could instil negative perception pertaining to the act of donating organ. These should reduce the negative attitude towards the act of donating organ.

In addition, social marketers should undertake a comprehensive review of the effectiveness of new media contents regarding organ donation. For example, The National Transplant Resource Centre is active in Facebook by updating their events as well as factual data about organ donation. As indicated in one of the previous studies by Wong (2010), it will be good to sustain relationship with those who have liked the Facebook page. For instance, they may consider displaying the pledgers' lists so that it would be a motivational factor for other visitors. They should create a promotional plan that can establish the feelings of being appreciated for their effort to sign up an organ donor. All these may increase awareness among the public and particularly youth.

## CONCLUSION

As one of the studies that explore the influence of socialization agents towards the intention of donating organ, the findings were more encouraging than expected. However, the sample of study was limited to one sub-cohort of population. Future research should be conducted among the other cohorts of population especially the aging cohort. It is expected the result of this cohort will differ greatly with the youth conception. Another possible limitation of this study is that it has not been determined whether the influence from the agents is permanent. It is likely that as youth mature, changes may happen in terms of their surroundings and therefore the informative and supportive influence may change too. Additional research could also be conducted on the rationality of agents' influence for their next phases in life.

Future analysis is clearly needed to look at how to maximize the positive health communication element on organ donation especially with regards to creative messages while minimizing its negative impact on the younger generation. This analysis is required to determine the type of health promotional tools needed to effectively motivate youth to sign up as an organ donor. As young demographics account for the greatest growth in recent years, future research could investigate how the socialization agents' influence varies across other cohorts of youth probably a generational study will be a great benefit to the expansion of knowledge. For instance, comparative studies may determine if differences play a part in the attitude and behavior towards organ donation.

In conclusion, this study confirms that parents are powerful socializers and still play a big role become in motivating youth to pledge as an organ donor. Their importance is due, among other factors, to their relation with the attitude and behavior of their children. This entails the need for the social marketers to realize that socialization agents affect youth's attitude and behavior through their informative and supportive influence. In the quest for achieving high health sustainable awareness, this study provides direction for social marketers and policy makers to develop a more integrated health communication strategies focusing on encouragement and inspiration. Advertising and promotion may play an important role spelling out the realities and meanings of organ donation. The use of social media is expected to contribute immensely towards the process of persuading the young individuals as a whole. By doing so, society at large would have better knowledge and perceptions, which in turn, result in higher levels of organ donation.

## REFERENCES

Ajzen, I., & Fishbein, M. (1980). *Understanding attitudes and predicting social behavior*. Englewood Cliffs, CA: Prentice Hall.

Aleti, T., Brennan, L., & Parker, L. (2015). Family communication for the modern era: A typology. *Young Consumers*, *16*(4), 367–384. doi:10.1108/YC-01-2015-00500

Alvarez, E. (2019). Information Sources That Influence the Financial Literacy of Puerto Rican College Students [Doctoral dissertation]. Walden University.

Arnett, J. J. (2006). Socialization in emerging adulthood: From the family to the wider world, from socialization to self-socialization. In *Grusec, J.E., & Hastings, P.D. Handbook of Socialization: Theory and Research*. New York: Guilford Press.

Arpita, K., Ceeba, P., & Ankita, M. (2012). Influence of normative and informative values on fashion clothing involvement of Indian women. *Journal of Customer Behaviour, 11*, 1, 9–32.

Barber, N. A. (2013). Investigating the potential influence of the internet as a new socialization agent in context with other traditional socialization agents. *Journal of Marketing Theory and Practice, 21*(2), 179–193. doi:10.2753/MTP1069-6679210204

Barkan, S. E. (2012). *Sociology: Comprehensive Edition*. Creative Commons.

Baskaran, N. D., Haron, H., & Valappil, M. (2019). Demographics of healthcare professionals' knowledge and attitude toward deceased organ donation: Survey of critical care areas in a tertiary hospital. *The Medical Journal of Malaysia, 74*(2), 109. PMID:31079120

Baumrind, D. (1989). Rearing competent children. In W. Damon (Ed.), *Child Development Today and Tomorrow*. San Francisco: Jossey-Bass.

Bearden, W. O., Netemeyer, R. G., & Teel, J. E. (1989). Measurement of consumer susceptibility to interpersonal influence. *The Journal of Consumer Research, 15*(4), 473–481. doi:10.1086/209186

Bukowski, W. M., Brendgen, M., & Vitaro, F. (2006). Peers and socialization: Effects on externalizing and internalizing problems. In *Grusec, J.E., & Hastings, P.D. Handbook of Socialization: Theory and Research*. New York: Guilford Press.

Carducci, B. J., & Deuser, P. S. (1984). The foot-in-the donor technique: Initial request and organ donation. *Basic and Applied Social Psychology, 5*(1), 75–81. doi:10.120715324834basp0501_5 PMID:11653697

Chilton, D. (2000). Teens excel at media multi-tasking. *Marketing Magazine, 105*(28), 4.

Delafrooz, N., Rahmati, Y., & Abdi, M. (2019). The influence of electronic word of mouth on Instagram users: An emphasis on consumer socialization framework. *Cogent Business & Management, 6*(1), 1606973. doi:10.1080/23311975.2019.1606973

Dubow, E. F., Huesmann, L. R., & Greenwood, D. (2006). Media and youth socialization: Underlying processes and moderators of effects. In *J.E. Grusec & P.D. Hastings (Eds.), Handbook of Socialization: Theory and Research*. New York: Guilford Press.

Dündar, Ş. (2017). The Effects of Conscious Consumerism-Related Parental Role Modeling, Parental Communication, and Social Studies on Children's Socialization as Conscious Consumers. *Mersin University Journal of the Faculty of Education, 13*(2).

Edmund, L. W. J., Shazlyiana, H. N., Nursyahirah, N. F., Fam, J. C., & Chua, J. (2018). Factors Affecting Attitude and Behaviour towards Organ Donation among Medical Students in Malaysia: A Cross-Sectional Study. *Journal of Advances in Medicine and Medical Research*, 1-13.

Foong, J. W., Sheng, O. J., Loon, O. W., Hossain, M. M., Baskaran, N. D., Haron, H., ... Varadarajan, R. (2019). Demographics of healthcare professionals' knowledge and attitude toward deceased organ donation: Survey of critical care areas in a tertiary hospital. *The Medical Journal of Malaysia, 74*(2), 109–115. PMID:31079120

Ghouse, S. M., Chaudhary, M., & Durrah, O. (2019). Socialization and the buying behaviour of the Arab child consumers: Insights from Oman. *Journal of Islamic Marketing*. doi:10.1108/JIMA-09-2018-0176

Grusec, J. E., & Davidov, M. (2006). Socialization in the family: The roles of parents. In *J.E. Grusec & P.D. Hastings (Eds.), Handbook of Socialization: Theory and Research*. New York: Guilford Press.

Harrison, T. R., Morgan, S. E., & Chewning, L. V. (2008). The challenges of social marketing of organ donation: News and entertainment coverage of donation and transplantation. *Health Marketing Quarterly*, *25*(1/2), 33–65. doi:10.1080/07359680802126079 PMID:18935879

Hota, M., & Bartsch, F. (2019). Consumer socialization in childhood and adolescence: Impact of psychological development and family structure. *Journal of Business Research*, *105*, 11–20. doi:10.1016/j.jbusres.2019.07.035

Huesmann, L. R. (1995). *Screen violence and real violence: Understanding the link*. Auckland, New Zealand: Media Aware.

Irving, M. J., Jan, S., Tong, A., Wong, G., Craig, J. C., Chadban, S., ... Howard, K. (2014). What factors influence people's decisions to register for organ donation? The results of a nominal group study. *Transplant International*, *27*(6), 617–624. doi:10.1111/tri.12307 PMID:24617320

Kamaruddin, A. R., & Mokhlis, S. (2003). Consumer socialization, social structural factors and decision-making styles: A case study of adolescents in Malaysia. *International Journal of Consumer Studies*, *27*(2), 145–156. doi:10.1046/j.1470-6431.2003.00297.x

Khare, A. (2014). Consumers' susceptibility to interpersonal influence as a determining factor of ecologically conscious behaviour. *Marketing Intelligence & Planning*, *32*(1), 2–20. doi:10.1108/MIP-04-2013-0062

Kopfman, J. E., Smith, S. W., Yun, J. K. A., & Hodges, A. (1998). Affective and cognitive reactions to narrative versus statistical evidence organ donation messages. *Journal of Applied Communication Research*, *26*(3), 279–300. doi:10.1080/00909889809365508

Kothari, C. R., & Garg, G. (2014). *Research methodology: Methods and techniques*. New Delhi: New Age International Publishers.

La Ferle, C., Edwards, S. M., & Lee, W. (2000). Teen's use of traditional media and the Internet. *Journal of Advertising Research*, *40*(3), 55–65. doi:10.2501/JAR-40-3-55-65

Lachance, M. J., & Legault, F. (2007). College students' consumer competence: Identifying the socialization sources. *Journal of Research for Consumers*, *13*, 1–21.

Lee, K. (2011). The green purchase behaviour of Hong Kong young consumers: The role of peer influence, local environmental involvement and concrete environmental knowledge. *Journal of International Consumer Marketing*, *23*(1), 21–44. doi:10.1080/08961530.2011.524575

Liu, R. R., & Zhang, W. (2010). Informational influence of online customer feedback: An empirical study. *Database Marketing & Customer Strategy Management*, *17*(2), 120–131. doi:10.1057/dbm.2010.11

Loch, A., Hilmi, I. N., Mazam, Z., Pillay, Y., & Choon, D. S. K. (2010). Differences in attitude towards cadaveric organ donation: Observations in a multiracial Malaysian society. *Hong Kong Journal of Emergency Medicine*, *17*(3), 236–243. doi:10.1177/102490791001700306

Luczak, C., & Younkin, N. (2012). Net generation: A conceptual framework of the consumer socialization process. *Academy of Marketing Studies Journal*, *16*(2), 47–51.

Lwin, M. O., Williams, J. D., & Lan, L. L. (2002). Social marketing initiatives: National Kidney Foundation's organ donation programs in Singapore. *Journal of Public Policy & Marketing*, *21*(1), 66–77. doi:10.1509/jppm.21.1.66.17613

Maccoby, E. E. (2006). Historical overview of socialization research and theory. In *J.E. Grusec & P.D. Hastings (Eds.), Handbook of Socialization: Theory and Research*. New York: Guilford Press.

Makmor, T., Abdillah, N., Raja Noriza, R. A., Nurulhuda, M. S., Soo-Kun, L., & Kok-Peng, N. (2014). Low organ donation rate in Malaysia: A survey. *Journal of Health and Translational Medicine*, *17*(1), 14–17. doi:10.22452/jummec.vol17no1.3

Malaysian Medical Council. (2006). Guideline on Organ Transplantation.

Marshall, H. M. (2006). The use of multidimensional scaling in the assessment of college students' attitudes toward organ donation [PhD Thesis]. University of New York.

Mascarenhas, O. A. J., & Higby, M. A. (1993). Peer, parent and media influences in teen apparel shopping. *Journal of the Academy of Marketing Science*, *21*(1), 53–58. doi:10.1177/0092070393211007

McGlade, D., & Pierscionek, B. (2013). Can education alter attitudes, behaviour and knowledge about organ donation? A pretest-post-test study. *BMJ Open*, *3*(12), 1–7. doi:10.1136/bmjopen-2013-003961 PMID:24381257

McMillan, S. J., & Morrison, M. (2006). Coming of age with the Internet: A qualitative exploration of how the Internet has become an integral part of young people's lives. *New Media & Society*, *8*(1), 73–95. doi:10.1177/1461444806059871

McQuail, D. (2010). *McQuail's Mass Communication Theory*. London: Sage Publications.

Moore, E. S., & Bowman, G. D. (2006). Of friends and family: How do peers affect the development of intergenerational influences? *Advances in Consumer Research. Association for Consumer Research (U. S.)*, *33*, 536–542.

Moore, J. N., Raymond, M. A., Mittelstaedt, J. D., & Tanner, J. F. Jr. (2002). Age and consumer socialization agent influences on adolescents' sexual knowledge, attitudes and behaviour: Implications for social marketing initiatives and public policy. *Journal of Public Policy & Marketing*, *21*(1), 37–52. doi:10.1509/jppm.21.1.37.17612

Moschis, G. P., & Churchill, G. A. Jr. (1978). Consumer socialization: A theoretical and empirical analysis. *JMR, Journal of Marketing Research*, *15*(4), 599–609. doi:10.1177/002224377801500409

Mostafa, M. M. (2010). Altruistic, cognitive and attitudinal determinants of organ donation intention in Egypt: A social marketing perspective. *Health Marketing Quarterly*, *27*(1), 97–115. doi:10.1080/07359680903519867 PMID:20155553

Nazni, N., Zaherawati, Z., Adnan, A., Mahazril, A. Y., Mohd Zool Hilmie, M. S., Mohd Shamsul, D., ... Kamarudin, N. (2012). Organ donation among Malaysian: The Malay dilemma toward social development. *Asian Social Science*, *8*(10), 8–15.

Nor Hatina Shuib dan Mahani Ishak. (2019). Eksklusif: Malaysia negara terendah derma organ, BH Online. Retrieved from https://www.bharian.com.my/wanita/keluarga/2019/01/517319/eksklusif-malaysia-negara-terendah-derma-organ

Norazlina, A. A., Muhammad, N. A., & Zuramaznum, S. (2006). Donor's consent in organ procurement process in Malaysia: Issues and suggestions for solutions. Research Paper, Institut Penyelidikan, Pembangunan dan Pengkomersilan, Universiti Teknologi Mara.

Normawati, A., Faridah, I., Normah, M., & Chang, P. K. (2011). Pencarian Maklumat Pendermaan Organ dan Tisu dalam Pengikrar Penderma Organ: Satu Tinjauan Awal. *Malaysian Journal of Communication*, *27*(1), 102–117.

Online, B. H. (2019). 80 peratus kes berpotensi derma organ ditolak ahli keluarga. Bharian. Retrieved from https://www.bharian.com.my/berita/nasional/2019/10/614745/80-peratus-kes-berpotensi-derma-organ-ditolak-ahli-keluarga

Pinto, M. B., & Mansfield, P. M. (2011). College students' attitudes toward the act of gambling: Influence from consumer socialization agents. *Services Marketing Quarterly*, *32*(3), 210–227. doi:10.1080/15332969.2011.581968

Quick, B. L., Morgan, S. E., LaVoie, N. R., & Bosch, D. (2014). Grey's Anatomy viewing and organ donation attitude formation: Examining mediators bridging this relationship among African Americans, Caucasians and Latinos. *Communication Research*, *41*(5), 690–716. doi:10.1177/0093650213475476

Rady, M. Y., McGregor, J. L., & Verheijde, J. L. (2012). Mass media campaigns and organ donation: Managing conflicting messages and interests. *Medicine, Health Care, and Philosophy*, *15*(2), 229–241. doi:10.100711019-011-9359-3 PMID:22020780

Ramaseshan, B., Wong, D., & Turner, B. (2011). Striking the right chord with organ donation, American Marketing Association, 22, 119.

Riyanti, S., Hatta, M., Norhafizah, S., Balkish, M. N., Siti, Z. M., Hamizatul Akmal, A. H., & Normawati, A. (2014). Organ donation by sociodemographic characteristics in Malaysia. *Asian Social Science*, *10*(4), 264–272.

Ryan, A. M. (2000). Peer groups as a context for the socialization of adolescents' motivation and education. *Educational Psychologist*, *35*, 2, 11–111. doi:10.1207/S15326985EP3502_4

Ryan, R. M., & Powelson, C. L. (1991). Autonomy and relatedness as fundamental to motivation and education. *Journal of Experimental Education*, *60*(1), 49–66. doi:10.1080/00220973.1991.10806579

Schultz, D., & Schultz, H. (2004). *IMC the next generation: Five steps for delivering value and measuring financial returns*. USA: McGraw Hill.

Shim, S. (1996). Adolescent consumer decision-making styles: The consumer socialization perspectives. *Psychology and Marketing, 13*(6), 547–569. doi:10.1002/(SICI)1520-6793(199609)13:6<547::AID-MAR2>3.0.CO;2-8

Sirois, B. C., Sears, S. F. Jr, & Marhefka, S. (2005). Do new drivers equal new donors? An examination of factors influencing organ donation attitudes and behaviors in adolescents. *Journal of Behavioral Medicine, 28*(1), 201–212. doi:10.100710865-005-3669-y PMID:15957575

Soh, P. C. H., Charlton, J. P., & Chew, K. W. (2014). *The influence of parental and peer attachment on Internet usage motives and addiction.* Peer reviewed Journal on the Internet, 19, *7.*

Šramová, B. (2017). *Children's Consumer Behavior.* Consumer Behavior - Practice Oriented Perspectives.

Sussman, S. W., & Siegal, W. S. (2003). Informational influence in organizations: An integrated approach to knowledge adoption. *Information Systems Research, 14*(1), 47–65. doi:10.1287/isre.14.1.47.14767

Swenson, R. R., & Prelow, H. M. (2005). Ethnic identity, self-esteem and perceived efficacy as mediators of the relation of supportive parenting to psychosocial outcomes among urban adolescents. *Journal of Adolescence, 28*(4), 465–477. doi:10.1016/j.adolescence.2004.09.005 PMID:16022882

Urbain, C., Gonzalez, C., & Le Gall-Ely, M. (2013). What does the future hold for giving? An approach using the social representations of generation Y. *International Journal of Nonprofit and Voluntary Sector Marketing, 18*(3), 159–171. doi:10.1002/nvsm.1448

Uzniene, R. (2010). Media – Agents of socialization. *Regional Formation and Development Studies, 3*(8), 231–239.

Wang, M., Zhang, W., & Wang, X. (2008). The principle of family determination in organ donation: The application of Confucian Ethics. *HEC Forum, 20*(2), 183–196. doi:10.100710730-008-9069-2 PMID:18642090

Wentzel, K. R. (1991). Social competence at school: Relations between social responsibility and academic achievement. *Review of Educational Research, 61*(1), 1–24. doi:10.3102/00346543061001001

Wentzel, K. R., & Looney, L. (2006). Socialization in school settings. In *J.E. Grusec & P.D. Hastings (Eds.), Handbook of Socialization: Theory and Research.* New York: Guilford Press.

Wiman, A. R. (1983). Parental influence and children's responses to television advertising. *Journal of Advertising, 12*(1), 12–18. doi:10.1080/00913367.1983.10672825

Wong, L. P. (2010). Information needs, preferred educational messages and channel of delivery, and opinion on strategies to promote organ donation: A multicultural perspectives. *Singapore Medical Journal, 51*(10), 790–795. PMID:21103814

Yunus, A., & Landau, E. (2019), Youth Now Defined as Those Between 15 to 30. New Straits Times. Retrieved from https://www.nst.com.my/news/nation/2019/07/501288/youth-now-defined-those-between-15-and-30

Zolkepli, I. A., & Kamarulzaman, Y. (2015). Social Media Adoption: The Role of Media Needs and Innovation Characteristics. *Computers in Human Behavior, 43*, 189–209. doi:10.1016/j.chb.2014.10.050

# Chapter 11
# The Role of Narrative Elements in Gamification Towards Value Co-Creation:
## A Case of Mobile App Users in Malaysia

**Cheah Wen Kit**
*Universiti Sains Malaysia, Malaysia*

**Izzal Asnira Zolkepli**
*Universiti Sains Malaysia, Malaysia*

## ABSTRACT

*This article discusses the role of narrative element in gamification towards value co-creation in a crowdsourcing application system. The discourse addresses the gap of knowledge to understand the user motivation and experience to co-create value in a gamified system. Value co-creation is an interactive engagement process that refers to the act of collaborating with a group of intended consumers through a crowdsourcing approach. As the decentralisation of the web enables participation of people to shape the future based on their contributions, understanding Internet users' motivation and experience to co-create value is crucial in ensuring that the initiatives are reciprocated by the intended parties. As gamification has been widely utilised in numerous contexts in order to encourage users to contribute their resources of knowledge and skills, the effectiveness of its elements, namely narrative, remains questionable.*

## INTRODUCTION

The gaming industry is one of the fastest growing and emerging industries that connects Internet users from all over the world. Anyone with Internet connectivity is able to experience an immersive platform that offers fun and enjoyable moments. Understanding the penetration that the gaming industry has on the players, firms are leveraging on numerous gaming and gamification platforms to ensure the sustenance of their business through the customers value co-creation.

DOI: 10.4018/978-1-7998-1566-2.ch011

As the fourth industrial revolution (termed as Industry 4.0), is transforming communication from all areas, understanding online user engagement is crucially important to achieve effective collaboration between the firm and its target market for any intended purpose. Given its rapid growth, research efforts to extricate the complexity of digital communication between humans to humans, humans to machines, and machines and machines is intensified more than ever. For instance, the deployment of artificial intelligence (A.I.) is changing the immersive state of user during connectivity. Each smartphone user nowadays is carrying a personal A.I. in his or her pocket, which makes acquisition and dissemination of knowledge conveniently and swiftly. Advanced technology tools that rely on rich interactive media to connect with online users are gradually blurring the line between the virtual and real-world experience. The information that is found on numerous online platforms are based on the contribution of the users to other users. As users are more likely to engage with the information provided from other users, firms and content providers are heavily relying on the knowledge possessed by other users. One of the effective ways that firms use to motivate the initiative of other users to contribute their knowledge and skills in a collaborative platform is through the use of gamification, a process that draws the potential of games in non-game contexts.

Understanding the potential of gamification would be beneficial for firms in leveraging the game design elements to encourage engagement and induce immersion among its users to co-create value in marketing communication approach. In brief, a game refers to physical or mental contest that has goal or objective, whereby an individual or a group of individuals will play according to a pre-determined framework or rules in the game world (Huzinga, 1970). The use of game elements in practical context is called gamification. The fundamental understanding of how and why certain users behave in a gamified environment is only at the tip of the iceberg. Therefore, researchers are calling for more in-depth research into this context that was derived from the rapid progress of the world wide web technology.

The advancement of Semantic Web, which is also referred as Web 3.0, that witnesses more intelligent and intuitive application system is empowering users towards seamless user experience to reach more targeted communication goal. Such democratisation of the Internet enables all users to connect dynamically with each other, including in the context of business to consumers (B2C) and consumers to consumers (C2C), in which the connectivity between users happens in split seconds to reach the intended objectives. At one time, countless of information exchange processes takes place that help firms to understand its users or consumers better. For instance, a search engine understands the thinking patterns of the users better than the users would aware. It is able to predict the user's search patterns thereby offering possible findings of consumer behaviours at a rapid pace. From the firm's perspective, developers and marketers are able to understand users in greater detail with the assistance of big data, which is intensified by the decentralisation of the Internet.

As the decentralisation of the web enables participation of the people to shape the future based on their very own behaviours, understanding Internet users' psychological aspects, namely their motivation and experience in the dissemination of knowledge is crucial to spread the message on pressing global issues, such as on sustainable development in the age of urbanisation in emerging countries such as Malaysia.

Malaysia is holding the number one position in terms of its economic standing among the emerging markets due to her stable economic growth according to Bloomberg (2018).

The engagement and contribution of skills and knowledge from the people through value co-creation are vital to ensure the success of sustainability development initiatives. Since Sustainable Development Goals (SDG) was introduced by the United Nations in September 2015, firms from the developing nations have initiated numerous initiatives to tackle environmental issues such as on reducing carbon footprint from the use of transportation.

Since the penetration of Internet stands at close to 50 percent in Asia, which is the highest in the world (Miniwatts Marketing Group, 2018), a research from an emerging market in Asia such as Malaysia to investigate the effectiveness of marketing communication messages to encourage engagement pertaining to green issues is highly needed. As Malaysia is gearing towards developed nation that is bound to experience even more pressing green issues, namely sustainable consumption, pro-environmental effort from all parties, including government, firms and the people, the collaboration between all parties is highly needed, especially that collaborative green behaviour is integral in combating sustainable development issues on local and global front (Klöckner, 2013). While policymakers in Malaysia are intensifying the efforts to inculcate green behaviours by introducing numerous green initiatives such as through Environmental Protection Plan (The Star, 2018), it is argued that the initiatives would continue to face behavioral response setbacks if the people do not co-create value with the government or any organisations that attempt to tackle this issue, which greatly requires collaborative efforts. It was reported that green attitude has been found to be one of the key factors in determining environmental practices (Choe & Yap, 2018). Although it was surveyed that the majority of Malaysians are giving their moral support to the zero single use plastic campaign by the government (Malay Mail, 2019), it is unknown if the people are behaviorally contributing to the cause. Past studies have shown that environmental awareness among Malaysians are negatively correlated with their environmental behaviour, an indicator which has been a concern for policymakers in order to ensure the success of the green initiatives that will soon to implement in the future (Mei, Wai, & Ahamad, 2016).

One of the approaches that can be leveraged to encourage response in the context of green initiative is through the implementation of gamification. Since there are a number of gamification elements, namely through the use of points, badges, leaderboards, performance graphs, avatars, teammates, and meaningful stories (also referred to as narrative), practitioners are able to design the most appropriate strategies to execute an effective approach to motivate online users to co-create value in a given context. It has been reported that the commonly used gamification element, which are points and leaderboards (Morschheuser, Hamari, Koivisto, & Maedche, 2017), have demonstrated positive results to encourage value co-creation in gamified crowdsourcing system. However, for a gamification strategy to be well-designed that ensures continuous user engagement, other gamification elements needs to be investigated to ensure its effectiveness. While the gaming industry has been utilising the story element, the use of gamification is still at its infancy stage and warrants for further investigation. Although narrative element is often used in numerous gamification approach for value co-creation, it requires further understanding on its effectiveness. As to date, there is no empirical research that has been conducted in this context. In addition, the psychological outcomes from the integration of story element needs to be investigated for the validation of this approach. Therefore, the present article aims to shed light on the discourse gap by discussing the effectiveness of the story element in gamification towards value co-creation, which focuses on the user experiences in driving collaboration.

## VALUE CO-CREATION

Value co-creation is a novel paradigm that is highly discussed by researchers from numerous disciplines, especially in marketing communication, management, and information system (Galvagno & Dalli, 2014). Value creation is vital to ensure the sustainability of a firm through service offer, which is a fundamental basis of exchange (Vargo & Lusch, 2004). Generally, researchers agree that value co-creation is a process that refers to a collaborative process between an initiator and its actors, an umbrella term that refers to individual consumers, groups of consumers, brand communities, employees, and suppliers, among others (Storbacka, Brodie, Böhmann, Maglio, & Nenonen, 2016). Service marketing researchers highlighted that it is a process that serves as a way to increase value between the service provider and the intended consumer (Deng, Lu, Wei, & Zhang, 2010; Hollebeek & Andreassen, 2018). To further explain, Brodie, Hollebeek, Jurić, and Ilić (2011) mentioned that value co-creation is an interactive, creative and social process between stakeholders, which include consumers, suppliers and investors. Galvagno and Dalli (2014) have explored the definition by mentioning that it is a "joint, collaborative, concurrent, peer-like process of producing new value, both materially and symbolically". In brief, it is agreed that value co-creation is a way to increase value between the service provider and the intended consumer (Payne, Storbacka, & Frow, 2008; Vargo & Lusch, 2004). From the definitions gathered, it signifies that value co-creation is in reference of an interactive process that involves several parties in a multidimensional dissemnination of message that goes beyond dyadic communication (Li, Juric, & Brodie, 2017).

The success of value co-creation is highly dependent on the voluntary participation of the intended stakeholders. The emergence of new technologies such as Web 3.0, blockchain technology, big data, and social media are offering numerous new opportunities for value co-creation on various online platforms (Geri, Gafni, & Bengov, 2017; Xu, He, & Li, 2014).

Each of the constructs used in the value co-creation concept needs to be further discussed. Value can be referred to the utilitarian or hedonic aspect of the outcome, which is the creation, whereas "co-" refers to the actors, which is also the customers, who are collaborating with the initiator (Islam, Rahman, & Hollebeek, 2018). The process is aimed towards developing beneficial collaborative behaviour as intended by the initiator.

In line with the fundamental understanding described by Storbacka et al. (2016), value co-creation is an approach that creates offerings through ideation, design and development through customer engagement. Based on the mentioned descriptions of value co-creation in past literatures, the researchers tend to view consumers from a capitalistic point of view. Customers are perceived as 'partner employee' that is able to contribute to their supply chain of business. Hence, the psychological characteristics of the co-creators needs to be explored as these would ensure a continuously healthy and interactive process between the parties involved (Bendapudi & Leone, 2003; France, Grace, Merrilees, & Miller, 2018; Heinonen, 2018).

A definite conceptualisation of value co-creation is still incomplete as it fails to capture the psychological state of the co-creator, which is termed as actor disposition (Chandler & Lusch, 2015; Li, Juric, & Brodie, 2018; Storbacka et al., 2016).

## Value Co-Creation and Crowdsourcing Explained

While the concept of both crowdsourcing and value co-creation are often discussed in overlapping manner, it needs to be clarified that the former refers to the approach performed by an initiator, which usually would be a firm or organisation, whereas the latter refers to the act of collaboration between two or more parties. Firms are leveraging on crowdsourcing system to co-create value with their consumers in order to build online brand community (Fuller, Hutter, & Faullant, 2011; Muniz & O'Guinn, Thomas, 2001; Schau, Muniz, & Arnould, 2009). Value co-creation is made possible especially as firms crowdsource resources from the intended consumers through an open call on the Internet. The purpose of carrying out a crowdsourcing approach is usually to innovate a product, service, or content by sourcing for ideas and solutions from the stakeholder's core competencies, such as expertise, knowledge and skills (Brabham, 2008; Duverger, 2015). Co-creation of value would be able to offer competitive advantage for the initiator as its collective outcome is catered to the consumers for a more personalised service.

Crowdsourcing was introduced by Jeff Howe in 2006 when he wrote an article on Wired magazine that caught the attention of many industry players (Howe, 2006). He sheds light on the rise of online platforms with the likes of iStockphoto, whereby publishers are able to source for user-submitted photos at a exceptionally low price rate instead of paying much higher to a photographer to capture the intended shots. The publisher ended up paying for only one dollar to the original owner of the photos instead of hundreds of dollars to a professional. These images are contributed by the Internet users on a dedicated website to showcase their talents and be discovered by industry players, aside from getting paid for their work. Another platform that rode on this dynamic approach was Innocentive, a website that allows industry players such as Procter and Gamble, DuPoint and Boeing to seek contributors' solution on the posted scientific, business and technical problems with a pre-determined prize money as incentive. Some of these problems seek to solve pressing global green issues. Hence, any Internet user who has solutions to the posted problems are able to contribute towards sustainable development while being rewarded for their resources, such as knowledge, skills, time and energy. Such phenomena have disrupted the market whereby a once confined industry becomes more permeable and receptive to anyone among the crowds. While not all the posted problems were solved, it was reported that about 30 percent of the problems were solved through the submitted ideas and solutions. However, 30 percent is still considered low compared to the high number of Internet users around the world who has untapped knowledge, skills and ideas. While offering prize money would be an easy solution to encourage value co-creation through user contribution, the downside would surface when the initiator does not possess sufficient financial resources for crowdsourcing. Incentivising the crowd with monetary value revolves around a carrot-and-stick method, whereby the crowd would be conditioned to only contribute ideas if there is financial reward, which many initiators, such as startups, nonprofit organisations and public-funded entities could not afford. Targeting the crowd's extrinsic motivation would be detrimental to the approach or campaign in the long run, whereby the consumers would only focus on the extrinsic reward. In order for an approach to provide the intended gameful experience, an experience that is unique to games, initiators would need to tap into the user's intrinsic motivation as well (Huotari & Hamari, 2012; Zhao & Zhu, 2014).

Realising the potential that the users possess, especially those who are enthusiastic towards contributing to a cause or brands, firms are beginning to engage their network of communities towards the company's initiatives through value co-creation that is made possible with the unique position that technology advancement offers.

With the proliferation of mobile applications (apps), websites, and social media due to the advancement of Web 3.0, the features and functionalities have been expanded with wide range of uses to encourage more users to co-create value. Understanding that the online users and consumers are demanding for more active role in an exchange of service (Labrecque, vor dem Esche, Mathwick, Novak, & Hofacker, 2013), individuals who engage with a particular Internet application are given the authority and autonomy to contribute their own ideas, skills and knowledge for a more personalised service experience (Füller & Bilgram, 2017). Such an approach takes steps forward to ensure that users are intrinsically motivated to co-create value. In addition, it leads to a win-win situation whereby the initiator is able to acquire the knowledge and skills from the users through their contributions and the users are able to enjoy a personalised service for a better experience. Without a doubt, value co-creation is growing at a rapid pace.

Researchers and marketers are beginning to understand the limitless harnessing potential that they are able to source from the online communities to ensure the success of their campaigns. For instance, a German-based skincare brand Nivea that is globally recognised for its extensive research and development efforts has been pushing boundaries with their technology-driven innovation capabilities. The firm has managed to hit the most successful sales from their 130-year history of product launch as they co-create a new deodorant with the consumers. Japanese consumer brand called Muji has also reported that the sales from its co-created products are higher than the internally developed ones.

## Theoretical Underpinning of Value Co-Creation: Service-Dominant Logic

According to Vargo and Lusch (2004), value is no longer determined only when the delivery of goods or tangible products reaches the stakeholders. Goods are viewed as the transmitter of knowledge, but not as the end product, in a value creation process. That is, any form of products and services delivered to the stakeholders is categorised as service. The novel marketing theory builds on the fundamental understanding that service is uniquely derived from the delivery of goods (Hansen, 2017). This new understanding of logic is viewed as the Service-Dominant (S-D) Logic, which is stimulating discussions from marketing and management researchers (Baron, Patterson, Warnaby, & Harris, 2010; Cheah & Zolkepli, 2018; Grönroos, 2006; Payne et al., 2008).

Vargo and Lusch (2008) defined service as the application of competence, such as through knowledge and skills, in benefiting the service recipients. Hence, regardless if it is a goods-based (tangible) or service-based (intangible) product, the logic postulates that the firm is delivering value.

S-D Logic is viewed as a transcendence of goods-dominant (G-D) logic that conservatively focuses on the delivery of goods (Vargo & Lusch, 2008). The logic that has been viewed to tie the loose ends in the fragments of marketing theories appears to have broadened the understanding of values which benefit both the firms and stakeholders (Saarijärvi, Kannan, & Kuusela, 2013). In line with the paradigm shift, firms need to acknowledge the importance and potential of value in service as they communicate with their customers or prospective consumers.

Through this understanding, service is viewed as a unit of exchange, whereas goods and service products are the mediator of trade (Payne et al., 2008; Saarijärvi, Kannan, & Kuusela, 2013). In other words, consumers who are utilising, adapting, and maintaining the goods or products are determining the value created. It is no longer that the firms dictating the value of a goods or service delivery, but the customers. The basis of this customer-oriented communication gives rise to the understanding that value creation is a collaborative function between the service provider and the receiver. Hence, a service, being the common denominator, is always co-created and is never a one-way process when a trade is done.

There are five axioms or principles that form the understanding of S-D Logic which needs to be met, according to Vargo and Lusch (2008). The first one is that service is the integral unit of exchange. The second axiom states that customers or the prospective customers are always the co-creater of value. An update in the authors' foundational premise (FP) in almost a decade later suggested that value is co-created by multiple actors (Vargo & Lusch, 2016). This shows that the authors have taken into account the multidimensional communication that exists in the network, a view that is in-line with the systems perspective. The third axiom states that all actors, including industry players, service initiators, and even customers in the social and economic context are the resource integrators. In other words, they are the ones who do not merely provide the service, but consolidate and integrate the available resources through various means as well. Resources, specifically operant resources, primarily refers to knowledge and skills (Vargo & Lusch, 2016). The fourth axiom states that value is always determined by the beneficiary, who is also the co-creator, of the service in phenomenological sense. In other words, the customers of a service are always determining the value in a trade process, as mentioned earlier. While Vargo and Lusch (2008), have only mentioned four axioms in their original conceptualisation of S-D Logic, the scholars have revised the understanding of S-D Logic through the injection of the fifth axiom in the theory's foundational premise (FP), which states that "value co-creation is coordinated through actor-generated institutions and institutional arrangements" (Vargo & Lusch, 2016, p. 8). It needs to be noted that the term "institution" does not refer to an organisation but the rules, meanings, symbols, practice, and norms that are designed by humans, in enabling and constraining action in giving meaning to an individual's social life.

The central focus of the revised S-D Logic is to provide a holistic and meaning-ladden worldview of actor's experience that encourage co-creation of value (Vargo & Lusch, 2016). In line with the contemporary understanding in marketing, customer-centric firms are seeking ways to generate better engagement and experience for value co-creators (Vega-Vazquez, Revilla-Camacho, Cossío-Silva, & Cossı, 2013). The scenario is applied to the situation that involves any approach or initiative that is carried out by an organisation, including government bodies. Hence, the integrated study of experience and intrinsic motivation to co-create value with a crowdsourcing platform needs to be done to understand the antecendents of value co-creation from the actor's psychological perspective.

## Motivations for Value Co-Creation

User engagement has been described to be a motivationally driven resource integration function through the interaction with a firm or organisation (Hollebeek & Andreassen, 2018). Therefore, the investigation on the motivations of an actor that influence his or her intention to co-create value with a firm shall provide more in-depth understanding on the antecedents of value co-creation. While extrinsic motivation such as obtaining the reward of money or other tangible prizes would enhance participation and involvement for users to co-create value, that is not always the case as there is no contract that abides users to integrate their resources (Zhao & Zhu, 2014). In other words, financial rewards may not always work if the actors are not intrinsically motivated to co-create value. Since users would consciously evaluate their contribution of service in the form of skills, knowledge, time and energy, practitioners would need to tap into the user's intrinsic motivation. Among the intrinsic motivations that could encourage participation in a co-creation approah include user's hedonic and utilitarian experience, the need for recognition and competition, and other forms of personal satisfaction (Zheng, Li, & Hou, 2011). A study based on the systems perspective has found that the experience of actors would be enhanced if the platform that actors engage on is able to provide usability and sociability (Zhao & Zhu, 2014). The study carried out by Zhao and Zhu (2014) showed that the two main dimensions that positively motivated the participants in a crowdsourcing contest are the attributes of hedonic and utilitarian. Concurrently, these values are able to be met through the elements of games, which is referred to as gamification (Deterding, 2011; Morschheuser et al., 2017). Immersive gamification is able to provide these functionalities towards the users. Users who are engaged on a crowdsourcing gamification platform are able to experience hedonism and utilitarianism, while being recognised in their contribution. Understanding that intrinsic motivation is able to influence participation efforts, the present article argues that actors who co-create value would be positively driven by the concept of gamification that is introduced on an interactive platform.

## Gamification in Value Co-Creation

Gamification in value co-creation has been widely utilised in numerous contexts. For instance, a highly used community-based traffic and navigation mobile application called Waze is depending on the contribution of the motorist to report on the traffic conditions including congestions, road hazards, roadblocks and sharing place of interest, among others. It is reported that Waze has over 90 million global active users in 185 countries (Waze, 2018), whereas Malaysia is recorded for having the highest number of users in Asia and top 5 in the world (The Star, 2017). Realising the huge number of reach and engagement, brands are leveraging on the app for location-based marketing to advertise and target their marketing communication to the nearby users. In order to attract more users and ensure continuous usage, the app offers a variety of features and function, such as interactivity and gamification. Apart from being able to communicate with each other on the app, Waze users are granted points from every trip and contribution to the system that they make. These points would help them to level up, which makes more features available to the higher level users. The icon that represents each user would have their avatar changes as they climbs up the stage. The change in representation of the user is to signifiy that they have achieved a higher level and more features are unlocked. Users who are connected with each other on the app are also able to compare their rankings and marks on the system's leaderboard. The granting of points and the availability of leaderboards are part of the gamification affordances that are integrated in the app. Having said that, the app is not leveraging on other gamification affordance, namely story element, which is argued to be effective in sustaining usage rather than being used for pure navigation purpose. Such could be the reason as to why navigational app users in Malaysia seems to opt for Google Maps, as an online poll shows (Awani, 2019). Users do not seem to appreciate the gamification feature that the system offers, despite its high emphasis on points and leaderboard integration.

In addition, an e-hailing app in Malaysia called Grab is implementing the use of gamification features to reward loyal users. The feature is called GrabRewards membership. There are four levels in total, which are labelled as "member", "silver", "gold" and "platinum". The higher level that a user climbs enables him or her to unlock more rewards and benefits offered by the app. For instance, a "platinum" user is able to earn double the reward point through each booking, compared to a regular member. With more points earned, users are able to exchange it with a wide variety of offers such as discounts on their next Grab rides or with their partner merchants. While Grab has implemented such gamification features to ensure continuous usage among users, the use of extrinsic rewards such as financial benefit could exhaust user's motivation in the long run. Hence, firms and developers would need to tap into the use of intrinsic motivation in order for the users to continuously engaged in a long run, despite the presence of other competing players that would consume a larger market share.

In the context of green initiative, an app developer is collaborating with a non-profit environmental organisation to bring higher awareness to the public in preserving the environment through tree-planting. For instance, a Chinese-based gamification productivity app called "Forest" is in cooperation with Trees For The Future, a non-profit organisation based in Maryland of United States of America to plant trees in Africa. Apart from greening the earth, their mission is to elevate poverty. Based on the points in the form of virtual coins that users earned through their focus time during the app usage, they are able to co-create with the firm to exchange the value for tree-planting. The use of point-based system is an application of gamification that functions to encourage user's enjoyment and their focus time. The trees that they plant will wither if they happen to switch to other app when the app is in use, a phenomenon known as 'phubbing'. Such approach can be related to the use of story element whereby users are role-playing as the tree planters. It is also reported that the proceeds from the app sales has helped the organisation to purchase trees from their partners as well, apart from public donation on their green initiative. In the app and on the brand's social media profiles, Forest has integrated the narrative elements by actively describing about the trees that they offer the users to plant to ensure continuous use. However, it is not known if users are driven to use the app due to the story element, for its functional purpose, or both. It needs to be highlighted that the element of story is one of the affordances that form gamification.

## Gamification Affordances Explained

Gamification affordances are the game design elements that motivates gameful experiences. These elements are the basic building blocks for games and gamified systems, in which each of them, or collectively, serves to propel the game towards its intended experience. Given its relatively novel area of study, game researchers have not come to conclusion the exact number of these elements that exists. For instance, Werbach and Hunter (2012) have suggested that there are 15 elements or components of a game, whereas Revees and Read (2009) suggested that there are 10 'ingredients' that make up an effective gamification approach. Having said that, many gamification systems do not utilize all of the available elements that exist. Based on Werbach and Hunter (2012), Sailer et al. (2017) proposed that there are seven common elements that are present in the previously discussed game literatures. The justification for the authors' selection are that these elements have direct visibility to the players, are easily activated and deactivated in experimental setting, and the ease of use in manipulating each variable to study the motivational aspect of players. For the benefit of the present context that revolves around user experience and motivation, the current study subscribes to the proposal by Sailer et al. (2017). The seven game elements are points, badges, leaderboards, performance graphs, avatars, and meaningful stories (Sailer et al., 2017).

Points is the most basic game element that is rewarded to player's based on his or her successful or accomplished task (Sailer et al., 2017; Werbach & Hunter, 2012). It serves to represent and track a player's progress numerically while providing feedback to the journey in the game. There are various types of points that have been used by game designers such as achievement or reputation points, experience points, redeemable points. There are apps that have integrated a combination of types of points as service. For instance, Grab, Southeast Asian e-hailing app that allows users to book a private ride, rewards reputation and redeemable points to its users. These points can be exchanged for ride discounts and other sales promotion discounts with their partners. Being the most basic form of game element, past research have revealed that the use of points has been shown to be effective in motivating users to participate in gamified crowdsourcing system (Morschheuser et al., 2017).

Badges are the visual representation, a sign, or symbol that indicates the achievements of the players, who would collect them throughout the game environment (Sailer et al., 2017; Werbach & Hunter, 2012). Badges visibly show player's achievement, which would give them a sense of pride and satisfaction. The use of badges would encourage continuous usage as players would be motivated to earn the available badges. However, once all the badges are earned, the motivation of players would perhaps deteriorate. Earning badges do not pose any narrative meaning. It is also not a compulsory task for players to journey forward in the gamified environment (Sailer et al., 2017). For instance, the use of badges is evident in the Forest app, a co-creative productivity app that encourages users to focus on a self-determined time. Once users have achieved a targeted focus time, badges are awarded. However, these badges do not propel the gameplay journey forward as they are not necessary to be collected.

Leaderboards is a game element that quantifies players' success according to their achievements (Sailer et al., 2017). The quantification may be calculated through points or other forms or measurable outcomes. In other words, the utilisation of leaderboard is useful in ranking users against certain success criterion. It is able to create a competitive environment among the players while providing users a sense of pride and satisfaction in the gamified network. However, it might not be an effective gamification affordance as the usage motivation would deplete once a player reaches the highest level of the leaderboard. In the case of Waze the the gamified navigation app, a player or user would climb the leaderboard based on the distance that the person has travelled. The distance is converted into points for leaderboard quantification. There are five rankings that a user may reach. It is argued that once a user has reached the highest level in the leaderboard, the sense of excitement and hedonic experience of users who focus on this element would deplete as there will be no motivating factor in driving the gameplay journey.

Performance graphs contains information that is used to compare the player's current performance against their preceding performance (Sailer et al., 2017). It does not compare the said player's performance with other players but with his or her own previous performance so that the player is able to evaluate the achievement thus far. In a networked game environment that involves the participation of multi-players, it is argued that the sole use of performance graph might not be suitable as a main or focal game design element to encourage a strong motivation and optimal experience for continuous usage.

Avatar represents players visually in a gamification or gamified environment through the use of a virtual character (Sailer et al., 2017; Werbach & Hunter, 2012). Avatars are either assigned by the game designer or chosen by the player. Avatars can be depicted as a simple pictogram or complex three-dimensional animation (Werbach & Hunter, 2012). In certain games such as a popular simulation game *The Sims*, highly-rated strategy game *PlayerUnknown Battleground* or briefly known as *PUBG*, or even a networked simulation game called *Second Life*, players are allowed to customise their own avatars. The function of customisation is to set their avatar apart from other players in the network. The use of avatars would give users a sense of immersive experience in creating and adopting their own identity in becoming part of the game environment (Sailer et al., 2017). While the use of avatar is argued to be able to contribute towards user's optimal experience in the gameplay, it is not compulsory for game designers to introduce this element towards inducing a complete gameful experience.

*Teammates* refers to the use of other human or computer players to induce conflict, competition or cooperation (Kapp, 2012). The introduction of teams is able to foster collaboration among players towards a shared goal or objective (Werbach & Hunter, 2012). The use of teammates as game design element is apparent in most networked games. In fact, as most of the mobile games are connected over the Internet, it is easy for game designers to utilise this teammates to make a gameplay more interesting. While it is not the focus of the present study, it needs to be acknowledged that the use of teammates would influence a player's gameful experience.

Meaningful stories refers to the use of narrative element that gives and expresses the meaning of a game that goes beyond the quest for points and quantifiable achievements (Kapp, 2012; Sailer et al., 2017). While it does not relate to the player's performance, the use of meaningful stories as game element could enrich gameful experience and further motivate the players towards engagement (Sailer et al., 2017). Being an important element of gamification, the implementation of story is said to be more engaging if it was applied in a less stimulating non-game contexts. The overlay of narrative element is able to alter the real-world meaning while immersing the players towards value co-creation (Hansen, 2017; Sailer et al., 2017). The relationship between stories and human behaviour has been acknowledged, but not as cause and effect (Hansen, 2017). Stories is arguably to be able to influence behaviour as it organises perceptions and would symbolically restructure one's experience (Jackson, 2013). The understanding of how stories influence behaviour needs to be investigated.

Sailer et al. (2017) have conducted a study to investigate the effect of game elements on human motivation and performance using an online simulation. The authors have grouped the game design elements based on the focus of the study, which were psychological need satisfaction, which are the need for competence and autonomy in a gameplay. The need for competence revolves around the feedback that the players would be getting. On the other hand, the need for autonomy involves two aspects, which are experience of decision freedom and experience of task meaningfulness. It was found that badges, leaderboards, and performance graphs yield positive effect on competence need satisfaction and task meaningfulness. Whereas, avatars, teammates, and meaningful stories game elements have shown to affect experience and social relatedness. While the grouping of game design elements appeared to be relevant towards investigating both study variables, it is still unclear how individual game design element could afford specific outcomes, such as user motivation and optimal experiences as users participate in a gamified system. In the context value co-creation, it is proclaimed that value is deconstructed through stories (Hansen, 2017).

## Story Immersion as Gamification Affordances

In order to win a game, players need to draw information presented to them as they navigate the gaming world. The information are often times given through a form of storyline. Hence, video game creators would need to employ interesting and immersive storyline in the dynamic, role-playing medium of gameplay (Baranowski, Buday, Thompson, & Baranowski, 2008). As a narrative series of events, a story functions to give meaning to a wide range of context, including gamification (Deterding, 2011; Green & Brock, 2000; Lu, Baranowski, Thompson, & Budday, 2012). A story takes place at a particular time (such as before World War II, during graduation, at year end), particular place (in a starship, on a mountain, in the prison), and usually has a character that lead the story forward, called the protagonist (Baranowski et al., 2008). Other character that opposes the protagonist is called antagonist. The struggle between both protagonist and antagonist refers to conflict, which is the motivating factor behind a story's plot.

A story is able to engage individuals by forming empathy with the protagonist. The engagement is even stronger and more effective when there is a change of value in the protagonist, such as being failure to a successful, betrayal to loyalty, cowardice to courage, so that the viewers or audiences is able to relate and exemplify the lessons to be learned from the story (Baranowski et al., 2008). The changes in value is usually presented as the protagonist's internal conflict. This conflict would be the common barriers towards individual's behaviour change (Baranowski et al., 2008). The change of behaviour as argued by Baranowski et al., (2008) implies that an individual would respond behaviorally upon presented persuasive message contained in the story. In this vein, the engaged individuals are immersed by the story element. The behavioral response as influenced by the story is argued to have the potential in engaging individuals towards value co-creation (Green & Brock, 2000; Lu et al., 2012).

In the context of game, the utilisation of narrative element in a gamification is able to transcend fantasy to the real-world scenario. It is said to be able to serve as analogies in the virtual world for user's understanding of the physical world (Sailer et al., 2017). While it is an emerging trend, there are not many studies to investigate that effectiveness of story element in gamified application systems which employ the use of story element to immerse players for continuous engagement. Although story element has been mentioned as one of the gamification affordances in numerous gamification literature (Morschheuser et al., 2017; Sailer et al., 2017), little to none empirical findings is found on the effectiveness this element. It remains questionable if users are driven to co-create value through a presented story element in gamification (Lu et al., 2012). Such gap of knowledge needs to be investigated especially that the common component of games is the story element (McKee, 1997). Now that everyone is able to create stories through user-generated content as mentioned earlier, an investigation on this gamification affordance is integral for co-creation of value approach. Hence, the present article encourages researchers to investigate the effectiveness of story immersion as an enabler in gamification towards value co-creation among online users. The studies would be beneficial to find out how users who are immersed in the narrative element during the process of gamification would respond to a call for collaboration towards pressing global issues.

## WHAT'S IN A STORY?

Narrative, which is also known as story, derives from the Latin word *narrare*, which is "to recount". Narration is originated from the ancient oral culture that involves the sharing of life stories among people (Lu et al., 2012). It is the most basic form of human communication and is enjoyed by many. On societal level, a narrative element is seen as a pre-condition for community living to collaborate and unite with each other (Riessman, 2008). Present in every age and society, narrative is defined as the "quintessential form of customary knowledge" (Lyotard, 1984, p. 19) in postmodernism philosophy. As narrative or story is broadly defined, it may be fictional or non-fictional in any medium, from text to video (Green & Brock, 2000). In that sense, the story recipient or narratee could be termed as a viewer, reader or listener. A phenomenon worth noting is that a narratee would be immersed in the narration through story immersion, according to transportation theory, which will be discussed in the following section (Green & Brock, 2000; Lu et al., 2012).

As a person is immersed in a story, the individual would reach an optimal experience or flow (Csikszenmihalyi, 1990), which would then influence behaviour even in online environment (Hoffman & Novak, 2009). In line with the hypothesised argument that a narrative element would influence online behaviour, narrative might serve as a bridging element between gamification value co-creation. As a user is engaged on a gamified co-creation platform, there is a high chance that the individual will be exposed to a narrative element, especially as it sets the context for the game.

From managerial perspective, understanding the customers through their narrative worldview that is expressed through stories would allow the firms to strategise a more effective engagement approach, especially when value is always determined by the customers. According to the narrative theory is based on the assumption that human beings share stories to connect with the society, firms are able to harness the wisdom of the crowd through narration as medium. In this vein, narrative is seen to connect the dot of understanding in the context of gamification and value co-creation in investigating the motivation and experience of customers towards the intended engagement properties. Understanding the composition of a narrative element would further facilitate understanding for researchers and practitioners.

While it is generally known that a story is made up of a beginning, middle and end according to Aristotle, researchers agreed that it is made up of six elements, which include abstract, orientation, complication, evaluation resolution and coda (Labov, 1997; Rezadoost & Charvadeh, 2013). According to Labov (1997), 'abstract' refers a short summary of the story; 'orientation' denotes the who, where and when of the story that requires the story recipient to understand who the characters are, and situation of the time and place. The element of 'complication' tells the exact happenings of the story. This element is the core or integrating factor of the story. As the name suggests, 'evaluation' refers to the analysing of story if it is worth to retelling. 'Resolution' refers to the last part of the story that concludes the narration. The last element in Labov's model, which is 'coda', which refers to the brief summary that explicitly states the retelling of the story by the story receiver.

From serious games perspective, an area that gamification borrows its understanding from, the element of story is widely found in most, if not all, of the game genres. The game genre that is heavily dependable on story element are games that are categorised in adventure, role-playing, strategy, action games, to name a few. For instance, the *Final Fantasy* series, *The Elder Scrolls*, *Assassin's Creed*, *Resident Evil*, *Tomb Raider*, are some of the highest-grossing games that are made available on numerous platforms, from computer, console to mobile. In fact, the successes of these games have spun off to big screen as films that garnered high number of viewers globally. Currently, the electronic game across platforms that appear to garner one of the highest number of players is *Player Unknown Battleground*, or *PUBG* for short. Released just a year, the shooter game has 32 million total players worldwide (Kain, 2018). Developed by Microsoft and Tencent, the game that has sold more than 20 million copies becomes one of the most talk-about online from players and practitioners (BBC News, 2018). It is a community-based connectivity game or termed as massive multiplayer online role-playing game (MMORPG) involves players from all over the world to connect and fight with each other to locate weapons and supplies in a remote island to survive and be the last person standing. A player is represented by each unique avatar or character of their choice. The game on mobile platform allows players to communicate with each other through voice as well. The success of the game suggests that players find enjoyment in the immersive gameplay experience through its story element as they collaborate with each other towards a common goal.

As for gamified systems, the elements of stories have been widely used as well. For instance, a highly-rated mobile exergame (integrating exercise and games) called 'Zombies, Run!' appears to leverage the element of story element as well. Developed in 2012, the app integrates science fiction storyline of zombie apocalypse that requires users to "survive" by running away from the "zombies". The gamified app introduces a series of missions as the players run, whereby they would need to collect items that help them to "survive" in the game. The users are playing as the characters called "Runner 5", which are also their representation in the game. The elements of points, avatars and storyline are created to influence users to be in the immersive state as they use the app while exercising or running. While most gamified apps contain story element, game designers and technology communication practitioners have yet to leverage on the potential that it brings. For instance, a top-rated productivity app called Forest, an example which was mentioned earlier, that serves as a timer while enabling users to collaborate with each other in trees planting in real life from the virtual coins that they earned, has hardly used the element of story. Although it features a hint of narrative element, such as the use of 'complication', it does not seem to have a fully utilise the concept of story that encourages users to be in an immersive state as they use the app. Having said that, the app appears to garner a high number of users. The question arises if being immersive in the story, or story immersion, is an essential element in a gamified app?

Story immersion as a construct has been proposed by Lu et al. (2012) in their conceptual framework. According to the authors who conducted a multidisciplinary literature reviews psychology, human computer interaction, communication to public health, it is proposed that story immersion would affect health behavioral change as mediated by intrapersonal psychological predictors such as attitude, observational learning, emotional arousal, self-efficacy, subjective norms and intrinsic motivation. However, the conceptual study did not show any empirical finding.

There are not many studies that reports the effectiveness of story element in the context of gamification. As to date, there are only seven studies that investigates storytelling as gamification affordances (Morschheuser et al., 2017). Therefore, such finding calls for more in-depth research on this dimension as it would facilitate gamified application designers, communication practitioners and researchers to understand the suitable elements that yields the strongest engagement towards co-creation of value.

However, the effectiveness of the story element approach in the context of gamification remains questionable. It is also unclear what motivates the users to integrate their resources of time, ideas and knowledge, since there is a lack of theoretical foundation to this novel phenomenon. Although it has found that stories have influenced beliefs (Green & Brock, 2000) and behaviours (Baranowski et al., 2008) through immersion, researchers have yet to conclude if individuals who are immersed in a story would co-create value with the initiator. It is not known if individuals in the virtual context or online users would carry out the behaviours as intended by the firm that initiates a crowdsourcing approach through the Internet. While previous findings have shown that gamification would serve as an effective strategy to motivate users to co-create value (Morschheuser et al., 2017; Piligrimiene, Dovaliene, & Virvilaite, 2015; Sailer et al., 2017), the effectiveness of the vital elements in gamification, which is its narrative element, remains unclear. In order to investigate the context of story that persuades and moves users towards an immersive state, its fundamental understanding is discussed further in the following section.

## Theoretical Underpinnings of Story Immersion: Transportation Theory

Narrative has always been known to change beliefs. Its persuasive potential is leveraged by practitioners in social marketing to influence and induce certain behaviours among its intended target market (Davis, 2002; Kotler & Zaltman, 1971). For instane, public service announcements would use a form of narrative as a strategy to communicate its intended message for change in social marketing campaigns. Some of the examples of social marketing messages that utilise narrative element include anti-smoking campaigns, save the environment campaigns, road safety campaigns, to name a few. Similar approach is witnessed in green marketing as well, whereby brands would use narrative element to communicate its environmental concern while promoting their environmentally friendly products (Kalafatis, Pollard, East, & Tsogas, 1999). These messages could either be presented in the form print advertisement or video documentaries. Similar approach is seen in the context of video games as mentioned earlier. As more sophisticated technology systems, applications, and equipments are facilitating the widespread adoption towards virtual worlds such as Second Life virtual game, the use of story element to provide an immersive experience is witnessed as well. The utilisation of narrative element has been effective in engaging consumers through offline and online environment. However, the use of narrative element does not guarantee success in continuous engagement for value co-creation, which requires active customer collaboration, in the context of gamification. The transcendence of story element from a serious game to non-game contexts is questionable due to its adaptability potential. For instance, it is unclear if the use of story element would be effective in influencing participation and engagement, which are the basic component that ensures gamification success, in functional gamified platform such as Waze or Forest (Berkling & Thomas, 2013). As such, the current study proposes to investigate the effectiveness of story element in influencing user's optimal experience during engagement.

As a story raises questions, builds conflicts, presents a journey of activities that is related to the characters and storyline, it has the potential to encapsulate a user in an immersive experience, which is accounted as the construct of story immersion (Green & Brock, 2000; Lu et al., 2012; Rubio-Tamayo, Barrio, & García, 2017). The use of story element has the potential to shift a user's current real-world context to the environment portrayed in the game. Such is a phenomenon that can be explained through transportation theory as proposed by Green and Brock (2000). According to transportation theory, a person who is immersed in a story would be "transported" into the narrative world and become involved with the characters in the storyline. As an individual is transported into the narrative environment, he or she would undergo certain mental processes, such as a change of beliefs and attitudes, that has effect on their real-world context (Green & Brock, 2000). The psychological experiences are the effects that resulted from the integration of attention, imagery and feelings due to the exposure of the recipient.

The transportation theory is originated from the accounts described by Gerrig (1993) in describing one's narrative experience. As an individual travels upon performing certain actions, the traveller is on a journey that changes his or her original perspective. During the course of transportation, such experience makes some aspects of the original world becomes inaccessible, which is similar to the literal experience of traveling as conceptualised by Gerrig (1993). The action perform could be reading a text or watching a video that has communicates the narrative element. As a transported narratee would experience a loss of accessibility to the real-world environment, the individual may be unaware of his or her surroundings. For instance, on a physical level, if a person is transported into a story as he is watching a movie in the living room, he may not notice others entering the room. On psychological perspective, the person would have "subjective distancing from reality" (Green & Brock, 2000, p. 702). Transcending from the loss of access to real-world facts, the narratee would experience strong experience and motivation even when the individual knows that the world is unreal. Such engagement in the narrative context explains the occurrences whereby a viewer would cry when watching a tragic scene in a movie. It also explains a situation where viewers would actively think what could have happened next if an outcome was to be changed when an individual is presented with unhappy ending in a movie, a phenomenon that (Gerrig, 1993) termed as 'anomalous replotting'.

According to the transportation theory, there are three conditions that determine a transported individual who is immersed in a story, namely suspension of disbelief, personal experience and affection towards the character (Green & Brock, 2000; Lu et al., 2012). Methodologically, these are the variables that make up the construct of story immersion.

Suspension of disbelief refers to the situation where the narratee would be less likely to counter-argue a claim in the story. In other words, the person would be easily influenced in accordance to the story context. This argument is supported by Gilbert (1991) whereby people tend to believe anything that they read or hear. As disbelieving appears to be an effortful corrective mental process, it is found that transported individuals would be less motivated to exert such effort (Green & Brock, 2000). The individual is highly absorbed into the story that he or she is reluctant to critically analyse any reasonings presented in the narrative. Therefore, it is argued that the attitude and belief of the individual would be easily influenced through story immersion. To test this hypothesis, Green and Brock (2000) conducted an experiment that requires the participants to read a textual content and identify the statements that they believe to be false. The result shows that the highly immersed participants identified fewer false statements and showed greater acceptance towards the story content. This shows that the participants who are more immersed tend to suspend their disbelief towards false content and enjoy the story.

Personal experience refers to the situation when the events portrayed in the story feels more self-directed and real (Green & Brock, 2000; Lu et al., 2012). In other words, narrative events would become more of personal experience when a person is immersed in a story. As the event seems closer to reality, it is argued that the attitude of a transported individual would be changed according to the narrative since direct experience is a powerful measure of forming attitude (Slater & Rouner, 2002).

Affection for the character refers to the situation where the narratee develops strong feelings towards the character portrayed in the story (Green & Brock, 2000; Lu et al., 2012). As a viewer or reader develops a strong sense of attachment with the protagonist, the values of the characters would have influence on the immersion level of the narratee as he or she journeys through the story. It is likely for a person who is high on story immersion to have affection for the character as it drives a story forward. Similar to passive narration, a character can either be portrayed through an actual character or an avatar, which was discussed earlier. However, it is unclear if an avatar in a gamified context would yield similar result of transportation effect as compared to a situation where a person is exposed to textual or motion narrative, since an avatar is already a representation of the self or the player in the game context. One might question if the avatar that is controlled by the storyline would show similar effect of persuasiveness or would it be that the player imposes his or her value on the avatar, if given the freedom? The current study intends to investigate this uncertainty.

As transportation theory discusses about attitude change through one's cognitive exposure to narrative element, it needs to be clarified that it differs from an attitude change theory called elaboration likelihood model (ELM) (Petty & Cacioppo, 1981). The main distinguishing factor is that transportation is considered as a convergent process whereby ELM is a divergent process. In other words, a person who is engaged in elaboration process might be accessing his or her own mental schema such as previous knowledge, thoughts or opinions in evaluating a situation. On the contrary, an individual who is on high level of transportation would temporarily disregard or distance his or her previous schemas and experience (Green & Brock, 2000) While suspension of disbelief, personal experience and affection towards the character may be true for passive narratee as suggested by Green and Brock (2000) through the experiment conducted, individuals who are exposed to the story element in an active engagement context such as gamification is questionable. In addition, since there is no empirical evidence in the conceptual model proposed by Lu et al. (2012), the argument calls for further clarification. Therefore, the present study intends to extend the construct by Lu et al. (2012) by investigating the influence of story immersion as fundamentally supported by transportation theory.

*Table 1. Characteristics of flow*

| No. | Construct | Description |
|---|---|---|
| 1. | Balance | When a person achieves a match between the perception of skills and challenges at high level. Else, boredom or anxiety may occur. Challenges refers to demanding tasks whereas skills refer to the capacities that one has to reach the desired outcome. |
| 2. | Control | When one feels that he or she is in control of the situation at any time, without a need to actively exert it. |
| 3. | Feedback | When one continuously receives feedback over the achievements or activities that one makes immediately and unambiguously. |
| 4. | Goals | When the objective of the activity is clearly defined distinctively. The person needs to be aware of requirements of the respective activity that he or she engages in. |
| 5. | Concentration | When the person is completely focused on the task at hand. There will be no distracting thoughts and the attention is high. |
| 6. | Merging | The element refers to the merging between awareness and action. When the individual has acts spontaneously without conscious effort, he or she experiences a unified consciousness of awareness and action. |
| 7. | Loss of self-consciousness | When an individual's concern about the self is out of the way. Any disturbing thoughts that interferes with the activity fades away during engagement. |
| 8. | Timelessness | When one feels that the perception of time is altered. Hours will be felt as though minutes has just passed by. The person finds that time moves slowly when he or she is experiencing timelessness. |
| 9. | Enjoyment | When one believes that the situation is highly enjoyable. The person's state of mind gratifies. Therefore, he or she has the desire to repeat the activity for its own sake. In other words, the activity becomes autotelic. |

## Theoretical Underpinning of Story Immersion: Flow

In order to investigate the experience of users who are in story immersion, it is essential to discuss the flow theory as its fundamental understanding to facilitate value co-creation. As mentioned, it is argued that immersed users would undergo flow, a state that Csikszenmihalyi (1990) described as a process of optimal experience. Flow is used to account for an individual's experience of pleasure and immersion in numerous activities, from mundane tasks to learning a new skill (Jin, 2011). It is a state of mind in which a person is fully immersed in performing a particular activity that characterised by deep focus, complete involvement, intense attention, and achieves an equilibrium perception of skill and challenge (Csikszenmihalyi, 1990).

In total, there are nine elements or constructs that conceptualised flow, which are balance, control, feedback, goals, concentration, merging, loss of self-consciousness, timelessness, and enjoyment. Table 1 summarises nine flow characteristics as theorised by Csikszenmihalyi and how it applies to gamification.

The concept of flow has been applied in a wide variety of discipline, from social psychology, consumer research, human-computer interaction to user experience, including gamification (Choi, Kim, & Kim, 2007; Green & Brock, 2000; Hansen, 2017; Henke, 2013; Huang, Backman, & Backman, 2012; Jin, 2011; Lu et al., 2012; Topu, Reisoğlu, Yılmaz, & Göktaş, 2018). The nine characteristics of flow is able to explain the state of mind of an individual who engages on a gamification platform. As a user plays a game, the person is bound to be immersed in the world that the game creates (Jennett et al., 2008). In other words, the user or player would lose himself or herself in the game that nothing else seems to matter.

As noted, the flow theory was conceptualised by Csikszenmihalyi since 1975. Recently, there has been a development who proposes that the theory of flow to be revisited (Drengner, Jahn, & Furchheim, 2018). According to the authors, flow should be viewed as a parsimonious process. The author argued flow does not apply only to a person who engages in a active functioning but when he or she engages in passive activities such as watching television or meditating as these activities would require awareness and being in a state of mindfulness. In fact, academics are very much shifting their attention to researching the effect of mindfulness on well-being. To support the argument, they have found that there is a process that takes place whenever a person experiences flow.

Firstly, one would experience the constructs of flow antecedents, which is categorised as the task related conditions. These include the constructs of balance, control, feedback and goals. Then, a person would thereafter experience the flow process, which is engrossment that leads to enjoyment. According to the researchers who challenged the traditional understanding of flow theory (Drengner et al., 2018), four of the nine flow characteristics, which are Concentration, Loss of self-consciousness Merging and Timelessness, should be considered as the flow core characteristics in a revised conceptualisation of flow. These characteristics are conceptualised as flow.

Thereafter, an individual would experience the flow outcome, which refers to the intended consequence of the activity that the person hopes to experience or as influenced by the service provider. This suggests that when one experiences flow, it is likely for a person to co-create value with the system. For instance, a research has found that players who are in Second Life, the networked simulation game that builds on its players' contribution, are more likely to co-create value when they are in a flow state of mind (Kohler, Fueller, Matzler, & Stieger, 2011).

As like many other games, Second Life rides on the power of narrative to hook its players to collaborate in building the virtual world. Second Life is a true depiction of gamified value co-creation that sources for the wisdom of the crowd. The use of story element in games seems to be the deciding factor if players would continue to be engaged or disengage. The implementation of narrative element would encourage users to co-create value with the system.

The revisitation of flow theory on a recent finding sheds new light on the understanding of user experience from the perspective of flow. As gamification requires skills to overcome the challenge imposed, a user would need to achieve a balance between skill and challenge to navigate and stay engaged in the system. The elements of gamification, namely narrative element, encapsulates the user in a context so as to give an understanding of the worldview that he or she is in. As the user have a better understanding, he or she would know how to utilise the skills needed for the challenge. It is argued that once balance is achieved, it is likely for the user to be engrossed in the gamified environment.

As a user is engaged on a gamified platform, it is likely that the user takes the role of a character or an avatar, which is the virtual representation of the user in the gamified environment. It is common for the user to be in control over other gamification affordances as well, such as the story element presented in the gamified environment. As the user navigates in the gamified environment with the avatar, a user is said to have the control in creating the story. It was reported that users who are in control would be in engrossment and achieve optimal experience.

A game often provides feedback to the player if they have correctly navigated through the story. As a user experiences story immersion through the narrative presented in the gamified environment, he or she would have stronger engagement for a longer duration, resulting in the user receiving more feedback. The feedback that the user receives would translate to higher engrossment.

*Figure 1. The role of story immersion in gamification towards value co-creation*

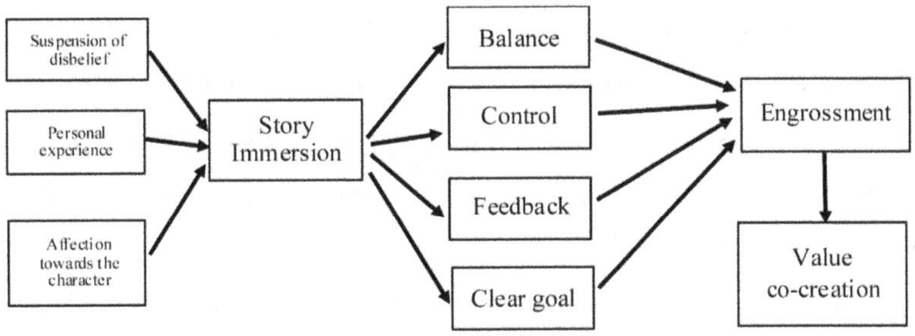


A user needs to have a clear goal as to why and how the game is played. The rules of the game is often stated at the start of the game so that the user have clear goal. The rules of a game is usually integrated in the story of a game. As a user navigates through a game, the user would need to know the finishing line. The motivation to reach the finishing line often influences the user to have high engrossment. When a user is highly engrossed, it is likely for the user to co-create value with the system, in which researchers refer to the flow outcome. The below framework shows a visual representation of the role of story immersion in gamification towards value co-creation, from the perspective of the revisitation of flow theory. Figure 1 shows the conceptual framework of the role of story immersion in gamification towards value co-creation.

## CONCLUSION

This article discussed the role of narrative element in gamification that motivates online users to co-create value. The derivation of gamification features in today's advanced technology ecosystem seems to be able to engage users for continued use. However, its effectiveness is still in question as discussed earlier. Despite the potential that gamification brings, the fact that Malaysia has the highest number of Waze users, but many seem to opt for Google Maps in a recent online poll brings questionable outcomes of gamification among Malaysian mobile app users.

To date, there has not been a study to validate the effectiveness of the gamification affordance, especially on narrative element. Hence the author encourages academics to conduct studies to fill the gap of knowledge so that practitioners are able to implement a more effective gamification approach. The study of user experience from a single gamification affordance as one engages on a gamified system would be a good start. As discussed, the potential that narrative element brings is worth considering, especially that it is able to transport its users into optimal experience of flow for continued usage and thereafter co-create value with the system. Its findings hold promising potential for businesses to engage with its prospects and customers for the sustenance of firm.

## REFERENCES

Awani, A. (2019). Rakyat Malaysia lebih gemar Google Maps? Astroawani. Retrieved from http://www.astroawani.com/gaya-hidup/rakyat-malaysia-lebih-gemar-google-maps-217093

Baranowski, T., Buday, R., Thompson, D., & Baranowski, J. (2008). Playing for real: Video games and Stories for Health-Related Behavior Change. *American Journal of Preventive Medicine*, *34*(1), 74–82. doi:10.1016/j.amepre.2007.09.027 PMID:18083454

Baron, S., Patterson, A., Warnaby, G., & Harris, K. (2010). Service-dominant logic: Marketing research implications and opportunities. *Journal of Customer Behaviour*, *9*(3), 253–264. doi:10.1362/147539210X533179

Bendapudi, N., & Leone, P. R. (2003). Psychological implications of customer participation in co-production. *Journal of Marketing*, *67*(1), 14–28. doi:10.1509/jmkg.67.1.14.18592

Berkling, K., & Thomas, C. (2013). Gamification of a software engineering course and a detailed analysis of the factors that lead to it's failure. *Proceedings of the 2013 International Conference on Interactive Collaborative Learning, ICL 2013* (pp. 525–530). Academic Press. 10.1109/ICL.2013.6644642

Bloomberg. (2018). Malaysia beats emerging market peers as Asia outshines. Retrieved from https://www.nst.com.my/business/2018/11/435823/malaysia-beats-emerging-market-peers-asia-outshines

Brabham, D. C. (2008). Crowdsourcing as a Model for Problem Solving: An Introduction and Cases. *Convergence (London)*, *14*(1), 75–90. doi:10.1177/1354856507084420

Brodie, R. J., Hollebeek, L. D., Jurić, B., & Ilić, A. (2011). Customer engagement: Conceptual domain, fundamental propositions, and implications for research. *Journal of Service Research*, *14*(3), 252–271. doi:10.1177/1094670511411703

Chandler, J. D., & Lusch, R. F. (2015). Service Systems: A Broadened Framework and Research Agenda on Value Propositions, Engagement, and Service Experience. *Journal of Service Research*, *18*(1), 6–22. doi:10.1177/1094670514537709

Cheah, W. K., & Zolkepli, I. A. (2018). Crowdsourcing and Service-Dominant Logic : The Interaction Effect of Brand Familiarity, Customer Experience an..... *Asia Pacific Journal of Advanced Business and Social Studies*, *4*(1), 154–162. doi:10.25275/apjabssv4i1bus16

Choe, K. H., & Yap, C. Y. (2018). Exploring Malaysian Attendee Attitude towards Green Practices in Exhibitions. *Proceedings of the 2017 Tokyo International Conference on Hospitality, Tourism, and Sports Management* (pp. 78–92). Academic Press.

Choi, D. H., Kim, J., & Kim, S. H. (2007). ERP training with a web-based electronic learning system: The flow theory perspective. *International Journal of Human-Computer Studies*, *65*(3), 223–243. doi:10.1016/j.ijhcs.2006.10.002

Csikszenmihalyi, M. (1990). *Flow: The Psychology of Optimal Experience*. New York: Harper & Row.

Da Xu, L., He, W., & Li, S. (2014). Internet of things in industries: A survey. *IEEE Transactions on Industrial Informatics*, *10*(4), 2233–2243. doi:10.1109/TII.2014.2300753

Davis, J. E. (2002). *Stories of change: Narrative and social movements*. New York: State University of New York Press.

Deng, Z., Lu, Y., Wei, K. K., & Zhang, J. (2010). Understanding customer satisfaction and loyalty: An empirical study of mobile instant messages in China. *International Journal of Information Management*, *30*(4), 289–300. doi:10.1016/j.ijinfomgt.2009.10.001

Deterding, S. (2011). Situated motivational affordances of game elements: A conceptual model. *Proceeding of ACM CHI Conference on Human Factors in Computing Systems*. ACM.

Drengner, J., Jahn, S., & Furchheim, P. (2018). *Flow revisited: process conceptualization and a novel application to service contexts. Journal of Service Management, 29*. doi:10.1108/JOSM-12-2016-0318

Duverger, P. (2015). Crowdsourcing innovative service ideas. *Journal of Hospitality and Tourism Technology*, *6*(3), 228–241. doi:10.1108/JHTT-10-2014-0063

France, C., Grace, D., Merrilees, B., & Miller, D. (2018). Customer brand co-creation behavior: Conceptualization and empirical validation. *Marketing Intelligence & Planning*, *36*(3), 334–348. doi:10.1108/MIP-10-2017-0266

Füller, J., & Bilgram, V. (2017). The moderating effect of personal features on the consequences of an enjoyable co-creation experience. *Journal of Product and Brand Management*, *26*(4), 386–401. doi:10.1108/JPBM-03-2016-1122

Fuller, J., Hutter, K., & Faullant, R. (2011). Why co-creation experience matters? Creative experience and its impact on the quantity and quality of creative contributions. *R & D Management*, *3*(41), 259–273. doi:10.1111/j.1467-9310.2011.00640.x

Galvagno, M., & Dalli, D. (2014). Theory of value co-creation: A systematic literature review. *Managing Service Quality*, *24*(6), 643–683. doi:10.1108/MSQ-09-2013-0187

Geri, N., Gafni, R., & Bengov, P. (2017). Crowdsourcing as a business model. *Journal of Global Operations and Strategic Sourcing*, *10*(1), 90–111. doi:10.1108/JGOSS-05-2016-0018

Gerrig, R. J. (1993). *Experiencing narrative worlds*. New Haven: Yale University Press.

Gilbert, D. T. (1991). How mental system believe. *The American Psychologist*, *46*(2), 107–199. doi:10.1037/0003-066X.46.2.107

Green, M. C., & Brock, T. C. (2000). The role of transportation in the persuasiveness of public narratives. *Journal of Personality and Social Psychology*, *79*(5), 701–721. doi:10.1037/0022-3514.79.5.701 PMID:11079236

Grönroos, C. (2006). Adopting a service logic for marketing. *Marketing Theory*, *6*(3), 317–333. doi:10.1177/1470593106066794

Hansen, A. V. (2017a). What stories unfold: Empirically grasping value co-creation. *European Business Review*, *29*(1), 2–14. doi:10.1108/EBR-08-2015-0080

Heinonen, K. (2018). Positive and negative valence influencing consumer engagement. *Journal of Service Theory and Practice*, *28*(2), 147–169. doi:10.1108/JSTP-02-2016-0020

Henke, L. L. (2013). Breaking through the clutter: The impact of emotions and flow on viral marketing. *Academy of Marketing Studies Journal, 17*(2), 111–118.

Hoffman, D. L., & Novak, T. P. (2009). Flow Online: Lessons Learned and Future Prospects. *Journal of Interactive Marketing, 23*(1), 23–34. doi:10.1016/j.intmar.2008.10.003

Hollebeek, L. D., & Andreassen, T. W. (2018). The S-D logic-informed "hamburger" model of service innovation and its implications for engagement and value. *Journal of Services Marketing, 32*(1), 1–7. doi:10.1108/JSM-11-2017-0389

Howe, J. (2006). The Rise of Crowdsourcing. *Wired Magazine, 14*(06), 1–5. doi:10.1086/599595

Huang, Y., Backman, S. J., & Backman, K. F. (2012). Exploring the impacts of involvement and flow experiences in Second Life on people's travel intentions. *Journal of Hospitality and Tourism Technology, 3*(1), 4–23. doi:10.1108/17579881211206507

Huotari, K., & Hamari, J. (2012). Definining Gamification - A Service Marketing Perspective. *Proceeding of the 16th International Academic MindTrek Conference* (pp. 17–22). Tampere: Mindtrek. 10.1145/2393132.2393137

Huzinga, J. (1970). *Homo Ludens: A Study of the Play Element in Culture*. New York: J & J Harper.

Islam, J. U., Rahman, Z., & Hollebeek, L. D. (2018). Consumer engagement in online brand communities: A solicitation of congruity theory. *Internet Research, 28*(1), 23–45. doi:10.1108/IntR-09-2016-0279

Jackson, M. (2013). *The politics of storytelling: Variations on a theme by Hannah Arendt*. Copenhagen: Museum Tusculanum Press.

Jennett, C., Cox, A. L., Cairns, P., Dhoparee, S., Epps, A., Tijs, T., & Walton, A. (2008). Measuring and defining the experience of immersion in games. *International Journal of Human-Computer Studies, 66*(9), 641–661. doi:10.1016/j.ijhcs.2008.04.004

Jin, S. A. A. (2011). "I feel present. therefore, I experience flow:" A structural equation modeling approach to flow and presence in video games. *Journal of Broadcasting & Electronic Media, 55*(1), 114–136. doi:10.1080/08838151.2011.546248

Kain, E. (2018). "PUBG" Tops 4 Million Players On Xbox One, Hands Out Free Stuff To Players. Forbes. Retrieved from https://www.forbes.com/sites/erikkain/2018/01/25/pubg-tops-4-million-players-on-xbox-one-hands-out-free-stuff-to-players/#5297c3d72621

Kalafatis, S. P., Pollard, M., East, R., & Tsogas, M. H. (1999). Green marketing and Ajzen's theory of planned behaviour: A cross-market examination. *Journal of Consumer Marketing, 16*(5), 441–460. doi:10.1108/07363769910289550

Kapp, K. M. (2012). *The gamification of learning and instruction: Game-based methods and strategies for training and education*. San Francisco, CA: Pfeiffer.

Klöckner, C. A. (2013). A Comprehensive Model of the Psychology of Environmental Behaviour – a Meta-Analysis. *Global Environmental Change, 23*(5), 1028–1038. doi:10.1016/j.gloenvcha.2013.05.014

Kohler, T., Fueller, J., Matzler, K., Stieger, D., & Füller. (2011). Co-Creation In Virtual Worlds: The Design Of The User Experience. *Management Information Systems Quarterly*, *35*(3), 773–788. doi:10.2307/23042808

Kotler, P., & Zaltman, G. (1971). Social Marketing: An Approach to Planned Social Change. *Journal of Marketing*, *35*(3), 3–12. doi:10.1177/002224297103500302 PMID:12276120

Labov, W. (1997). Some further steps in narrative analysis. *Journal of Narrative and Life History*, *7*(1/4), 395–415. doi:10.1075/jnlh.7.49som

Labrecque, L. I., vor dem Esche, J., Mathwick, C., Novak, T. P., & Hofacker, C. F. (2013). Consumer power: Evolution in the digital age. *Journal of Interactive Marketing*, *27*(4), 257–269. doi:10.1016/j.intmar.2013.09.002

Li, L. P., Juric, B., & Brodie, R. J. (2017). Dynamic multi-actor engagement in networks: The case of United Breaks Guitars. *Journal of Service Theory and Practice*, *27*(4), 738–760. doi:10.1108/JSTP-04-2016-0066

Li, L. P., Juric, B., & Brodie, R. J. (2018). Actor engagement valence: Conceptual foundations, propositions and research directions. *Journal of Service Management*, *29*(3), 491–516. doi:10.1108/JOSM-08-2016-0235

Lu, A. S., Baranowski, T., Thompson, D., & Budday, R. (2012). Story Immersion of Videogames for Youth Health Promotion: A Review of Literature. *Games for Health Journal*, *1*(3), 199–204. doi:10.1089/g4h.2011.0012 PMID:24416639

Lyotard, J. F. (1984). *The postmodern condition: A report on knowledge*. Minneapolis: University of Minnesota Press.

Malay Mail. (2019). Survey shows majority support "no single-use plastic" campaign.

McKee, R. (1997). *Story, substance, structure, style and the principle of screenwriting*. New York: HarperCollins.

Mei, N. S., Wai, C. W., & Ahamad, R. (2016). Environmental Awareness and Behaviour Index for Malaysia. *Procedia: Social and Behavioral Sciences*, *222*(07), 668–675. doi:10.1016/j.sbspro.2016.05.223

Miniwatts Marketing Group. (2018). Internet World Stats.

Morschheuser, B., Hamari, J., Koivisto, J., & Maedche, A. (2017). Gamified crowdsourcing: Conceptualization, literature review, and future agenda. *International Journal of Human-Computer Studies*, *106*(April), 26–43. doi:10.1016/j.ijhcs.2017.04.005

Muniz, A. M. J. Jr, & O'Guinn, T. C. (2001). Brand Community. *The Journal of Consumer Research*, *27*(4), 412–432. doi:10.1086/319618

BBC News. (2018). PubG: The story behind one of the world's most popular games. Retrieved from https://www.bbc.com/news/av/newsbeat-42660502/pubg-the-story-behind-one-of-the-world-s-most-popular-games

Payne, A. F., Storbacka, K., & Frow, P. (2008). Managing the co-creation of value. *Journal of the Academy of Marketing Science*, *36*(1), 83–96. doi:10.100711747-007-0070-0

Petty, R., & Cacioppo, J. T. (1981). *Attitudes and Persuasion: Classic and contemporary approaches*. Iowa: Brown.

Piligrimiene, Z., Dovaliene, A., & Virvilaite, R. (2015). Consumer engagement in value co-creation: What kind of value it creates for company? *The Engineering Economist*, *26*(4), 452–460. doi:10.5755/j01.ee.26.4.12502

Revees, B., & Read, J. L. (2009). *Total Engagement: Using Games and Virtual Worlds to Change the Way People Work and Businesses Compete*. Boston: Harvard Business School Press; doi:10.1016/j.chb.2010.03.035

Rezadoost, A., & Charvadeh, B. K. (2013). Studying narrative structure of one thousand and one night tales based on Labov's diamond- shaped model. *International Research Journal of Applied and Basic Sciences*, *4*(11), 3324–3327.

Riessman, C. K. (2008). *Narrative Methods for Human Sciences*. London: Sage Publications.

Rubio-Tamayo, J. L., Barrio, M. G., & García, F. G. (2017). Immersive Environments and Virtual Reality: Systematic Review and Advances in Communication, Interaction and Simulation. *Multimodal Technologies and Interaction*, *1*(4), 21. doi:10.3390/mti1040021

Saarijärvi, H., Kannan, P. K., & Kuusela, H. (2013). Value co-creation: Theoretical approaches and practical implications. *European Business Review*, *25*(1), 6–19. doi:10.1108/09555341311287718

Sailer, M., Hense, J. U., Mayr, S. K., & Mandl, H. (2017). How gamification motivates: An experimental study of the effects of specific game design elements on psychological need satisfaction. *Computers in Human Behavior*, *69*, 371–380. doi:10.1016/j.chb.2016.12.033

Schau, J. H., Muniz, A. M. Jr, & Arnould, E. J. (2009). How brand communities create value. *Journal of Marketing*, *73*(September), 30–51. doi:10.1509/jmkg.73.5.30

Slater, M. D., & Rouner, D. (2002). Entertainment-education and elaboration likelihood: Understanding the processing of narrative persuasion. *Communication Theory*, *2*(12), 173–191.

Storbacka, K., Brodie, R. J., Böhmann, T., Maglio, P. P., & Nenonen, S. (2016). Actor engagement as a microfoundation for value co-creation. *Journal of Business Research*, *69*(8), 3008–3017. doi:10.1016/j.jbusres.2016.02.034

The Star. (2017). Malaysia has the most Wazers.

The Star. (2018). Make green growth a priority. Retrieved from https://www.thestar.com.my/news/nation/2018/10/19/make-green-growth-a-priority-we-must-focus-on-making-sustainable-development-a-mainstream-goal/

Topu, F. B., Reisoğlu, İ., Yılmaz, T. K., & Göktaş, Y. (2018). Information retention's relationships with flow, presence and engagement in guided 3D virtual environments. *Education and Information Technologies*, *23*(4), 1621–1637. doi:10.100710639-017-9683-1

Vargo, S. L., & Lusch, R. F. (2004). Evolving to a New Dominant Logic for Marketing. *Journal of Marketing*, *68*(1), 1–17. doi:10.1509/jmkg.68.1.1.24036

Vargo, S. L., & Lusch, R. F. (2008). Service-dominant logic: Continuing the evolution. *Journal of the Academy of Marketing Science*, *36*(1), 1–10. doi:10.100711747-007-0069-6

Vargo, S. L., & Lusch, R. F. (2016). Institutions and axioms: An extension and update of service-dominant logic. *Journal of the Academy of Marketing Science*, *44*(1), 5–23. doi:10.100711747-015-0456-3

Vega-Vazquez, M., Revilla-Camacho, M. Á., Cossío-Silva, F. J., & Cossı, F. J. (2013). The value co-creation process as a determinant of customer satisfaction. *Management Decision*, *51*(10), 1945–1953. doi:10.1108/MD-04-2013-0227

Waze. (2018). Data, Driven. Retrieved from https://www.waze.com/brands/drivers/

Werbach, K., & Hunter, D. (2012). *For win: How game thinking can revolutionize your business*. Philadelphia: Wharton Digital Press.

Zhao, Y., & Zhu, Q. (2014). Effects of extrinsic and intrinsic motivation on participation in crowdsourcing contest. *Online Information Review*, *38*(7), 896–917. doi:10.1108/OIR-08-2014-0188

Zheng, H., Li, D., & Hou, W. (2011). Task Design, Motivation, and Participation in Crowdsourcing Contests. *International Journal of Electronic Commerce*, *15*(4), 57–88. doi:10.2753/JEC1086-4415150402

# Chapter 12

# Nexus Between Social Network, Social Media Use, and Loneliness:

## A Case Study of University Students, Bangladesh

**Md. Aminul Islam**
https://orcid.org/0000-0001-5636-3140
*University of Liberal Arts Bangladesh, Bangladesh*

**Bezon Kumar**
https://orcid.org/0000-0003-4489-431X
*Rabindra University, Bangladesh*

## ABSTRACT

*This paper mainly explores how real-life social network and social media use are related to loneliness among university students in Bangladesh. To carry out this paper, primary data and several methods are used. This paper uses Lubben Social Network Scale and UCLA loneliness scale to measure the level of real life social network and loneliness, respectively. Besides Pearson's partial correlation matrix is used to find out the correlation between social network, social media use, and loneliness. The study finds that students are averagely engaged in real-life social network and moderately lonely. The study also finds a significantly positive relationship between social media (Facebook) use and loneliness, and a significantly negative relationship between real life social network and loneliness. This paper calls for the students to be careful in using social media and be engaged more in real life social network to avoid loneliness.*

DOI: 10.4018/978-1-7998-1566-2.ch012

*Table 1. Internet users and 2019 population statistics for Asia*

| Asia Region | Population (2019 Est.) | Population. % World | Internet Users 30-June-2019 | Penetration (% Population) | Internet % Users | Facebook 31-Dec-2018 |
|---|---|---|---|---|---|---|
| Asia Only | 4,241,972,790 | 55.0 | 2,200,658,148 | 51.9 | 49.8 | 867,984,000 |
| Rest of World | 3,474,250,419 | 45.0 | 2,221,836,474 | 64.0 | 50.2 | 1,331,444,570 |
| All the World | 7,716,223,209 | 100.0 | 4,422,494,622 | 57.3 | 100.0 | 2,199,428,57 |
| Bangladesh | 168,065,920 | | 94,445,000 | 56.2 | | 28,000,000 |

Source: https://www.internetworldstats.com/stats3.htm

## INTRODUCTION

One of the most defining characteristics of the human being is their need for meaningful social connection and they feel pain without it. The need for interaction with others is deeply embedded in human being genetic code. Cacioppo and Patrick (2009) metaphorically termed loneliness as social pain but is very much similar to physical pain or hunger. By using functional magnetic resonance imaging (FMRI), they found that loneliness and social isolation or rejection activates the same parts of the brain as physical pain. Researchers often use the terms loneliness and social isolation interchangeably. Contrarily, Valtorta and Hanratty (2012) argued that these two terms are completely distinct concepts. For them, individuals can be lonely despite having numerous persons to interact with. On the other hand, an individual may not feel lonely despite being isolated from the social world or not having a huge number of active social connections. So, loneliness is a two dimensional phenomenon– both social and emotional. Social loneliness can be measured by a number of connections of a person has. On the other hand, emotional loneliness can be understood in terms of quality of the relationship and desired companionship. Lack of social contacts and interactions and lack of meaningful quality and quantity relationship lead to social isolation (Victor and Bond, 2009).

Latest modern communication technologies and social media platforms have connected to each other today than ever before. Internet, mobile phone and social Medias are the key factors to this connections. According to the statistics of International Telecommunication Union (ITU), about 51.2 per cent of the global populations (3.9 billion) used the internet actively by the end of 2018 (ITU, 2019). This figure has been changed to more than 56 percent (4.33 billion) by July 2019 (Statista, 2019). According to The Global State of Digital in 2019 Report, worldwide the number of social media users was 3.484 billion and mobile phone users was 5.112 billion in 2019 while on an average people had 7.6 social media accounts around the world and daily time spent on social was 142 minutes a day (Smith, 2019).

Bangladesh is a country of the third world with 16.8 million population. The access to and use of communication technologies are increasing rapidly in the country. The number of internet users in the country reached at 96.199 million and the number of mobile phone subscribers reached at 161.772 million at the end of June, 2019 in Bangladesh (BTRC, 2019). Several reports reveals that Bangladesh is in the third position in terms of ‘growing number of Facebook users’. About 92 million people in the country use internet in multiple devices. Among them, 34 million are active in using various social media platforms which is about 20 percent of the total population in 2019. The table below shows picture of total population, internet and Facebook users in Bangladesh in comparison to South Asia and Global perspective.

*Figure 1. Individuals using the internet per 100 inhabitants around the world, 2018*

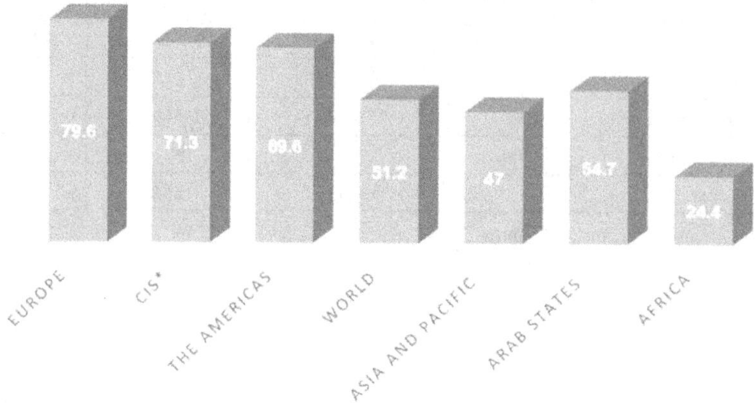


Bangladesh is now passing through the massive transformations and constant changes in terms of its economic development, penetration of technology, mobility of people and the structure and function of the society. Such development has increased urbanization which demanded new living spaces which resulted in increasing number of high rise building. In the past, Bangladesh mainly had individually owned houses but the economic development brought high rise buildings which has introduced new spheres of social relationships. The rise of mobile communication technologies and social media around the world has led changes in the forms of social interaction. In combination with urbanization, modernization and social media's popularity, the social interactions have been going through changes. Similar is the case in a developing country like Bangladesh. Over last few decades, Bangladeshi population migrated from rural areas to urban areas searching for better education and job purposes, which has increased the number of urban populations. Globalization and modernization also played a significant role in contributing to the transformation of urban life, especially to the forms of social interaction. As social institutions, the universities and the university students are not distant island of the society. Rather, they are the significant part of the society, who lead the transformation and changes.

People are now spending too much time communicating and interacting with others. Yet several studies show that people feel increasingly alone and the sense and feeling of loneliness are becoming an epidemic in modern society (Alberti, 2018; Killeen, 1998; Wood, 2013; Brown and Wood, 1953 and Kar-Purkayastha, 2010). However, a critical look into how technology-mediated communication affects users to offer mixed and conflicting results. Some studies show that use of social media such as Facebook is linked to depressive symptoms (Steers et al. 2014 and Alshammari et al. 2017), feelings of isolation (Song et al. 2014), self-esteem and sense of belonging (Tobin et al. 2014), sleep disturbance (Wolniczak et al. 2013 and Levenson et al. 2016) and frequency of social media use initially predicted decreased loneliness and increased happiness (Pittman, 2018). Another study found that there is a relationship between loneliness, depression and internet addiction, and loneliness and internet addiction are risk factors for depression (Demir, 2016). While some other studies indicate that Facebook didn't make people lonely, rather lonely people were more likely to use the social media site (Song et al. 2014). In an online social networking experiment, Deters and Mehl (2012) found that status updating activities on Facebook decrease loneliness. In a study, Lin et al. (2016) found that there association between the amount of time spent on social media and level of depression. Their results indicate that individuals who spend more time using social media are more likely to be depressed. Blanchnio et al. (2018) indicated that among young Polish people, friendship decreases loneliness and loneliness decreases Facebook intrusion. Scott, et al. (2018) found that there is a positive relationship between posting photo and narcissism. However, their results reveal that in the case of loneliness and shyness the relationship is negative. Meanwhile, narcissists are more frequently update about their achievements, diet, and exercise, and individuals with outgoing characteristics post more frequently update about their social activities. Another study found that Facebook 'Likes' less likely to affect the self-esteem of people with the purpose (Burrow and Rainone, 2017). However, Facebook profiles raise users' self-esteem and affect behavior (Toma, 2013).

Some contrasting views about the impact of social media on face-to-face connections are also found. Researchers note that more interactions on social media are replacing face-to-face connections. Meanwhile, other researchers argue that social media does not decrease face-to-face interactions (Hall et al. 2018). No statistically significant association was found between loneliness and Facebook use (Yavich et al., 2019). However, studies suggest that more reliance on social technology instead of face to face interaction create a feeling of social isolation. In a meta-analysis, Huang ((2010) investigated forty studies to find out the relationship between depression, loneliness, self-esteem, and life satisfaction. Its results show that there is a small detrimental effect of Internet use on psychological well-being. Another study finds that social interaction on Facebook may decline subjective well-being in young adults, while it is increased as a result of frequent interactions with supportive "offline" social networks (Kross et al. 2013).

A growing body of researches indicates that loneliness has multiple causes and effects on the human body and mind (Cohen, 2004 and Umberson and Montez, 2010). Results of several studies suggest that suffering from loneliness for a long time can lead to changes in the cardiovascular (Xia and Li, 2018 and Valtora et al. 2016), nervous systems (Cacioppo and Patrick, 2009 and Zelikowsky et al, 2018), health behavior (Winkel et al. 2017), immune and mortality risk (Cole et al. 2015; Holt-Lunstad, et al. 2015 and Steptoe et al. 2013) and loneliness deteriorates the productivity of human being which ultimately decelerates the economic growth of a country (Ahmed, 2018). In the cases, the determining factors are the structural characteristics and types of social networks (Medevene, et al. 2015), quality and quantity of the relationship (Wiseman et al. 2006; Schmidt and Sermat, 1983 and Valtorta and Hanratty, 2012). For example, Cohen (2004) explored three aspects of social relations and their associated health outcomes. The researcher found that social relationships-social support, social integration, and negative interaction-strongly affect both physical and mental health. Another study revealed the mechanism in which age, socioeconomic status, and other factors contribute to social isolation and poorer mental health (Alberti, 2018). Contrarily, Weeks, et al. (1980) argued the relationship between loneliness and depression is not causal, meaning that neither causes directly the other, though the origins of the both are same.

For a comprehensive understanding, the role of social media in mental health can be outlined as the interpersonal-connection-behaviors framework. For them, social media sites can benefit people when they use it for meaningful social connections. On the contrary, the use of the sites can cause harm through multiple ways such as isolation and social comparison. Hunt et al. (2018) found that the limited use group showed significant reductions in loneliness and depression over three weeks compared to the control group. Balchi and Olkun (2015) revealed a positive relationship between the loneliness of foreigners and Facebook usage that implies social network usage is increasing with rising loneliness level. Similarly, Karakose et al. (2016) revealed that students most commonly share photographs and videos on Facebook for spending leisure time. In addition, they have found that there is a statistically significant relationship between the time participants spend on Facebook and their Facebook addiction scores. Primack et al. (2017) found that young adults with high social media use seem to feel more socially isolated than their counterparts with lower social media use. In addition, Masthi et al. (2017) revealed that private school students were more prone to social media addiction and a multitude of physical, psychological and behavioral problems was observed among social media users. Shettar et al. (2017) found that more than one-fourth (26%) of the study participants had Facebook addiction and 33% had a possibility of Facebook addiction. Besides, they have also found that there was a significant positive correlation between the severity of Facebook addiction and extent of experience of loneliness.

In another theoretical model, Nowland et al. (2017) the researchers note that the relationship between loneliness and social internet use is bidirectional and dynamic. According to the model, social technologies can be a strong tool for reducing loneliness when it is used to enhance existing relationships and forge new social connections. On the contrary, the technologies can increase the 'social pain' of interaction and feelings of loneliness when people use it as a scope for escaping from the reality of life and social world. So, it can be argued that the impact of Facebook use is two dimensional- it can deteriorate and improve mental health conditions.

A causal association between social network and health is well established by the wealth of empirical shreds of evidence in various recent cross-sectional, longitudinal, experimental and quasi-experimental studies, most from western perspectives. But the mechanisms through which social relationships affect mental health remain to be explored from the context of developing countries like Bangladesh. Many factors are found in the earlier studies which affects loneliness among the people of Bangladesh. For instance, Pervin and Ferdowshi (2016) examined the relationship between suicidal ideation and depression, loneliness, hopelessness among University students in Bangladesh. They found that suicidal ideation was positively correlated with depression, loneliness and hopelessness. In the context of Bangladesh, mostly young people are found depressed and lonely after the significant visible inclusion of internet particularly after using of various social Medias like Facebook. In addition, Kabir et al. (2018) found that students feel lonely for anxiety, stress and so on. Although very few study has been carried out on loneliness in Bangladesh, to the best of knowledge of the researchers, this is the first study which explores the relationship between mental health (loneliness) and real-life social network and virtual social media use (Facebook) in the context of private university students in Bangladesh. This study is an effort to fill up the gap and this study investigates the structure of real and social media and their relationship with loneliness.

## METHODOLOGY

### Study Area and Sample Selection

This paper is mainly based on primary data. For carrying out this paper, Rajshahi district among 64 districts of Bangladesh is selected randomly as the study area. There are four universities in Rajshahi district of which two are public university and the rest two are private. As this study focuses only on the private university, public universities are ignored here. Among two private universities, one university is selected randomly and the selected university is Varendra University. The university has about 5000 students in 3 faculties such as Arts and Social Sciences, Business, and Science and Engineering faculties. From the university registrar office, the list of faculty wise students is collected and these faculties are assigned as a stratum. Using the stratified sampling method, sample is determined. The number of sample size is determined using the following formula stated by Taro Yamane.

$$n = \frac{N}{1 + Ne^2},$$

where, n = sample size, N= population size and e = rate of precision (0.05).

$$n = \frac{5000}{1 + 5000\left(0.05^2\right)} = 370$$

Data are collected from 370 students of different faculties randomly with a well-structured questionnaire from January to June 2019 through face to face interview.

## Empirical Methods

This study uses three distinct empirical methods. To measure the level of real life social network and loneliness among university students, 6 items Lubben Social Network Scale (LSNS-6) is used following Jobe and White (2007) and Cecen (2008). The LSNS-6 total score is an equally weighted sum of these six items. The scale consisted of 6 items, and scores for each question ranged from zero to five. The score zero indicated minimal social integration and five indicated substantial social integrations. The total score was an equally weighted sum of the 6 items. Scores ranged from 0 to 30 with higher scores indicated a greater level of social support and low risk for isolation. A score less than 12 indicated a person with an extremely limited social network and high risk for isolation. Authors of this study also tried to understand the level of social network of the respondents by dividing total score in three categories. A score of 0-10.00 indicated extremely lower or limited social network, 10.01-20.00 indicated average or moderate social network and 20.01- 30.00 indicated higher or strong social network.

Besides measuring the level of real life social network, the study measures the level of loneliness. To measure the level of loneliness, this study uses twenty items the University of California and Los Angeles (UCLA-20) loneliness scale consisting with 10 negatively stated (lonely) and 10 positively stated (non-lonely) items following Jobe and White (2007) and Cecen (2008). The estimated score of this index ranges from 0 to 60 (Russell et al. 1980). The score zero means student is not lonely while the score 60 means student is highly lonely. More specifically, the score ranges from 0 to 20.00 reveals the lower level of loneliness, the score ranges from 20.01 to 40.00 and 40.01 to 60.00 reveal the moderate level and severe level of loneliness, respectively.

After being informed with the level of real life social network and the level of loneliness, the study uses Pearson's partial correlation matrix to find out the correlation between real life social network, social media use (Facebook use) and loneliness among the university students. Variables considered in the Pearson's partial correlation matrix are described in Table 2.

*Table 2. Description of the variables included in the Pearson's correlation test*

| Variables | Types | Measurement |
|---|---|---|
| Age (AG) | Continuous | Age of the students (years) |
| Gender (GN) | Dummy | 1 if the student is male, 0 otherwise |
| Romantic relationship (RR) | Dummy | 1 if the student has romantic relationship, 0 otherwise |
| Study year (SY) | Continuous | Student's studying year |
| Time spent on Facebook (TSF) | Continuous | Total hours spent in Facebook by the student in a day |
| Real life social network (RLSN) | Continuous | Measured by Lubben Social Network Scale |
| Loneliness (LN) | Continuous | Measured by UCLA Loneliness Scale |

## RESULTS AND DISCUSSION

### Descriptive Analysis

This section presents the students' demographic and Facebook use related features like age, gender, study year, hours spent on Facebook per day, years of Facebook use and so on. The result is analyzed with SPSS 23.00 and presented in Table 3.

Table 3 shows that about 43 percent of students are aged between 20 to 22 years while 62 percent of students are male. It is also found that most of the students read in the fourth year and use Facebook from 1 to 2 years. Majority of the students (58 percent) use Facebook for keeping contact with others and mostly they spent time on Facebook for viewing friends' update. From this analysis, it is found that most of the students (32 percent) spent time on Facebook from 2 to 3 hours a day. At the same time, the study also finds that most of the students (68 percent) share various links on their Facebook timeline. Interestingly, the study finds that the majority of the students (32 percent) never changed their profile picture in last three months.

### Reasons behind Updating Facebook Status

Reasons behind updating the Facebook status of the students are analyzed through SPSS 23.00 and presented in Table 4. This analysis has been done for checking whether students update status when they feel lonely.

The above table shows the causes of updating Facebook status. The study finds eleven causes behind updating Facebook status and they are ranked according to the order of their mean value. It is found that to share memorable moments is the topmost reason behind updating Facebook status. The study finds that about 50 of percent students update their Facebook status just for this reason. On the other hand, to express feelings, to make people aware of any important issue are the second and third reasons behind updating Facebook status. Similarly, to earn name and fame is the lowest reason behind updating Facebook status. About 3 percent of students update their Facebook status for this reason. From this analysis, it is found that students mainly update Facebook status not only to express loneliness but also to share memorable moments.

*Table 3. Demographic and Facebook use related features of students*

| Variables | Categories | Percentage |
|---|---|---|
| Age (years) | 18.00-20.00 | 18.50 |
| | 20.01-22.00 | 43.00 |
| | 22.01-24.00 | 32.50 |
| | 24.01 and above | 6.00 |
| Gender | Male | 62.00 |
| | Female | 38.00 |
| Year of Study | First | 23.50 |
| | Second | 17.90 |
| | Third | 24.80 |
| | Fourth | 33.80 |
| Years of Facebook use | 1.00 and below | 32.10 |
| | 1.01 2.00 | 33.30 |
| | 2.01-3.00 | 23.50 |
| | 3.01 and above | 11.10 |
| Purpose of using Facebook | Passing time | 11.5 |
| | A kind of addiction | 9.4 |
| | Keeping contact with others | 58.1 |
| | Entertainment | 20.9 |
| Major Facebook activities | Making new friendship | 15.4 |
| | Viewing friends' updates | 79.5 |
| | Uploading photo/selfie | 5.1 |
| Time spent on Facebook per day (hours) | 1.00 and below | 15.40 |
| | 1.01 -2.00 | 24.80 |
| | 2.01-3.00 | 32.50 |
| | 3.01 and above | 27.40 |
| Sharing of link | Yes | 67.9 |
| | No | 32.1 |
| Frequency of changing profile picture in last three months | Never | 32.1 |
| | Once | 5.1 |
| | Twice | 22.2 |
| | Thrice | 15.4 |
| | Fourth and above | 25.2 |

*Table 4. Reasons behind updating a Facebook status*

| Reasons | Percentage of total | Mean | Rank |
|---|---|---|---|
| To share memorable moments | 50.00 | 0.50 | 1 |
| To express feelings | 48.70 | 0.49 | 2 |
| To make people aware about various issues | 40.20 | 0.40 | 3 |
| To express joys | 38.00 | 0.38 | 4 |
| To express rudeness, anger and egotism indirectly | 19.70 | 0.20 | 5 |
| To express depression | 15.80 | 0.16 | 6 |
| To get rid of loneliness | 13.70 | 0.14 | 7 |
| To express personality | 11.10 | 0.11 | 8 |
| To bring to light the skill, knowledge and appearance | 10.30 | 0.10 | 9 |
| To get like and comments | 6.80 | 0.07 | 10 |
| To earn name and fame | 3.40 | 0.03 | 11 |

Source: Field survey, 2019

*Table 5. Level of real life social network of students*

| Levels of social network | Frequency | Percentage |
|---|---|---|
| Lower level (0.00 to 10.00) | 71 | 19.23 |
| Average level (10.01 to 20.00) | 213 | 57.69 |
| Higher level (20.01 to 30.00) | 86 | 23.08 |
| Total | 370 | 100 |

Note: Average level of real life social network of all students is average i.e. 15.02.
Source: Field survey, 2019

*Table 6. Level of loneliness of university students in Bangladesh*

| Levels of loneliness | Frequency | Percentage |
|---|---|---|
| Lower level (0 to 20.00) | 131 | 35.47 |
| Moderate level (20.01 to 40.00) | 190 | 51.28 |
| Severe level (40.01 to 60.00) | 49 | 13.25 |
| Total | 370 | 100 |

Note: Average level of loneliness of all students is moderate i.e. 26.86.
Source: Field survey, 2019

*Table 7. Correlations among different characteristics and loneliness*

| | AG | GN | RR | SY | HF | RLSN | LN |
|---|---|---|---|---|---|---|---|
| AG | 1 | | | | | | |
| GN | -.157** | 1 | | | | | |
| RR | -.036 | -.057 | 1 | | | | |
| SY | .126* | -.372** | .183** | 1 | | | |
| TSF | .187** | -.038 | -.114* | -.339** | 1 | | |
| RLSN | -.095 | -.219** | -.094 | .111* | .051 | 1 | |
| LN | -.040 | .466** | -.186** | -.689** | .443** | -.234** | 1 |

Note: ** and * means 1 percent and 5 percent levels of significance (two-tailed test)
Source: Survey data, 2019

## Lubben Social Network Scale

Lubben Social Network Scale (LSNS) is measured through the MS Excel 2013 and the result is presented in tabular form in the following table.

Table 5 shows that about 19 percent of students in the study area are lower engaged in real life social network. On the other hand, the majority of the students (58 percent) in the study area are averagely engaged in real life social network while 23 of percent students are highly engaged in real life social network. From the analysis, it is also found that the level of real life social network among the university students in the study area is 15. This interprets that students are averagely engaged in real life social network in the study area.

## UCLA Loneliness Scale

UCLA Loneliness Scale is measured through the MS Excel 2013 and the result is presented in tabular form in the following table.

Table 6 shows only 35 percent of students in the study area suffered from lower level loneliness. On the other hand, it is found that most of the students (51 percent) in the study area suffered from the moderate level of loneliness while 13 percent of students severely suffered from loneliness. From this analysis, it is also found that the level of loneliness among the university students in the study area is 27. This interprets that students suffered from loneliness moderately in the study area.

## Correlations among Different Characteristics and Loneliness

Correlations among various characteristics and loneliness are analyzed by Pearson's partial correlation matrix though SPSS 23.00 and presented in Table 7.

Pearson's partial correlation test reveals the correlation among variables. The above table shows the correlation of some variables with loneliness. All tests were two-tailed and conducted at 5% level of significance. From the above table, it is found that there is a significant correlation between all variables and loneliness except age. The above table reveals that there is a perfect correlation of a particular variable with that variable and the correlation coefficient is one. The estimated correlation coefficient

0.466** reveals that there is a positive correlation between gender and loneliness which is significant at 1 percent level of significance. The estimated correlation coefficient of the romantic relationship and study year with loneliness are -0.186** and -0.689**, respectively. Similarly, 0.443** and -0.234** shows the correlation of Facebook use and real life social network with loneliness. These values interpret that there is a positive and significant relationship between Facebook use and loneliness while a negative relationship exists between real life social network and loneliness.

Although there is a negative relationship between the age of students and the level of loneliness, the relationship is not statistically significant. This finding is consistent with the results of many studies related to the loneliness of students (Cassidy and Asher, 1992). The findings of the study also shows that there is a positive and significant relationship between gender and level of loneliness which reveals that female students are lonelier than male students. This finding may be interpreted by the fact that female students in the study area cannot move freely everywhere anytime with anyone and cannot share their feeling with others, consequently, they have to stay at home in most cases. For this reason, they got depressed and feel lonelier than male students. This study also finds that there is a negative and significant relationship between the romantic relationship and the level of loneliness. This can be interpreted by the fact that students who are in a romantic relationship are less lonely than students who are not in any romantic relation as students enjoying romantic relationship can share their both positive and negative feelings with their partners but others cannot do that. Thus, romantic relationship holder students do not suffer from loneliness but the others. Weiss (1973) and Ozdemir and Tuncay (2008) have found the similar findings. On the other hand, a negative and significant relationship has been found between study year and loneliness. This finding explains that if a student reads in the higher class, the level of loneliness of those students will be relatively lower than students who read in the lower class. Because students read in higher class has many friends and acquainted persons in the university and they can adapt themselves in their university life. As a result, they do not get depressed and do not suffer from loneliness. The study also finds that the level of loneliness among university students differentiates over the time spent on Facebook. This result is positively and significantly correlated which implies that the more the students spent time on Facebook, the more they are lonelier. This finding may be explained by the fact that students' views of others improvements or updates on social media for a long times and staying without family and friends result mental dissatisfaction which makes students lonely. This result is in line with Deters and Mehl (2012). Contrarily, it is found that the students who engaged mostly in real life social network are less lonely than others. The finding is consistent with the result found by Holt-Lunstad et al. (2015).

## CONCLUSION

This paper investigates two separate research questions. First, what is the level of real life social network and the level of loneliness among the university students in Bangladesh? Second, what is the correlation between real life social network, social media, and loneliness? To find out the solution to these questions, this paper uses primary data and several methods, and finds some interesting results. Results are interesting in two different aspects. First, the study finds that university students of Bangladesh have an average level of real life social network and a moderate level of loneliness. Second, the study finds that gender and hours on Facebook are positively and significantly while romantic relationship, study year and real life social network are negatively and significantly related to the level of loneliness of university students. Therefore, the study recommends that students should be more engaged in real life social network and romantic relationship, and be careful in using social media to get rid of the effects of loneliness. As researchers of this paper were constraint by time and money, small sample was taken. Authors here suggest taking a big sample in doing further study on it.

## ACKNOWLEDGMENT

This research has not received grant from any funding agency in the public, commercial or non-profit sectors.

## REFERENCES

Ahmed, O. (2018). Relationship between loneliness and mental health among first-year undergraduate students: Mediating role of timeline browsing and chatting on Facebook. *International Journal of Contemporary Education*, *1*(2), 86–94. doi:10.11114/ijce.v1i2.3625

Alberti, B. F. (2018). This "modern epidemic": Loneliness as an emotion cluster and a neglected subject in the history of emotions. *Emotion Review*, *10*(3), 242–254. doi:10.1177/1754073918768876

Alshammari, A., Al-Harbi, M., Alhamidah, A., Alreshidi, Y., & Abdulrahim, O. (2017). Association between social media use and depression among Saudi population. *Annals of International Medical and Dental Research*, *4*(1). doi:10.21276/aimdr.2018.4.1.ME7

Balchi, S., & Olkun, E. O. (2015). The relation between loneliness in social life and Facebook usage. *Proceedings of the 16th International Academic Conference*. Academic Press. doi:10.20472/IAC.2015.016.015

Blachnio, A., Przepiorka, A., Wolonciej, M., Mahmoud, A. B., Holdos, J., & Yafi, E. (2018). Loneliness, friendship, and Facebook intrusion. a study in Poland, Slovakia, Syria, Malaysia, and Ecuador. *Studia Psychologica*, *60*(3), 183–194. doi:10.21909p.2018.03.761

Brown, L., & Wood, M. (1953). Paths of Loneliness: The individual isolated in modern society. *American Sociological Review*, *18*(6), 712. doi:10.2307/2088145

BTRC. (2019). Internet Subscribers in Bangladesh June, 2019. Retrieved from http://www.btrc.gov.bd/telco/internet

Burrow, A., & Rainone, N. (2017). How many likes did I get?: Purpose moderates links between positive social media feedback and self-esteem. *Journal of Experimental Social Psychology*, *69*, 232–236. doi:10.1016/j.jesp.2016.09.005

Cacioppo, J., & Patrick, W. (2009). *Loneliness: Human Nature and the Need for Social Connection*. New York: Norton. Retrieved from https://psycnet.apa.org/record/2008-07755-000

Cassidy, J., & Asher, S. R. (1992). Loneliness and peer relations in young children. *Child Development*, *63*(2), 350–365. doi:10.2307/1131484 PMID:1611939

Cecen, A. R. (2008). The effects of gender and loneliness levels on ways of coping among university students. *College Student Journal*, *42*, 510–516. Retrieved from https://eric.ed.gov/?id=EJ816934

Cohen, S. (2004). Social relationships and health. *The American Psychologist*, *59*(8), 676–684. doi:10.1037/0003-066X.59.8.676 PMID:15554821

Cole, S., Capitanio, J., Chun, K., Arevalo, J., Ma, J., & Cacioppo, J. (2015). Myeloid differentiation architecture of leukocyte transcriptase dynamics in perceived social isolation. *Proceedings of the National Academy of Sciences of the United States of America*, *112*(49), 15142–15147. doi:10.1073/pnas.1514249112 PMID:26598672

Demir, Y. (2016). The relationship between loneliness and depression: Mediation role of internet addiction. *Educational Process International Journal*, *5*(2), 97–105. doi:10.12973/edupij.2016.52.1

Deters, F., & Mehl, M. (2012). Does posting Facebook status update increase or decrease loneliness? An online social networking experiment. *Social Psychological & Personality Science*, *4*(5), 579–586. doi:10.1177/1948550612469233

Hall, J., Kearney, M., & Xing, C. (2018). Two tests of social displacement through social media use. *Information Communication and Society*, *10*, 1–18. doi:10.1080/1369118x.2018.1430162

Holt-Lunstad, J., Smith, T., Baker, M., Harris, T., & Stephenson, D. (2015). Loneliness and social isolation as risk factors for mortality. *Perspectives on Psychological Science*, *10*(2), 227–237. doi:10.1177/1745691614568352 PMID:25910392

Huang, C. (2010). Internet use and psychological well-being: A meta-analysis. *Cyberpsychology, Behavior, and Social Networking*, *13*(3), 241–249. doi:10.1089/cyber.2009.0217 PMID:20557242

Hunt, M. G., Marx, R., Lipson, C., & Young, J. (2018). No more FOMO: Limiting social media decreases loneliness and depression. *Journal of Social and Clinical Psychology*, *37*(10), 751–768. doi:10.1521/jscp.2018.37.10.751

ITU. (2019). ITU World Telecommunication /ICT Indicators database. Retrieved from https://www.itu.int/en/ITU-D/Statistics/Pages/stat/default.aspx

Jobe, L. E., & White, S. W. (2007). Loneliness, social relationships, and a broader autism phenotype in college students. *Personality and Individual Differences*, *42*(8), 1479–1489. doi:10.1016/j.paid.2006.10.021

Kabir, M. A., Sultana, F., Ibrahim, M., Islam, M. M., Rahman, M. M., Islam, M. J., & Bayen, S. (2018). Evaluation of stress, loneliness and depression among residential and non-residential students of Dhaka University: Case-control study. *Pharmaceutical Science and Technology*, *2*(1), 1–6. doi:10.11648/j.pst.20180201.11

Kar-Purkayastha, I. (2010). An epidemic of loneliness. *Lancet*, *376*(9758), 2114–2115. doi:10.1016/S0140-6736(10)62190-3 PMID:21167590

Karakose, T., Yirci, R., Uygun, H., & Ozdemir, T. Y. (2016). Relationship between high school students' Facebook addiction and loneliness status. *Eurasia Journal of Mathematics. Science and Technology Education*, *12*, 2419–2429. doi:10.12973/eurasia.2016.1557a

Killeen, C. (1998). Loneliness: An epidemic in modern society. *Journal of Advanced Nursing*, *28*(4), 762–770. doi:10.1046/j.1365-2648.1998.00703.x PMID:9829664

Kross, E., Verduyn, P., Demiralp, E., Park, J., Lee, D., Lin, N., ... Ybarra, O. (2013). Facebook use predicts declines in subjective well-being in young adults. *PLoS One*, *8*(8), e69841. doi:10.1371/journal.pone.0069841 PMID:23967061

Levenson, J., Shensa, A., Sidani, J., Colditz, J., & Primack, B. (2016). The Association between social media use and sleep disturbance among young adults. *Preventive Medicine*, *85*, 36–41. doi:10.1016/j.ypmed.2016.01.001 PMID:26791323

Lin, L., Sidani, J., Shensa, A., Radovic, A., Miller, E., Colditz, J., ... Primack, B. A. (2016). Association between social media use and depression among US young adults. *Depression and Anxiety*, *33*(4), 323–331. doi:10.1002/da.22466 PMID:26783723

Masthi, N. R. R. M., Pruthvi, S., & Mallekavn, P. (2017). A comparative study on social media addiction between public and private high school students of urban Bengaluru, India. *ASEAN Journal of Psychiatry*, *18*, 206–215.

Medvene, L., Nilsen, K., Smith, R., Ofei-Dodoo, S., DiLollo, A., Webster, N., ... Nance, A. (2015). Social networks and links to isolation and loneliness among elderly HCBS clients. *Aging & Mental Health*, *20*(5), 485–493. doi:10.1080/13607863.2015.1021751 PMID:25808879

Nowland, R., Necka, E., & Cacioppo, J. (2017). Loneliness and social internet use: Pathways to reconnection in a digital world? *Perspectives on Psychological Science*, *13*(1), 70–87. doi:10.1177/1745691617713052 PMID:28937910

Ozdemir, U., & Tuncay, T. (2008). Correlates of loneliness among university students. *Child and Adolescent Psychiatry*, *2*, 1–6. doi:10.11866/1753-2000-2-29 PMID:18851744

Pervin, M. M., & Ferdowshi, N. (2016). Suicidal ideation in relation to depression, loneliness and hopelessness among university students. *Dhaka University Journal of Biological Sciences*, *25*(1), 57–64. doi:10.3329/dujbs.v25i1.28495

Pittman, M. (2018). Happiness, loneliness, and social media: perceived intimacy mediates the emotional benefits of platform use. *The Journal of Social media in Society*, 7, 164-176.

Primack, B. A., Shensa, A., Sidani, J. E., Whaite, E. O., Lin, L. Y., Rosen, D., ... Miller, E. (2017). Social media use and perceived social isolation among young adults in the U.S. *American Journal of Preventive Medicine*, *53*(1), 1–8. doi:10.1016/j.amepre.2017.01.010 PMID:28279545

Russell, D., Peplau, L. A., & Cutrona, C. E. (1980). The revised UCLA loneliness scale– concurrent and discriminate validity evidence. *Journal of Personality and Social Psychology*, *39*(3), 472–480. doi:10.1037/0022-3514.39.3.472 PMID:7431205

Schmidt, N., & Sermat, V. (1983). Measuring loneliness in different relationships. *Journal of Personality and Social Psychology*, *44*(5), 1038–1047. doi:10.1037/0022-3514.44.5.1038

Scott, G., Boyle, E., Czerniawska, K., & Courtney, A. (2018). Posting photos on Facebook: The impact of narcissism, social anxiety, loneliness, and shyness. *Personality and Individual Differences, 133*, 67–72. doi:10.1016/j.paid.2016.12.039

Shettar, M., Karkal, R., Kakunje, A., Mendonsa, R. D., & Chandran, V. V. M. (2017). Facebook addiction and loneliness in the post-graduate students of a university in southern India. *The International Journal of Social Psychiatry, 63*(4), 325–329. doi:10.1177/0020764017705895 PMID:28504040

Smith, K. (2019). *126 Amazing Social Media Statistics and Facts*. Retrieved from https://www.brandwatch.com/blog/amazing-social-media-statistics-and-facts/#section-10

Song, H., Zmyslinski-Seelig, A., Kim, J., Drent, A., Victor, A., Omori, K., & Allen, M. (2014). Does Facebook make you lonely?: A meta analysis. *Computers in Human Behavior, 36*, 446–452. doi:10.1016/j.chb.2014.04.011

Statista, (2019). Global digital population as of July 2019 (in millions). Retrieved from https://www.statista.com/statistics/617136/digital-population-worldwide/

Steers, M., Wickham, R., & Acitelli, L. (2014). Seeing everyone else's highlight reels: How Facebook usage is linked to depressive symptoms. *Journal of Social and Clinical Psychology, 33*(8), 701–731. doi:10.1521/jscp.2014.33.8.701

Steptoe, A., Shankar, A., Demakakos, P., & Wardle, J. (2013). Social isolation, loneliness, and all-cause mortality in older men and women. *Proceedings of the National Academy of Sciences of the United States of America, 110*(15), 5797–5801. doi:10.1073/pnas.1219686110 PMID:23530191

Tobin, S., Vanman, E., Verreynne, M., & Saeri, A. (2014). Threats to belonging on Facebook: Lurking and ostracism. *Social Influence, 10*(1), 31–42. doi:10.1080/15534510.2014.893924

Toma, C. (2013). Feeling better but doing worse: Effects of Facebook self-presentation on implicit self-esteem and cognitive task performance. *Media Psychology, 16*(2), 199–220. doi:10.1080/15213269.2012.762189

Umberson, D. K., & Montez, J. (2010). Social relationships and health: A flashpoint for health policy. *Journal of Health and Social Behavior, 51*(1 suppl), S54–S66. doi:10.1177/0022146510383501 PMID:20943583

Valtorta, N., & Hanratty, B. (2012). Loneliness, isolation and the health of older adults: Do we need a new research agenda? *Journal of the Royal Society of Medicine, 105*(12), 518–522. doi:10.1258/jrsm.2012.120128 PMID:23288086

Valtorta, N., Kanaan, M., Gilbody, S., & Hanratty, B. (2016). Loneliness, social isolation and social relationships: What are we measuring? A novel framework for classifying and comparing tools. *BMJ Open, 6*(4), e010799. doi:10.1136/bmjopen-2015-010799 PMID:27091822

Victor, C., Scambler, S., & Bond, J. (2009). *The social world of older people: Understanding Loneliness and Social Isolation in Later Life*. Maidenhead: Open University Press.

Weeks, D., Michela, J., Peplau, L., & Bragg, M. (1980). Relation between loneliness and depression: A structural equation analysis. *Journal of Personality and Social Psychology*, *39*(6), 1238–1244. doi:10.1037/h0077709 PMID:7205551

Weiss, R. S. (1973). *Loneliness: The experience of emotional and social isolation London*. MIT Press. Retrieved from https://mitpress.mit.edu/ books/loneliness

Winkel, V. M., Wichers, M., Collip, D., Jacobs, N., Derom, C., & Thiery, E. (2017). Unraveling the role of loneliness in depression: The relationship between daily life experience and behavior. *Psychiatry*, *80*, 104–117. doi:10.1080/00332747.2016.1256143

Wiseman, H., Mayseless, O., & Sharabany, R. (2006). Why are they lonely? Perceived quality of early relationships with parents, attachment, personality predispositions and loneliness in first-year university students. *Personality and Individual Differences*, *40*(2), 237–248. doi:10.1016/j.paid.2005.05.015

Wolniczak, I., Cáceres-DelAguila, J., Palma-Ardiles, G., Arroyo, K., Solís-Visscher, R., Paredes-Yauri, S., ... Bernabe-Ortiz, A. (2013). Association between Facebook dependence and poor sleep quality: A study in a sample of undergraduate students in Peru. *PLoS One*, *8*(3), e59087. doi:10.1371/journal.pone.0059087 PMID:23554978

Wood, C. (2013). Loneliness: A silent epidemic. *Independent Nurse*, *6*. doi:10.12968/indn.2013.17.6.99114

Xia, N., & Li, H. (2018). Loneliness, social isolation, and cardiovascular health. *Antioxidants & Redox Signalling*, *28*(9), 837–851. doi:10.1089/ars.2017.7312 PMID:28903579

Yavich, R., Davidivutch, N., & Frenkel, Z. (2019). Social media and loneliness - forever connected? *Higher Education Studies*, *9*(2), 10–21. doi:10.5539/hes.v9n2p10

Zelikowsky, M., Hui, M., Karigo, T., Choe, A., Yang, B., Blanco, M., ... Anderson, D. J. (2018). The neuropeptide controls a distributed brain state induced by chronic social isolation stress. *Cell*, *173*(5), 1265–1279. doi:10.1016/j.cell.2018.03.037 PMID:29775595

# Chapter 13
# A Comparative Historical Analysis of the Evolution of the Venture Capital Industry in the Economic Regions of the United States of America, Europe, and China

**Som Sekhar Bhattacharyya**
https://orcid.org/0000-0003-3259-0319
*National Institute of Industrial Engineering, India*

**Bibhash Laik**
*National Institute of Industrial Engineering, India*

**Divya Sharma**
*National Institute of Industrial Engineering, India*

**Tirthankar Bose**
*National Institute of Industrial Engineering, India*

## ABSTRACT

*Venture capital (VC) provides a platform to empowered individuals with financial constraints to transform their ideas into business models and attain commercial success. This article reviewed the growth and trends of VC industry across various regions such as the United States of America (USA), Europe, China, and India. Initially, VC firms flourished and developed in the USA and still it harbors the largest VC industry. From the USA, VC firms spread to Europe and then much later to emerging economies like China and India. Although the VC ecosystem had started late in China, it had registered higher growth when compared to Europe in terms of VC investment. China has become the second largest VC market. It was backed by government initiatives, vast market opportunities, and home-grown technology firm investments. India has started observing growth in VC space later than China but had ample opportunities to allow for a surge in VC activities.*

DOI: 10.4018/978-1-7998-1566-2.ch013

## INTRODUCTION

Global economy post second world war registered growth especially in the western world (Maddison, 1995). Global economy fluctuated during the1970s-1990s (Inglehart & Abramson, 1994). This was because of a sharp decline in global economy during the 1970s due to rising oil prices, fluctuating exchange rates and political instability around the world (Barsky & Kilian, 2004). A sharp rise in global economic activities was observed during the 1990s due to introduction of Information Technology (IT) as an industry (Oliner & Sichel, 2000). Till the 1990s, Venture Capital (VC) industry remained mostly confined within United States of America (USA) then it spread to other parts like Europe and Asia (Bottazzi and Da Rin, 2002). VC firms witnessed a sharp rise in developing nations in the 2000s and it influenced industrialist, governments and most significantly entrepreneurs (Megginson, 2004). Multiple perspectives regarding the description of VC as a construct has been defined (Sapienza, 1992). According to Bottazzi and Rin (2002), VC could be defined as a support to entrepreneurs in the form of investment and turning their ideas into product and services. Kaplan and Stromberg (2003), defined VC as a long-term investment for Small to Medium sized Enterprise (SME), in the form of capital and providing intangible benefits like network connection and business expertise. VC funds had been able to build companies from scratch to mature organizations (Rice et al., 2000).VC investors engaged actively with the startup firm by placing its member in the board of the startup and provided valuable assistance (Hansen et al., 2000). Over the years, VC firms concentrated on management, revenue model, ownership, market conditions, potential of the firm to be funded, amount of capital required by a firm to overcome losses (if any) (Metrick & Yasuda, 2010). VC firms aspired to secure strategic fit to VC fund's objectives (De Clercq et al., 2006). VC firms, subsequent to the funding, over the next four to seven years generally worked with the owners and helped it become a mature organization (Sahlman, 1990). Once the invested company reached maturity and became successful, VC firms exited from the company by taking it public through an Initial Public Offering (IPO) or selling its shares to big companies in the form of acquisition or liquidation (Barry et al., 1990). This process helped the VC firms fund to be free from its previous investment and allowed it to invest in new companies (Gompers, 1997). VC firms typically lived by such cycles of entry to exit (Teker, Teker & Teraman, 2016). So, for new small sized companies having limited access to capital markets VC firms had become an essential source of capital (Martin, Sunley & Turner, 2002) The variation of VC at different stages while funding a new firm has been tabulated in table 1.

Despite all of these, VC firms remained as a popular source for fund raising as funding through bank loans and other methods was not readily available and had relatively strict repayment schedules (Jeng & Wells, 2000). Highly competent companies with higher valuations were usually targeted by VC firms (Zider, 1998). VC firms which provided for new and small business in emerging industries were termed as angel investors (DeGennaro, 2010).

*Table 1. Variation of VC and it's different stages*

| Stage | Description | Applicability | Associated risk |
|---|---|---|---|
| Seed | Funds provided to entrepreneurs or inventors for the early development stage of new product or service, (Burzacka & Gąsiorowska, 2016). | This stage focussed on developing business plan to building management team. It further focusses on product development (Burzacka & Gąsiorowska, 2016). | This stage generally has been very risky as at this stage the commercial operations haven't yet started (Burzacka & Gąsiorowska, 2016). |
| Early | Funds were provided to firms which were in activities which ranged from operational level to sales. Small firms running business for less than three years and having a initial prototype of product or service were considered (Sobanska & Sieradzan, 2004). | Commercial manufacturing and sales would start during this stage (Sobanska & Sieradzan, 2004). | Risk would be associated in this stage as sales of product or service remained uncertain as the product was new (Sobanska & Sieradzan, 2004). |
| Formative | This included seed stage and early stage financing (Gompers, 1994). | In this stage, capital would be provided for major expansion like plant expansion and for product marketing and operational improvements (Gompers, 1994). | Risk would be associated with sales which would depend on market conditions (Gompers, 1994). |

## LITERATURE REVIEW

In this section the importance, advantages and disadvantages of VC firms have been discussed respectively.

### Importance of Venture Capital

Countries having strong economy would become global economic leaders (Kindleberger, 1981). Funding of entrepreneurial ventures had been looked upon as an important aspect in this regard (Beugelsdijk & Smeets, 2008). Promotion of entrepreneurship and growth of individuals endowed with rich business acumen would result in the growth of a country economically (Baumol, 1993). Venture Capital (VC) provided a platform that encouraged individuals from different fields to showcase their talent and convert it into a commercial success with requisite financial viability (Chishti & Barberis, 2016). VC funds could be used as a tool to encourage entrepreneurs from different fields (Henderson, 2002). Transformation of the innovation process in many countries had been driven by the rise of fast growing & high technology companies, which were backed by VC finance (Deorah, 2015). The result was the coming up of giants in technology such as Intel, Digital Equipment Corporation, Apple, Microsoft, Sun Microsystems (Florida, & Kenney,1988). Even technology companies like Google or taxi aggregators like Uber had been backed by VC funds which brought these dream companies into reality (Deorah, 2015). Often business ideas failed due to lack of ample funding and this increased the rate of unemployment (Belke et al., 2003). Such problems could be solved by VC funding (Belke et al., 2003). Small and medium sized companies could play a major role in the development of economy and VC funding would help in the development of these companies (Agbeibor, 2006).

*Table 2. VC funding challenges*

| S. No. | Author | Challenges |
|---|---|---|
| 1 | Storey (1985) | Raising fund had not been an easy task. |
| 2 | Ruhnka & Young (1991) | Investor preferred to retain cash as compared to long term investment in new ventures. |
| 3 | Gompers (1995) | Confusion among VC to invest in established brand or non-established brand. |
| 4 | Kaiser, & Westarp (2010) | Financial crisis and dot com bubble burst have made fundraising tough. |
| 5 | Kohler (2016) | Launching startups has been easy in present day resulting in high end competitive pressures. |

## Challenges

Fund raising in the capital market had not been an easy task for most entrepreneurs (Storey, 1985). There had been multiple challenges, one of which was securing capital for an idea (Buttner & Rosen, 1992). Getting a deal struck out of multiple competitors (competing ideas of many entrepreneurs) in the market also posed hurdle (Kim, & Mauborgne, 2014). It had been a tough decision for a VC fund manager to decide whether to invest in a firm with an established brand name or one with none (Gompers, 1995). It was noticed in a few cases that many pre-unicorns with no established brand name did well in later stages (Block et al., 2018). Execution of new startups also posed a difficult challenge (Kohler, 2016). In present day context, there had been great ease with which startups were launched and the cost associated with it had dropped (Kohler, 2016). This resulted in high end competitive pressure for both VC firms as well as startups (Block & Sandner, 2011). The returns by the VC industry after the internet bubble burst had been uncertain and created a doubt in the minds of investors and fundraising had been tough ever since (Kaiser, & Westarp, 2010). Considering uncertainties created in global markets, securing VC fund had been challenging (Murray, 1992). This was because VC investors preferred to retain cash rather than making long term investment in new ventures (Ruhnka & Young, 1991). Robust stock market development, poor contract and tax legislation, inhibiting labor market regulations were some of the other factors which posed challenges in VC activities development (Lerner & Tag, 2013). In table 2, the challenges of a VC firm as well that of a startup has been tabulated chronologically.

## Advantages of Venture Capital

After VC firms came into existence, it transformed the way startups started & operated (Colombo & Grilli, 2010). It altered multiple industries & sectors, such as manufacturing and services which were technology intensive (Feldman et al., 2005). The major reason for going through VC financing route has been because VC funding was not a loan scheme (Kortum & Lerner, 2001). There was also an absence of strict repayment schedule in VC financing as compared to bank loans or other debt instruments (Gompers, 1996). Also, the VC firms provided expertise to a startup in the form of consulting in areas like financial management & strategic management (Ehrlich et al., 1994). This was valuable and affected the growth of firm (Ehrlich et al., 1994). It was noticed that VC fund managers were well connected in the business community and one single VC firm interest would attract other funding ventures to participate in funding (Hochberg et al., 2007). Also, VC firms' route for exit (for return & liquidity) aided startups

*Table 3. VC funding advantages*

| S. No. | Author | Advantages |
|---|---|---|
| 1 | Ehrlich et al. (1994) | VC firms provided expertise for startup in areas like financial management & strategic management. |
| 2 | Gompers (1996) | VC funding is not a debt. VC financial route was better than banks as it had absence of strict repay schedule. |
| 3 | Feldman et al. (2005) | It opened scope of investment for multiple sectors. |
| 4 | Hochberg et al. (2007) | Connection of VC firm managers led to participation of other funding ventures. |
| 5 | DeTienne et al. (2015) | VC firms helped the firms to achieve IPO. |
| 6 | Kalpan & Lerner (2016) | VC firms helped in creating more job opportunities, providing proper scope to new entrepreneurs and also bringing profit for investors. |

to achieve IPOs (DeTienne et al., 2015). On a broader perspective, the advantages of VC firms could be classified within three elements namely investor, entrepreneur and economy (Kalpan & Lerner, 2016). Economy related benefits helped in job creation and industrialization of a nation (Tzannatos, 1999). This was because VC firms have been investing in newly budding companies or start-ups (Manchanda & Muralidharan,2014). Entrepreneurial advantages provided by VC firms could be through providing a proper platform to execute new ideas and support entrepreneurial spirit despite bearing considerable risks along with the improvement of technological advances in the market (Sapsed et al., 2007). Investor benefits have been related to profit, as profit to VC firms has been profit to investor (Callison & Vestal, 2010). The advantages of VC funding both for VC firms as well as startups has been tabulated in table 3 in chronological manner.

## Disadvantages of VC

The major disadvantage that has been reported with regard to VCs was the intervention of venture capitalists in the operational and strategic decision making of a startup (Gompers & Lerner, 2001). Further, there was a dilution of control mechanisms for the top management positions of a startup firm (invested by VC firms) (Gompers & Lerner, 2001). The role played by VC firms at every stage from birth to maturity had been varying (Sahlman, 1990). VC firms held a significant percentage of the company share in the form of equity (Mehran, 1995). In many cases VC firms placed a member of their team in the startups lead team with the intention to provide quality assistance in managerial decision (Isaksson, 2006). VC firm managers believed that these kinds of structure resulted in higher performance for a startup (Isaksson, 2006). However, it was noticed that such governance structure manifested into a tension between the two parties and hampered the entrepreneur's work (Arthurs & Busenitz, 2003). The advent of internet wave created a lot of opportunities for several startups to grow in this area (Miller & Bound, 2011). It has also been noticed that despite support and nurturing by VC fund managers, majority of startups failed and very few had turned into successful ventures (Barry, 1994). Thus, the notion that VC fund managers always made money and every deal had been profitable was a wrong one (Fried & Hisrich, 1994). The disadvantages associated with VC funding has been presented chronologically in table 4.

*Table 4. VC funding disadvantages*

| S. No. | Author | Disadvantages |
|---|---|---|
| 1 | Barry (1994) | Majority of startup failed & few turned successful despite the backup by VC firm managers expertise. |
| 2 | Gompers & Lerner (2001) | Intervention of VC firms in strategic & operational decision making of firm resulted in dilution of control mechanism. |
| 3 | Arthurs & Busenitz, (2003) | Changes in governance structure resulted in hindrance of entrepreneurs working and gave rise to tension between parties. |

The authors had undertaken Literature Review (LR) and for this the authors conducted the LR search on EBSCO, Proquest, Emerald Insight, Jistor and such other journals databases for the keyword of 'Venture Capital'. A detailed analysis of the VC literature has been provided in a chronological order in table 5.

## HISTORICAL ANALYSIS

In this section, an evolutionary historical discussion regarding the VC industry in the contexts of USA, Europe, China & India has been provided respectively. Finally, a comparative analysis in the recent decade of 2010s has been provided.

### VC Industry in USA

The existence of VC ecosystem has been ascribed to its founder George Doriot, who was chosen to lead American Research & Development Corporation (ARDC) (Liles, 1977). This was the first public VC firm in the year 1946 (Myhrvold, 2010). The need of entrepreneurial firms to raise funds without collateral, borrowing risk capital in the form of equity gave rise to the formation of ARDC (Scherer, 2000). During the 1960- 1980 period, a considerable upliftment in VC atmosphere was made by the enactment of Small Business Investment Act 1958 (Gilbertson, 1959). This act provided tax advantages and permitted banks to invest in Small Business Investment Companies (SBICs) (Dossani & Kenney, 2002). During the 1981- 1990 period, there had been rise in investing activities till 1983 (amounting to $3Bn) and throughout the 1980s stable investment happened due to the relaxation of investment rules for U.S.A. based pension funds (Metrick & Yasuda, 2011). During the 1991- 2000 period, major changes in investment patterns had been observed because of the dawn of internet era (Metrick & Yasuda, 2010). This period was also called as the boom period and multifold returns were witnessed in this period (Kaplan & Lerner, 2010). In the 2001- 2010 period, the crash of NASDAQ and technology slump degraded the valuation of startup companies which shook the entire VC industry (Goodnight & Green, 2010). The investments during this period were at its lowest but the numbers were higher than that prior to the year 1995 (Metrick & Yasuda, 2011). In the 2011- 2016 period, VC activity reduced in angel and seed stage as software industry matured into different business models (Kolmakov et al., 2015). VC investors made considerable investments in less risky stages such as the late stage instead of seed stage (Agrawal et al., 2014).

*Table 5. Systematic 'VC' literature review*

| S.No. | Author | Remarks |
|---|---|---|
| 1 | Jeng & Wells (2000) | The authors focused on the determinants of VC investment like Initial Public offerings (IPOs), Gross domestic Product (GDP), market capitalisation growth, labour market rigidities across 21 countries. |
| 2 | Dehesa (2002) | The author discussed the micro and macro aspects of VC ecosystem in EU for its large gap with US. There has been huge gap in terms of volume of venture capital raised and investment between the two countries. To reduce the gap, it was advocated that VC firms in Europe must focus on the volume of venture capital investment as well as quality of venture capital investment. |
| 3 | Lockett, Wright, Sapienza & Pruthi, (2002) | The authors focused on the source of information and approach to investee valuation used by VC investors of US, Hong Kong, India, and Singapore. The authors found out that there had been commonality between the countries for valuation methods but there had been significant differences in the source of information used for these valuations. Source of information had been more culturally determined and differed from country to country. |
| 4 | Manigart, Waele & Wright (2002) | The authors focused on the determinants of required return in VC investments of US, United Kingdom (UK), France, Belgium, and the Netherlands. The authors demonstrated the risk reward trade-off in VC industry. Stage diversification had not been a risk reduction strategy by the VC firms. The required return for independent VC companies were higher as compared to public supported firms for early stage and expansion stage investments. |
| 5 | Cumming & Macintosh (2003) | The authors studied on the differences between US & Canada VC industries & impact of legal & institutional factors on different exits. The degree of information asymmetry between selling VC and purchaser of VC investments was a major factor for determining full exit or partial exit. In Canada, the hold period requirement was more stringent, so it affected the partial exits more as compared to the USA. |
| 6 | Megginson (2004) | The author studied on the emergence of global market on VC. The author found out that there had been difficulty in integrating global VC market due to the differences existing among legal system across countries. The author also provided reasons in favor of global VC market by demonstrating the increasing fundraising and investment activities across countries. |
| 7 | Bruton, Fried & Manigart (2005) | The article discussed on the impact and influence of various institutions on VC firm's behavior in different countries. The authors found three categories of institutional influence namely normative, regulatory & cognitive that affected the expansion of venture capitalist in different countries. |
| 8 | Cornelius (2005) | The author did a comparative study of VC firms from Australia, US, China, and Europe. The author found out that in US, the emphasis was on early stage investment whereas in Europe the emphasis was on buy-out, turn-around and in late stage investment. In China and Australia, the focus was on expansion stage development. |
| 9 | Mayer, Schoors & Yafeh (2005) | The authors compared the investment activities and sources of funds of VC funds in UK, Israel, Japan, and Germany. The authors found out that VC investments differed across countries in terms of geographical focus, sectors and stage. Sources of VC funds were different across countries like pension funds were important in UK, corporations in Israel, banks have been important in Germany & Japan. It was also found out that bank backed funds were more focused for late-stage investment, while individual & corporate backed funds were focused on early stage investment. |
| 10 | Grichnik & Hisrich, (2006) | The authors investigated investment behaviors of VC firms of Germany & Israel in comparison with USA. The authors considered board involvement, strategy formation, entrepreneur capabilities & industry performance as some of the parameters to judge the similarities and differences across the VC firms. |
| 11 | Zacharakis, McMullen & Shepherd (2007) | The authors studied the influence of economic institutions upon VC firms decision policies in USA (mature economy), South Korea (emerging economy), and China (transitional economy). The authors found out that mature market economy relied upon market information to a greater extent as compared to emerging economies. Transitional and emerging economies relied upon human capital factors to a greater extent. |
| 12 | Aizenmann & Kendall (2008) | The authors investigated the increasing internationalization of VC investments and the factors behind it. High-end human capital, better business environment, deeper financial markets and military expenditure were some of the significant local factors that appeared to attract international VC firms. |
| 13 | Bruton, Ahlstrom & Puky (2009) | The article examined the impact of institutional clusters on the VC industry. The authors did a comparative study of VC firms in Asia, and Latin America. The authors found out that the VC industry exhibited consistency across multiple dimensions in both these regions despite having significant differences in industry practice in both these regions. |
| 14 | Liu (2010) | The paper studied on VC investment behaviour pattern differences between China, and US. The author considered four investment stages namely selecting, structure and monitoring, value-adding and exiting and tried to find out the differences. The author focused on VC activities in each stage and skill & instrument applied in each activity. |
| 15 | McCahery & Vermeulen (2010) | The authors discussed on the corporate venture role in boosting new entrepreneurial clusters after financial crisis in US, Germany & Russia. The authors found out that after the financial crisis greater attention were paid to Corporate Venture Capital (CVC) initiatives. The authors suggested that government must act as a facilitator of CVC alliances and help in starting entrepreneurial growth similar to Silicon Valley. |
| 16 | Block, Vries & Sandner (2010) | The authors deliberated on the impact of financial crisis on VC market across industries and countries. The authors found out that there was a sharp decrease in in the number of initial funding rounds along with the amount of funding raised in later funding rounds. The effect of crisis on USA was stronger as compared to other countries. |
| 17 | Ning, Wang & Yu (2015) | The authors discussed the driving forces of VC investment in US between the years of 1995 to 2011. The authors found out the major drivers were macroeconomic factors and public market signals. The major macroeconomic factors like higher GDP growth rate, lower unemployment rate and greater industry production index had positive impact on the VC industry. |
| 18 | Teker & Teraman (2016) | The authors did a cross country analysis of VC markets of US, Europe, Israel, Canada, China and India. The authors found out that the most attractive sector for Europe, China & India had been consumer product while for USA, Israel, and Canada has been information technology. |

## VC Industry in Europe

VC activities started in Europe during the decade of 1980's but it was underdeveloped due to lack of robust legal policies (Bruton et al., 2005). During the 1991-2000 period, especially in the second half of 1990s there had been significant growth of VC industry due to supportive government policies which led to an aggregate investment of 12 billion euros by the year 1999 (Schertler, 2001). Some of the eminent companies like Cambridge silicon radio (1998) and MySQL (1995) were founded through VC investments (Hege et al., 2003). During 2002-2004, a small number of deals (as low as twelve) were completed because of lack of interests of VCs in fundraising and not indulging in new investments (Kräussl, & Krause, 2014). But during the 2004-2005 period after the entry of Noble Venture Finance (NVF) there was a steep rise in investment levels from 8 million euros to 119 million euros (Bottazzi et al., 2004). During this period, noted companies like SKYPE, Last.fm were started and these events improved the growth opportunities as around two million people were employed by the year 2010 in such sectors backed by VC investments (Bertoni et al., 2015). During these periods it was also noticed that there were continued disinterest of VC firms towards young companies (Cherif & Gazdar, 2011). As new investments struggled to lead young portfolio company into profitability (Cherif & Gazdar, 2011). During the 2011-2014 period, the participation rate of angel investors or incubators increased from 9% in 2010 to around 27% in the year 2013 (Nunes et al., 2014). It was also noticed that there was a lack of private funding in this period (Nunes et al., 2014). During this time, the European VC funds were dependent on government financing which had led to stripping away of risk- taking attitude and entrepreneurial culture (Mason, 2009). During the 2015-2016 period, there was increased uncertainties due to UK's Brexit and geopolitical unrest in other European countries (Harvey, & Hubbard, 2016). However, there was a positive sign as successful IPO exits had increased (Cumming et al., 2017). This provided a successful exit route for the venture capitalist and investors (Cumming et al., 2017). Further, Google venture aimed to invest a substantial amount of $125 million in European startups in the year 2015 (Zhang et al., 2015). Thus, the closing decade of 2010s was positive for Europe.

## VC Industry in China

The first VC firm of China was China New Technology Venture Capital Investment Corporation (CNTVI) (Haitian et al., 2007). It was set up in the year 1985 to boost up economic activity in China (Haitian et al., 2007). Not much investment activity happened during the period between 1985-2000 (Zeng, 2004). During the mid-1990s, technology development became a pressing government policy but there were strict government regulation and lack of awareness regarding VC industry working in China (Zeng, 2004). The relaxation of mandatory government's permission for private fund raising led to a wave of substantial foreign direct investment across industries in China, (this was around 95% of the total VC funding in the1990s) (Wong 2011). During the late 1990s, Innovation Fund for Small and Medium Technology-based Enterprises (Innofund) was introduced when CNTVI was shut down due to its inability to commercialize (Wang, 2009). During the period of 2001-2005, there had been a slump in VC activities due to dot com bubble burst in the year 2001 and economic slowdown in the early 2000s (Lin, 2015). Government's efforts towards constructive regulation policies and making the atmosphere VC friendly helped boost investments from 5% in 2003 to 43.7% in the year 2006 (Ahlstrom et al., 2007). During 2006-2013, three major expenditure programs to develop VC industry in China were introduced namely Industrial Technology Research and Development Budgets Funded VC Funds (ITRDF), Government Directory Venture Capital Fund (GDVCF), and Industrial Technology Research and Development Budget Participated Venture Capital Funds (ITRDPF) (Hu, 2010). During the period of 2013-2016, a noticeable shift from a government directed approach to a government led as well as market operation model had been witnessed (Lin, 2015). This boosted up the new ventures and contributed in increased seed funding in order to provide leverage to the growing innovation manifesting in China (Zhang & Mayes, 2018). Overall in China VC landscape was both healthy as well as booming.

## VC Industry in India

India was extremely conservative towards its economy after independence and it affected many industries in spite of having abundant talent (Pandey, 1998). During the year 1988, government issued guidelines to legalize venture capital operations (Ramesh & Gupta, 1995). In 1988, the Technology Development & Information Company of India Ltd. (TDICI), was established in Bangalore as a subsidiary of the Industrial Credit & Investment Corporation of India, Ltd. (ICICI) (Dossani & Kenney, 2002). This was the first organization which could be identified as a venture capital operator (Dossani & Kenney, 2002). The success of Indian entrepreneurs in Silicon Valley during the 1990s encouraged the notion in USA that India might have a good talented pool of entrepreneurs (Pandey, 1998). There had been an involvement of overseas private sector in the Indian venture capital industry since then (Subhash, 2004). During the late 1996, Walden Group's, Walden International Investment Group (WIIG) initiated its India-focused venture capital operation focusing on early and late stage companies (Dossani & Kenney, 2002). In the early 2000s, the focus of VC investment shifted from technology-oriented business to commercially viable industries due to dot com bubble burst (Thacker & Kaul, 2017). A sharp drop was noticed in early stage financing and VC money was getting invested in late stage development in order to minimize risks (Da Rin et al., 2006). India felt the effect of internet bubble burst as during that period a substantial portion of country's private equity investments originated from USA (Lin, 2008). The average deal size almost doubled while the number of early stage deals reduced sharply during the 2000s (Sengupta, 2006). By the early 2004, emerging markets like India and China began to stabilize (Sengupta, 2006). After 2005, early stage investing started to make a comeback (Gohil, 2014). Few notable deals during these periods, included $12 million investment from Sasken, Makemytrip.com bagging $10 million from Softbank Asia Infrastructure Fund (SAIF) (Gohil, 2014). Overall India also became a hot bed for VC investments by the 2010s.

## VC Industry Comparative Findings and Analysis in the 2010s for the Four Regions

To undertake a comparative analysis of VC industry in India with respect to USA, Europe, China and India data was gathered from the following sources by the authors. This has been tabulated in table 6.

The authors having presented the historical perspectives on VC landscapes in the four regions in this section deliberated on the developments registered in the decade of 2010s. VC activities analysis across different economic geographies has been carried out from the data available in table 6. The authors have analyzed the data and have tabulated the comparison below.

*Table 6. Research reports consulted for data analysis*

| | |
|---|---|
| CB Insights | PWC & CB Insights Moneytree Report, Q4 2018<br>PWC & CB Insights Healthcare Moneytree Report, Q4 2018 |
| Bain & IVCA | Perspectives of Indian VC Ecosystem 2018, December 2018 |
| Preqin | Venture Capital in India, September 2015<br>Venture Capital deals & exits, Q4 2018<br>Asia's VC Eclipse: A Preqin & Vertex Ventures Study, Oct 2018<br>Quarterly Asian Private Equity & Venture Capital, 2014-18 |
| Pitchbook | Annual Global League Tables 2013-2018<br>Annual European Venture Report, 2015-2018<br>Venture Ecosystem Factbook, London & DACH |
| Dow Jones | Quarterly Venture Capital Report, China, 2013-2018 |

## Comparative Analysis Of VC Landscape In USA, Europe, China And India During The Decade Of 2010s

### USA

- The total capital investment in was substantive, this indicated that the market was considered safe and recovered completely.
- USA consisted of 50% of the global VC deals on value basis in the year 2018 which highlighted the size and importance of USA among VC market.
- The angel/seed stage indicated decline in the percentage size wise, but the deal value had increased in this stage. This implied that the VC firms have been spending more on early and later stages.
- The deal value has increased in the Early stage compared to previous years which implied that VC firm managers thought that nurturing a firm right from this stage might help it to become a unicorn at a later stage.
- The angel/ seed stage funding had been in its peak during the years 2010-2015, but by end of 2010s it has been indicating a declining trend. On the other hand, late stage activities ($50M+ deal values) had increased significantly which implied VC industry in USA had been moving towards more mature phase with larger chunks of later stage deals.
- The deal size having value of more than $50M+ grew from $13.19B (27% of total VC deal value) in 2013 to $81.1 B (61% of total VC deal value) in 2018 which implied VC firms were investing significantly in later stage deals. This implied that VC firms were prospecting for more secured deals.
- The Internet, healthcare and software sectors continued to dominate the count of USA VC deals and it covered a major portion of capital investments in USA.
- The software sector had been growing steadily since last decade while healthcare and pharma had been steady sectors since last decade. The least growing sectors were IT hardware, media and energy.
- The Corporate VC (CVC) participation witnessed a sharp increase in 2010s. It could be interpreted that CVC had been targeting the later and larger stage deals as count-wise it was less but value wise it contributed almost 50%.

- Corporate rate tax cut gave boost to CVC funding as the participation of investors like Softbank increased in recent years.

## Europe

- In 2018, the deal counts decreased but the investment value had increased sharply.
- The contribution of deals over $100 M was less but the deals of 10M-15M pounds witnessed the biggest year on year increase based on deal value. This signified that a greater number of start-ups were able to secure bigger deals.
- There had been a rapid increase in the median deal size because of higher expectation from selective investors.
- Different geographic regions of Europe witnessed similar year on year decline of VC deals on count basis, but the deal value had increased. This implied that VC firms were focused on fewer number of larger deals in Europe.
- Countries like Israel, Germany, Austria and Switzerland (DACH) which had strong backbone in healthcare and technology. Only these countries of Europe witnessed an increase of more than 2% in proportion of total deals.
- Recently, in Europe there had been multi-billion dollar exits like Spotify, iZettle and Supercell which impacted the VC ecosystem and the VC firms in late 2010s focused on larger deals after its success and therefore the number of deals reduced.
- Foreign capital played a major role in Europe VC market, especially through the investors from USA. Despite the downward trend of broader market, the involvement of corporate and USA investors reached a new high.
- The percentage of USA investor capital in total deals reached a new high. The availability of funding and increased valuation of USA startups had initiated the search for lucrative assets in Europe and other regions, which resulted in an upward trend.
- The Corporate VC participation had increased steeply compared to the last decade. The CVC's participated in larger and fewer deals. The VC ecosystem in Europe had been maturing which could be observed from CVC investment. Count of later stage deals had seen a steep year on year increase while angel/seed stage and early stage deal count there was decline.
- The median deal size of CVC backed deals had increased. CVC's had provided substantial amount to startups to secure stakes in leading technology companies in the market. CVC firms' had invested more in late stage and more in mature companies. This provided the CVC's valuable insights about emerging technologies. Early stage startups were considered a bit risky in nature to invest as its chances of success was unreliable by VC's and CVC's in Europe.

## China

- China witnessed substantial growth in 2010s both in terms of deal count as well as deal value compared to the last decade.
- From the period 20005-2013 the year on year growth in number of deals in all stages had remained flat but after 2014 there has been a steep growth in first and second stage deals signifying growth stage of the VC ecosystem in China.

- The number of later stage deals had also witnessed a steep rise compared to the previous year's indicating that the Chinese VC ecosystem had been maturing than the previous years.
- Viewed from a sector wise perspective, consumer services & financial services business were the major sectors that contributed to VC deal count and deal value significantly.
- Healthcare industry was the emerging industry as in funding and deal counts it had increased significantly as compared to previous year though it contributed to only 8.5% of total VC activity. However, there was a clear trend of high investor interest.
- The major reasons behind the increase of healthcare industry in China was due to rising and active middle-class population, shortage of quality healthcare, growth of urban population and technology adoption from US firms.
- Initial public offering (IPO) remained as the most significant option in China throughout the decade of 2010s. The number of IPO had increased significantly compared to last couples of decades.
- Along with IPO, Mergers & Acquisition (M&A) gained significance for the exit option. Compared to previous year the number of M&A deals had increased significantly on an year on year basis.
- China's easy domestic exit policy and favorable political regulation had been the major factors in VC market growth.
- China had the second highest number of High Net Worth Individuals (HNWI) in Asia. This factor had helped in economic prosperity and growth of VC market in China.

## India

- The VC investments in India had grown by nearly five times within a span of 10 years from $0.88B in 2009 to $4 B in 2018.
- The number of deals or deal counts had increased steadily till 2016 signifying a growth in VC ecosystem. However, in the last two years the number of deals had declined.
- The median VC deal size had increased sharply from $3.7M in 2013 to $10M in 2018.
- The number of seed stage deals had increased sharply from the year 2013 to the year 2016 but it dipped after that time period. This was a bad sign for the VC ecosystem in India because if there would be less seed stage deals, the number of new startups would be drying up in the future because of lack of funding.
- The later stage deals had increased to some extent, but the deal size had risen sharply. This indicated that VC firms were interested in lesser number of deals but of larger value.
- The number of active VC firms in India had increased from 157 in the year 2013 to 270 in the year 2018. This indicated that VC ecosystem in India has been growing.
- The CVC investment in Indian deals had increased by 20% from 59 deals in 2017 to 71 deals in the year 2018. The largest deal which was significant portion of CVC deals was $1B series E to OYO rooms. Globally renowned CVC like Google, Intel had increased direct investment in India.
- In sector wise VC investment in India, the prominent sectors were consumer technology followed by IT and consumer retail. In consumer technology there had been a decline in horizontal e-commerce and vertical e-commerce investment. This was because investors had become more focused on profits and viable operation.
- Financial technology sector had witnessed substantial growth in investment due to surge of cashless transactions after RBI's demonetization initiative.

- The emerging sector for VC investment in India were agriculture-food sector and healthcare technology. The major reason for the growth of agriculture-food sector in India was due to factors like presence of substantial market size, higher margins availability, unavailability of large national food brands and the growing middle-class section.
- The reason for the growth of health care technology and technology space in general were due to factors like increased internet penetration, increase in digital payment post demonetization, gaps in healthcare due to quality, affordability, and improved government initiatives focused on digitization in social initiatives.

## CONCLUSION

The authors from the references for report mentioned in table-6 and analysis in table-7 prepared over the years 2010 to 2018 conclude that the year 2018 was a remarkable year in US VC industry history. While VC market had experienced a sharp decline in the count of angel & seed deals, but larger deals constituted a majority of total capital investment. This resulted in the increase of median deal sizes and valuations across all stages. The capital invested in companies working on life sciences and artificial intelligence had grown to a decade high level. Chinese VC companies invested heavily on the life sciences sector. In Europe also, there was an upsurge of VC investment in both deal values and the total number of deals in the year 2018. The significant thing to observe was that the number of large value deals decreased but the number of small deals increased. From this it could be implied that larger number of startups in Europe secured early and medium stage deals however for the later stage the investment dried up. Region wise, DACH recorded the largest relative growth on the basis of number of deals while UK retained the best spot like previous years. As there was not abundance of local professional investors, foreign capital had played a major role through years. In 2018, US investors had participated in a large number of VC deals relative to the last few years. This was due to the increased valuation of US startups and the target to capture attractive assets in Europe and other regions. In case of CVC investments, later stage deals had observed healthy growth while there were decline in seed stages and in early stages compared to earlier years. This signified the trend of CVCs capturing the later stage & larger rounds of successful startups. Europe witnessed decent growth and has been second largest VC market after USA. Though USA and Europe have been similar sized economies, VC activity of Europe was only one fifth of USA. European and Indian VC market observed decent growth in past decade with the help of its encouraging government regulatory policies. Recent Brexit deal had created political instability in VC and private equity activity in Europe. IT and health care startups received majority of VC investments in Europe. Countries in Europe like Sweden, Switzerland and Netherland had been amongst the top 10 innovative markets across the globe that attracted VC investments. India could follow a similar path by creating innovative platforms and hubs for entrepreneurship to flourish.

China had witnessed a significant growth in terms of VC investment and deals. China was the second biggest VC ecosystem in the world in 2018 when a large portion of global VC had been invested in Chinese startups. This happened majorly due to several focused programs to develop Chinese economic and technology infrastructure such as Torch (one of the largest and most successful incubators), Innofund (Government Initiative to provide seed funding), state-sponsored VC funding and foreign Rainminbi (RMB) funds (Foreign VCs who were raising money using local currency Rainminbi). China had under witnessed robust growth in VC funding. Major reasons for this growth could be attributed to good policy regulation, high GDP growth rate and easy exit policies. The conducive role played by the Chinese government in the growth of VC funding was also a factor. Increased economic prosperity in China had also helped in the growth of VC firms. On the contrary, India though improved the ease of doing business, but the regulatory policies of Indian government had not been encouraging. India government should provide supportive business landscape like that of China by making good policies. With 'Digital India', 'Startup India' & 'Make in India' campaign, India has attracted the attention of global VC players and with successful brand building activities, India could improve its position in the market. In India although VC investment had been in a nascent stage, yet it had grown five times in last decade in 2018. One could observe that larger VC firms had been doing investments mostly in later stage deals & helping more startups to become unicorns. This also pointed towards the risk averting nature of VC firms in India. Influx of CVCs and its investment diversification across various sectors also helped Indian VC ecosystem. Consumer technology had been emerged as the most significant sector and health-tech, education technology sectors have been growing steadily. It has been observed that VC investment had been rising in India YOY and significant contribution has been from US institutional investors (like mutual funds, insurance firms, banks and others). So political and economic sentiments of USA affect the Indian VC market. It has also been observed that political and instructional policies affect all markets of VC. Regulation compliance and Foreign Direct Investment (FDI) rules had been major hindrance factor in growth of Indian VC market. Though, India stands nowhere close to developed markets in terms of volume of investment through VC, Indian startup ecosystem has been growing at a good pace. A lot of technology-based startup attracted VC funding. IT sector and consumer products were the most attractive sector for VC funding in India. Nations with higher GDP growth has attracted higher VC funds. Economies of countries like USA, China, India, Europe have been playing a significant role in growth of VC market. It has been observed that developing nations have become more favorable for VC investment as the developed markets reached its saturation state. However, despite of saturation stage, USA leads in number of deals in VC and volume compared to other markets. To conclude the VC ecosystem has evolved in all four regions differently and are still evolving differentially.

## REFERENCES

Agbeibor, W. Jr. (2006). Pro-poor economic growth: Role of small and medium sized enterprises. *Journal of Asian Economics*, *17*(1), 35–40. doi:10.1016/j.asieco.2006.01.005

Agrawal, A., Catalini, C., & Goldfarb, A. (2014). Some simple economics of crowdfunding. *Innovation Policy and the Economy*, *14*(1), 63–97. doi:10.1086/674021

Ahlstrom, D., Bruton, G. D., & Yeh, K. S. (2007). Venture capital in China: Past, present, and future. *Asia Pacific Journal of Management*, *24*(3), 247–268. doi:10.100710490-006-9032-1

Aizenman, J., & Kendall, J. (2008). *The internationalization of venture capital and private equity*. National Bureau of Economic Research. doi:10.3386/w14344

Arthurs, J. D., & Busenitz, L. W. (2003). The boundaries and limitations of agency theory and stewardship theory in the venture capitalist/entrepreneur relationship. *Entrepreneurship Theory and Practice*, *28*(2), 145–162. doi:10.1046/j.1540-6520.2003.00036.x

Bain & IVCA. (2018). Perspective of Indian VC ecosystem. Retrieved from https://ivca.in/wp-content/uploads/2018/12/IVCA-Bain-India-VC-report-2018.pdf

Barry, C. B. (1994). New directions in research on venture capital finance. *Financial Management, 23*(3), 3–15. doi:10.2307/3665617

Barry, C. B., Muscarella, C. J., Peavy Iii, J. W., & Vetsuypens, M. R. (1990). The role of venture capital in the creation of public companies: Evidence from the going-public process. *Journal of Financial Economics, 27*(2), 447–471. doi:10.1016/0304-405X(90)90064-7

Barsky, R. B., & Kilian, L. (2004). Oil and the Macroeconomy since the 1970s. *The Journal of Economic Perspectives, 18*(4), 115–134. doi:10.1257/0895330042632708

Baumol, W. J. (1993). *Entrepreneurship, management, and the structure of payoffs*. MIT Press.

Belke, A. H., Fehn, R., & Foster-McGregor, N. (2003). Does venture capital investment spur employment growth? Centre for European policy studies.

Bertoni, F., Colombo, M. G., & Quas, A. (2015). The patterns of venture capital investment in Europe. *Small Business Economics, 45*(3), 543–560. doi:10.100711187-015-9662-0

Beugelsdijk, S., & Smeets, R. (2008). Entrepreneurial Culture and Economic Growth: Revisiting McClelland's Thesis. *American Journal of Economics and Sociology, 67*(5), 915-939.

Block, J. H., Colombo, M. G., Cumming, D. J., & Vismara, S. (2018). New players in entrepreneurial finance and why they are there. *Small Business Economics, 50*(2), 239–250. doi:10.100711187-016-9826-6

Block, J. H., De Vries, G., & Sandner, P. G. (2010). *Venture capital and the financial crisis: An empirical study across industries and countries*. Oxford.

Block, J. H., & Sandner, P. G. (2011). Venture capital funding in the middle of the year 2011: Are we back to pre-crisis boom levels? *Strategic Change: Briefings in Entrepreneurial Finance, 20*(5-6), 161–169. doi:10.1002/jsc.893

Bottazzi, L., & Da Rin, M. (2002). Venture capital in Europe and the financing of innovative companies. *Economic Policy, 17*(34), 229–270. doi:10.1111/1468-0327.00088

Bottazzi, L., Da Rin, M., & Hellmann, T. (2004). The changing face of the European venture capital industry: Facts and analysis. *Journal of Private Equity, 7*(2), 26–53. doi:10.3905/jpe.2004.391048

Bruton, G. D., Ahlstrom, D., & Puky, T. (2009). Institutional differences and the development of entrepreneurial ventures: A comparison of the venture capital industries in Latin America and Asia. *Journal of International Business Studies, 40*(5), 762–778. doi:10.1057/jibs.2009.3

Bruton, G. D., Fried, V. H., & Manigart, S. (2005). Institutional influences on the worldwide expansion of venture capital. *Entrepreneurship Theory and Practice, 29*(6), 737–760. doi:10.1111/j.1540-6520.2005.00106.x

Burżacka, M., & Gąsiorowska, E. (2016). The importance of venture capital financing of start-up companies. *Forum Scientiae Oeconomia, 4*(3), 141–150.

Buttner, E. H., & Rosen, B. (1992). Rejection in the loan application process: Male and female entrepreneurs' perceptions and subsequent intentions. *Journal of Small Business Management*, *30*(1), 58.

Callison, J. W., & Vestal, A. W. (2010). The L3C Illusion: Why Low-Profit Limited Liability Companies Will Not Stimulate Socially Optimal Private Foundation Investment in Entrepreneurial Ventures. *Vermont Law Review*, *35*, 273.

Carlos Nunes, J., Gomes Santana Félix, E., & Pacheco Pires, C. (2014). Which criteria matter most in the evaluation of venture capital investments? *Journal of Small Business and Enterprise Development*, *21*(3), 505–527. doi:10.1108/JSBED-10-2013-0165

Cherif, M., & Gazdar, K. (2011). What drives venture capital investments in Europe? New results from a panel data analysis. *Journal of Applied Business and Economics*, *12*(3), 122–139.

Chishti, S., & Barberis, J. (2016). *The FinTech book: the financial technology handbook for investors, entrepreneurs and visionaries*. John Wiley & Sons. doi:10.1002/9781119218906

Colombo, M. G., & Grilli, L. (2010). On growth drivers of high-tech start-ups: Exploring the role of founders' human capital and venture capital. *Journal of Business Venturing*, *25*(6), 610–626. doi:10.1016/j.jbusvent.2009.01.005

Cornelius, B. (2005). The institutionalization of venture capital. *Technovation*, *25*(6), 599–608. doi:10.1016/j.technovation.2003.12.001

Cumming, D. J., Grilli, L., & Murtinu, S. (2017). Governmental and independent venture capital investments in Europe: A firm-level performance analysis. *Journal of Corporate Finance*, *42*, 439–459. doi:10.1016/j.jcorpfin.2014.10.016

Cumming, D. J., & MacIntosh, J. G. (2003). A cross-country comparison of full and partial venture capital exits. *Journal of Banking & Finance*, *27*(3), 511–548. doi:10.1016/S0378-4266(02)00389-8

Da Rin, M., Nicodano, G., & Sembenelli, A. (2006). Public policy and the creation of active venture capital markets. *Journal of Public Economics*, *90*(8-9), 1699–1723. doi:10.1016/j.jpubeco.2005.09.013

De Clercq, D., Fried, V. H., Lehtonen, O., & Sapienza, H. J. (2006). An entrepreneur's guide to the venture capital galaxy. *The Academy of Management Perspectives*, *20*(3), 90–112. doi:10.5465/amp.2006.21903483

De la Dehesa, G. (2002). *Venture capital in the United States and Europe*. Group of Thirty.

DeGennaro, R. P. (2010). Angel Investors: Who They Are and What They Do; Can I Be One, Too? *The Journal of Wealth Management*, *13*(2), 55–60.

Deorah, K. (2015). *The Golden Tap: The Inside Story of Hyper-funded Indian Start-ups*. Lotus Collection, Roli Books.

DeTienne, D. R., McKelvie, A., & Chandler, G. N. (2015). Making sense of entrepreneurial exit strategies: A typology and test. *Journal of Business Venturing*, *30*(2), 255–272. doi:10.1016/j.jbusvent.2014.07.007

Dossani, R., & Kenney, M. (2002). Creating an environment for venture capital in India. *World Development*, *30*(2), 227–253. doi:10.1016/S0305-750X(01)00110-3

Dow Jones. (2018). Venture Capital Report, China Q2. Retrieved from https://images.dowjones.com/wp-content/uploads/sites/43/2018/07/13022934/2Q18-VS-China-final1.pdf

Ehrlich, S. B., De Noble, A. F., Moore, T., & Weaver, R. R. (1994). After the cash arrives: A comparative study of venture capital and private investor involvement in entrepreneurial firms. *Journal of Business Venturing*, *9*(1), 67–82. doi:10.1016/0883-9026(94)90027-2

Feldman, M., Francis, J., & Bercovitz, J. (2005). Creating a cluster while building a firm: Entrepreneurs and the formation of industrial clusters. *Regional Studies*, *39*(1), 129–141. doi:10.1080/0034340052000320888

Florida, R. L., & Kenney, M. (1988). Venture capital-financed innovation and technological change in the USA. *Research Policy*, *17*(3), 119–137. doi:10.1016/0048-7333(88)90038-8

Fried, V. H., & Hisrich, R. D. (1994). Toward a model of venture capital investment decision making. *Financial Management*, *23*(3), 28–37. doi:10.2307/3665619

Gilbertson, L. D. (1959). Small Business Financing Under the Small Business Act and the Small Business Investment Act of 1958. *U. Kan. L. Rev.*, *8*, 538.

Gohil, R. (2014). The Performance of Private Equity Funds in India. *Journal of Private Equity*, *18*(1), 63–72. doi:10.3905/jpe.2014.18.1.063

Gompers, P. (1994). The rise and fall of venture capital. *Business and Economic History*, *23*(2), 1–26.

Gompers, P., & Lerner, J. (2001). The venture capital revolution. *The Journal of Economic Perspectives*, *15*(2), 145–168. doi:10.1257/jep.15.2.145

Gompers, P. A. (1995). Optimal investment, monitoring, and the staging of venture capital. *The Journal of Finance*, *50*(5), 1461–1489. doi:10.1111/j.1540-6261.1995.tb05185.x

Gompers, P. A. (1996). Grandstanding in the venture capital industry. *Journal of Financial Economics*, *42*(1), 133–156. doi:10.1016/0304-405X(96)00874-4

Gompers, P. A. (1997). *Ownership and control in entrepreneurial firms: an examination of convertible securities in venture capital investments*. Unpublished working paper, Harvard Business School.

Goodnight, G. T., & Green, S. (2010). Rhetoric, risk, and markets: The dot-com bubble. *The Quarterly Journal of Speech*, *96*(2), 115–140. doi:10.1080/00335631003796669

Grichnik, D., & Hisrich, R. D. (2006). Strategic and investment behaviour in the German and Israeli venture capital industries: A comparison with the USA. *International Journal of Technology Management*, *34*(1-2), 88–104. doi:10.1504/IJTM.2006.009449

Haitian, L., Yi, T., & Gongmeng, C. (2007). Venture capital and the law in China. *Hong Kong Law Journal*, *37*, 229.

Hansen, M. T., Chesbrough, H. W., Nohria, N., & Sull, D. N. (2000). Networked incubators. *Harvard Business Review*, *78*(5), 74–84. PMID:11143156

Harvey, D., & Hubbard, C. (2016). Why Brexit. *Centre for Rural Economy Discussion Paper Series*, (35).

Hege, U., Palomino, F., & Schwienbacher, A. (2003). *Determinants of venture capital performance: Europe and the United States*. HEC School of Management.

Henderson, J. (2002). Building the rural economy with high-growth entrepreneurs. *Economic Review-Federal Reserve Bank of Kansas City*, *87*(3), 45–75.

Hochberg, Y. V., Ljungqvist, A., & Lu, Y. (2007). Whom you know matters: Venture capital networks and investment performance. *The Journal of Finance*, *62*(1), 251–301. doi:10.1111/j.1540-6261.2007.01207.x

Hu, Y. (2010). *Improving Chinese Expenditure Incentive Programs for Venture Capital Investment: A Comparative Study of Government Expenditure Supporting Policies of Venture Capital Investment in the United States, Canada and China* [Doctoral dissertation].

Inglehart, R., & Abramson, P. R. (1994). Economic security and value change. *The American Political Science Review*, *88*(2), 336–354. doi:10.2307/2944708

CB Insights. (2018). Venture Capital Funding Report Q4. Retrieved from https://www.cbinsights.com/research/report/venture-capital-q4-2018/

Isaksson, A. (2006). *Studies on the venture capital process*[Doctoral dissertation]. Handelshögskolan vid Umeå universitet.

Jeng, L. A., & Wells, P. C. (2000). The determinants of venture capital funding: Evidence across countries. *Journal of Corporate Finance*, *6*(3), 241–289. doi:10.1016/S0929-1199(00)00003-1

Kaiser, K. M., & Westarp, C. (2010). Value Creation in the Private Equity: And Venture Capital Industry.

Kaplan, S. N., & Lerner, J. (2010). It ain't broke: The past, present, and future of venture capital. *Journal of Applied Corporate Finance*, *22*(2), 36–47. doi:10.1111/j.1745-6622.2010.00272.x

Kaplan, S. N., & Lerner, J. (2016). *Venture capital data: Opportunities and challenges (No. w22500)*. National Bureau of Economic Research. doi:10.3386/w22500

Kaplan, S. N., & Strömberg, P. (2003). Financial contracting theory meets the real world: An empirical analysis of venture capital contracts. *The Review of Economic Studies*, *70*(2), 281–315. doi:10.1111/1467-937X.00245

Kim, W. C., & Mauborgne, R. A. (2014). *Blue ocean strategy, expanded edition: How to create uncontested market space and make the competition irrelevant*. Harvard business review Press.

Kindleberger, C. P. (1981). Dominance and leadership in the international economy: Exploitation, public goods, and free rides. *International Studies Quarterly*, *25*(2), 242–254. doi:10.2307/2600355

Kohler, T. (2016). Corporate accelerators: Building bridges between corporations and startups. *Business Horizons*, *59*(3), 347–357. doi:10.1016/j.bushor.2016.01.008

Kolmakov, V. V., Polyakova, A. G., & Shalaev, V. S. (2015). An analysis of the impact of venture capital investment on economic growth and innovation: Evidence from the USA and Russia. *Ekonomski Anali/Economic Annals, 60*(207).

Kortum, S., & Lerner, J. (2001). Does venture capital spur innovation? In *Entrepreneurial inputs and outcomes: New studies of entrepreneurship in the United States* (pp. 1–44). Emerald Group Publishing Limited. doi:10.1016/S1048-4736(01)13003-1

Kräussl, R., & Krause, S. (2014). Has Europe been catching up? An industry level analysis of venture capital success over 1985–2009. *European Financial Management, 20*(1), 179–205. doi:10.1111/j.1468-036X.2013.12012.x

Lerner, J., & Tåg, J. (2013). Institutions and venture capital. *Industrial and Corporate Change, 22*(1), 153–182. doi:10.1093/icc/dts050

Liles, P. R. (1977). *Sustaining the venture capital firm.* Management Analysis Center.

Lin, J. Y. (2008). The impact of the financial crisis on developing countries.

Lin, L. (2015). *Re-Engineering a Venture Capital Market: The Case of China* (Vol. 4). NUS Centre for Law & Business Working Paper 15.

Liu, C. (2010). *A Comparison of the venture capitalists´ investment behavior pattern between China and America.*

Lockett, A., Wright, M., Sapienza, H., & Pruthi, S. (2002). Venture capital investors, valuation and information: A comparative study of the US, Hong Kong, India and Singapore. *Venture Capital: An International Journal of Entrepreneurial Finance, 4*(3), 237–252. doi:10.1080/13691060213715

Maddison, A. (1995). *Monitoring the world economy, 1820-1992* (p. 238). Paris: Development Centre of the Organisation for Economic Co-operation and Development.

Manchanda, K., & Muralidharan, P. (2014, January). Crowdfunding: a new paradigm in start-up financing. *Proceedings of the Global Conference on Business & Finance Proceedings* (p. 369). Institute for Business & Finance Research.

Manigart, S., De Waele, K., Wright, M., Robbie, K., Desbrières, P., Sapienza, H. J., & Beekman, A. (2002). Determinants of required return in venture capital investments: A five-country study. *Journal of Business Venturing, 17*(4), 291–312. doi:10.1016/S0883-9026(00)00067-7

Martin, R., Sunley, P., & Turner, D. (2002). Taking risks in regions: The geographical anatomy of Europe's emerging venture capital market. *Journal of Economic Geography, 2*(2), 121–150. doi:10.1093/jeg/2.2.121

Mason, C. M. (2009). Public policy support for the informal venture capital market in Europe: A critical review. *International Small Business Journal, 27*(5), 536–556. doi:10.1177/0266242609338754

Mayer, C., Schoors, K., & Yafeh, Y. (2005). Sources of funds and investment activities of venture capital funds: Evidence from Germany, Israel, Japan and the United Kingdom. *Journal of Corporate Finance, 11*(3), 586–608. doi:10.1016/j.jcorpfin.2004.02.003

McCahery, J. A., & Vermeulen, E. P. (2010). Venture capital beyond the financial crisis: How corporate venturing boosts new entrepreneurial clusters (and assists governments in their innovation efforts). *Capital Markets Law Journal, 5*(4), 471–500. doi:10.1093/cmlj/kmq018

Megginson, W. L. (2004). Toward a global model of venture capital? *Journal of Applied Corporate Finance, 16*(1), 89–107. doi:10.1111/j.1745-6622.2004.tb00599.x

Mehran, H. (1995). Executive compensation structure, ownership, and firm performance. *Journal of Financial Economics, 38*(2), 163–184. doi:10.1016/0304-405X(94)00809-F

Metrick, A., & Yasuda, A. (2010). Venture capital and the finance of innovation. In A. Metrick and A. Yasuda (Eds.), Venture capital and the finance of innovation (2nd ed.) John Wiley and Sons, Inc.

Metrick, A., & Yasuda, A. (2011). *Venture Capital & The Finance of Innovation*. John Wiley & Sons.

Miller, P., & Bound, K. (2011). The startup factories. *NESTA*. Retrieved from http://www.nesta.org.uk/library/documents/StartupFactories.pdf

Murray, G. C. (1992). A challenging marketplace for venture capital. *Long Range Planning, 25*(6), 79–86. doi:10.1016/0024-6301(92)90173-Y

Myhrvold, N. (2010). The big idea: Funding eureka! *Harvard Business Review, 88*(2), 40–50.

Ning, Y., Wang, W., & Yu, B. (2015). The driving forces of venture capital investments. *Small Business Economics, 44*(2), 315–344. doi:10.100711187-014-9591-3

Oliner, S. D., & Sichel, D. E. (2000). The resurgence of growth in the late 1990s: Is information technology the story? *The Journal of Economic Perspectives, 14*(4), 3–22. doi:10.1257/jep.14.4.3

Pandey, I. M. (1998). The process of developing venture capital in India. *Technovation, 18*(4), 253–261. doi:10.1016/S0166-4972(98)00003-0

Pitchbook. (2018). Annual European Venture Report. Retrieved from https://files.pitchbook.com/website/files/pdf/2018_Annual_European_Venture_Report.pdf

Preqin. (2018). Venture Capital Deals and Exit Q4. Retrieved from https://docs.preqin.com/reports/Preqin-Venture-Capital-Deals-and-Exits-Q4-2018.pdf

Ramesh, S., & Gupta, A. (1995). *Venture capital and the Indian financial sector*. USA: Oxford University Press.

Rice, M. P., O'Connor, G. C., Leifer, R., McDermott, C. M., & Standish-Kuon, T. (2000). Corporate venture capital models for promoting radical innovation. *Journal of Marketing Theory and Practice, 8*(3), 1–10. doi:10.1080/10696679.2000.11501870

Ruhnka, J. C., & Young, J. E. (1991). Some hypotheses about risk in venture capital investing. *Journal of Business Venturing, 6*(2), 115–133. doi:10.1016/0883-9026(91)90014-5

Sahlman, W. A. (1990). The structure and governance of venture-capital organizations. *Journal of Financial Economics, 27*(2), 473–521. doi:10.1016/0304-405X(90)90065-8

Sapienza, H. J. (1992). When do venture capitalists add value? *Journal of Business Venturing, 7*(1), 9–27. doi:10.1016/0883-9026(92)90032-M

Sapsed, J., Grantham, A., & DeFillippi, R. (2007). A bridge over troubled waters: Bridging organisations and entrepreneurial opportunities in emerging sectors. *Research Policy*, *36*(9), 1314–1334. doi:10.1016/j.respol.2007.05.003

Scherer, F. M. (2000). The size distribution of profits from innovation. In *The Economics and Econometrics of Innovation* (pp. 473–494). Boston, MA: Springer. doi:10.1007/978-1-4757-3194-1_19

Schertler, A. (2001). *Venture capital in Europe's common market: a quantitative description* (No. 1087). Kiel working paper.

Sengupta, S. (2006). The New World of Private Equity. *Business World*, 16-17.

Sieradzan, P., & Sobańska, K. (2004). Risk of venture capitale / private equity investments. Basic conditions for the construction of the private equity fund's portfolio Part 1. Studies and Works Collegium of Management and Finance / Warsaw School of Economics, (52), 62-75.

Storey, D. J. (1985). The Problems Facing New Firms. *Journal of Management Studies*, *22*(3), 327–345. doi:10.1111/j.1467-6486.1985.tb00079.x

Subhash, K. B. (2004). Globalisation and Indian venture capital system.

Teker, D., Teker, S., & Teraman, Ö. (2016). Venture capital markets: A cross country analysis. *Procedia Economics and Finance*, *38*, 213–218. doi:10.1016/S2212-5671(16)30192-7

Thacker, S., & Kaul, M. (2017). The Evolution of Private Equity in India. *Ct. Uncourt*, *4*, 5.

Tzannatos, Z. (1999). Women and labor market changes in the global economy: Growth helps, inequalities hurt and public policy matters. *World Development*, *27*(3), 551–569. doi:10.1016/S0305-750X(98)00156-9

Wang, A. (2009). Government's role as public venture capitalist in high-technology small and medium sized enterprises.

Wong, A. (2011). The Evolution of the Venture Capital Market in China: Current Trends in Venture Capital Financing Strategies and Investment Preferences. *Journal of Investing*, *20*(4), 16–26. doi:10.3905/joi.2011.20.4.016

Zacharakis, A. L., McMullen, J. S., & Shepherd, D. A. (2007). Venture capitalists' decision policies across three countries: An institutional theory perspective. *Journal of International Business Studies*, *38*(5), 691–708. doi:10.1057/palgrave.jibs.8400291

Zeng, F. (2004). *Venture capital investments in China*. Santa Monica, CA: Rand Corp.

Zhang, B., Wardrop, R., Rau, P. R., & Gray, M. (2015). Moving mainstream: Benchmarking the European alternative finance market. *Journal of Financial Perspectives*, *3*(3).

Zhang, Y., & Mayes, D. G. (2018). The performance of governmental venture capital firms: A life cycle perspective and evidence from China. *Pacific-Basin Finance Journal*, *48*, 162–185. doi:10.1016/j.pacfin.2018.02.002

Zider, B. (1998). How venture capital works. *Harvard Business Review*, *76*(6), 131–139. PMID:10187243

# Chapter 14
# Development and Access to Finance of Small and Medium-Sized Enterprises in Mongolia

**Gan-Ochir Doojav**
*Bank of Mongolia, Mongolia*

**Davaajargal Luvsannyam**
*Bank of Mongolia, Mongolia*

**Bilguun Sukhbaatar**
*Bank of Mongolia, Mongolia*

**Bilguunzul Sodnomdarjaa**
*Bank of Mongolia, Mongolia*

**Tsolmon Otgonbat**
*Bank of Mongolia, Mongolia*

**Khuslen Batmunkh**
*Bank of Mongolia, Mongolia*

**Munkhbayar Gantumur**
*Bank of Mongolia, Mongolia*

**Elbegjargal Enkh-Amgalan**
*Bank of Mongolia, Mongolia*

## ABSTRACT

*This chapter presents a recent survey on the development and accessibility for finance of small and medium-size enterprises (SMEs) in Mongolia. The survey covers a sample of 1922 SMEs from Ulaanbaatar, the capital city, and 21 provinces. We find that banks and local government administrations are the most supportive institutions for SME development. Political instability, corruption, and labour supply, a high lending rate, short maturity loans, and service fees are perceived as the major obstacles that SMEs face in the business environment. Our results also suggest that SMEs in Mongolia are less likely to have access to external finance because of tight credit condition, potentially explaining the lack of SME growth. Implementing country-specific reform strategy for SME development covering key building blocks is needed to promote SME financial inclusion and facilitate SMEs to contribute to the economic growth.*

DOI: 10.4018/978-1-7998-1566-2.ch014

## INTRODUCTION

Small and medium-sized enterprises (SMEs) play an important role in most countries, particularly in developing countries. A number of papers find that (i) SMEs are closely linked with economic growth (e.g., Beck et al., 2005, Ayyagari et al., 2007), and (ii) the majority of jobs created by SMEs, specifically in low-income countries (e.g., Beck et al., 2008, Dietrich, 2010, Ayyagari et al., 2011). Therefore, SMEs are the engine of economic development, reduced economic vulnerability for individual households, poverty alleviation, and SME financial inclusion is at the core of the economic diversification and growth challenges. Improving SME financial inclusion can help increase economic growth and job creation. However, SMEs are financially more constrained than large firms and are less likely to have access to formal finance (Beck and Demirgüç-Kunt, 2006). Improving SME financial inclusion can help increase economic growth, job creation and the effectiveness of fiscal and monetary policy and could also contribute to financial stability. Blancher et al. (2019) argue that economic fundamentals and financial sector characteristics (i.e., macroeconomic stability, limited public sector size, financial sector soundness, a competitive banking system) and institutional factors (i.e., strong governance and financial regulatory and supervisory capacity, credit information availability, and supportive business environment and legal systems that allow to adequately enforce property rights and contracts) can help scale up SME bank credit.

This paper presents a recent survey on development and access to finance of small and medium-size enterprises (SMEs) in Mongolia, developing and commodity dependent economy. This survey aims to examine the current development, access to finance and other challenges faced by SMEs. The target group of respondents is business owners and/or management level employees (CEO, general manager and accountant) of the SMEs. According to Doing Business Report 2019 published by World Bank, Mongolia is ranked in 74 out of 190 economies. Though ease of doing business score (67.74) is higher than East Asia and Pacific regional average (63.41), the score is far below from other economies (i.e., China, Japan, Kazakhstan and South Korea). As the economy is heavily dependent on the resource sector accounting for 90% of total exports and 25% of budget revenues, there is essential need of economic diversification. The economy also faces challenges to strengthen its resilience to transform its natural resource wealth into assets that support sustainable growth and prosperity. There is no doubt that SME development can contribute to economic diversification and resilience. Hence, evidence (and lessons) from the Mongolian case would be of high relevance formulating policies in scaling-up SME access to financial services and lowering barriers to SME development for resource-rich developing countries.

The present survey is conducted in October 2018, and the Bank of Mongolia has successfully conducted nationwide SMEs survey for the last 7 consecutive years in order to create the SME database. Without reliable SME data, it is difficult for policy makers to implement policies aimed at expending and strengthening the SME sector. Highlights of this year's survey are as follows: i) Increased coverage of SMEs operating in slums within 50 km from the province center, ii) The World Bank tablet-based CAPI (Computer Assisted Personal Interview) system was first used nationwide, and iii) SMEs producing value-added products were targeted. The survey creates database for SME development in Mongolia, and the survey questionnaire follows international practices. Sampling takes location, number of employees into account in terms of representation. Total sample is determined with 95% confidence and with 5% significance level. Population for the survey is determined by the overall number of businesses (83,086) registered at the National Statistics Office's database as of 2018. According to the statistical calculations, optimal sample size is 850-1000 for Ulaanbaatar and 30-50 for each province, respectively.

However, realization of sampling has been altered during the process of interviewing due to quality of responses. As a consequence, 1922 samples, exceeding critical number by 5%, is sampled in this survey. Geographically, 54% of the participation is sampled from Ulaanbaatar and the rest is from the provinces.

The reminder of the paper is structured as follows. Section 2 discusses the characteristics of SMEs including types and age of business operation, number of employee and ownership structure. Section 3 discusses the business environment in terms of macroeconomic, social, political, financial, market and infrastructure conditions. Section 4 discusses SMEs' access to finance and financing obstacles. Section 5 explores operational challenges faced by SMEs and discusses SMEs' financial performances. Finally, section concludes the paper with some policy implications to foster SMEs' development and access to finance in Mongolia.

## LITERATURE REVIEW

Development and access to finance of SMEs has been focus of researchers and policy makers over last three decades. A number of papers (e.g., Levy 1993, Beck et al., 2005, Ayyagari et al., 2007 and Beck et al. 2013) find that SMEs are the engine of economic development, reduced economic vulnerability for individual households, poverty alleviation and economic diversification. SMEs also create majority of jobs, specifically in low-income. Harwood and Konidaris (2015) find that SMEs account for more than 60% of all jobs in developing countries. Ayyagari et al. (2007) emphasize that employment in SMEs constitutes over 60% of total employment in manufacturing in developed economies. According to the IFC (2013), the private sector is the main engine of job creation and the source of 9 out of every 10 jobs in the world. Within the private sector, SMEs account for more than half of all jobs worldwide. However, Beck et al. (2005) do not find any evidence for any association of a large SME sector with faster income growth of the lowest income quintile and faster rates of poverty reduction.

Our paper is closely related the growing literature on firms' access to external finance. The existing literature emphasizes that access to finance of SMEs has become critical in many developing countries. Lack of access to finance emerges as the binding constraint for smaller, less established firms in Sri Lanka and for all of Tanzania's SMEs (Levy 1993). SMEs are financially more constrained than large firms and are less likely to have access to formal finance (Beck and Demirgüç-Kunt, 2006, Ayyagari et al., 2007). Schiffer and Weder (2001) show that small firms consistently report higher growth obstacles than medium-size or large firms. Demirgüç-Kunt and Maksimovic (1998) find that financing constraints are lower in countries with more efficient legal systems. Ardic et al. (2012) find income per, private credit to GDP, the legal and business environment, and the efficiency of the banking system are among the factors that influence SME lending. Berger and Udell (2006) show that specific financing forms such as leasing or factoring have been promoted as conducive to easing financing constraints of SMEs, as they are based on the underlying assets and cash flows rather than borrower's financial history. Taiwo et al. (2012) reveal that the most common constraints hindering small and medium scale business growth in Nigeria are lack of financial support, poor management, corruption, lack of training and experience, poor infrastructure, insufficient profits, and low demand for product and services. Beck et al. (2013) find that (i) dominance of the financial system by banks is associated with lower use of financial services by firms of all sizes, and (ii) there is no evidence that smaller institutions are better in providing access to finance. Financial development allows existing firms to exploit growth and investment opportunities

(Beck et al. 2005), and SMEs do not only report higher financing obstacles, they are also more adversely affected by these obstacles in their operation and growth (Beck et al., 2008).

Market structure and innovative lending tools also drives SME-friendly financial development. Improving financial market infrastructure greatly reduces costs for financial institutions to access SME ability and willingness to repay, and broadens the range of assets SMEs can use to support their borrowing (IFC 2013). Frame et al. (2001) show that the use of credit-scoring techniques has increased small business lending by banks in the United States. There are mixed results concerning the effect of bank concentration and competitiveness on the availability of SME financing (Berger et al. 2004). Clarke et al. (2003) show that the entry of foreign banks is mostly associated with greater SME credit availability. By competing with domestic banks for large corporate clients, they can force domestic banks to go down market to cater to SMEs (De Haans and Naaborg, 2005).

There are few studies focused on the SME issues in Mongolia. Lkhagvasuren (2014) finds that branding is especially crucial in Mongolia where the population is heterogeneous in terms of composition and needs, and this could be beneficial for various business sectors in Mongolia. Tuul and Bing (2019) find that SMEs in Mongolia face many challenges including high employee turnover, shortage of funds, lack of facilities and inferior training policy. OECD report (2016) states that access to finance is one of the major barriers to micro, small and medium enterprise (MSME) competitiveness in Mongolia, and recommends five policy measures to help mitigate market failures affecting MSME access to finance: (i) Enhancing the quality of data on SMEs, (ii) Improving the product offer and scope of the SME Development Fund (SMEDF), (iii) Improving the Mongolian Credit Guarantee Fund's risk assessment techniques and operational reliability, (iv) Establishing one-stop shops for MSMEs and streamlining the process between SMEDF and banks, and (v) Joining the OECD International Network for financial education and adding MSME specific actions to the National strategy for financial literacy. Jargalsaikhan (2019) argues that supporting SMEs by developing hard and soft infrastructures is necessary for SMEs more than loans in Mongolia. Hard infrastructures such as roads, telecommunication, and electricity play a crucial role. In addition, he claims that soft infrastructures can be improved, for instance, by measuring the transaction cost. If a transaction is slow and cost is high, or even if it fails, the circulation of money will cease.

## CHARACTERISTICS OF SMES IN MONGOLIA

In this section, we present general characteristics of SMEs. The majority of businesses surveyed (75%) operate in one sector, whereas remaining (25%) operate in multiple sectors. Thus, the questionnaire is further interviewed only for the main operation of SMEs. Figure 1 shows the number of employees in the core business of the surveyed entities.

SMEs having 1-10 employees constitute the clear majority (86%) followed by SMEs with 11-30 employees (10%). Only 2% of SMEs can be regarded as medium-sized enterprises with more than 50 employees (Figure 1).

About half of the SMEs have annual sales revenue of 50-499 million MNT, and 36% of SMEs has sales revenue of 10-49 million MNT. 12% of SMEs surveyed has sales revenue of below 9.9 million MNT (Figure 2). Out of total SMEs surveyed, 49% of them are private business owners, 42% are limited liability companies (Figure 3). Regarding to the ownership type, a clear majority of the SMEs (98%) have domestically invested enterprises, while remaining are foreign and mixed invested enterprises (Figure 4).

*Figure 1. Number of employees*
*Source: SME survey 2018*

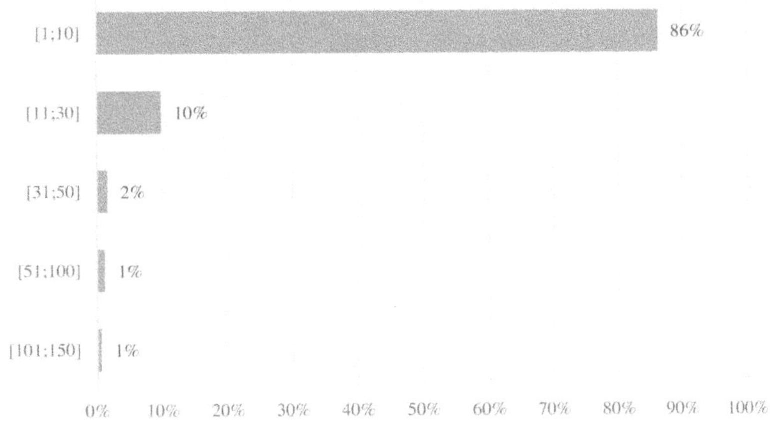


*Figure 2. Sales income, in millions MNT*
*Source: SME survey 2018*

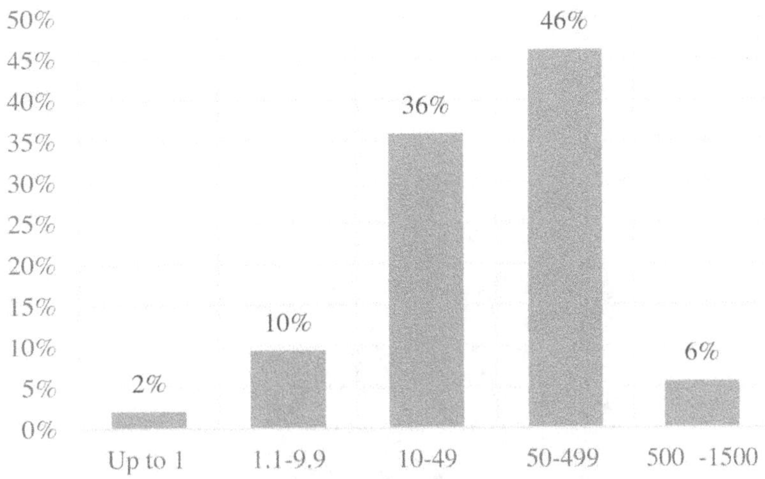


Beck et al. (2006) show that size, age and ownership are most reliable predictors of firms' financing obstacles. Schiffer and Weder (2001) find that small firms consistently report higher growth obstacles then medium-size or large firms. As the surveyed SMEs are young and their size is relatively small in terms of number of employees and sales revenues, it may be expected that higher growth and financing obstacles are reported within the survey. However, we believe that the identified growth and financing obstacles would reflect the real facts faced by the SMEs.

*Figure 3. Corporate structure*
*Source: SME survey 2018*

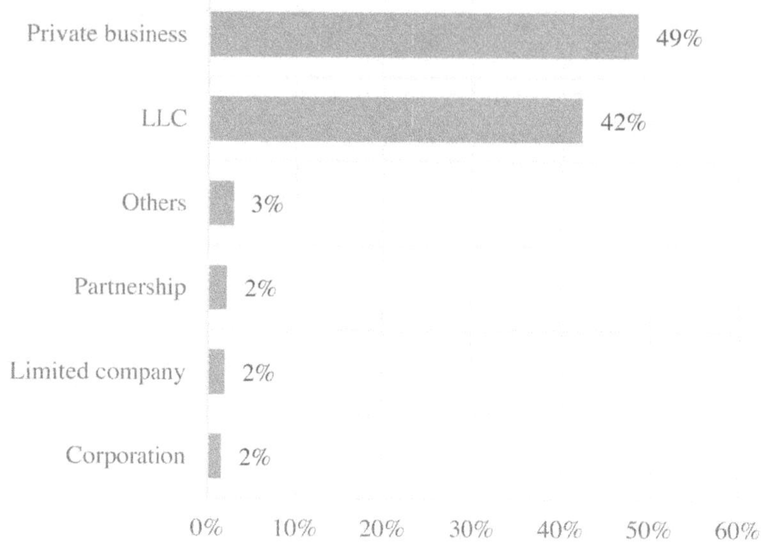


*Figure 4. Ownership structure*
*Source: SME survey 2018*

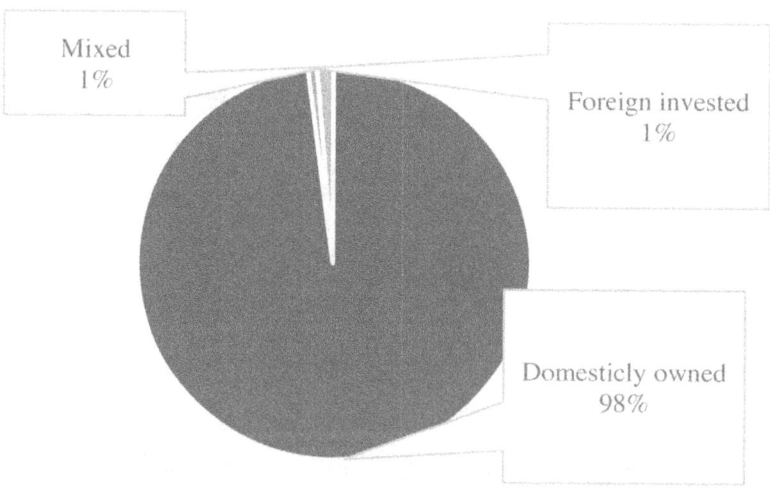


The survey for 2018 aims to cover more value-added producers (manufacturing), making up 21% of the surveyed SMEs. Coverage of manufacturing sector is increased by 8 percentage point from last year's survey. Majority of SMEs operate in retail trade (37%) and service (34%) sectors, while 5% of them operate in the agricultural sectors (Figure 5). The surveyed SMEs are asked how long they had

*Figure 5. Sector of main activities*
*Source: SME survey 2018*

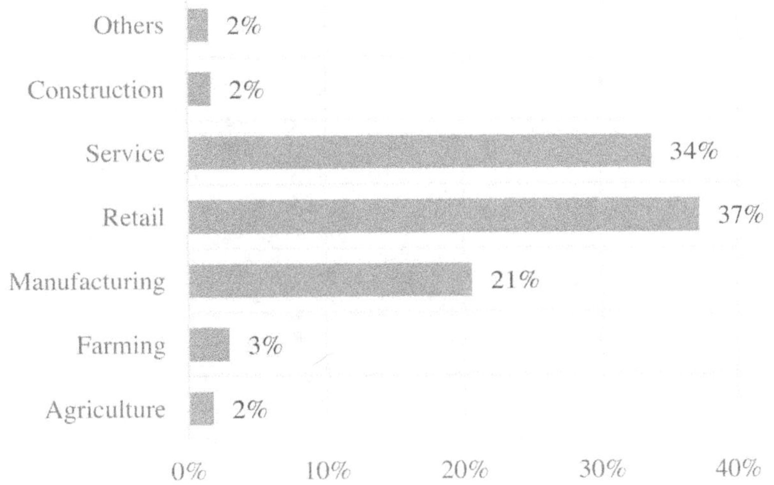


*Figure 6. Years in operation*
*Source: SME survey 2018*

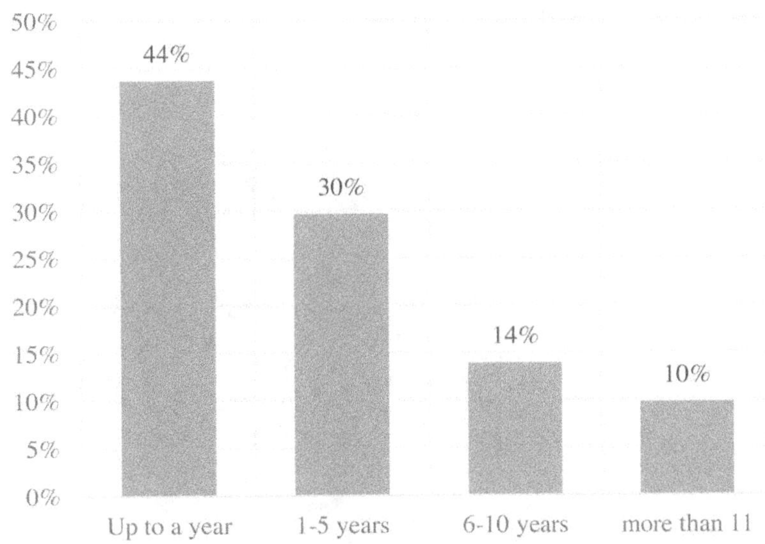

*Figure 7. Holdings of licenses*
*Source: SME survey 2018*

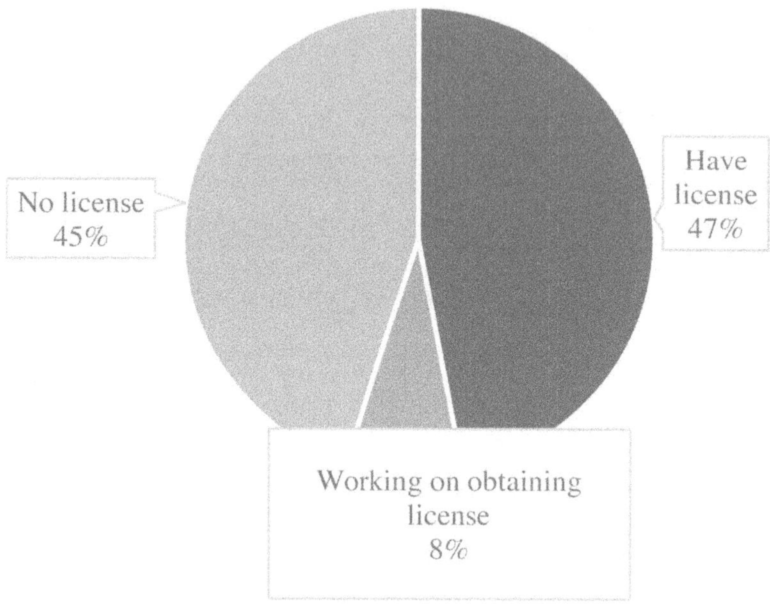


*Figure 8. Preparation of financial statements*
*Source: SME survey 2018*

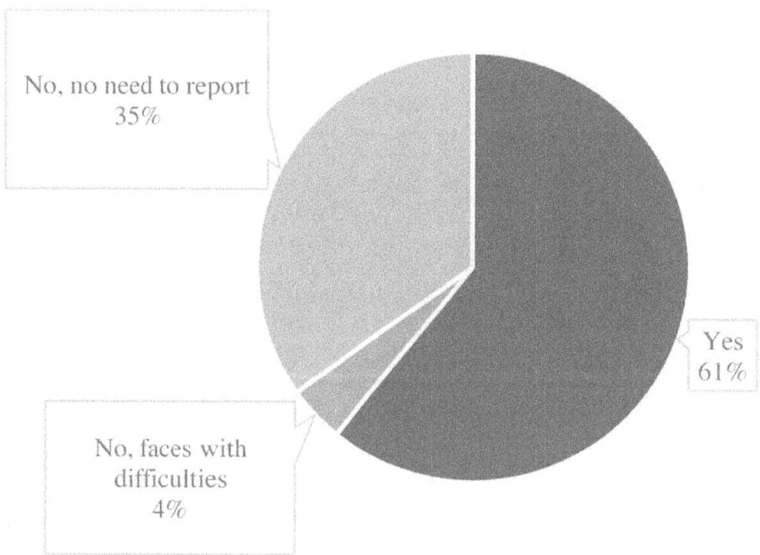

*Figure 9. Types of supports to SMEs*
*Source: SME survey 2018*

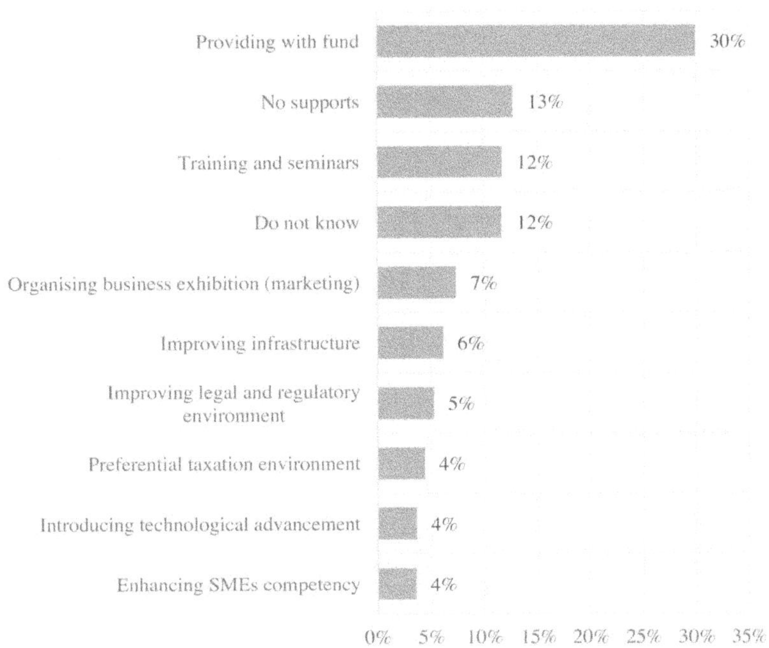


been operating, and results are as follows: 44% of SMEs have been operating for less than 1 year, 30% for 1-5 years, another 14% for 6-10 years, and remaining 10% for more than 11 years (Figure 6). The results may imply that SMEs are young and operate on retail and service sectors, which require a relatively low level of investment and expertise.

Of all the surveyed SMEs, 47% has obtained a business license, while 45% have not and 8% are in application procedure (Figure 7). We also ask whether SMEs keep records of financial statements. The results show that 61% of the SMEs surveyed prepare general-purpose financial statements, whereas 39% do not. Of those enterprises which do not, 35% of them reports that they do not have to report, and only 4% of are incapable of preparing the statements (Figure 8). We also find that majority of surveyed SMEs (95%) do not export, and only 5% export their products and services to foreign markets.

When asking how the government and other organizations support SMEs, they answer that much support is provided by commercial banks to SMEs (29%) followed by local government (12%), other financial institutions (7%), chamber of commerce (7%) and the government (6%). However, SMEs report that supports from consulting firms and international organizations are not satisfactory. They highlight the importance of the establishment of SME promotion agency. International practices show that establishing SME promotion agency is more efficient than spreading resources over numerous supporting organizations.

SMEs mostly receive supports in the form of financing, training, consultancy, organizing business exhibition and providing infrastructure. However, they receive less support in the forms of enhancing business competitiveness, introducing technological advances and exemption from VAT (Figure 9). Regarding to necessary policy measures and actions demanded to support SMEs, they place a higher priority on reducing government bureaucracy, creating a favorable tax regime, reducing corruption and

*Figure 10. Necessary measures to support SMEs*
*Source: SME survey 2018*

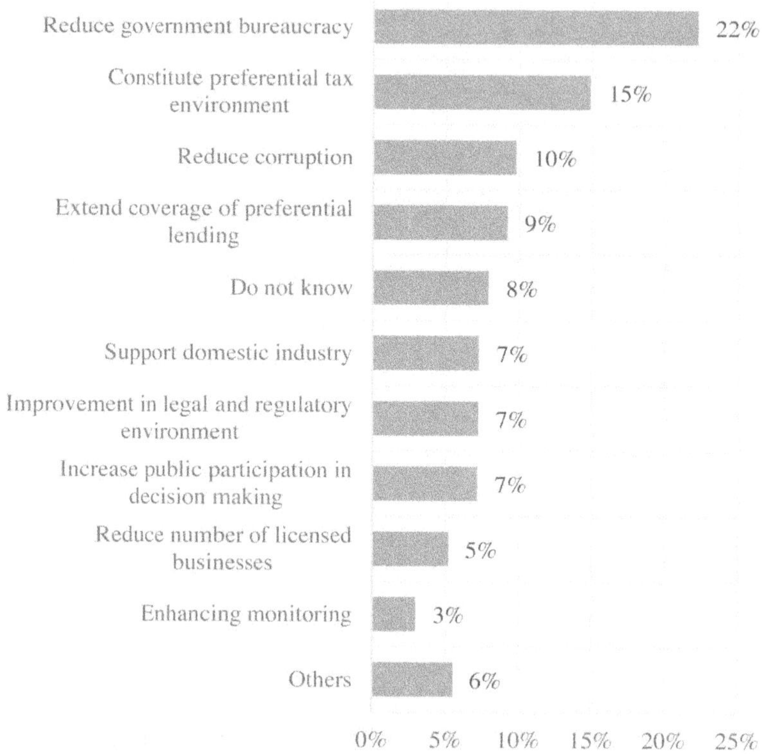


extending coverage of concessional lending (Figure 10). SMEs demand more effective and transparent regulatory environment as they face with more difficulties compared to larger companies. In line with other studies conducted in the case of Mongolia, over 30% of responders highlight that bureaucracy of government agencies and corruption at all levels are standing issues in doing business. This implies that the government's recent measures on tackling corruption and introducing e-governance were in right direction, but the implementation of the measures is highly demanded.

## BUSINESS ENVIRONMENT

This section focuses on business environment for SMEs. Factors potentially determining the business environment are classified into 6 subgroups:

- Social, political environment
- Macroeconomic environment
- Financial environment
- Legal and institutional environment
- Market condition
- Infrastructure condition

*Figure 11. SME business environment*
*Source: SME survey 2018*

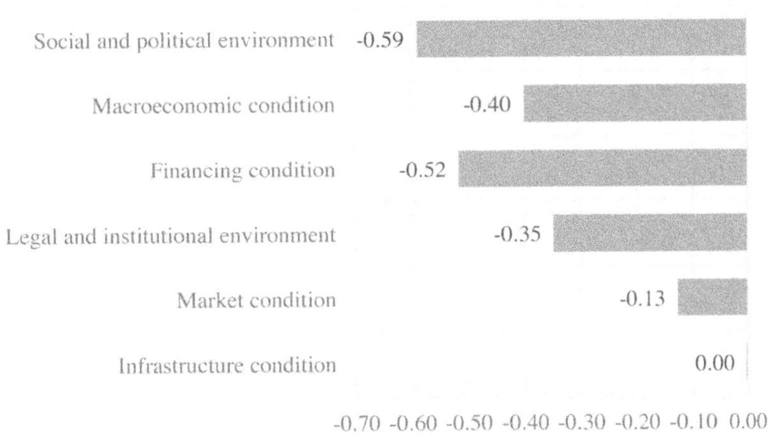


*Figure 12. Social and political environment index*
*Source: SME survey 2018*

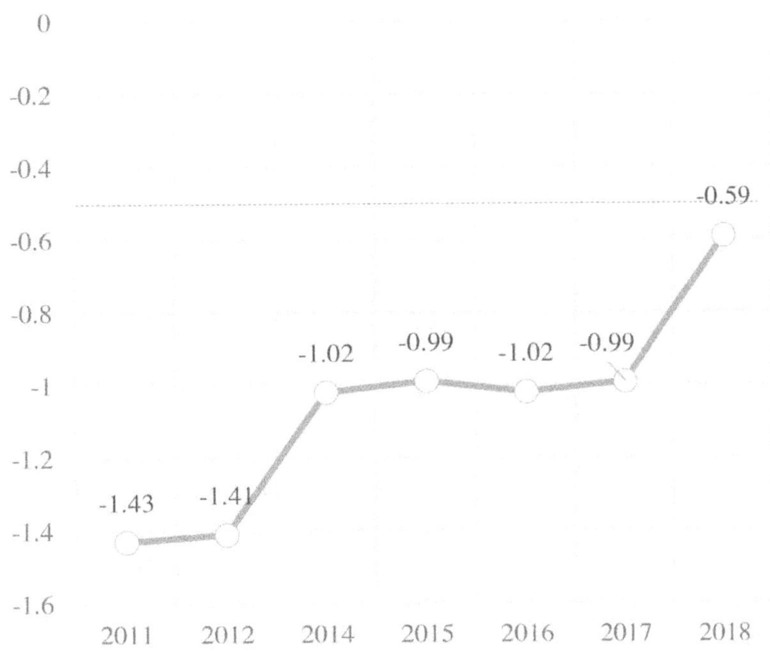


In total, 33 factors are considered under the subgroups. For each of 33 factors, indexes are calculated based on qualitative answers ranging from very bad (-2) to very good (+2) and subgroup index is estimated as average index of individual factors within the sub-group.

*Figure 13. Social and political environment by factors*
*Source: SME survey 2018*

*Figure 14. Macroeconomic environment index*
*Source: SME survey 2018*

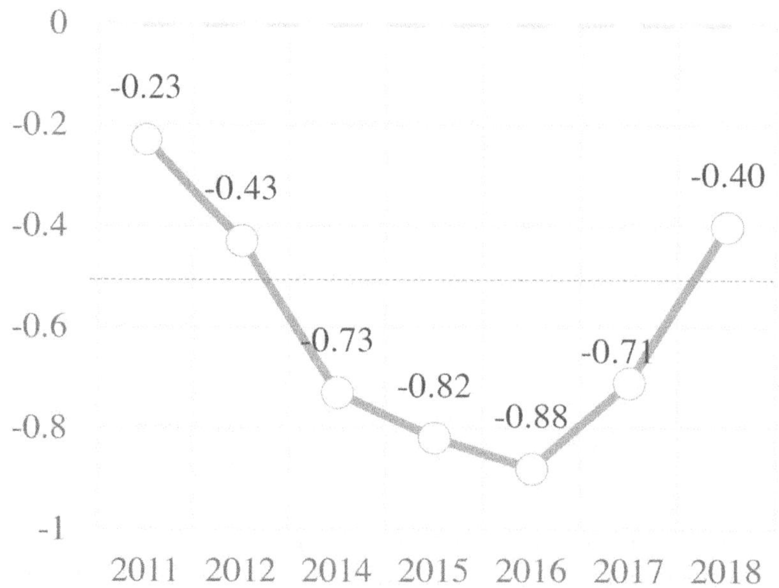

*Figure 15. Macroeconomic environment by factors*
*Source: SME survey 2018*

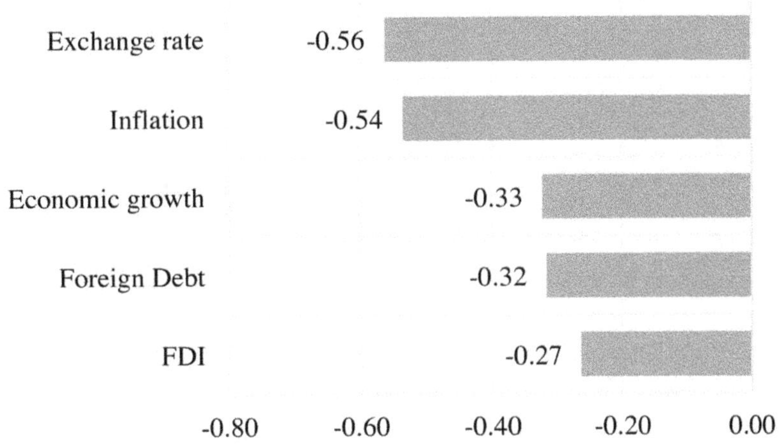


*Figure 16. Financial environment index*
*Source: SME survey 2018*

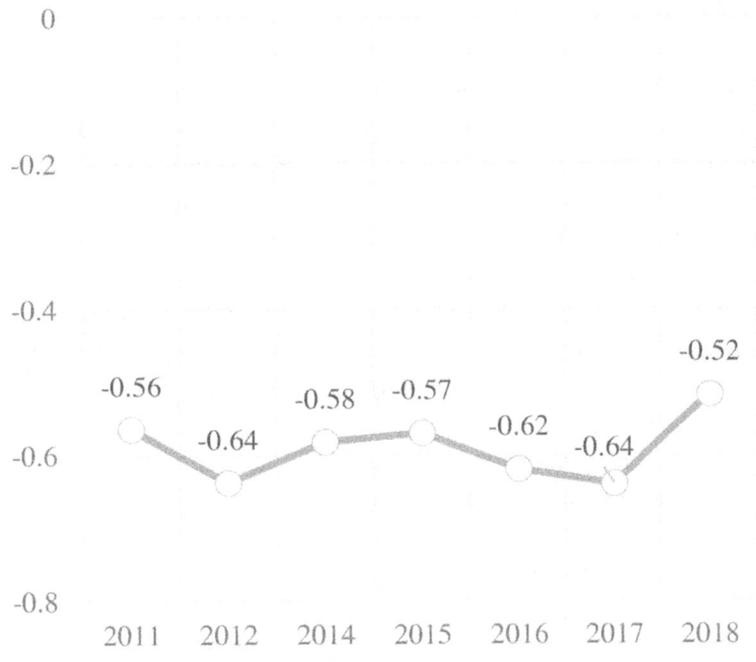


Results show that social and political environment (-0.59), macroeconomic environment (-0.40), financing environment (-0.52), legal and institutional condition (-0.35) and market condition (-0.13) are evaluated as reasonable (but vulnerable to become unfavorable) for SME business environment (Figure 11). There are also some feedback effects: legal frameworks and the business environment are important factors affecting the level of SME financing (i.e., Chavis et al., 2010 and Ardic et al., 2012).

*Figure 17. Financial environment by factors*
*Source: SME survey 2018*

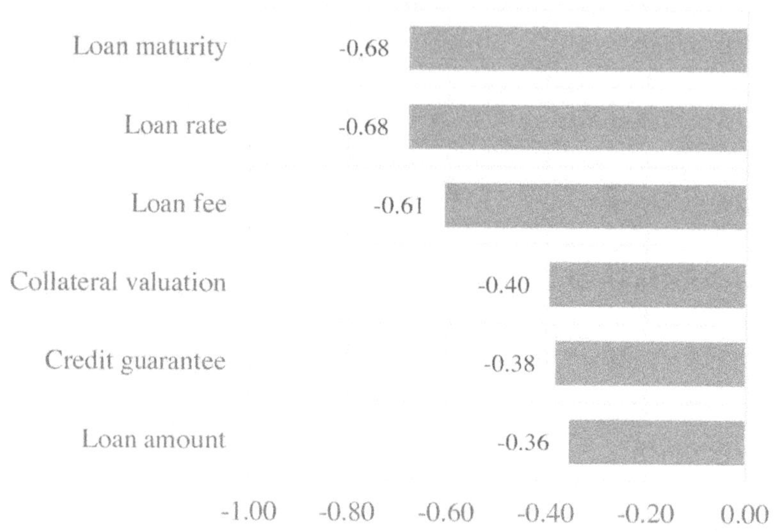


Social and political environment index is calculated by broad range of factors namely poverty, unemployment, corruption, crime and political situation that may hinder the SME operation. The overall evaluation of social and political environment is relatively unfavorable (-0.59) compared to other subgroups, but with significant improvement compared to previous 7 years (Figure 12). Political situation and corruption stand out as the most challenging factors in the social and political environment for 2018. According to the Transparency International Survey (2017), Mongolia ranked in 103 out of 180 countries in terms of the corruption perception index. Poverty and unemployment are found as the leading concern for SMEs, and both factors undermine sales, revenue (Figure 13).

In this survey, macroeconomic environment is represented by five factors, namely economic growth, inflation, exchange rate, foreign direct investment (FDI) and foreign debt. Overall macroeconomic condition for 2018 is evaluated as reasonable (-0.40), but with little improvement compared to the last year (Figure 14). Result of the survey for 2018 suggests that exchange rate and inflation are the most challenging factors, whereas economic growth, foreign debt level and FDI are referred to as having less negative impact on the SME operation (Figure 15).

The financial environment index has improved to (-0.52) in 2018, however it still indicates unfavorable financing condition (Figure 16). It is also observed that access to finance is a significant obstacle for SMEs during the economic recession. Among the factors, high loan rate and short loan maturity are the worst factors for the financial environment. SMEs find that it difficult to obtain financing, especially loan-term loans, for several reasons, including high risk premiums, high loan fee, lack of collateral and credit guarantees (inadequate credit history) and inadequate loan amount (Figure 17).

Even though numerous efforts targeting to strengthen SMEs' legal and institutional environment In terms of external financing, 48% of SMEs operating in manufacturing sector, 45% of agriculture SMEs have obtained external financing, and the shares are higher compared to that of others sectors. This result may also reflect the fact that financial need for agriculture and manufacturing sectors are higher. Moreover, finance for Agri-SMEs faces special challenges since farmers are risky borrowers and all agriculture has

*Figure 18. Legal and institutional environment index*
*Source: SME survey 2018*

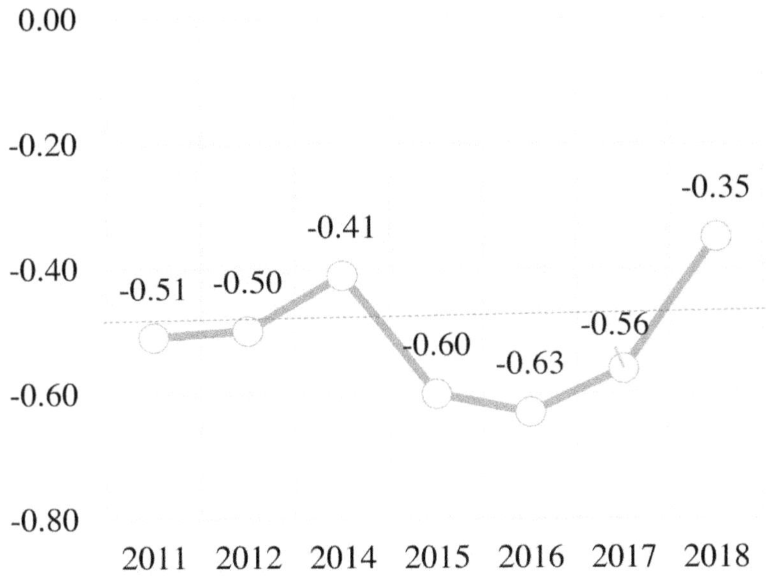


*Figure 19. Legal and institutional environment by factors*
*Source: SME survey 2018*

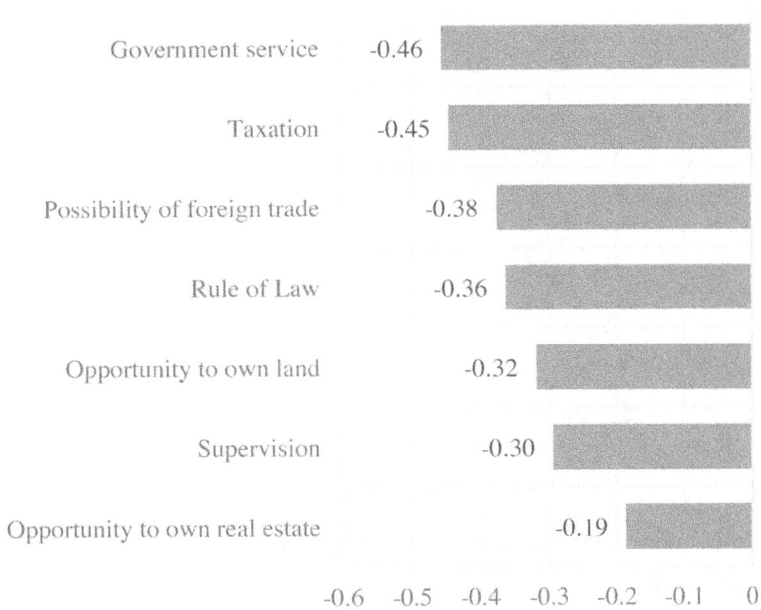

*Figure 20. Market condition index*
*Source: SME survey 2018*

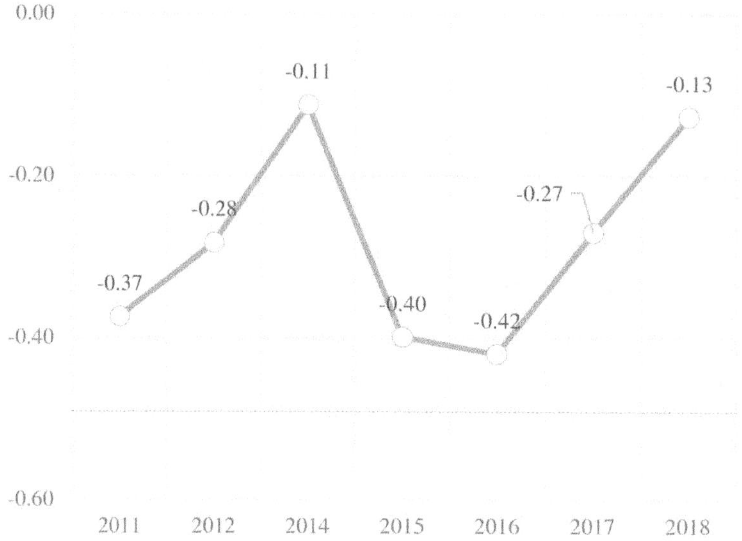


*Figure 21. Market condition by factors*
*Source: SME survey 2018*

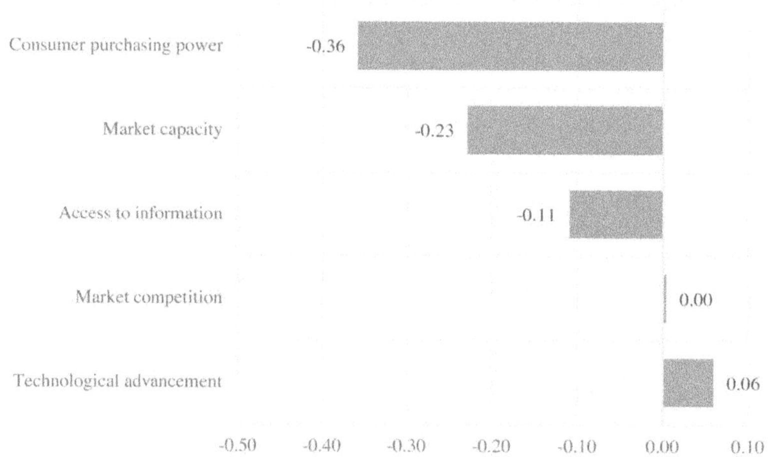


seasonal ebbs and flows, with irregular cash flows. Majority of SMEs had access to funding through banks, and 12% of them were able to fund with subsidized rates (Figure 24). The major share of external funding (57%) is used to purchase current asset, and 17% is used as machinery investment (Figure 25).

over the past years, the enterprises expect further improvements in this regard. Legal and institutional environment index was evaluated at a reasonable level (-0.35) with significant improvement from previous

*Figure 22. Infrastructure condition index*
*Source: SME survey 2018*

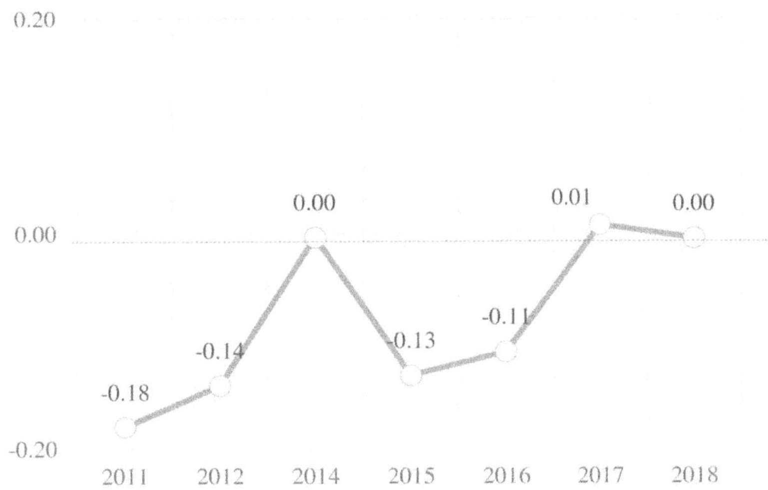


*Figure 23. infrastructure condition by factors*
*Source: SME survey 2018*

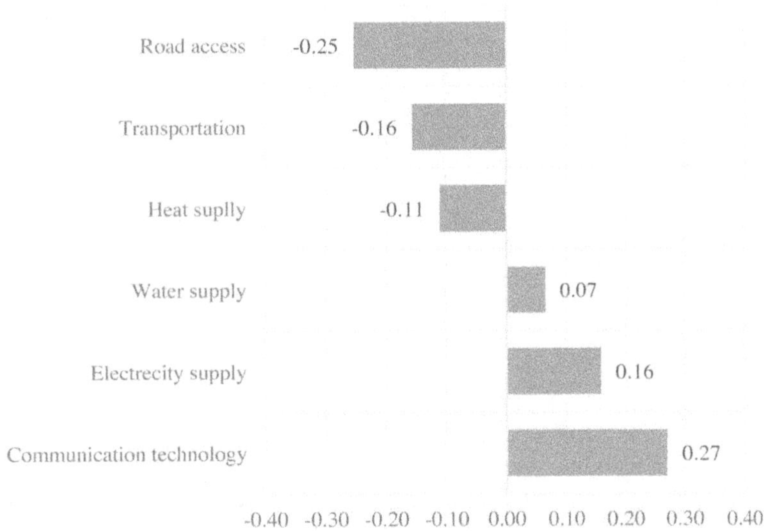


year (Figure 18). A deeper look into the legal and institutional environment suggests that public service, taxation framework and possibility of foreign trade stands out as the greatest concerns (Figure 19).

Market condition index has improved in 2018 and valued as the most favorable in last 4 years (Figure 20). Results of the survey for 2018 points out that low purchasing power of consumers and market capacity are worst market factors affecting SME operation, while market competition and technological advancements are at satisfactory level (Figure 21).

*Figure 24. Sources of financing (share in total responses)*
*Source: SME survey 2018*

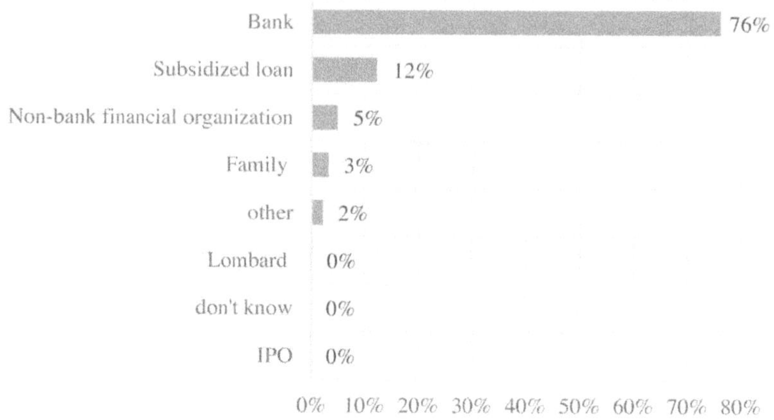


*Figure 25. Purpose of financing*
*Source: SME survey 2018*

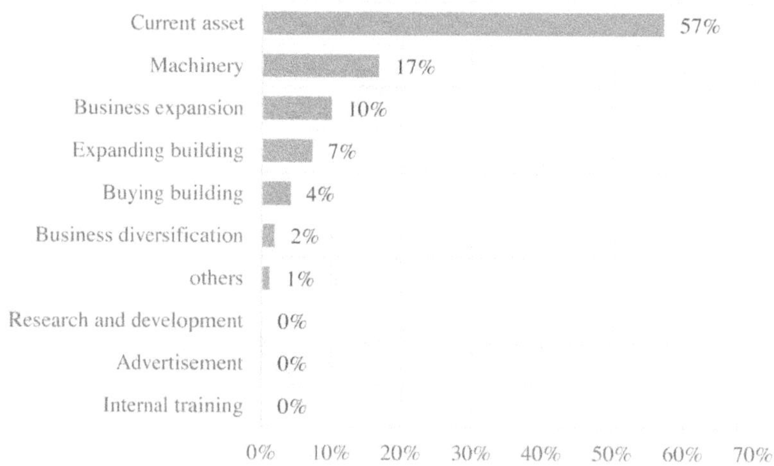


Infrastructure is no doubt a key component for business development. In this survey, infrastructure condition is represented by road access, heat supply, transportation, water supply, electricity supply and communication technology. Amongst other group of factors infrastructure is the least troublesome in Mongolia (Figure 22). The weakest factors of infrastructure seem to be road access and transportation, while there is no problem for communication technology and electricity supply (Figure 23).

*Figure 26. Reasons for tight credit condition (share in total responses)*
*Source: SME survey 2018*

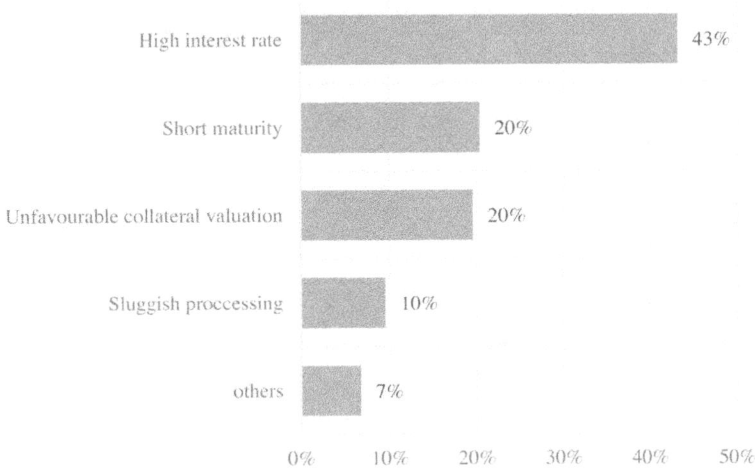


*Figure 27. Reasons for rejected applications*
*Source: SME survey 2018*

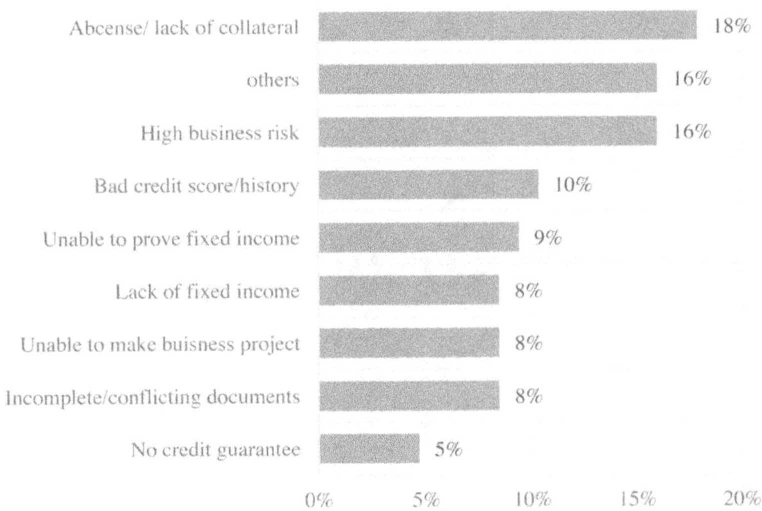


## ACCESS TO FINANCE AND FINANCING OBSTACLES

In line with the literature (i.e., Berger and Udell, 1998; Calindo and Schiantarelli, 2003), SMEs in Mongolia have less access to external finance and to be more constrained in their operation and growth. For instance, amongst 1864 SMEs responded, 41% of them obtain external financing, whereas 54% did not get any external financing. The low level of financial inclusion for SMEs is a result of weak economic

*Figure 28. Sources of financing (share in total responses)*
*Source: SME survey 2018*

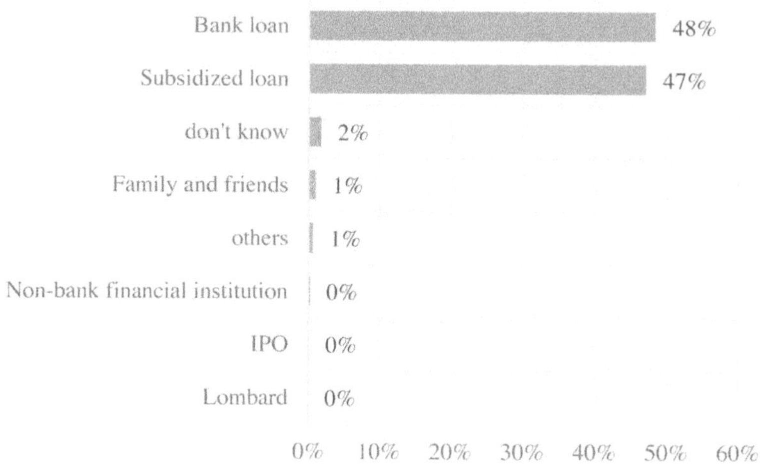


*Figure 29. Amount of necessary funding*
*Source: SME survey 2018*

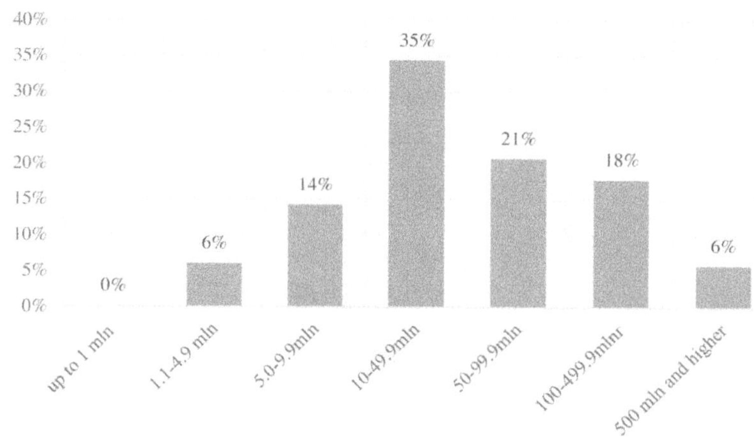


fundamentals, weak governance, weak financial regulation and supervision, undiversified and uncompetitive financial system in Mongolia.

SMEs mostly take loans from commercial banks, and 12% of SMEs obtain financing from the SME development fund of the government (Figure 24). For last couple of years, the structure of financing purposes has remained similar with slight tilt toward machinery investment from current asset financing. In terms of repayment, 13% of the SMEs surveyed have delayed their loan repayment.

*Figure 30. Amount of necessary financing by years in operation (share in total responses)*
*Source: SME survey 2018*

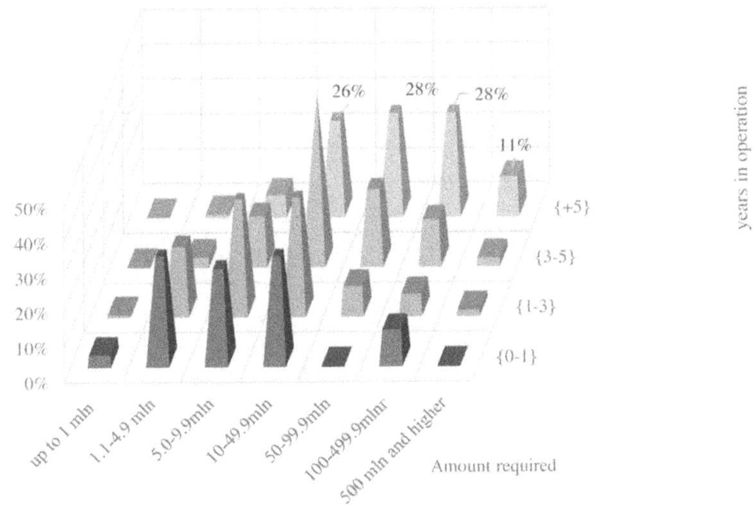


From the SMEs that have not obtained external finance, we also ask reasons: 21% of SMEs respond that they did not obtain external financing because of tight credit condition, and 9% of them could not meet the credit criteria. The remaining SMEs use their own capital. The result is in line with the view that entrepreneurs in transition economies are more likely to reinvest their profits (i.e., Johnson et al. 2002, Cull and Xu 2005).

SMEs operating in trade and service sectors take more external financing compared to those working in manufacturing sector. This may be partially explained by the fact that the firms need high level of financial resources to operate in manufacturing and agriculture sectors. In addition, the share of responses for not obtaining external financing because of tight credit condition is the highest in manufacturing and agricultural sectors. SMEs in agriculture have more limited collateral, and operate in high risks since they tend to be very concentrated in one activity or have a portfolio of activities all exposed to similar risks (drought, disease, etc.). Therefore, credit condition for Agri-SMEs is tight compared to those in operating other sectors.

It is crucial to understand obstacles to SMEs' operation and growth. For responders that did not obtain external financing because tight credit condition, we also asked why credit condition is evaluated as tight for them. Most pressing issue in tight credit condition is high interest rate. The 43% of them points that lending rate is high, and 20% of them responds that maturity is short and collateral valuation is unfair (Figure 26). This result is in line with the fact that SMEs worldwide consider access to finance their greatest obstacle to growth (i.e., IFC, 2013). In the bank dominated financial system, SMEs have difficulty accessing cheap finance.

For SMEs (106 of the total sample) applying for external financing, but got rejected, we also asked reasons. The most common reason for rejection is lack of or absence of collateral (18%) followed by high business risk (16%) and bad credit history (16%). In terms of need for financing, half of the SMEs

*Figure 31. Operational challenges by factors*
*Source: SME survey 2018*

*Figure 32. Amount of necessary funding*
*Source: SME survey 2018*

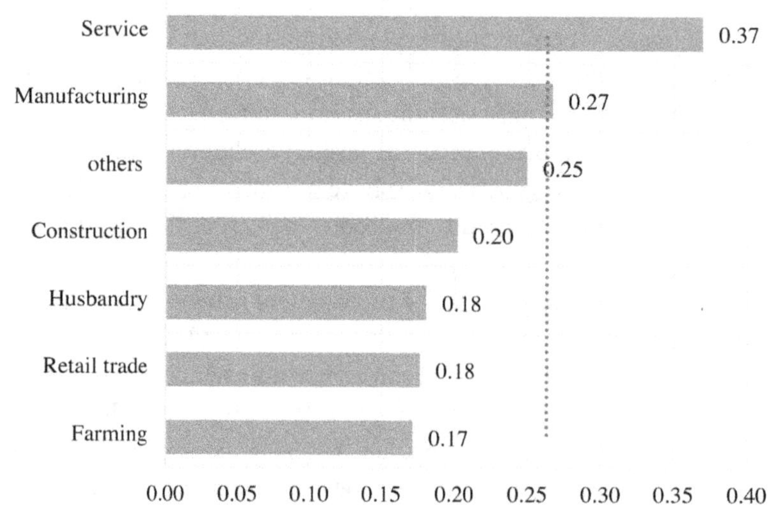


are certain to seek for external financing for upcoming year's operation, whereas 39% of them claim that there is no need for it. Therefore, improving financial market infrastructure, particularly credit information, movable collateral and secured transactions systems greatly reduce costs for financial institutions to assess SME ability and willingness to repay and broadens the range of assets SMEs can use to support their borrowing.

*Figure 33. Annual revenue (share in total responses)*
*Source: SME survey 2018*

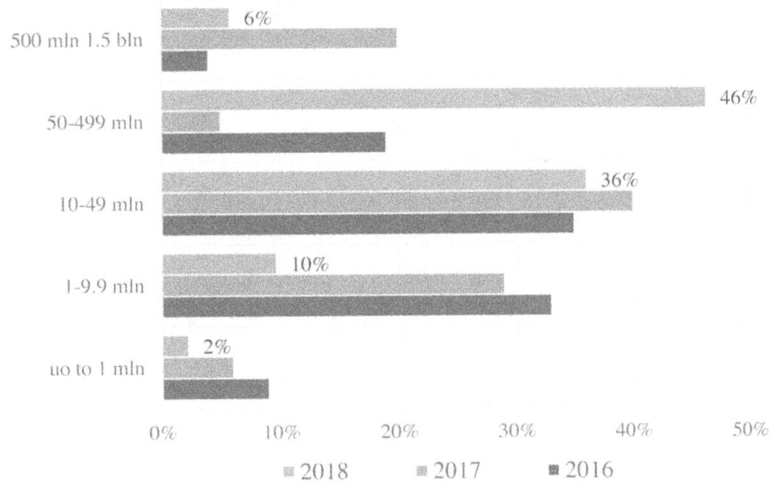


*Figure 34. Changes in annual revenue*
*Source: SME survey 2018*

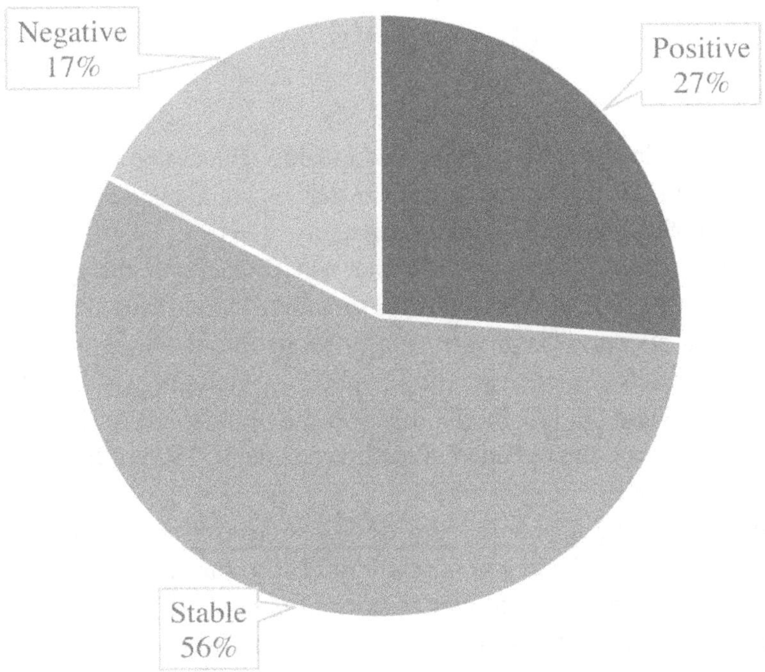

*Figure 35. SMEs' total cost index*
*Source: SME survey 2018*

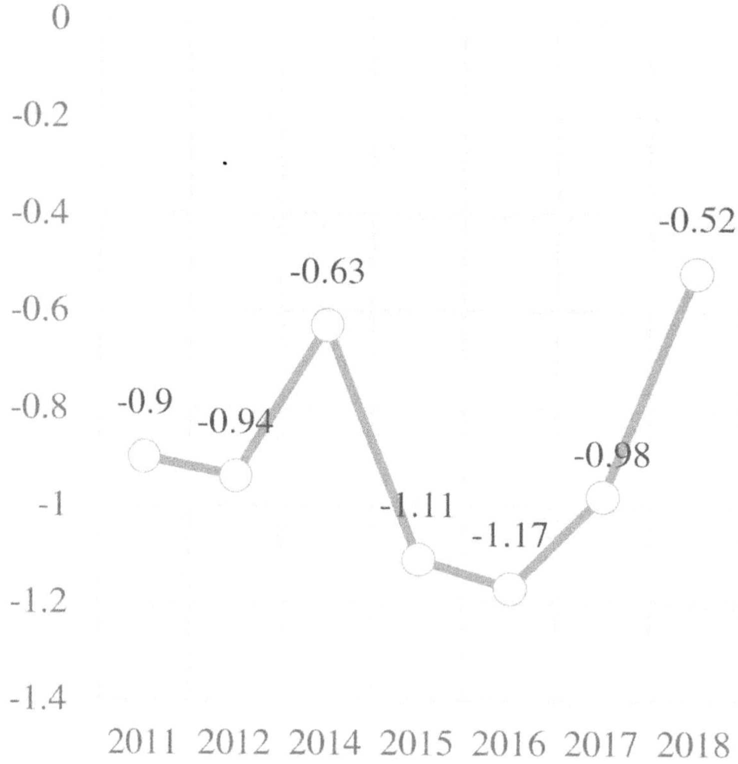


Share of business owners in agriculture and manufacturing sectors that require financing is considerably larger than in trade and service sectors. It is clear that financial necessity is considerably larger in agriculture and manufacturing compared with other sectors.

Regarding to source of financing, 48% of SMEs that seek financing in the upcoming year is planning to take bank loans and another 47% of them looks for subsidized loans from the SME development fund (Figure 28). In last couple of years, the government of Mongolia has implemented partial approaches such as policies focusing solely on direct public financing or guarantees for SMEs, which have been ineffective and unlikely to yield larger benefits. Instead, the policies make the SMEs more dependent on the public supports, which are limited and not inclusive enough. The public financing for SMEs also led to political corruption issues.

In terms of required financing for SMEs, 56% of SMEs covered in the survey seeks for financing between 10-100 million MNT (equivalent to 4000-40000 USD) for their operation. The clients are too large for microfinance institutions, and too small, risky and remote for commercial banks. Only 24% of them need more than 100 million MNT of financing (Figure 29). Closing the SME finance gap can bring about significant gains in growth potential as well. Hence, public interventions (ensuring that the legal system can quickly and reliably resolve issues of contract and property rights strengthens value chain financing options, providing alternative collateral possibilities) will be crucial in scaling up promising

*Figure 36. Total cost index by factors*
*Source: SME survey 2018*

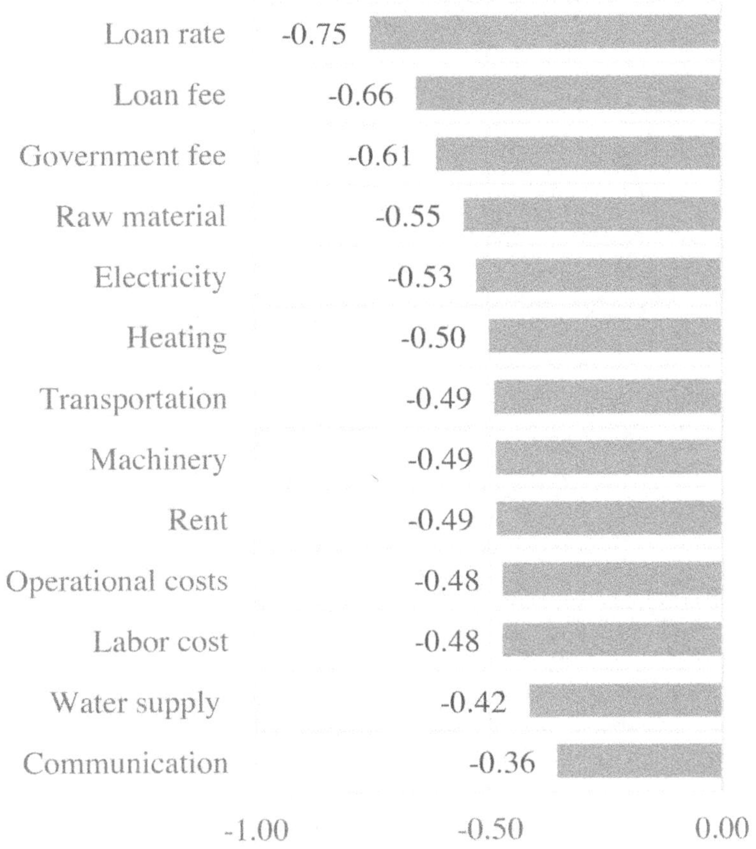


SME-finance innovations. Supporting guarantee funds, particularly those focused on SMEs with massive potential cam complement improved risk management in banks and other financial institutions.

There is considerable positive correlation between obtained amount of financing and SMEs' years in operation. This implies that banks are cautious in SME lending and more concerned about risks associated with SMEs. SMEs that are relatively younger (under 3 years in operation) seek for financing under 50 million MNT, while more mature businesses seek for loan up to 500 million MNT (Figure 30). Younger SMEs are financially more constrained since creditors have not had enough time to monitor such SMEs (i.e., the information asymmetries are likely to be especially large for them).

## SME OPERATIONAL CHALLENGES

Operational challenges are evaluated based on a scoring with a range from -2.0 (the worst) to +2.0 (the best). Surveyed SMEs respond that the most difficult operational challenges are labour supply, building supply and sales (Figure 31).

*Figure 37. Total cost decomposition (share in total responses)*

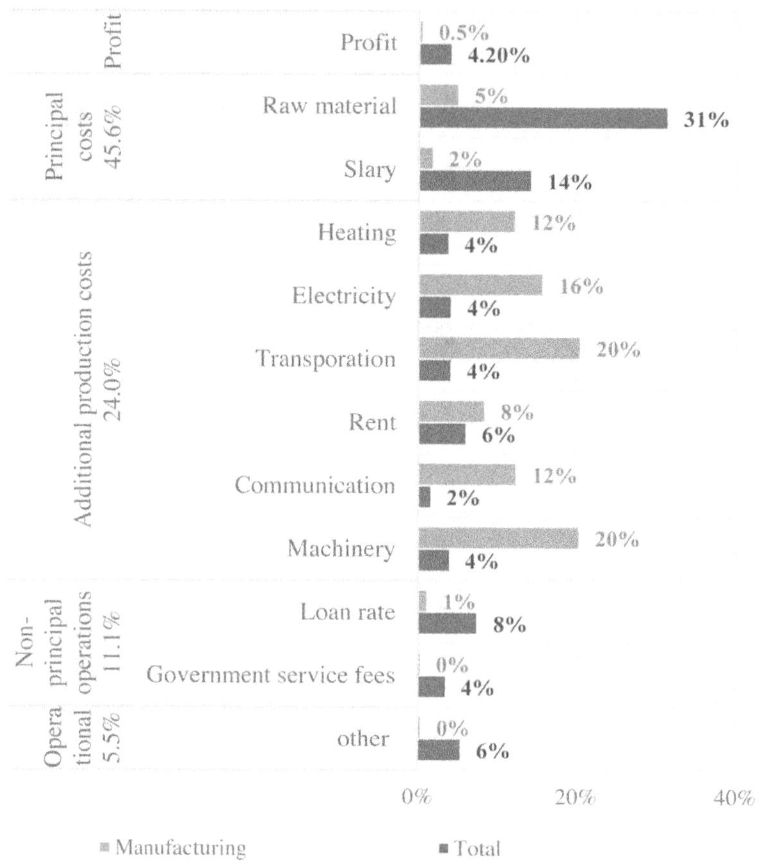


*Figure 38. Profitability for 2018*

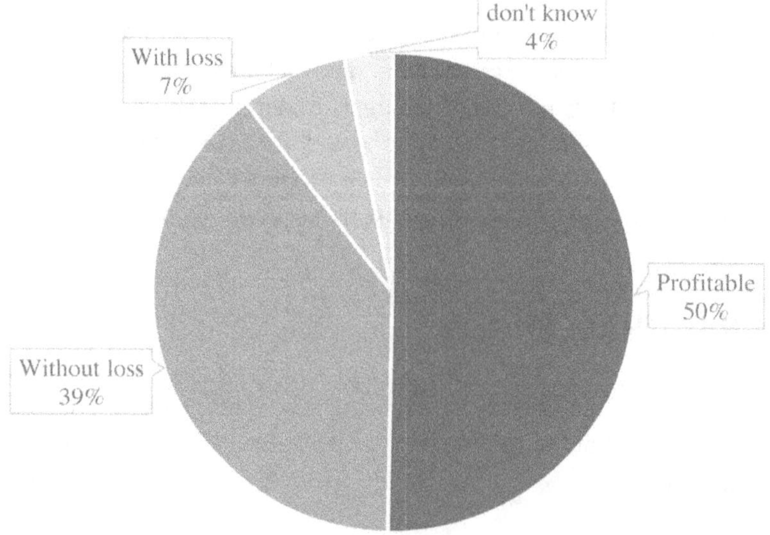

Regarding to sector-level operational challenges for SMEs, there are relatively less operational obstacles in service, manufacturing and construction sectors. However, trade and farming sectors tend to face more difficulties, but in manageable levels (Figure 32).

From all SMEs surveyed, 61% of them respond that they produce a balance sheet in their operation (Figure 8). Figure 33 shows annual revenues of surveyed SMEs for last three years. Almost half (of all SMEs 48%) has revenue up to 50 million MNT in which 36% of them has 10-49 million MNT revenues, 10% of them has 1.1- 9.9 million MNT revenues and 2% of them has only up to 1 million MNT revenues. The most frequently observed revenue range for 2018 (46%) is 50-499 million MNT with an increase of 41 percentage point compared to previous year.

In terms of changes in annual revenue, 27% of surveyed SMEs have positive revenues growth and 17% of them has negative revenues growth, while 56% of them has not changed for 2018 (Figure 34). Manufacturing sector leads with highest positive revenues growth followed by service and agriculture.

Costs related to business operations of SMEs and their pressing issues are also considered in this survey. The cost index for 2018 is evaluated at (-0.52), gaining 0.46 points from the previous year's index value, but still remaining in concerning zone (Figure 35).

A deeper look suggests that interest rate costs are the most pressing one followed by loan fees, government service fees, raw material costs, heating costs and transportation costs. Communication cost, water supply cost and labor costs seem to be the least troubling costs (Figure 36).

Regarding to cost decomposition, main production costs like labour costs and raw material costs are taking the largest portion, 45.6% of all costs for SMEs. Other operational costs are accounting for 24.0%, while non-operational costs like interest payment and service fees takes up to 11.5% of total costs (Figure 37). As for manufacturers, 90% of the revenues is dedicated to additional operational costs such as machinery, rent, transportation, electricity, heat, water supply, and communication.

In terms of profitability, 50% of the SMEs have reported profit and 7% of them have operated with loss for 2018. Meantime 39% of SMEs have operated with neither loss nor profit (Figure 38). 57% of SMEs in manufacturing sector have operated with profit, while 44% of SMEs in trade sector have reported profit.

## CONCLUSION

This paper has presented results of the survey on development and access to finance of SMEs in Mongolia. A number of important findings stand out. First, supports from government, international organization, financial organizations and non-governmental organizations play a crucial role in the SME development. In particular, banks and local government administration are the most supportive institutions. The institutions provide financing, training, assistance for SMEs. Second, SMEs are more constrained by different obstacles. They demand the government to take some policy measures in cutting out bureaucracy, creating a favorable tax regime and combating corruption. Social and political environment, particularly political instability and corruption are perceived as one of the major obstacles that SMEs face in their business environments. Third, SMEs think that business environment has improved for 2018. For instance, macroeconomic, legal and market conditions tend to favor the SME operation. Fourth, like many other emerging economies, SMEs in Mongolia are less likely to have access to formal finance. Only 40% of SMEs has been able to access external financing through bank loans or subsidized financing. External financing is used mainly on current asset, expanding operation and purchasing machinery. SMEs argue that financing condition is tight, hence does not sufficiently support the SME operation. In particular, lending rate is too high, loan maturity is short, and fees are still unfavorable. The SME survey conducted by the Bank of Mongolia (BOM) for the last 7 years constantly shows that the high lending rate is the worst financial factor affecting business environment in Mongolia. About 10% of SMEs obtained external financing have faced with difficulties for paying debt on time. Fifth, more than half of the SMEs need external financing in the upcoming years. For most SMEs, the amount of required financing is 10-50 million MNT (4000-8000 USD) with maturity of 4-5 years. Relatively younger SMEs (under 3 years in operation), for instance, seek for financing under 50 million MNT, while more mature businesses seek for loan up to 500 million MNT. Finally, for operational challenges, labor supply stands out as the biggest obstacle, while competitiveness of the businesses, skill of labor force and quality of machinery are at reasonable level. A deeper look at the costs of SMEs suggests that interest rate cost is the most troubling one followed by loan and government service fees. Main production costs such as labour costs and raw material costs are taking the largest portion (46%), while non-operational costs like interest payment and service fees account for 11.5% of total costs for SMEs. About 50% of the SMEs have reported profit and 7% of them has operated with loss for 2018.

These findings provide some policy implications. The measures and reforms may include (i) improving institutions (including low corruption and political stability) and the overall business environment is the most effective way of relaxing the growth constraints SMEs face, (ii) building diversified and competitive financial system, higher levels of financial intermediary development, adequate financial infrastructure and more efficient legal system would lower financing obstacles, and (iii) Designing country-specific reform strategy for SME development and SME access to financing is important to promote safe and stable SME financial inclusion and facilitate SMEs to contribute the economic growth.

Though the results have yielded significant insights about development and access to finance of SMEs in Mongolia, future research should further investigate the impact of access to finance on growth of SMEs based on the accumulated SME survey data using a formal econometric analysis. It is also possible to conduct a survey to deepen the understanding of what types of financing (i.e., trade credit, industry specific financing forms such as leasing or factoring, etc.) is really needed for SMEs as an alternative to bank financing. Nevertheless, these were out of the scope of this paper.

## REFERENCES


Ardic, O., Mylemko, N., & Saltane, V. (2012). Access to finance by small and medium enterprises: A cross-country analysis with a new data set. *Pacific Economic Review*, *17*(4), 491–513. doi:10.1111/j.1468-0106.2012.00596.x

Ayyagari, M., Beck, T., & Demirgüç-Kunt, A. (2007). Small and medium enterprises across the globe. *Small Business Economics*, *29*(4), 415–434. doi:10.100711187-006-9002-5

Ayyagari, M., Demirguc-Kunt, A., & Maksimovic, V. (2011). *Small vs. young firms across the world: contribution to employment, job creation, and growth*. The World Bank.

Beck, T., & Demirgüç-Kunt, A. (2006). Small and medium-size enterprises: Access to finance as a growth constant. *Journal of Banking & Finance*, *30*(11), 2931–2943. doi:10.1016/j.jbankfin.2006.05.009

Beck, T., Demirgüç-Kunt, A., Laeven, L., & Maksimovic, V. (2006). The determinants of financing obstacles. *Journal of International Money and Finance*, *25*(6), 932–952. doi:10.1016/j.jimonfin.2006.07.005

Beck, T., Demirgüç-Kunt, A., & Levine, R. (2005). SMEs, growth, poverty: Cross-country evidence. *Journal of Economic Growth, 10*(3), 197–227. doi:10.100710887-005-3533-5

Beck, T., Demirguc-Kunt, A., & Martinez Peria, M. S. (2008). *Bank financing for SMEs around the world: Drivers, obstacles, business models, and lending practices.* The World Bank.

Beck, T., Demirgüç-Kunt, A., & Singer, D. (2013). Is small Beautiful? Financial structure, size and access to finance. *World Development, 52*, 19–33. doi:10.1016/j.worlddev.2013.05.014

Berger, A., Demirgüç-Kunt, A., Haubrich, J., & Levine, R. (2004). Bank concentration and competition: An evolution in the making. *Journal of Money, Banking and Credit, 36*(3b), 433–453. doi:10.1353/mcb.2004.0040

Berger, A., & Udell, G. (1998). The economics of small business finance: The role of private equity and debt markets in the financial growth cycle. *Journal of Banking & Finance, 22*(6-8), 613–673. doi:10.1016/S0378-4266(98)00038-7

Blancher, N., Appendino, M., Bibolov, A., Fouejieu, A., Li, J., Ndoye, A., ... Sydorenko, T. (2019). *Financial inclusion of small and medium-sized enterprises in the Middle East and Central Asia.* Middle East and Central Asia Department, International Monetary Fund. doi:10.5089/9781484383124.087

Calindo, A., & Schiantarelli, F. (Eds.). (2003). Credit constraints and investment in Latin America. Inter-American Development Bank, Washington, DC.

Chavis, L., Klapper, L., & Love, I. (2010). The impact of the business environment on young firm financing. World Bank.

Clarke, G., Cull, R., Soledad, M., Peria, M., & Sanchez, S. (2003). Foreign bank entry: Experience, implications for developing economies and agenda for further research. *The World Bank Research Observer, 18*(1), 25–59. doi:10.1093/wbro/lkg002

Cull, R., & Xu, L. (2005). Institutions, ownership, and finance: The determinants of profit reinvestment among Chinese firms. *Journal of Financial Economics, 77*(1), 117–146. doi:10.1016/j.jfineco.2004.05.010

De Haas, R., & Naaborg, I. (2005). Does foreign bank entry reduce small firms' access to credit? Evidence from European transition economies. De Nederlandsche Bank.

Demirgüç-Kunt, A., & Maksimovic, V. (1998). Law, finance, and firm growth. *The Journal of Finance, 53*(6), 2107–2137. doi:10.1111/0022-1082.00084

Harwood, A., & Konidaris, T. (2015). SME exchanges in emerging market economies: A stocktaking of development practices. The World Bank.

IFC. (2013). *Small and medium enterprise finance: new findings, trends and G-20/global partnership for financial inclusion progress. International Finance Corporation (IFC) and Global Partnership for Financial Inclusion.* GPFI.

Jargalsaikhan, D (2019). SMEs are the reflection of the development. The Facto Gazette.

Johnson, S., McMillan, J., & Woodruff, C. (2002). Property rights and finance. *The American Economic Review, 92*(5), 1335–1356. doi:10.1257/000282802762024539

Levy, B. (1993). Obstacles to Developing Indigenous Small and Medium Enterprises: An Empirical Assessment. *The World Bank Economic Review*, *7*(1), 65–83. doi:10.1093/wber/7.1.65

Lkhagvasuren, T. (2014). Marketing of small and medium-sized enterprises in Mongolia. *International Journal of Technology Enhancements and Emerging Engineering Research*, *2*(5), 33–35.

OECD. (2016). Enhancing access to finance for micro, small and medium-sized enterprises in Mongolia.

Schiffer, M., & Weder, B. (2001). Firm size and the business environment: Worldwide survey results. International Finance Cooperation.

Taiwo, M., Ayodeji, A., & Yusuf, B. (2012). Impact of small and medium enterprises on economic growth and development. *American Journal of Business and Management*, *1*(1), 18–22. doi:10.11634/21679606170644

Tuul, O., & Bing, S. (2019). Employee training in small and medium-sized enterprises in Mongolia. *Proceedings of the Mongolian Academy of Sciences*, *59*(1), 75–84.

World Bank. (2019). *Doing Business 2019* (16th ed.). World Bank Group.

# Chapter 15
# The Role of Internet of Things, Knowledge Management, and Open Innovation in SME Sustainability

**Muhammad Imran**
https://orcid.org/0000-0001-7126-2831
*The Islamia University of Bahawalpur, Pakistan*

**Jawad Iqbal**
*The Islamia University of Bahawalpur, Pakistan*

**Hassan Mujtaba Nawaz Saleem**
*The Islamia University of Bahawalpur, Pakistan*

## ABSTRACT

*The main objective of the chapter is to discuss the relationship between internet of things and knowledge management; knowledge management and open innovation; open innovation and SMEs sustainability. The relationship between the constructs developed and discuss on the behalf of past studies. The present chapter found that Internet of Things is playing an important role in knowledge generation and management, further, knowledge management is very important for open innovation environment in SMEs. Moreover, the open innovation sustains the SMEs performance. In respect of implications, the owner / managers of SMEs should consider the Internet of Things, knowledge management, and open innovation capabilities during the decision making for SME sustainability. Moreover, this is a process framework which brings the effect of one variable to other variables. However, the future studies should empirically validate the proposed research framework.*

DOI: 10.4018/978-1-7998-1566-2.ch015

## INTRODUCTION

In this digital world, firms are adopting and adapting the latest technologies to speed up the business process to achieve the higher efficiency in the business as well as enhance the innovation within in the firms through external and internal knowledge and information gathering (Frishammar, Richtnér, Brattström, Magnusson, & Björk, 2019). In other words, it is called knowledge and technology driven economy. Basically, large number of scholarly researchers reported the issue regarding management of knowledge, which is crucial for company's better performance. Essentially, the knowledge management is very important to organize and leverage the collective knowledge of firms which can improve the innovativeness in the firms and achieve the sustainable performance of the firms (Garcia-Morales, Martín-Rojas, & Lardón-López, 2018). However, the knowledge management is supporting to open innovation environment in the firms as well. Furthermore, proper knowledge management responds to dynamic environment changes.

Amazingly, a few researchers have been conducted the researches on the role of information & communication technology (ICT) in management of knowledge. However, this is significant research gap, because there are many firms working on knowledge management, in order to create, share and store the information. As discuss previously, knowledge management is important for firm efficiency which leads to higher firm performance as well as improve the open innovation within the firm (Papa, Dezi, Gregori, Mueller, & Miglietta, 2018). More importantly, considering the internet of things (IoT) occurrence, this is a global network in which millions of devices are interconnected to exchange the data and interact to spread their utility beyond the physical world (Miraz, Ali, Excell, & Picking, 2018). Basically, firms can get the competitive advantage in order to gathering and exchange of data through IoT capability. Furthermore, itself IoT capability is changing the nature of innovation.

According to Santoro, Vrontis, Thrassou, and Dezi (2018) suggested that firms can developed the IoT which can facilitate or sustain to knowledge management system through exploration of advance external knowledge, which converted into higher innovation performance. The innovation performance explains in this chapter as introduction of new product/service, process and opening a new market (Expósito, Fernández-Serrano, & Liñán, 2019). Mainly, as past researchers identified that knowledge management is exploring the internal & external knowledge and retaining the knowledge to adopting the open innovation (Lifshitz-Assaf, 2018). Furthermore, past researches presented that different practices related with open innovation, such as looking for opportunity to using external knowledge for innovation process within the firm beyond the firm boundary is not new for firms. Consequently, the main concern of the firms to build the network through they can collaborate and outsource the knowledge which can help to manage the innovation strategies (Tidd & Bessant, 2018). This open innovation behavior is very important small and medium enterprises (SMEs), since SMEs are lying on external information and research collaboration in order to attain the competitive advantage, which leads to sustain their performance (de Zubielqui, Lindsay, Lindsay, & Jones, 2019).

Moreover, the only few studies have investigated the role of open innovation in context of SMEs sustainability. Additionally, the past researches have main focused on the differences between open innovation in SMEs and large firms. However, this research gap is motivating the current research for further investigation. Therefore, this chapter is going to explore the role of IoT in knowledge management and open innovation domain. Hence, this research work aims to contribute into knowledge management and open innovation literature to highlight how IoT can facilitate to knowledge management and further how knowledge management can improve the open innovation. Moreover, this chapter also focus that how open innovation sustain the performance of SMEs.

The chapter will answer the following question.

Q1: Are Internet of things generating and managing knowledge?
Q2: Is Knowledge management creating open innovation environment in firms?
Q3: Is open innovation sustaining the SMEs performance?

After the presentation of study introduction, the remaining sections of the chapter as follows. The section will cover the chapter literate review and will propose the proposition, this section will cover the research framework as well. The third section will present the research methodology of the chapter. The fourth section of the chapter will cover the discussion of the study. The last section will cover the conclusion, implications, limitations and future recommendation of the chapter.

## LITERATURE REVIEW

### Internet of Things and Knowledge Management

According to Ardolino et al. (2018) stated that "a global network in which billions of devices can be heterogeneously interconnected to exchange data and interact to extend their functions beyond the physical world and reach common goals without direct human intervention". One more study explained that "IoT is the integration of sensors and computing in internet environment through wireless communication" (Premsankar, Di Francesco, & Taleb, 2018). However, the present development in internet is allowing the communications of many objects. Moreover, this concept is reducing the cost of sensors, which can sense the any kind of object and connection to global network (Tao, Qi, Liu, & Kusiak, 2018). Furthermore, the IoT considered as truculent technologies, which is influencing the daily life of company and individual. Basically, in this modern era, the firms are developing and implementing the truculent ICTs technologies for better efficiency and innovation through knowledge and data / information (Ardolino et al., 2018). As a result, the management of knowledge can be supported by ICTs.

As previous researchers stated that the IoT is the ICTs part, IoT is combinations of billion devices, which are generating data (Ardolino et al., 2018). However, the firms can integrate with technologies through the open and collaborative IoT environment (Tao et al., 2018). The technological integration respect to integration of website with back-end system and data base; secondly integrate with internal database and also with external database of stockholders as well (Miraz et al., 2018). This whole process can manage the knowledge effectively. Moreover, the IoT in respect of information, aims to providing commercial information to several stakeholders, secondly IoT in respect of communication allows the cost reduction and interaction with several business partner and competitors and lastly IoT in term of workflow, can established the electronic process through corporate technologies (Santoro et al., 2018). These all process can manage the knowledge effectively. Hence on the behalf of past literature discussion, the present chapter is proposing the following proposition.

P1: The Internet of Things is generating and managing knowledge

## Knowledge Management and Open Innovation

Knowledge management is a firm capability to manage the different knowledge processes. Furthermore, according to Frishammar et al. (2019) explained the KM is the degree to create, share and utilize the knowledge between functional entities. Moreover, according to Garcia-Morales et al. (2018) KM includes knowledge transfer, sharing, integration and creation. However, according to Santoro et al. (2018) KM is a function in which knowledge is created, shared, codified and utilized to make the environment for increase in innovation which leads to higher firm performance. More specifically, according to Papa et al. (2018) KM can motivate to firms that knowledge is a key sources for the firm which can bring accuracy and speed in innovation, secondly KM can provide the information about customers, products, markets and technologies, this kind of information can improve firm performance. KM provides the new and advance knowledge to accelerate the innovation processes.

However, the present chapter found the two main dimensions of KM such as knowledge acquisition and knowledge sharing. Knowledge acquisition refers to firm capability which cans acquire the external knowledge (Huang, Bhattacherjee, & Wong, 2018). Additionally, knowledge acquisition is an integrated contrivance, which helps to absorb the important knowledge and create the latest knowledge from internal and external source (Sun, Liu, & Ding, 2019). Secondly, the share of knowledge from individual to group is called sharing of knowledge. Basically, the knowledge sharing is a complex and unique development of internal and external knowledge, and also included the dissemination of different organization knowledge (Connell, Kriz, & Thorpe, 2014).

Ultimately, open innovation incorporates the inside and outside innovation concepts, which brings more easy approach to knowledge accumulation, further, improve the firm capability to handle the knowledge and motivates firms to achieve the knowledge acquisition (Sun et al., 2019). On the other hand, more of studies stated that useful knowledge sharing between companies can developed the open innovation environment (Connell et al., 2014). However, the recent literature suggests that the KM of the firms can support to open innovation, in which firms can utilized the internal and external knowledge to produce the new products and services. Hence the current chapter is proposing the following proposition for further investigation.

P2: The knowledge management is supporting open innovation in SMEs

## Open Innovation and SMEs Sustainability

Open innovation is the use of internal and external knowledge to quicken the internal innovation and spread the markets for external use of innovation (Expósito et al., 2019). In other words, the open innovation is opposite of closed innovation. However, the most of studies divided the open innovation into two dimensions, such as inbound and outbound. The inbound open innovation explained that the firms explores and attain the latest knowledge and technologies from external sources such as customers, suppliers, distributor, competitors, universities, Government (Popa, Soto-Acosta, & Martinez-Conesa, 2017). Moreover, the inbound dimensions changing the firms behavior that firms can think beyond the boundaries and enriching the own knowledge as well (Li-Ying, 2018). However, the firms are benefited through inbound open innovation behavior because firms are generating new ideas, new markets and effective handling of problems. On the other hand, the outbound open innovation refers to the manipulation of knowledge regarding internal ideas and technologies that run out through licensing, patents, agreements in respect to gain financial and non-financial benefits (Popa et al., 2017).

However, according to past literature, most of studies reveled that firms focused on inbound open innovation rather out bound open innovation. Additionally, both dimensions of open innovation are not mutually exclusive. Anyhow, some literature suggested that firms should practice the both type of open innovation, since, some of researchers said that implementation of both types can achieve high value from their knowledge and technology resources (Popa et al., 2017). More specifically, the main concern of the chapter is the sustainability of small and medium enterprises (SMEs). In this regard, researchers said that the close innovation is strictly bounded within firm, it is using their own internal knowledge and creating ideas, then develop the product, distribute it into market and support it (Van de Vrande, De Jong, Vanhaverbeke, & De Rochemont, 2009). This model is very tight and bounded only within the firms; it's very difficult to share with other stockholders. On the other hand, as we know, open innovation is very flexible and easy to acquire and share the knowledge between stakeholders. Basically, small and medium business have no ability to build the close innovation environment, hence, the open innovation can support the SMEs due to their nature (Lee, Park, Yoon, & Park, 2010).

According to Brunswicker and Vanhaverbeke (2015) open innovation recognize the SMEs that can play the prominent role in innovation background. However, the innovation in SMEs is very low due their scant financial position, less skill employees and low risk taking (Greco, Grimaldi, & Cricelli, 2016). Accordingly, SMEs need to collaborate with stakeholder to find the missing innovation resources (Scuotto, Del Giudice, Bresciani, & Meissner, 2017). Furthermore, SMEs required shortening the product life cycle with the help of stockholders (Van de Vrande et al., 2009). However, these all strategic step cannot meet up without the positive role networks. Hence, we can say, the open innovation practices not effective for large firms, these are equally important for SMEs. Therefore, the following proposition proposed.

P3: The open innovation is sustaining SME performance

## Series Mediation

As past literature stated that management would use the different mechanism to influence SMEs sustainability through diverse processes and levels. According to IoT considered advance and fastest technology in this world, which is influencing daily life of firm and as well as individual. As many researchers presented that companies are implementing IoT technologies to generates the knowledge. As of now, KM is source of innovation. KM is the key source which can bring accuracy and speed in innovation. However, there are different researches found the positive and significant relationship between KM and open innovation. Hence, according to Hayes (2013) that causal relationship between both mediator such as KM and open innovation should be high. As on the behalf of past researches which reported the strong relationship between KM and innovation, the current study is believing that the series mediation relationship between IoT and SMEs sustainability. Hence the present study proposed the following proposition.

P4: The relationship between IoT and SMEs sustainability is serially mediated by knowledge management and open innovation.

*Figure 1.*

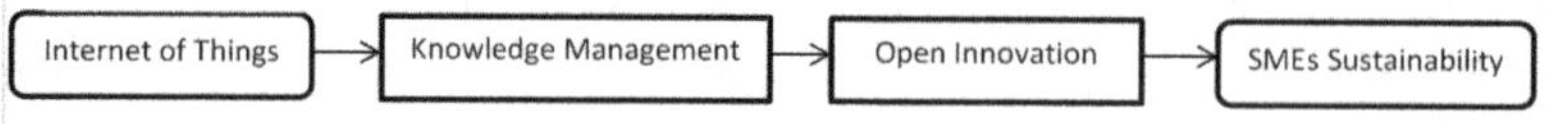


## RESEARCH FRAMEWORK

The research framework is presenting the independent variable such as internet of things, secondly the focus of the study is series or chain mediation role of knowledge management and open innovation between internet of things and dependent variable which is SMEs sustainability.

## RESEARCH METHODOLOGY

The past researches are reviewed related with internet of things, knowledge management, open innovation and SMEs sustainability. The articles are only taken from famous publishers' journals such as Emerald, ScienceDirect, Taylor and Francis, Springer, Sage and Wiley. However, the present has reviewed the internet of things, knowledge management, open innovation and SMEs sustainability to present the study propositions.

## DISCUSSION AND CONCLUSION OF THE STUDY

The current chapter has raised the three questions, such as: Are internet things generating knowledge management? Is KM creating open innovation? Is open innovation sustaining the SMEs performance? However, on the basis of past literature discussion and suggestions, the present chapter conclude to answer the above questions. Definitely, in this modern World, we cannot ignore the ICTs used in the firms, as previous studies revealed that IoT are connecting with billions of devices, which are providing the data on daily basis. The firms can use this data for knowledge management and innovation. However, the IoT are playing main vital role in knowledge management. In response of second question, proper KM is providing the internal and external knowledge to firms, which is very important for innovation, as per past literature, open innovation required updated knowledge for idea generation, which convert into product development, then searching for new market for products selling and distribution. Furthermore, open innovation bringing the string networks inside the firms and outside the firms. This network is providing the opportunity to firms to exchange the knowledge, ideas and technologies. These all strategic steps cannot achieve without proper knowledge management. Hence, KM is playing main role in open innovation environment.

The answer of third question, SMEs are facing lot of challenges due to their small size, low capital and low skill employees. These barriers are bringing low intention towards innovation. Basically, it's very difficult to adopt the innovation process, which required high capital and advance technologies. However, the open innovation is providing the opportunity to SMEs to gain the strong networks with stockholder, inside and outside, such as employees, suppliers, government, universities, and competitors.

These stockholders can help to SMEs to sustain their performance through open innovation. Basically, these stakeholders are providing them load, ideas, latest technologies and trainings to employees, which can generate new ideas, can develop the new product and can find the new market foe products sales.

Therefore, the current study concludes that the internet of things generating and managing the knowledge for open innovation. The open innovation is sustaining the SMEs performance. Furthermore, the present chapter suggested to owner / manager of SMEs should consider the internet things, knowledge management, and open innovation during the decision making for SMEs sustainability. In respect of future recommendations, the present chapter suggested that the future studies should validate the proposed framework empirically in developing countries where the less use of ICTs.

## LIMITATION AND FUTURE RECOMMENDATIONS

The current study has only covered the past literature and proposed the research framework due to time constraint, hence, the future study would validate the proposed research framework empirically, specifically in context of Pakistan, whereas the SMEs are reporting low performance. Secondly, this chapter has discussed the about small and medium firm performance sustainability, the future study can cover the large firms. The present study is discussed the series mediation role between IoT and SMEs sustainability, the future study should converse the parallel mediation as well.

## REFERENCES

Ardolino, M., Rapaccini, M., Saccani, N., Gaiardelli, P., Crespi, G., & Ruggeri, C. (2018). The role of digital technologies for the service transformation of industrial companies. *International Journal of Production Research*, *56*(6), 2116–2132. doi:10.1080/00207543.2017.1324224

Brunswicker, S., & Vanhaverbeke, W. (2015). Open innovation in small and medium-sized enterprises (SMEs): External knowledge sourcing strategies and internal organizational facilitators. *Journal of Small Business Management*, *53*(4), 1241–1263. doi:10.1111/jsbm.12120

Connell, J., Kriz, A., & Thorpe, M. (2014). Industry clusters: An antidote for knowledge sharing and collaborative innovation? *Journal of Knowledge Management*, *18*(1), 137–151. doi:10.1108/JKM-08-2013-0312

de Zubielqui, G. C., Lindsay, N., Lindsay, W., & Jones, J. (2019). Knowledge quality, innovation and firm performance: A study of knowledge transfer in SMEs. *Small Business Economics*, *53*(1), 145–164. doi:10.100711187-018-0046-0

Expósito, A., Fernández-Serrano, J., & Liñán, F. (2019). The impact of open innovation on SMEs' innovation outcomes: New empirical evidence from a multidimensional approach. *Journal of Organizational Change Management*, *32*(5), 558–577. doi:10.1108/JOCM-09-2018-0253

Frishammar, J., Richtnér, A., Brattström, A., Magnusson, M., & Björk, J. (2019). Opportunities and challenges in the new innovation landscape: Implications for innovation auditing and innovation management. *European Management Journal*, *37*(2), 151–164. doi:10.1016/j.emj.2018.05.002

Garcia-Morales, V. J., Martín-Rojas, R., & Lardón-López, M. E. (2018). Influence of social media technologies on organizational performance through knowledge and innovation. *Baltic Journal of Management, 13*(3), 345–367. doi:10.1108/BJM-04-2017-0123

Greco, M., Grimaldi, M., & Cricelli, L. (2016). An analysis of the open innovation effect on firm performance. *European Management Journal, 34*(5), 501–516. doi:10.1016/j.emj.2016.02.008

Huang, M., Bhattacherjee, A., & Wong, C.-S. (2018). Gatekeepers' innovative use of IT: An absorptive capacity model at the unit level. *Information & Management, 55*(2), 235–244. doi:10.1016/j.im.2017.06.001

Lee, S., Park, G., Yoon, B., & Park, J. (2010). Open innovation in SMEs—An intermediated network model. *Research Policy, 39*(2), 290–300. doi:10.1016/j.respol.2009.12.009

Li-Ying, J. (2018). Dual boundary spanning: Toward a typology of outside-in open innovation in the Canadian context. *Canadian Journal of Administrative Sciences [Revue Canadienne des Sciences de l'Administration], 35*(3), 429-443.

Lifshitz-Assaf, H. (2018). Dismantling knowledge boundaries at NASA: The critical role of professional identity in open innovation. *Administrative Science Quarterly, 63*(4), 746–782. doi:10.1177/0001839217747876 PMID:30443046

Miraz, M., Ali, M., Excell, P., & Picking, R. (2018). Internet of nano-things, things and everything: Future growth trends. *Future Internet, 10*(8), 68. doi:10.3390/fi10080068

Papa, A., Dezi, L., Gregori, G. L., Mueller, J., & Miglietta, N. (2018). Improving innovation performance through knowledge acquisition: The moderating role of employee retention and human resource management practices. *Journal of Knowledge Management*. doi:10.1108/JKM-09-2017-0391

Popa, S., Soto-Acosta, P., & Martinez-Conesa, I. (2017). Antecedents, moderators, and outcomes of innovation climate and open innovation: An empirical study in SMEs. *Technological Forecasting and Social Change, 118*, 134–142. doi:10.1016/j.techfore.2017.02.014

Premsankar, G., Di Francesco, M., & Taleb, T. (2018). Edge computing for the Internet of Things: A case study. *IEEE Internet of Things Journal, 5*(2), 1275–1284. doi:10.1109/JIOT.2018.2805263

Santoro, G., Vrontis, D., Thrassou, A., & Dezi, L. (2018). The Internet of Things: Building a knowledge management system for open innovation and knowledge management capacity. *Technological Forecasting and Social Change, 136*, 347–354. doi:10.1016/j.techfore.2017.02.034

Scuotto, V., Del Giudice, M., Bresciani, S., & Meissner, D. (2017). Knowledge-driven preferences in informal inbound open innovation modes. An explorative view on small to medium enterprises. *Journal of Knowledge Management*, *21*(3), 640–655. doi:10.1108/JKM-10-2016-0465

Sun, Y., Liu, J., & Ding, Y. (2019). Analysis of the relationship between open innovation, knowledge management capability and dual innovation. *Technology Analysis and Strategic Management*, 1–14. doi:10.1080/09537325.2019.1632431

Tao, F., Qi, Q., Liu, A., & Kusiak, A. (2018). Data-driven smart manufacturing. *Journal of Manufacturing Systems*, *48*, 157–169. doi:10.1016/j.jmsy.2018.01.006

Tidd, J., & Bessant, J. R. (2018). *Managing innovation: integrating technological, market and organizational change*. John Wiley & Sons.

Van de Vrande, V., De Jong, J. P., Vanhaverbeke, W., & De Rochemont, M. (2009). Open innovation in SMEs: Trends, motives and management challenges. *Technovation*, *29*(6-7), 423–437. doi:10.1016/j.technovation.2008.10.001

# Compilation of References


Abbas, W., & Asghar, I. (2010). The role of leadership in organizatinal change: relating the successful organizational change with visionary and innovative leadership.

Abdennadher, E., & Hellara, S. (2018). Causality and contagion in emerging stock markets. *Borsa Istanbul Review*, *18*(4), 300–311. doi:10.1016/j.bir.2018.07.001

Abel, A. T. L. (2008). *The Development of A Conceptual Framework and Taxonomy for Defining and Classifying Corporate Universities*. New York University. doi:10.1017/CBO9781107415324.004

Adjei, N. K., Jonsson, K. R., & Brand, T. (2018). Time spent on work-related activities, social activities and time pressure as intermediary determinants of health disparities among elderly women and men in 5 European countries: A structural equation model. *International Journal for Equity in Health*, *17*(1), 121. doi:10.118612939-018-0840-y PMID:30115073

Agarwal, R., & Helfat, C. E. (2009). Strategic renewal of organizations. *Organization Science*, *20*(2), 281–293. doi:10.1287/orsc.1090.0423

Agbeibor, W. Jr. (2006). Pro-poor economic growth: Role of small and medium sized enterprises. *Journal of Asian Economics*, *17*(1), 35–40. doi:10.1016/j.asieco.2006.01.005

Aggarwal, I., Woolley, A. W., Chabris, C. F., & Malone, T. W. (2019). The impact of cognitive style diversity on implicit learning in teams. *Frontiers in Psychology*, *10*, 112. doi:10.3389/fpsyg.2019.00112 PMID:30792672

Agrawal, A., Catalini, C., & Goldfarb, A. (2014). Some simple economics of crowdfunding. *Innovation Policy and the Economy*, *14*(1), 63–97. doi:10.1086/674021

Ahlstrom, D., Bruton, G. D., & Yeh, K. S. (2007). Venture capital in China: Past, present, and future. *Asia Pacific Journal of Management*, *24*(3), 247–268. doi:10.100710490-006-9032-1

Ahmed, O. (2018). Relationship between loneliness and mental health among first-year undergraduate students: Mediating role of timeline browsing and chatting on Facebook. *International Journal of Contemporary Education*, *1*(2), 86–94. doi:10.11114/ijce.v1i2.3625

Ahn, J. Y., Kim, S. K., & Han, K. S. (2003). On the design concepts for CRM system. *Industrial Management & Data Systems*, *103*(5), 324–331. doi:10.1108/02635570310477370

Aizenman, J., & Kendall, J. (2008). *The internationalization of venture capital and private equity*. National Bureau of Economic Research. doi:10.3386/w14344

Ajzen, I., & Fishbein, M. (1980). *Understanding attitudes and predicting social behavior*. Englewood Cliffs, CA: Prentice Hall.

Akhavan,P., & Jafari, M .(2006). Learning organizations. the necessity of knowledge era. *Journal of Tadbir*, *169*.

Alagaraja, M., & Li, J. (2015). Utilizing institutional perspectives to investigate the emergence, rise, and (relative) decline of corporate universities. *Human Resource Development International*, *18*(1), 4–23. doi:10.1080/13678868.2014.979003

Al-Azzam, Z. F., Irtaimeh, H. J., & Khaddam, A. A. H. (2017). Examining the Mediating Effect of Strategic Agility in the Relationship between Intellectual Capital and Organizational Excellence in Jordan Service Sector. *The Journal of Business*, *6*(1), 7–15.

Alberti, B. F. (2018). This "modern epidemic": Loneliness as an emotion cluster and a neglected subject in the history of emotions. *Emotion Review*, *10*(3), 242–254. doi:10.1177/1754073918768876

Aldrich, H., Reese, P. R., & Dubini, P. (1989). Women on the verge of a breakthrough: Networking among entrepreneurs in the United States and Italy. *Entrepreneurship and Regional Development*, *1*(4), 339–356. doi:10.1080/08985628900000029

Aleti, T., Brennan, L., & Parker, L. (2015). Family communication for the modern era: A typology. *Young Consumers*, *16*(4), 367–384. doi:10.1108/YC-01-2015-00500

Allen, M. (Ed.). (2002). *The Corporate University Handbook: designing, managing and growing a successful program.* New York: Amacom.

Almirall, E., & Casadesus-Masanell, R. (2010). Open versus closed innovation: A model of discovery and divergence. *Academy of Management Review*, *35*(1), 27–47.

Alotaibi, A. R., & Mishra, A. V. (2017). Time varying international financial integration for GCC stock markets. *The Quarterly Review of Economics and Finance*, *63*, 66–78. doi:10.1016/j.qref.2016.03.001

Aloui, R., Aïssa, M. S. B., & Nguyen, D. K. (2011). Global financial crisis, extreme interdependences, and contagion effects: The role of economic structure? *Journal of Banking & Finance*, *35*(1), 130–141. doi:10.1016/j.jbankfin.2010.07.021

Alshammari, A., Al-Harbi, M., Alhamidah, A., Alreshidi, Y., & Abdulrahim, O. (2017). Association between social media use and depression among Saudi population. *Annals of International Medical and Dental Research*, *4*(1). doi:10.21276/aimdr.2018.4.1.ME7

Alvani, M. (2008). Public Management (32th ed.). Tehran: Ney publication.

Alvarez, E. (2019). Information Sources That Influence the Financial Literacy of Puerto Rican College Students [Doctoral dissertation]. Walden University.

Amar, A. D. (2004). Motivating knowledge workers to innovate: A model integrating motivation dynamics and antecedents. *European Journal of Innovation Management*, *7*(2), 89–101. doi:10.1108/14601060410534366

Amoroso, S., & Link, A. N. (2019). *Intellectual Property Protection Mechanisms and the Characteristics of Founding Teams*. Joint Research Centre.

Andresen, M., & Lichtenberger, B. (2007). The corporate university landscape in Germany. *Journal of Workplace Learning*, *19*(2), 109–123. doi:10.1108/13665620710728484

Aragón-Correa, J. A., García-Morales, V. J., & Cordón-Pozo, E. (2007). Leadership and organizational learning's role on innovation and performance: Lessons from Spain. *Industrial Marketing Management*, *36*(3), 349–359. doi:10.1016/j.indmarman.2005.09.006

Ara, I. (2004). Is Pakistan's manufacturing sector competitive? *Pakistan Development Review*, *43*(4II), 685–706. doi:10.30541/v43i4IIpp.685-706

Ardic, O., Mylemko, N., & Saltane, V. (2012). Access to finance by small and medium enterprises: A cross-country analysis with a new data set. *Pacific Economic Review*, *17*(4), 491–513. doi:10.1111/j.1468-0106.2012.00596.x

Ardolino, M., Rapaccini, M., Saccani, N., Gaiardelli, P., Crespi, G., & Ruggeri, C. (2018). The role of digital technologies for the service transformation of industrial companies. *International Journal of Production Research*, *56*(6), 2116–2132. doi:10.1080/00207543.2017.1324224

Argote, L. (2011). Organizational learning research: Past, present and future. *Management Learning*, *42*(4), 439–446. doi:10.1177/1350507611408217

Argyris, C., & Schon, D. A. (1978). *Organizational Learning: A theory of action perspective*. Reading, MA: Addison Wesley.

Arifio, A. (2003). Measures of Strategic Alliances Performance: An Analysis of Construct Validity. *Journal of International Business Studies*, *34*(1), 66–79. doi:10.1057/palgrave.jibs.8400005

Armstrong, K. (2019). Mastering Motivation. *APS Observer*, *32*(7).

Armstrong, M., & Taylor, S. (2014). *Armstrong's handbook of human resource management practice*. Kogan Page Publishers.

Arnett, J. J. (2006). Socialization in emerging adulthood: From the family to the wider world, from socialization to self-socialization. In *Grusec, J.E., & Hastings, P.D. Handbook of Socialization: Theory and Research*. New York: Guilford Press.

Arpita, K., Ceeba, P., & Ankita, M. (2012). Influence of normative and informative values on fashion clothing involvement of Indian women. *Journal of Customer Behaviour*, *11*, 1, 9–32.

Arthurs, J. D., & Busenitz, L. W. (2003). The boundaries and limitations of agency theory and stewardship theory in the venture capitalist/entrepreneur relationship. *Entrepreneurship Theory and Practice*, *28*(2), 145–162. doi:10.1046/j.1540-6520.2003.00036.x

Asian Development Bank Annual Report. (2018). Asian Development Bank. Retrieved from https://www.adb.org/documents/adb-annual-report-2018

Assunção, A. A., & Abreu, M. N. S. (2019). Pressure to work, health status, and work conditions of schoolteachers in Basic Education in Brazil. *Cadernos de Saude Publica*, 35. PMID:31141023

Astley, W. G. (1984). Toward an appreciation of collective strategy. *Academy of Management Review*, *9*(3), 526–535. doi:10.5465/amr.1984.4279700

Awani, A. (2019). Rakyat Malaysia lebih gemar Google Maps? Astroawani. Retrieved from http://www.astroawani.com/gaya-hidup/rakyat-malaysia-lebih-gemar-google-maps-217093

Ayyagari, M., Beck, T., & Demirgüç-Kunt, A. (2007). Small and medium enterprises across the globe. *Small Business Economics*, *29*(4), 415–434. doi:10.100711187-006-9002-5

Ayyagari, M., Demirguc-Kunt, A., & Maksimovic, V. (2011). *Small vs. young firms across the world: contribution to employment, job creation, and growth*. The World Bank.

Baele, L., Ferrando, A., Hördahl, P., Krylova, E., & Monnet, C. (2004). *Measuring financial integration in the euro area* (No. 14). ECB occasional paper.

Baghel, R., & Nüsser, M. (2015). Securing the heights: The vertical dimension of the Siachen conflict between India and Pakistan in the Eastern Karakoram. *Political Geography*, *48*, 24–36. doi:10.1016/j.polgeo.2015.05.001

Bagozzi, R. P., & Baumgartner, H. (1994). The evaluation of structural equation models and hypothesis testing In R. P. Bagozzi (Ed.), Principles of Marketing Research (pp. 386–422). Cambridge, MA: Blackwell.

Bagozzi, R. P., & Yi, Y. (1988). On the evaluation of structural equation models. *Journal of the Academy of Marketing Science*, *16*(1), 74–94. doi:10.1007/BF02723327

Bain & IVCA. (2018). Perspective of Indian VC ecosystem. Retrieved from https://ivca.in/wp-content/uploads/2018/12/IVCA-Bain-India-VC-report-2018.pdf

Bain, P. G., Mann, L., & Pirola-Merlo, A. (2001). The innovation imperative: The relationships between team climate, innovation, and performance in research and development teams. *Small Group Research*, *32*(1), 55–73. doi:10.1177/104649640103200103

Baker, W. E., & Sinkula, J. M. (2002). Market orientation, learning orientation and product innovation: delving into the organization's black box. *Journal of market-focused management, 5*(1), 5-23.

Baker, T., & Nelson, R. E. (2005). Creating something from nothing: Resource construction through entrepreneurial bricolage. *Administrative Science Quarterly*, *50*(3), 329–366. doi:10.2189/asqu.2005.50.3.329

Baker, W. E., & Sinkula, J. M. (1999). The synergistic effect of market orientation and learning orientation on organizational performance. *Journal of the Academy of Marketing Science*, *27*(4), 411–427. doi:10.1177/0092070399274002

Bakker, M., Leendes, T. A. J., Gabby, S. M., Krazer, J., & Engelen, J. M. L. V. (2006). Is trust really social capital? Knowledge sharing in product development projects. *The Learning Organization*, *13*(6), 594–605. doi:10.1108/09696470610705479

Balasingham, K. (2016). *Industry 4.0: Securing the Future for German Manufacturing Companies*. University of Twente.

Balchi, S., & Olkun, E. O. (2015). The relation between loneliness in social life and Facebook usage. *Proceedings of the 16th International Academic Conference*. Academic Press. doi:10.20472/IAC.2015.016.015

Balli, F., Balli, H. O., & Luu, M. N. (2014). Diversification across ASEAN-wide sectoral and national equity returns. *Economic Modelling*, *41*, 398–407. doi:10.1016/j.econmod.2014.05.035

Banerjee, S., & Mehta, P. (2016). Determining the antecedents of job stress and their impact on job performance: A study among faculty members. *IUP Journal of Organizational Behavior*, *15*(2), 7.

Baporikar, N. (2014). Corporate university edification in knowledge society. *International Journal of Strategic Change Management*, *5*(2), 125–139. doi:10.1504/IJSCM.2014.060258

Baranowski, T., Buday, R., Thompson, D., & Baranowski, J. (2008). Playing for real: Video games and Stories for Health-Related Behavior Change. *American Journal of Preventive Medicine*, *34*(1), 74–82. doi:10.1016/j.amepre.2007.09.027 PMID:18083454

Barber, N. A. (2013). Investigating the potential influence of the internet as a new socialization agent in context with other traditional socialization agents. *Journal of Marketing Theory and Practice*, *21*(2), 179–193. doi:10.2753/MTP1069-6679210204

Barkan, S. E. (2012). *Sociology: Comprehensive Edition.* Creative Commons.

Barlas, B., & Izci, F. B. (2018). Individual and workplace factors related to fatal occupational accidents among shipyard workers in Turkey. *Safety Science*, *101*, 173–179. doi:10.1016/j.ssci.2017.09.012

Barley, K. (2002). Corporate University Structures that Reflect Organizational Structures. In M. Allen (Ed.), *The Corporate University Handbook: designing, managing and growing a successful program.* New York: Amacom.

Baron, S., Patterson, A., Warnaby, G., & Harris, K. (2010). Service-dominant logic: Marketing research implications and opportunities. *Journal of Customer Behaviour*, *9*(3), 253–264. doi:10.1362/147539210X533179

Barry, C. B. (1994). New directions in research on venture capital finance. *Financial Management, 23*(3), 3–15. doi:10.2307/3665617

Barry, C. B., Muscarella, C. J., Peavy Iii, J. W., & Vetsuypens, M. R. (1990). The role of venture capital in the creation of public companies: Evidence from the going-public process. *Journal of Financial Economics, 27*(2), 447–471. doi:10.1016/0304-405X(90)90064-7

Barsky, R. B., & Kilian, L. (2004). Oil and the Macroeconomy since the 1970s. *The Journal of Economic Perspectives, 18*(4), 115–134. doi:10.1257/0895330042632708

Baskaran, N. D., Haron, H., & Valappil, M. (2019). Demographics of healthcare professionals' knowledge and attitude toward deceased organ donation: Survey of critical care areas in a tertiary hospital. *The Medical Journal of Malaysia, 74*(2), 109. PMID:31079120

Bas, T. K., & Teng, B. S. (2000). A Resource-Based Theory of Strategic Alliances. *Journal of Management, 26*(1), 31–61. doi:10.1177/014920630002600105

Bateson, G. (1972). The logical categories of learning and communication. In *Steps to an Ecology of Mind* (pp. 279-308). Academic Press.

Batten, J. A., Kinateder, H., Szilagyi, P. G., & Wagner, N. F. (2019). Time-Varying Energy and Stock Market Integration in Asia. *Energy Economics, 80*, 777–792. doi:10.1016/j.eneco.2019.01.008

Bauer, W., Hämmerle, M., Schlund, S., & Vocke, C. (2015). Transforming to a hyper-connected society and economy–towards an "Industry 4.0". *Procedia Manufacturing, 3*, 417–424.

Baumol, W. J. (1993). *Entrepreneurship, management, and the structure of payoffs*. MIT Press.

Baumrind, D. (1989). Rearing competent children. In W. Damon (Ed.), *Child Development Today and Tomorrow*. San Francisco: Jossey-Bass.

BBC News. (2018). PubG: The story behind one of the world's most popular games. Retrieved from https://www.bbc.com/news/av/newsbeat-42660502/pubg-the-story-behind-one-of-the-world-s-most-popular-games

Beamer, L. (1992). Learning intercultural communication competence. *The Journal of Business Communication, 29*(3), 285-303.

Bearden, W. O., Netemeyer, R. G., & Teel, J. E. (1989). Measurement of consumer susceptibility to interpersonal influence. *The Journal of Consumer Research, 15*(4), 473–481. doi:10.1086/209186

Beck, T., & Demirgüç-Kunt, A. (2006). Small and medium-size enterprises: Access to finance as a growth constant. *Journal of Banking & Finance, 30*(11), 2931–2943. doi:10.1016/j.jbankfin.2006.05.009

Beck, T., Demirgüç-Kunt, A., Laeven, L., & Maksimovic, V. (2006). The determinants of financing obstacles. *Journal of International Money and Finance, 25*(6), 932–952. doi:10.1016/j.jimonfin.2006.07.005

Beck, T., Demirgüç-Kunt, A., & Levine, R. (2005). SMEs, growth, poverty: Cross-country evidence. *Journal of Economic Growth, 10*(3), 197–227. doi:10.100710887-005-3533-5

Beck, T., Demirguc-Kunt, A., & Martinez Peria, M. S. (2008). *Bank financing for SMEs around the world: Drivers, obstacles, business models, and lending practices*. The World Bank.

Beck, T., Demirgüç-Kunt, A., & Singer, D. (2013). Is small Beautiful? Financial structure, size and access to finance. *World Development, 52*, 19–33. doi:10.1016/j.worlddev.2013.05.014

Beirne, J., Caporale, G. M., Schulze-Ghattas, M., & Spagnolo, N. (2013). Volatility spillovers and contagion from mature to emerging stock markets. *Review of International Economics, 21*(5), 1060–1075. doi:10.1111/roie.12091

Bekaert, G., Harvey, C., & Lundblad, C. (2006). Growth volatility and financial liberalization. *Journal of International Money and Finance, 25*(3), 370–403. doi:10.1016/j.jimonfin.2006.01.003

Belke, A. H., Fehn, R., & Foster-McGregor, N. (2003). Does venture capital investment spur employment growth? Centre for European policy studies.

Bendapudi, N., & Leone, P. R. (2003). Psychological implications of customer participation in co-production. *Journal of Marketing, 67*(1), 14–28. doi:10.1509/jmkg.67.1.14.18592

Bengtsson, M., & Kock, S. (2000). "Coopetition" in business Networks—To cooperate and compete simultaneously. *Industrial Marketing Management, 29*(5), 411–426. doi:10.1016/S0019-8501(99)00067-X

Ben-Hur, S., Jaworski, B., & Gray, D. (2015). Aligning Corporate Learning With Strategy. *MIT Sloan Management Review, 57*(1), 1–12.

Berger, A., Demirgüç-Kunt, A., Haubrich, J., & Levine, R. (2004). Bank concentration and competition: An evolution in the making. *Journal of Money, Banking and Credit, 36*(3b), 433–453. doi:10.1353/mcb.2004.0040

Berger, A., & Udell, G. (1998). The economics of small business finance: The role of private equity and debt markets in the financial growth cycle. *Journal of Banking & Finance, 22*(6-8), 613–673. doi:10.1016/S0378-4266(98)00038-7

Berkling, K., & Thomas, C. (2013). Gamification of a software engineering course and a detailed analysis of the factors that lead to it's failure. *Proceedings of the 2013 International Conference on Interactive Collaborative Learning, ICL 2013* (pp. 525–530). Academic Press. 10.1109/ICL.2013.6644642

Berry, L. L., & Carbone, L. (2007). Building loyalty through experience management. *Quality Progress, 40*(9), 26–32.

Bertoni, F., Colombo, M. G., & Quas, A. (2015). The patterns of venture capital investment in Europe. *Small Business Economics, 45*(3), 543–560. doi:10.100711187-015-9662-0

Beugelsdijk, S., & Smeets, R. (2008). Entrepreneurial Culture and Economic Growth: Revisiting McClelland's Thesis. *American Journal of Economics and Sociology, 67*(5), 915-939.

Beverland, M., & Bretherton, P. (2001). The uncertain search for opportunities: Determinants of strategic alliances. *Qualitative Market Research, 4*(2), 88–99. doi:10.1108/13522750110388581

Bhattacharyya, S. S. (2019b). Mergers and Acquisition by Indian Firms: Managerial Perspectives on Challenges and Solutions. In Dynamic Perspectives on Globalization and Sustainable Business in Asia (pp. 41-55). Hershey, PA: IGI Global.

Bhattacharyya, S. S. (2011). Reflections on strategic insights for winning in the complex emerging market of India. *International Journal of Business Excellence, 4*(1), 15–43. doi:10.1504/IJBEX.2011.037247

Bhattacharyya, S. S. (2018). A holistic perspective on strategic alliances for Indian managers. *Journal of Services Research, 18*(1).

Bhattacharyya, S. S. (2019a). Development of an Integrated Model of Strategic Alliance. *Indian Journal of Industrial Relations, 54*(3).

Bhattacharyya, S. S., Rangarajan, R., & Vyas, K. G. (2011). Reflections on mapping chaos in the business organisational landscape. *International Journal of Business Innovation and Research, 6*(1), 76–116. doi:10.1504/IJBIR.2012.044258

Bhattacharyya, S. S., & Shaik, J. (2009). Topic–strategic alliances: Perspectives from Indian life insurance industry. *International Journal of Indian Culture and Business Management*, *3*(1), 98–120. doi:10.1504/IJICBM.2010.029531

Blachnio, A., Przepiorka, A., Wolonciej, M., Mahmoud, A. B., Holdos, J., & Yafi, E. (2018). Loneliness, friendship, and Facebook intrusion. a study in Poland, Slovakia, Syria, Malaysia, and Ecuador. *Studia Psychologica*, *60*(3), 183–194. doi:10.21909p.2018.03.761

Blancher, N., Appendino, M., Bibolov, A., Fouejieu, A., Li, J., Ndoye, A., ... Sydorenko, T. (2019). *Financial inclusion of small and medium-sized enterprises in the Middle East and Central Asia*. Middle East and Central Asia Department, International Monetary Fund. doi:10.5089/9781484383124.087

Blattberg, R. C., Getz, G., & Thomas, J. S. (2001). *Customer Equity: Building and Managing Relationships as Valuable Assets*. Boston, MA: Harvard Business School Press.

Bloch, C., & Bugge, M. M. (2013). Public sector innovation—From theory to measurement. *Structural Change and Economic Dynamics*, *27*, 133–145. doi:10.1016/j.strueco.2013.06.008

Block, J. H., De Vries, G., & Sandner, P. G. (2010). *Venture capital and the financial crisis: An empirical study across industries and countries*. Oxford.

Block, J. H., Colombo, M. G., Cumming, D. J., & Vismara, S. (2018). New players in entrepreneurial finance and why they are there. *Small Business Economics*, *50*(2), 239–250. doi:10.100711187-016-9826-6

Block, J. H., & Sandner, P. G. (2011). Venture capital funding in the middle of the year 2011: Are we back to pre-crisis boom levels? *Strategic Change: Briefings in Entrepreneurial Finance*, *20*(5-6), 161–169. doi:10.1002/jsc.893

Bloomberg. (2018). Malaysia beats emerging market peers as Asia outshines. Retrieved from https://www.nst.com.my/business/2018/11/435823/malaysia-beats-emerging-market-peers-asia-outshines

Bogers, M., Burcharth, A., & Chesbrough, H. (2019). Open innovation in Brazil: Exploring. *Int. J. Innov*, *7*(2), 178–191. doi:10.5585/iji.v7i2.417.2318-9975

Bogers, M., Chesbrough, H., & Moedas, C. (2018). Open innovation: Research, practices, and policies. *California Management Review*, *60*(2), 5–16. doi:10.1177/0008125617745086

Bojica, A. M., Istanbouli, A., & Fuentes-Fuentes, M. D. M. (2014). Bricolage and growth strategies: Effects on the performance of Palestinian women-led firms. *Journal of Developmental Entrepreneurship*, *19*(4). doi:10.1142/S108494671450023X

Bollerslev, T. (1986). Generalized autoregressive conditional heteroskedasticity. *Journal of Econometrics*, *31*(1), 307–327. doi:10.1016/0304-4076(86)90063-1

Borgatti, S. P., & Foster, P. C. (2003). The network paradigm in organizational research: A review and typology. *Journal of Management*, *29*(6), 991–1013. doi:10.1016/S0149-2063(03)00087-4

Bottazzi, L., & Da Rin, M. (2002). Venture capital in Europe and the financing of innovative companies. *Economic Policy*, *17*(34), 229–270. doi:10.1111/1468-0327.00088

Bottazzi, L., Da Rin, M., & Hellmann, T. (2004). The changing face of the European venture capital industry: Facts and analysis. *Journal of Private Equity*, *7*(2), 26–53. doi:10.3905/jpe.2004.391048

Boubakri, S., & Guillaumin, C. (2015). Regional integration of the East Asian stock markets: An empirical assessment. *Journal of International Money and Finance*, *57*, 136–160. doi:10.1016/j.jimonfin.2015.07.011

Bouncken, R. B., & Fredrich, V. (2012). Coopetition: Performance implications and management antecedents. *International Journal of Innovation Management, 16*(05), 1–28. doi:10.1142/S1363919612500284

Bower, J. L. (2001). *Not All M&As Are Alike – and That Matters* (pp. 92–101). Harvard Business School Publishing.

Boyer, K. K., & McDermott, C. (1999). Strategic consensus in operations strategy. *Journal of Operations Management, 17*(3), 289–305. doi:10.1016/S0272-6963(98)00042-4

Brabham, D. C. (2008). Crowdsourcing as a Model for Problem Solving: An Introduction and Cases. *Convergence (London), 14*(1), 75–90. doi:10.1177/1354856507084420

Bradshaw, D., & Brash, C. (2001). Management consumer relationships in the e-business world: How to personalise computer relationships for increased profitability. *International Journal of Retail & Distribution Management, 29*(12), 520–530. doi:10.1108/09590550110696969

Braunscheidel, M. J., & Suresh, N. C. (2009). The organizational antecedents of a firm's supply chain agility for risk mitigation and response. *Journal of Operations Management, 27*(2), 119–140. doi:10.1016/j.jom.2008.09.006

Brettel, M., Friederichsen, N., Keller, M., & Rosenberg, M. (2014). How virtualization, decentralization and network building change the manufacturing landscape: An Industry 4.0 Perspective. *International Journal of Mechanical. Industrial Science and Engineering, 8*(1), 37–44.

Brodie, R. J., Hollebeek, L. D., Jurić, B., & Ilić, A. (2011). Customer engagement: Conceptual domain, fundamental propositions, and implications for research. *Journal of Service Research, 14*(3), 252–271. doi:10.1177/1094670511411703

Bronder, C., & Pritzi, R. (1992). Developing strategic alliances: A conceptual framework for successful co-operation. *European Management Journal, 10*(4), 412–421. doi:10.1016/0263-2373(92)90005-O

Brown, L., & Wood, M. (1953). Paths of Loneliness: The individual isolated in modern society. *American Sociological Review, 18*(6), 712. doi:10.2307/2088145

Brueller, D., & Carmeli, A. (2011). Linking capacities of high-quality relationships to team learning and performance in service organizations. *Human Resource Management, 50*(4), 455–477. doi:10.1002/hrm.20435

Brunswicker, S., & Chesbrough, H. (2018). The Adoption of Open Innovation in Large Firms: Practices, Measures, and Risks A survey of large firms examines how firms approach open innovation strategically and manage knowledge flows at the project level. *Research Technology Management, 61*(1), 35–45. doi:10.1080/08956308.2018.1399022

Brunswicker, S., & Vanhaverbeke, W. (2015). Open innovation in small and medium-sized enterprises (SMEs): External knowledge sourcing strategies and internal organizational facilitators. *Journal of Small Business Management, 53*(4), 1241–1263. doi:10.1111/jsbm.12120

Bruton, G. D., Ahlstrom, D., & Puky, T. (2009). Institutional differences and the development of entrepreneurial ventures: A comparison of the venture capital industries in Latin America and Asia. *Journal of International Business Studies, 40*(5), 762–778. doi:10.1057/jibs.2009.3

Bruton, G. D., Fried, V. H., & Manigart, S. (2005). Institutional influences on the worldwide expansion of venture capital. *Entrepreneurship Theory and Practice, 29*(6), 737–760. doi:10.1111/j.1540-6520.2005.00106.x

BTRC. (2019). Internet Subscribers in Bangladesh June, 2019. Retrieved from http://www.btrc.gov.bd/telco/internet

Bucklin, L. P., & Sengupta, S. (1993). Organizing successful co-marketing alliances. *Journal of Marketing, 57*(2), 32–46. doi:10.1177/002224299305700203

Bukowski, W. M., Brendgen, M., & Vitaro, F. (2006). Peers and socialization: Effects on externalizing and internalizing problems. In *Grusec, J.E., & Hastings, P.D. Handbook of Socialization: Theory and Research*. New York: Guilford Press.

Burrow, A., & Rainone, N. (2017). How many likes did I get?: Purpose moderates links between positive social media feedback and self-esteem. *Journal of Experimental Social Psychology*, *69*, 232–236. doi:10.1016/j.jesp.2016.09.005

Burżacka, M., & Gąsiorowska, E. (2016). The importance of venture capital financing of start-up companies. *Forum Scientiae Oeconomia*, *4*(3), 141–150.

Butler, C. J. (2008). Problems in global strategic alliance management for European defence manufacturing firms. *Management Decision*, *46*(2), 330–341. doi:10.1108/00251740810854186

Butler, C. J. (2010). Internal and lateral communication in strategic alliance decision making. *Management Decision*, *48*(5), 698–712. doi:10.1108/00251741011043885

Butler, J. K. Jr. (1991). Toward understanding and measuring conditions of trust: Evolution of a conditions of trust inventory. *Journal of Management*, *17*(3), 643–663. doi:10.1177/014920639101700307

Buttner, E. H., & Rosen, B. (1992). Rejection in the loan application process: Male and female entrepreneurs' perceptions and subsequent intentions. *Journal of Small Business Management*, *30*(1), 58.

Cacioppo, J., & Patrick, W. (2009). *Loneliness: Human Nature and the Need for Social Connection*. New York: Norton. Retrieved from https://psycnet.apa.org/record/2008-07755-000

Calindo, A., & Schiantarelli, F. (Eds.). (2003). Credit constraints and investment in Latin America. Inter-American Development Bank, Washington, DC.

Callison, J. W., & Vestal, A. W. (2010). The L3C Illusion: Why Low-Profit Limited Liability Companies Will Not Stimulate Socially Optimal Private Foundation Investment in Entrepreneurial Ventures. *Vermont Law Review*, *35*, 273.

Canary, D. J., Cody, M. J., & Manusov, V. L. (2008). Interpersonal communication competence. In *Interpersonal communication: A goals-based approach* (pp. 502-520). Academic Press.

Canzaniello, A., Hartmann, E., & Fifka, M. S. (2017). Intra-industry strategic alliances for managing sustainability-related supplier risks: Motivation and outcome. *International Journal of Physical Distribution & Logistics Management*, *47*(5), 387–409. doi:10.1108/IJPDLM-01-2016-0034

Capon, N., Farley, J. U., & Hoenig, S. (1990). Determinants of financial performance: A meta-analysis. *Management Science*, *36*(10), 1143–1159. doi:10.1287/mnsc.36.10.1143

Cappiello, L., Engle, R. F., & Sheppard, K. (2006). Asymmetric dynamics in the correlations of global equity and bond returns. *Journal of Financial Econometrics*, *4*(4), 537–572. doi:10.1093/jjfinec/nbl005

Carducci, B. J., & Deuser, P. S. (1984). The foot-in-the donor technique: Initial request and organ donation. *Basic and Applied Social Psychology*, *5*(1), 75–81. doi:10.120715324834basp0501_5 PMID:11653697

Carlos Nunes, J., Gomes Santana Félix, E., & Pacheco Pires, C. (2014). Which criteria matter most in the evaluation of venture capital investments? *Journal of Small Business and Enterprise Development*, *21*(3), 505–527. doi:10.1108/JSBED-10-2013-0165

Cassidy, J., & Asher, S. R. (1992). Loneliness and peer relations in young children. *Child Development*, *63*(2), 350–365. doi:10.2307/1131484 PMID:1611939

Cavusgil, S. T., Calantone, R. J., & Zhao, Y. (2003). Tacit knowledge transfer and firm innovation capability. *Journal of Business and Industrial Marketing*, *18*(1), 6–21. doi:10.1108/08858620310458615

CB Insights. (2018). Venture Capital Funding Report Q4. Retrieved from https://www.cbinsights.com/research/report/venture-capital-q4-2018/

Cecen, A. R. (2008). The effects of gender and loneliness levels on ways of coping among university students. *College Student Journal*, *42*, 510–516. Retrieved from https://eric.ed.gov/?id=EJ816934

Černe, M., Jaklič, M., & Škerlavaj, M. (2013). Authentic leadership, creativity, and innovation: A multilevel perspective. *Leadership*, *9*(1), 63–85. doi:10.1177/1742715012455130

Chae, H.-C., Koh, C. E., & Prybutok, V. R. (2014). Information technology capability and firm performance: Contradictory findings and their possible causes. *Management Information Systems Quarterly*, *38*(1), 305–326. doi:10.25300/MISQ/2014/38.1.14

Chandler, J. D., & Lusch, R. F. (2015). Service Systems: A Broadened Framework and Research Agenda on Value Propositions, Engagement, and Service Experience. *Journal of Service Research*, *18*(1), 6–22. doi:10.1177/1094670514537709

Chand, M., & Katou, A. A. (2012). Strategic determinants for the selection of partner alliances in the Indian tour operator industry: A cross-national study. *Journal of World Business*, *47*(2), 167–177. doi:10.1016/j.jwb.2011.04.003

Chan, S. H., Kensinger, J. W., Keown, A. J., & Martin, J. D. (1997). Do strategic alliances create value? *Journal of Financial Economics*, *46*(2), 199–221. doi:10.1016/S0304-405X(97)00029-9

Chan, Y. E. (2000). IT value: The great divide between qualitative and quantitative and individual and organizational measures. *Journal of Management Information Systems*, *16*(4), 225–261. doi:10.1080/07421222.2000.11518272

Chaudhuri, A., & Holbrook, M. B. (2001). The chain of effects from brand trust and brand affect to brand performance: The role of brand loyalty. *Journal of Marketing*, *65*(2), 81–93. doi:10.1509/jmkg.65.2.81.18255

Chavis, L., Klapper, L., & Love, I. (2010). The impact of the business environment on young firm financing. World Bank.

Cheah, W. K., & Zolkepli, I. A. (2018). Crowdsourcing and Service-Dominant Logic : The Interaction Effect of Brand Familiarity, Customer Experience an..... *Asia Pacific Journal of Advanced Business and Social Studies*, *4*(1), 154–162. doi:10.25275/apjabssv4i1bus16

Cheng, J. H., Yeh, C. H., & Tu, C. W. (2008). Trust and knowledge sharing in green supply chains. *Supply Chain Management*, *13*(4), 283–295. doi:10.1108/13598540810882170

Chen, P. T., & Cheng, J. Z. (2010). Unlocking the promise of mobile value-added services by applying new collaborative business models. *Technological Forecasting and Social Change*, *77*(4), 678–693. doi:10.1016/j.techfore.2009.11.007

Chen, S. H., Lee, H. T., & Wu, Y. F. (2008). Applying ANP approach to partner selection for strategic alliance. *Management Decision*, *46*(3), 449–465. doi:10.1108/00251740810863889

Chen, Z., Duan, L. Y., Wang, S., Lou, Y., Huang, T., Wu, D. O., & Gao, W. (2019). Towards knowledge as a service over networks: A deep learning model communication paradigm. *IEEE Journal on Selected Areas in Communications*, *37*(6), 1349–1363. doi:10.1109/JSAC.2019.2904360

Cherif, M., & Gazdar, K. (2011). What drives venture capital investments in Europe? New results from a panel data analysis. *Journal of Applied Business and Economics*, *12*(3), 122–139.

Chesbrough, H. (2003). *Open innovation*. Boston: Harvard Business School Press.

Chesbrough, H. (2012). Open innovation: Where we've been and where we're going. *Research Technology Management*, *55*(4), 20–27. doi:10.5437/08956308X5504085

Chesbrough, H. W. (2006). *Open innovation: The new imperative for creating and profiting from technology*. Harvard Business Press.

Chevallier, J., Nguyen, D. K., Siverskog, J., & Uddin, G. S. (2018). Market integration and financial linkages among stock markets in Pacific Basin countries. *Journal of Empirical Finance*, *46*, 77–92. doi:10.1016/j.jempfin.2017.12.006

Chilton, D. (2000). Teens excel at media multi-tasking. *Marketing Magazine*, *105*(28), 4.

Chishti, S., & Barberis, J. (2016). *The FinTech book: the financial technology handbook for investors, entrepreneurs and visionaries*. John Wiley & Sons. doi:10.1002/9781119218906

Chiu, Y. J., Chen, H. C., Tzeng, G. H., & Shyu, J. Z. (2006). Marketing strategy based on customer behaviour for the LCD-TV. *International Journal and Decision Making*, *7*(2/3), 143–165. doi:10.1504/IJMDM.2006.009140

Choe, K. H., & Yap, C. Y. (2018). Exploring Malaysian Attendee Attitude towards Green Practices in Exhibitions. *Proceedings of the 2017 Tokyo International Conference on Hospitality, Tourism, and Sports Management* (pp. 78–92). Academic Press.

Choi, D. H., Kim, J., & Kim, S. H. (2007). ERP training with a web-based electronic learning system: The flow theory perspective. *International Journal of Human-Computer Studies*, *65*(3), 223–243. doi:10.1016/j.ijhcs.2006.10.002

Choi, T. Y., & Krause, D. R. (2006). The supply base and its complexity: Implications for transaction costs, risks, responsiveness, and innovation. *Journal of Operations Management*, *24*(5), 637–652. doi:10.1016/j.jom.2005.07.002

Cho, S., Mathiassen, L., & Robey, D. (2007). Dialectics of Resilience: A Multi-level Analysis of a Telehealth Innovation. *Journal of Information Technology*, *22*(1), 24–35. doi:10.1057/palgrave.jit.2000088

Chowdhury, M. A. F., Haque, M. M., & Islam, M. N. (2017). Contagion effects on stock market of Bangladesh: An empirical study on Dhaka stock exchange Shariah (DSES) index. *International Journal of Asian Business and Information Management*, *8*(2), 114. doi:10.4018/IJABIM.2017040101

Christopher, M. (2000). The agile supply chain: Competing in volatile markets. *Industrial Marketing Management*, *29*(1), 37–44. doi:10.1016/S0019-8501(99)00110-8

Chung, H. F. (2011). Market orientation, guanxi, and business performance. *Industrial Marketing Management*, *40*(4), 522–533. doi:10.1016/j.indmarman.2010.12.008

Chung, H. F., Yang, Z., & Huang, P.-H. (2015). How does organizational learning matter in strategic business performance? The contingency role of guanxi networking. *Journal of Business Research*, *68*(6), 1216–1224. doi:10.1016/j.jbusres.2014.11.016

Chung, Q. B., Luo, W., & Wagner, W. P. (2006). Strategic alliance of small firms in knowledge industries: A management consulting perspective. *Business Process Management Journal*, *12*(2), 206–233. doi:10.1108/14637150610657549

Clarke, G., Cull, R., Soledad, M., Peria, M., & Sanchez, S. (2003). Foreign bank entry: Experience, implications for developing economies and agenda for further research. *The World Bank Research Observer*, *18*(1), 25–59. doi:10.1093/wbro/lkg002

Claro, D. P., Claro, P. B. D. O., & Hagelaar, G. (2006). Coordinating collaborative joint efforts with suppliers: The effects of trust, transaction specific investment and information network in the Dutch flower industry. *Supply Chain Management*, *11*(3), 216–224. doi:10.1108/13598540610662112

Cleland, D. I. (1999). *Project management: strategic design and implementation*. McGraw-Hill.

Coffeng, J. K., Hendriksen, I. J., Duijts, S. F., Twisk, J. W., van Mechelen, W., & Boot, C. R. (2014). Effectiveness of a combined social and physical environmental intervention on presenteeism, absenteeism, work performance, and work engagement in office employees. *Journal of Occupational and Environmental Medicine*, *56*(3), 258–265. doi:10.1097/JOM.0000000000000116 PMID:24603201

Cohen, S. (2004). Social relationships and health. *The American Psychologist*, *59*(8), 676–684. doi:10.1037/0003-066X.59.8.676 PMID:15554821

Cohen, W. M., & Levinthal, D. A. (1989). Innovation and learning: The two faces of R & D. *Economic Journal (London)*, *99*(397), 569–596. doi:10.2307/2233763

Cohen, W. M., & Levinthal, D. A. (2000). Absorptive capacity: A new perspective on learning and innovation. In *Strategic Learning in a Knowledge economy* (pp. 39–67). Elsevier. doi:10.1016/B978-0-7506-7223-8.50005-8

Coleman, J. S. (1994). *Foundations of social theory*. Harvard university press.

Cole, S., Capitanio, J., Chun, K., Arevalo, J., Ma, J., & Cacioppo, J. (2015). Myeloid differentiation architecture of leukocyte transcriptase dynamics in perceived social isolation. *Proceedings of the National Academy of Sciences of the United States of America*, *112*(49), 15142–15147. doi:10.1073/pnas.1514249112 PMID:26598672

Colombo, M. G., & Grilli, L. (2010). On growth drivers of high-tech start-ups: Exploring the role of founders' human capital and venture capital. *Journal of Business Venturing*, *25*(6), 610–626. doi:10.1016/j.jbusvent.2009.01.005

Comrey, A., & Lee, H. (1992). A First Course in Factor Analysis (2nd ed.). Hillsdale, NJ: Lawrence Earlbaum Associates.

Conboy, K., & Morgan, L. (2011). Beyond the customer: Opening the agile systems development process. *Information and Software Technology*, *53*(5), 535–542. doi:10.1016/j.infsof.2010.10.007

Connell, J., Kriz, A., & Thorpe, M. (2014). Industry clusters: An antidote for knowledge sharing and collaborative innovation? *Journal of Knowledge Management*, *18*(1), 137–151. doi:10.1108/JKM-08-2013-0312

Cornelius, B. (2005). The institutionalization of venture capital. *Technovation*, *25*(6), 599–608. doi:10.1016/j.technovation.2003.12.001

Costa, A. C. (2002). The role of trust for the functioning of teams in organisations. Delft University.

Craven, J. (2019). *A Primer on Green Architecture and Green Design*. ThoughtCo. Retrieved from https://www.thoughtco.com/what-is-green-architecture-and-green-design-177955

Crossan, M. M., Lane, H. W., & White, R. E. (1999). An organizational learning framework: From intuition to institution. *Academy of Management Review*, *24*(3), 522–537. doi:10.5465/amr.1999.2202135

Csikszenmihalyi, M. (1990). *Flow: The Psychology of Optimal Experience*. New York: Harper & Row.

Cull, R., & Xu, L. (2005). Institutions, ownership, and finance: The determinants of profit reinvestment among Chinese firms. *Journal of Financial Economics*, *77*(1), 117–146. doi:10.1016/j.jfineco.2004.05.010

Cumming, D. J., Grilli, L., & Murtinu, S. (2017). Governmental and independent venture capital investments in Europe: A firm-level performance analysis. *Journal of Corporate Finance*, *42*, 439–459. doi:10.1016/j.jcorpfin.2014.10.016

Cumming, D. J., & MacIntosh, J. G. (2003). A cross-country comparison of full and partial venture capital exits. *Journal of Banking & Finance*, *27*(3), 511–548. doi:10.1016/S0378-4266(02)00389-8

Cummings, J. L., & Holmberg, S. R. (2012). Best-fit alliance partners: The use of critical success factors in a comprehensive partner selection process. *Long Range Planning*, *45*(2-3), 136–159. doi:10.1016/j.lrp.2012.01.001

Cunha, M. (2005). Bricolage in organizations. *SSRN Electronic Journal.*

Cunha, M. P., Rego, A., Oliveira, P., Rosado, P., & Habib, N. (2014). Product innovation in resource-poor environments: Three research streams. *Journal of Product Innovation Management, 31*(2), 202–210. doi:10.1111/jpim.12090

Cunningham, R. (1956). Brand loyalty- what where how much? *Harvard Business Review, 34*(1), 116–128.

Da Rin, M., Nicodano, G., & Sembenelli, A. (2006). Public policy and the creation of active venture capital markets. *Journal of Public Economics, 90*(8-9), 1699–1723. doi:10.1016/j.jpubeco.2005.09.013

Da Xu, L., He, W., & Li, S. (2014). Internet of things in industries: A survey. *IEEE Transactions on Industrial Informatics, 10*(4), 2233–2243. doi:10.1109/TII.2014.2300753

Daellenbach, U. S., McCarthy, A. M., & Schoenecker, T. S. (1999). Commitment to innovation: The impact of top management team characteristics. *Research Management, 29*(3), 199–208.

Daft, R. L., & Weick, K. E. (1984). Toward a model of organizations as interpretation systems. *Academy of Management Review, 9*(2), 284–295. doi:10.5465/amr.1984.4277657

Dainton, M., & Zelley, E. D. (2004). Explaining theories of interpersonal communication.

Darroch, J., & McNaughton, R. (2002). Examining the link between knowledge management practices and types of innovation. *Journal of Intellectual Capital, 3*(3), 210–222. doi:10.1108/14691930210435570

Das, T. K., & Teng, B. S. (1998). Resource and risk management in the strategic alliance making process. *Journal of Management, 24*(1), 21–42. doi:10.1177/014920639802400103

Das, T. K., & Teng, B. S. (2001). Trust, control and risk in strategic alliances: An integrated framework. *Organization Studies, 22*(2), 251–283. doi:10.1177/0170840601222004

Dastmalchian, A., & Blyton, P. (1998). Organizational flexibility in cross-national perspective: An introduction. *International Journal of Human Resource Management, 9*(3), 437–444. doi:10.1080/095851998340991

Davis, J. E. (2002). *Stories of change: Narrative and social movements.* New York: State University of New York Press.

Davoudi, S. M. M., Fartash, K., Venera, G. Z., Asiya, M. B., Rashad, A. K., Anna, V. B., & Zhanna, M. S. (2018). Testing the Mediating Role of Open Innovation on the Relationship between Intellectual Property Rights and Organizational Performance: A Case of Science and Technology Park. *EURASIA Journal of Mathematics Science and Technology Education, 14*(4), 1359–1369.

De Clercq, D., Fried, V. H., Lehtonen, O., & Sapienza, H. J. (2006). An entrepreneur's guide to the venture capital galaxy. *The Academy of Management Perspectives, 20*(3), 90–112. doi:10.5465/amp.2006.21903483

De Haas, R., & Naaborg, I. (2005). Does foreign bank entry reduce small firms' access to credit? Evidence from European transition economies. De Nederlandsche Bank.

De la Dehesa, G. (2002). *Venture capital in the United States and Europe.* Group of Thirty.

de Sousa Jabbour, A. B. L., Jabbour, C. J. C., Godinho Filho, M., & Roubaud, D. (2018). Industry 4.0 and the circular economy: A proposed research agenda and original roadmap for sustainable operations. *Annals of Operations Research*, 1–14.

de Zubielqui, G. C., Lindsay, N., Lindsay, W., & Jones, J. (2019). Knowledge quality, innovation and firm performance: A study of knowledge transfer in SMEs. *Small Business Economics, 53*(1), 145–164. doi:10.100711187-018-0046-0

Deeds, D. L., & Hill, C. W. (1996). Strategic alliances and the rate of new product development: An empirical study of entrepreneurial biotechnology firms. *Journal of Business Venturing*, *11*(1), 41–55. doi:10.1016/0883-9026(95)00087-9

DeGennaro, R. P. (2010). Angel Investors: Who They Are and What They Do; Can I Be One, Too? *The Journal of Wealth Management*, *13*(2), 55–60.

Del Vecchio, P., Secundo, G., Rubino, M., Garzoni, A., & Vrontis, D. (2019). Open innovation in family firms: Empirical evidence about internal and external knowledge flows. *Business Process Management Journal.*

Delafrooz, N., Rahmati, Y., & Abdi, M. (2019). The influence of electronic word of mouth on Instagram users: An emphasis on consumer socialization framework. *Cogent Business & Management*, *6*(1), 1606973. doi:10.1080/23311975.2019.1606973

Demerouti, E., Bakker, A. B., & Bulters, A. J. (2004). The loss spiral of work pressure, work–home interference and exhaustion: Reciprocal relations in a three-wave study. *Journal of Vocational Behavior*, *64*(1), 131–149. doi:10.1016/S0001-8791(03)00030-7

Demirgüç-Kunt, A., & Maksimovic, V. (1998). Law, finance, and firm growth. *The Journal of Finance*, *53*(6), 2107–2137. doi:10.1111/0022-1082.00084

Demir, Y. (2016). The relationship between loneliness and depression: Mediation role of internet addiction. *Educational Process International Journal*, *5*(2), 97–105. doi:10.12973/edupij.2016.52.1

Deng, Z., Lu, Y., Wei, K. K., & Zhang, J. (2010). Understanding customer satisfaction and loyalty: An empirical study of mobile instant messages in China. *International Journal of Information Management*, *30*(4), 289–300. doi:10.1016/j.ijinfomgt.2009.10.001

Deorah, K. (2015). *The Golden Tap: The Inside Story of Hyper-funded Indian Start-ups*. Lotus Collection, Roli Books.

DeOrtentiis, P. S., Summers, J. K., Ammeter, A. P., Douglas, C., & Ferris, G. R. (2013). Cohesion and satisfaction as mediators of the team trust–team effectiveness relationship: An interdependence theory perspective. *Career Development International*, *18*(5), 521–543. doi:10.1108/CDI-03-2013-0035

Deterding, S. (2011). Situated motivational affordances of game elements: A conceptual model. *Proceeding of ACM CHI Conference on Human Factors in Computing Systems*. ACM.

Deters, F., & Mehl, M. (2012). Does posting Facebook status update increase or decrease loneliness? An online social networking experiment. *Social Psychological & Personality Science*, *4*(5), 579–586. doi:10.1177/1948550612469233

DeTienne, D. R., McKelvie, A., & Chandler, G. N. (2015). Making sense of entrepreneurial exit strategies: A typology and test. *Journal of Business Venturing*, *30*(2), 255–272. doi:10.1016/j.jbusvent.2014.07.007

Dev, C. S., Klein, S., & Fisher, R. A. (1996). A market-based approach for partner selection in marketing alliances. *Journal of Travel Research*, *35*(1), 11–17. doi:10.1177/004728759603500103

Dewandaru, G., Masih, R., & Masih, A. M. M. (2016). Contagion and interdependence across Asia-Pacific equity markets: An analysis based on multi-horizon discrete and continuous wavelet transformations. *International Review of Economics & Finance*, *43*, 363–377. doi:10.1016/j.iref.2016.01.002

Dewandaru, G., Masih, R., & Masih, M. (2018). Unraveling the Financial Contagion in European Stock Markets During Financial Crises: Multi-Timescale Analysis. *Emerging Markets Finance & Trade*, *54*(4), 859–880. doi:10.1080/1540496X.2016.1266614

Dhingra, V., & Dhingra, M. (2017). Industry Framework of Indian Art-Metalware Handicraft Industry: A Case Study. *IUP Journal of Business Strategy, 14*(1), 52.

Di Domenico, M., Haugh, H., & Tracey, P. (2010). Social bricolage: Theorizing social value creation in social enterprises. *Entrepreneurship Theory and Practice, 34*(4), 681–703. doi:10.1111/j.1540-6520.2010.00370.x

Dick, A. S., & Basu, K. (1994). Customer loyalty: Toward an integrated conceptual framework. *Journal of the Academy of Marketing Science, 22*(2), 99–113. doi:10.1177/0092070394222001

Dickey, D. A., & Fuller, W. A. (1979). Distribution of the estimates for autoregressive time series with a unit root. *Journal of the American Statistical Association, 74*, 427–482.

Diestel, S., Wegge, J., & Schmidt, K. H. (2014). The impact of social context on the relationship between individual job satisfaction and absenteeism: The roles of different foci of job satisfaction and work-unit absenteeism. *Academy of Management Journal, 57*(2), 353–382. doi:10.5465/amj.2010.1087

Digan, S. P., Sahi, G. K., Mantok, S., & Patel, P. C. (2019). Women's Perceived Empowerment in Entrepreneurial Efforts: The Role of Bricolage and Psychological Capital. *Journal of Small Business Management, 57*(1), 206–229. doi:10.1111/jsbm.12402

Dodgson, M. (1993). Learning, trust, and technological collaboration. *Human Relations, 46*(1), 77–95. doi:10.1177/001872679304600106

Dossani, R., & Kenney, M. (2002). Creating an environment for venture capital in India. *World Development, 30*(2), 227–253. doi:10.1016/S0305-750X(01)00110-3

Dove, R. (1991). *The 21s Century Manufacturing Enterprise Strategy. Iacocca Institute*, Lehigh University.

Dove, R. (2002). *Response ability: the language, structure, and culture of the agile enterprise*. John Wiley & Sons.

Dow Jones. (2018). Venture Capital Report, China Q2. Retrieved from https://images.dowjones.com/wp-content/uploads/sites/43/2018/07/13022934/2Q18-VS-China-final1.pdf

Downs, G. W. Jr, & Mohr, L. B. (1979). Toward a theory of innovation. *Administration & Society, 10*(4), 379–408. doi:10.1177/009539977901000401

Doz, Y. L. (1996). The evolution of cooperation in strategic alliance: Initial conditions or learning processes? *Strategic Management Journal, 17*(1), 55–83.

Doz, Y. L., & Kosonen, M. (2010). Embedding strategic agility: A leadership agenda for accelerating business model renewal. *Long Range Planning, 43*(2-3), 370–382. doi:10.1016/j.lrp.2009.07.006

Drengner, J., Jahn, S., & Furchheim, P. (2018). *Flow revisited: process conceptualization and a novel application to service contexts. Journal of Service Management, 29*. doi:10.1108/JOSM-12-2016-0318

Drucker, P. F. (1999). Knowledge-Worker Productivity: The Biggest Challenge. *California Management Review, 41*(2), 79–94. doi:10.2307/41165987

Dubow, E. F., Huesmann, L. R., & Greenwood, D. (2006). Media and youth socialization: Underlying processes and moderators of effects. In *J.E. Grusec & P.D. Hastings (Eds.), Handbook of Socialization: Theory and Research*. New York: Guilford Press.

Dündar, Ş. (2017). The Effects of Conscious Consumerism-Related Parental Role Modeling, Parental Communication, and Social Studies on Children's Socialization as Conscious Consumers. *Mersin University Journal of the Faculty of Education, 13*(2).

Duverger, P. (2015). Crowdsourcing innovative service ideas. *Journal of Hospitality and Tourism Technology*, *6*(3), 228–241. doi:10.1108/JHTT-10-2014-0063

Dwayne, B., Pedro, C. S., & Manuel, V. J. (2006). Service personalization and loyalty. *Journal of Services Marketing*, *20*(6), 391–403. doi:10.1108/08876040610691284

Dyre, J. H., Kale, P., & Singh, H. (2014). When to Ally and When to Acquire. *Harvard Business Review*, 3–10.

Edmondson, A., & Moingeon, B. (1999). Learning, trust and organizational change. *Organizational learning and the learning organization*, 157-175.

Edmondson, A. C., & Mogelof, J. P. (2006). Explaining psychological safety in innovation teams: organizational culture, team dynamics, or personality? In *Creativity and innovation in organizational teams* (pp. 129–156). Psychology Press.

Edmondson, A., & Moingeon, B. (1999). Learning trust and organizational change. In M. Easterby-Smith, J. Burgoyne, & L. Araujo (Eds.), *Organizational Learning and the Learning Organisation* (pp. 157–175). Sage.

Edmund, L. W. J., Shazlyiana, H. N., Nursyahirah, N. F., Fam, J. C., & Chua, J. (2018). Factors Affecting Attitude and Behaviour towards Organ Donation among Medical Students in Malaysia: A Cross-Sectional Study. *Journal of Advances in Medicine and Medical Research*, 1-13.

Ehrlich, S. B., De Noble, A. F., Moore, T., & Weaver, R. R. (1994). After the cash arrives: A comparative study of venture capital and private investor involvement in entrepreneurial firms. *Journal of Business Venturing*, *9*(1), 67–82. doi:10.1016/0883-9026(94)90027-2

Eisenberger, R., Fasolo, P., & Davis-LaMastro, V. (1990). Perceived organizational support and employee diligence, commitment, and innovation. *The Journal of Applied Psychology*, *75*(1), 51–59. doi:10.1037/0021-9010.75.1.51

Eisenhardt, K. M., & Martin, J. A. (2000). Dynamic capabilities: What are they? *Strategic Management Journal*, *21*(10-11), 1105–1121. doi:10.1002/1097-0266(200010/11)21:10/11<1105::AID-SMJ133>3.0.CO;2-E

Elia, S., Messeni Petruzzelli, A., & Piscitello, L. (2019). The impact of cultural diversity on innovation performance of MNC subsidiaries in strategic alliances. *Journal of Business Research*, *98*, 204–213. doi:10.1016/j.jbusres.2019.01.062

Elmuti, D., & Kathawala, Y. (2001). An overview of strategic alliances. *Management Decision*, *39*(3), 205–217. doi:10.1108/EUM0000000005452

El-Tannir, A. A. (2002). The corporate university model for continuous learning, training and development. *Education + Training*, *44*(2), 76–81. doi:10.1108/00400910210419973

Engle, R. (2002). Dynamic conditional correlation: A simple class of multivariate generalized autoregressive conditional heteroskedasticity models. *Journal of Business & Economic Statistics*, *20*(3), 339–350. doi:10.1198/073500102288618487

Enkel, E., Gassmann, O., & Chesbrough, H. (2009). Open R&D and open innovation: Exploring the phenomenon. *Research Management*, *39*(4), 311–316.

Erdem, F., & İşbaşı, J. Ö. (2000). Takım Çalışmalarında Güven ve Güvensizlik: Performans için Koşulsuz Güven mi, Optimum Güven mi? *Erciyes Üniversitesi*, *8*, 25–27.

Erdem, F., Ozen, J., & Atsan, N. (2003). The relationship between trust and team performance. *Work Study*, *52*(7), 337–340. doi:10.1108/00438020310502633

Ernst and Young. (2018). Wellness Industry Report. Retrieved from https://www.indiaservices.in/medical

Ewer, G., & Russ-Eft, D. (2017). 35 Corporate University Theory and Practice: the Case of Platt University, USA. *International Journal of Human Resource Development: Practice. Policy & Research, 2*(1), 35–49. doi:10.22324/ijhrdppr.2.104

Expósito, A., Fernández-Serrano, J., & Liñán, F. (2019). The impact of open innovation on SMEs' innovation outcomes: New empirical evidence from a multidimensional approach. *Journal of Organizational Change Management, 32*(5), 558–577. doi:10.1108/JOCM-09-2018-0253

Fairbank, J. F., & Williams, S. D. (2001). Motivating creativity and enhancing innovation through employee suggestion system technology. *Creativity and Innovation Management, 10*(2), 68–74. doi:10.1111/1467-8691.00204

Faraj, S., & Sproull, L. (2000). Coordinating expertise in software development teams. *Management Science, 46*(12), 1554–1568. doi:10.1287/mnsc.46.12.1554.12072

Feldman, M., Francis, J., & Bercovitz, J. (2005). Creating a cluster while building a firm: Entrepreneurs and the formation of industrial clusters. *Regional Studies, 39*(1), 129–141. doi:10.1080/0034340052000320888

Ferris, G. R., Treadway, D. C., Perrewé, P. L., Brouer, R. L., Douglas, C., & Lux, S. (2007). Political skill in organizations. *Journal of Management, 33*(3), 290–320. doi:10.1177/0149206307300813

Fey, C. F., & Beamish, P. W. (2000). Joint venture conflict: The case of Russian international joint ventures. *International Business Review, 9*(2), 139–162. doi:10.1016/S0969-5931(99)00034-7

Fiol, C. M., & Lyles, M. A. (1985). Organizational learning. *Academy of Management Review, 10*(4), 803–813. doi:10.5465/amr.1985.4279103

Fisher, G. (2012). Effectuation, causation, and bricolage: A behavioral comparison of emerging theories in entrepreneurship research. *Entrepreneurship Theory and Practice, 36*(5), 1019–1051. doi:10.1111/j.1540-6520.2012.00537.x

Florida, R. L., & Kenney, M. (1988). Venture capital-financed innovation and technological change in the USA. *Research Policy, 17*(3), 119–137. doi:10.1016/0048-7333(88)90038-8

Følstad, A., & Kvale, K. (2018). Customer journeys: A systematic literature review. *Journal of Service Theory and Practice, 28*(2), 196–227. doi:10.1108/JSTP-11-2014-0261

Fong, K. H., & Snape, E. (2015). Empowering leadership, psychological empowerment and employee Outcomes: Testing a multi-level mediating model. *British Journal of Management, 26*(1), 126–138. doi:10.1111/1467-8551.12048

Fontela, E., & Gabus, A. (Eds.). (1976). *The DEMATEL Observe*. Battelle Institute, Geneva Research Center.

Foong, J. W., Sheng, O. J., Loon, O. W., Hossain, M. M., Baskaran, N. D., Haron, H., ... Varadarajan, R. (2019). Demographics of healthcare professionals' knowledge and attitude toward deceased organ donation: Survey of critical care areas in a tertiary hospital. *The Medical Journal of Malaysia, 74*(2), 109–115. PMID:31079120

Fornell, C., & Larcker, D. F. (1981). Evaluating structural equation models with unobservable variables and measurement error. *JMR, Journal of Marketing Research, 18*(1), 39–50. doi:10.1177/002224378101800104

Fournerino, M., Helme-Guizon, A., & de Gaudemaris, C. (2006). Mesurer L'immersion dans une experience de consommation: Premiers developpements. *Proceedings of the XXIIth Congress de l'AFM*. Academic Press.

France, C., Grace, D., Merrilees, B., & Miller, D. (2018). Customer brand co-creation behavior: Conceptualization and empirical validation. *Marketing Intelligence & Planning, 36*(3), 334–348. doi:10.1108/MIP-10-2017-0266

Franco, A. M., & Filson, D. (2006). Spin-outs: Knowledge diffusion through employee mobility. *The Rand Journal of Economics, 37*(4), 841–860. doi:10.1111/j.1756-2171.2006.tb00060.x

Frankel, J. A., & Rose, A. K. (1998). The endogenity of the optimum currency area criteria. *Economic Journal (London), 108*(449), 1009–1025. doi:10.1111/1468-0297.00327

Franken, A., & Thomsett, H. (2013). When it takes a network: Creating strategy and agility through wargaming. *California Management Review, 55*(3), 107–133. doi:10.1525/cmr.2013.55.3.107

Fried, V. H., & Hisrich, R. D. (1994). Toward a model of venture capital investment decision making. *Financial Management, 23*(3), 28–37. doi:10.2307/3665619

Frishammar, J., Richtnér, A., Brattström, A., Magnusson, M., & Björk, J. (2019). Opportunities and challenges in the new innovation landscape: Implications for innovation auditing and innovation management. *European Management Journal, 37*(2), 151–164. doi:10.1016/j.emj.2018.05.002

Frone, M. R., Russell, M., & Cooper, M. L. (1997). Relation of work–family conflict to health outcomes: A four-year longitudinal study of employed parents. *Journal of Occupational and Organizational Psychology, 70*(4), 325–335. doi:10.1111/j.2044-8325.1997.tb00652.x

Füller, J., & Bilgram, V. (2017). The moderating effect of personal features on the consequences of an enjoyable co-creation experience. *Journal of Product and Brand Management, 26*(4), 386–401. doi:10.1108/JPBM-03-2016-1122

Fuller, J., Hutter, K., & Faullant, R. (2011). Why co-creation experience matters? Creative experience and its impact on the quantity and quality of creative contributions. *R & D Management, 3*(41), 259–273. doi:10.1111/j.1467-9310.2011.00640.x

Galvagno, M., & Dalli, D. (2014). Theory of value co-creation: A systematic literature review. *Managing Service Quality, 24*(6), 643–683. doi:10.1108/MSQ-09-2013-0187

Garcia-Morales, V. J., Lloréns-Montes, F. J., & Verdu-Jover, A. J. (2007). Influence of personal mastery on organizational performance through organizational learning and innovation in large firms and SMEs. *Technovation, 27*(9), 547–568. doi:10.1016/j.technovation.2007.02.013

Garcia-Morales, V. J., Martín-Rojas, R., & Lardón-López, M. E. (2018). Influence of social media technologies on organizational performance through knowledge and innovation. *Baltic Journal of Management, 13*(3), 345–367. doi:10.1108/BJM-04-2017-0123

GBI. (2019). *What and Why Green Buildings?* Retrieved from https://new.greenbuildingindex.org/whatandwhy

Gehani, R. R. (1995). Time-based management of technology: A taxonomic integration of tactical and strategic roles. *International Journal of Operations & Production Management, 15*(2), 19–35. doi:10.1108/01443579510080391

Gelhard, C., von Delft, S., & Gudergan, S. P. (2016). Heterogeneity in dynamic capability configurations: Equifinality and strategic performance. *Journal of Business Research, 69*(11), 5272–5279. doi:10.1016/j.jbusres.2016.04.124

Gentile, C., Spiller, N., & Noci, G. (2007). How to sustain the customer experience: an overview of experience components that co-create value with the customer. *European Management Journal, 25*(5), 395-410. Retrieved from http://www.sciencedirect.com/science/article/pii/S0263237307000886

Geri, N., Gafni, R., & Bengov, P. (2017). Crowdsourcing as a business model. *Journal of Global Operations and Strategic Sourcing, 10*(1), 90–111. doi:10.1108/JGOSS-05-2016-0018

Gerrig, R. J. (1993). *Experiencing narrative worlds*. New Haven: Yale University Press.

Gerwin, D. (2004). Coordinating new product development in strategic alliances. *Academy of Management Review, 29*(2), 241–257. doi:10.5465/amr.2004.12736085

Gerwin, D., & Ferris, J. S. (2004). Organizing new product development projects in strategic alliances. *Organization Science*, *15*(1), 22–37. doi:10.1287/orsc.1030.0052

Ghasemi, F., Kalatpour, O., Moghimbeigi, A., & Mohhamadfam, I. (2018). A path analysis model for explaining unsafe behavior in workplaces: The effect of perceived work pressure. *International Journal of Occupational Safety and Ergonomics*, *24*(2), 303–310. doi:10.1080/10803548.2017.1313494 PMID:28399726

Ghouse, S. M., Chaudhary, M., & Durrah, O. (2019). Socialization and the buying behaviour of the Arab child consumers: Insights from Oman. *Journal of Islamic Marketing*. doi:10.1108/JIMA-09-2018-0176

Gil-Alana, L., Carcel, H., & Abakah, E. J. A. (2018). On the linkages between Africa's emerging equity markets and global markets: Evidence from fractional integration and cointegration. *Review of development finance, 8*(2), 96-105.

Gilbert, D. T. (1991). How mental system believe. *The American Psychologist*, *46*(2), 107–199. doi:10.1037/0003-066X.46.2.107

Gilbertson, L. D. (1959). Small Business Financing Under the Small Business Act and the Small Business Investment Act of 1958. *U. Kan. L. Rev.*, *8*, 538.

Gohil, R. (2014). The Performance of Private Equity Funds in India. *Journal of Private Equity*, *18*(1), 63–72. doi:10.3905/jpe.2014.18.1.063

Gompers, P. A. (1997). *Ownership and control in entrepreneurial firms: an examination of convertible securities in venture capital investments*. Unpublished working paper, Harvard Business School.

Gompers, P. (1994). The rise and fall of venture capital. *Business and Economic History*, *23*(2), 1–26.

Gompers, P. A. (1995). Optimal investment, monitoring, and the staging of venture capital. *The Journal of Finance*, *50*(5), 1461–1489. doi:10.1111/j.1540-6261.1995.tb05185.x

Gompers, P. A. (1996). Grandstanding in the venture capital industry. *Journal of Financial Economics*, *42*(1), 133–156. doi:10.1016/0304-405X(96)00874-4

Gompers, P., & Lerner, J. (2001). The venture capital revolution. *The Journal of Economic Perspectives*, *15*(2), 145–168. doi:10.1257/jep.15.2.145

Gondal, A. M., & Khan, A. (2008). Impact of team empowerment on team performance: Case of the telecommunications industry in Islamabad. *International Review of Business Research Papers*, *4*(5), 138–146.

Goodnight, G. T., & Green, S. (2010). Rhetoric, risk, and markets: The dot-com bubble. *The Quarterly Journal of Speech*, *96*(2), 115–140. doi:10.1080/00335631003796669

Gottschling, J., Hahn, E., Maas, H., & Spinath, F. M. (2016). Explaining the relationship between personality and coping with professional demands: Where and why do optimism, self-regulation, and self-efficacy matter. *Personality and Individual Differences*, *100*, 49–55. doi:10.1016/j.paid.2016.03.085

Gourinchas, P.-O., & Jeanne, O. (2006). The elusive gains from international financial integration. *The Review of Economic Studies*, *73*(3), 715–741. doi:10.1111/j.1467-937X.2006.00393.x

Govindan, K., Khodaverdi, R., & Vafadarnikjoo, A. (2015). Intuitionistic fuzzy based DEMATEL method for developing green practices and performances in a green supply chain. *Expert Systems with Applications*, *42*(20), 7207–7220. doi:10.1016/j.eswa.2015.04.030

Granger, C. (1969). Investigating causal relations by econometric models and cross-spectral methods. *Econometrica*, *37*(3), 424–438. doi:10.2307/1912791

Grant, R. (2005). *Contemporary strategy analysis*. Malden, MA: Blackwell Publishers.

Greco, M., Grimaldi, M., & Cricelli, L. (2016). An analysis of the open innovation effect on firm performance. *European Management Journal*, *34*(5), 501–516. doi:10.1016/j.emj.2016.02.008

Greenberg, P. (2012). Customer Experience Finally…. and once again. *Customer Relationship Management*, (October), 38-39. Retrieved from www.destinationCRM.com

Green, K. M., Covin, J. G., & Slevin, D. P. (2008). Exploring the relationship between strategic reactiveness and entrepreneurial orientation: The role of structure–style fit. *Journal of Business Venturing*, *23*(3), 356–383. doi:10.1016/j.jbusvent.2007.01.002

Green, K. W. Jr, Whitten, D., & Inman, R. A. (2012). Aligning marketing strategies throughout the supply chain to enhance performance. *Industrial Marketing Management*, *41*(6), 1008–1018. doi:10.1016/j.indmarman.2012.02.003

Green, M. C., & Brock, T. C. (2000). The role of transportation in the persuasiveness of public narratives. *Journal of Personality and Social Psychology*, *79*(5), 701–721. doi:10.1037/0022-3514.79.5.701 PMID:11079236

Gremler, D. D., & Brown, S. (1996). Service Loyalty: its nature, importance and implications In B. Edvardsson, S.W. Brown, R. Johnston, & E.E. Schewing (Eds), Advancing Service Quality: A Global Perspective (pp. 171-180). International Service Quality Association.

Grichnik, D., & Hisrich, R. D. (2006). Strategic and investment behaviour in the German and Israeli venture capital industries: A comparison with the USA. *International Journal of Technology Management*, *34*(1-2), 88–104. doi:10.1504/IJTM.2006.009449

Grönroos, C. (2006). Adopting a service logic for marketing. *Marketing Theory*, *6*(3), 317–333. doi:10.1177/1470593106066794

Grunwald, R., & Kieser, A. (2007). Learning to Reduce Interorganizational Learning: An Analysis of Architectural Product Innovation in Strategic Alliances. *Journal of Product Innovation Management*, *24*(4), 369–391. doi:10.1111/j.1540-5885.2007.00256.x

Grusec, J. E., & Davidov, M. (2006). Socialization in the family: The roles of parents. In *J.E. Grusec & P.D. Hastings (Eds.), Handbook of Socialization: Theory and Research*. New York: Guilford Press.

Guerrero Baca, L., & Soria López, F. (2018). Traditional Architecture and Sustainable Conservation. *Journal of Cultural Heritage Management and Sustainable Development*, *8*(2), 194–206. doi:10.1108/JCHMSD-06-2017-0036

Gumusluoğlu, L., & Ilsev, A. (2009). Transformational leadership and organizational innovation: The roles of internal and external support for innovation. *Journal of Product Innovation Management*, *26*(3), 264–277. doi:10.1111/j.1540-5885.2009.00657.x

Gundry, L. K., Kickul, J. R., Griffiths, M. D., & Bacq, S. C. (2011). Creating social change out of nothing: The role of entrepreneurial bricolage in social entrepreneurs' catalytic innovations. In *Social and sustainable entrepreneurship* (pp. 1–24). Emerald Group Publishing Limited. doi:10.1108/S1074-7540(2011)0000013005

Hagedoorn, J., & Ridder, A.-K. (2012). Open innovation, contracts, and intellectual property rights: an exploratory empirical study.

Hagedoorn, J., & Duysters, G. (2002). External sources of innovative capabilities: The preference for strategic alliances or mergers and acquisitions. *Journal of Management Studies*, *39*(2), 167–188. doi:10.1111/1467-6486.00287

Hair, J. F., Black, W. C., Babin, B. J., & Anderson, R. E. (2010). *Multivariate Data Analysis* (7th ed.). Upper Saddle River, NJ: Prentice Hall.

Haitian, L., Yi, T., & Gongmeng, C. (2007). Venture capital and the law in China. *Hong Kong Law Journal*, *37*, 229.

Hakanen, M., & Soudunsaari, A. (2012). Building trust in high-performing teams. *Technology Innovation Management Review*, *2*(6).

Håkansson, H., & Ford, D. (2002). How should companies interact in business networks? *Journal of Business Research*, *55*(2), 133–139. doi:10.1016/S0148-2963(00)00148-X

Hall, J., Kearney, M., & Xing, C. (2018). Two tests of social displacement through social media use. *Information Communication and Society*, *10*, 1–18. doi:10.1080/1369118x.2018.1430162

Hameed, W. U., Altaf, M., & Ahmed, A. (2019). Open Innovation Challenges: Empirical Evidence from Malaysian Small and Medium-Sized Enterprises (SME's). *Journal of Management Research*, *6*(1), 28–53.

Hameed, W. U., Basheer, M. F., Iqbal, J., Anwar, A., & Ahmad, H. K. (2018). Determinants of Firm's open innovation performance and the role of R & D department: An empirical evidence from Malaysian SME's. *Journal of Global Entrepreneurship Research*, *8*(1), 1–20. doi:10.118640497-018-0112-8

Hameed, W., & Altaf, M. (2019). Open Innovation Challenges: Empirical Evidence from Malaysian Small And Medium-Sized Enterprises (SME's). *Journal of Management Research*, *5*(2).

Hameed, W., & Naveed, F. (2019). Coopetition-Based Open-Innovation and Innovation Performance: Role of Trust and Dependency Evidence from Malaysian High-Tech SMEs. *Pakistan Journal of Commerce and Social Sciences*, *13*(1), 209–230.

Hameed, W.-U., Nadeem, S., Azeem, M., Aljumah, A. I., & Adeyemi, R. A. (2018). Determinants of E-Logistic Customer Satisfaction: A Mediating Role of Information and Communication Technology (ICT). *International Journal of Supply Chain Management*, *7*(1), 105–111.

Hamel, G. (1991). Competition for competence and interpartner learning within international strategic alliances. *Strategic Management Journal*, *12*(1), 83–103. doi:10.1002mj.4250120908

Hamilton, J. (1989). A new approach to the economic analysis of nonstationary time series and the business cycle. *Econometrica*, *57*(2), 357–384. doi:10.2307/1912559

Hamilton, J. D. (2005). *Regime-Switching Models* (2nd ed.). The New Palgrave Dictionary of Economics.

Han, H., & Kim, W. (2009). Outcomes of relational benefits: Restaurant customers' perspective. *Journal of Travel & Tourism Marketing*, *26*(8), 820–835. doi:10.1080/10548400903356236

Hansen, A. V. (2017a). What stories unfold: Empirically grasping value co-creation. *European Business Review*, *29*(1), 2–14. doi:10.1108/EBR-08-2015-0080

Hansen, M. T., Chesbrough, H. W., Nohria, N., & Sull, D. N. (2000). Networked incubators. *Harvard Business Review*, *78*(5), 74–84. PMID:11143156

Harrison, T. R., Morgan, S. E., & Chewning, L. V. (2008). The challenges of social marketing of organ donation: News and entertainment coverage of donation and transplantation. *Health Marketing Quarterly*, *25*(1/2), 33–65. doi:10.1080/07359680802126079 PMID:18935879

Hartmann, A. M. (2018). *Negotiating for Strategic Alliances*. In *The Palgrave Handbook of Cross-Cultural Business Negotiation* (pp. 53–70). Palgrave.

Harvey, D., & Hubbard, C. (2016). Why Brexit. *Centre for Rural Economy Discussion Paper Series*, (35).

Harwood, A., & Konidaris, T. (2015). SME exchanges in emerging market economies: A stocktaking of development practices. The World Bank.

Hayashi, T., Kobayashi, Y., Yamaoka, K., & Yano, E. (1996). Effect of overtime work on 24-hour ambulatory blood pressure. *Journal of Occupational and Environmental Medicine*, *38*(10), 1007–1011. doi:10.1097/00043764-199610000-00010 PMID:8899576

Hearn, D. R. (2002) Education in the Workplace: An Examination of Corporate University Models. New Foundations. Retrieved from http://www.newfoundations.com/OrgTheory

Hege, U., Palomino, F., & Schwienbacher, A. (2003). *Determinants of venture capital performance: Europe and the United States*. HEC School of Management.

Heinonen, K. (2018). Positive and negative valence influencing consumer engagement. *Journal of Service Theory and Practice*, *28*(2), 147–169. doi:10.1108/JSTP-02-2016-0020

Henderson, J. (2002). Building the rural economy with high-growth entrepreneurs. *Economic Review-Federal Reserve Bank of Kansas City*, *87*(3), 45–75.

Henke, L. L. (2013). Breaking through the clutter: The impact of emotions and flow on viral marketing. *Academy of Marketing Studies Journal*, *17*(2), 111–118.

Heywood, J. S., Jirjahn, U., & Wei, X. (2008). Teamwork, monitoring and absence. *Journal of Economic Behavior & Organization*, *68*(3-4), 676–690. doi:10.1016/j.jebo.2008.09.004

Hillman, A. J., Withers, M. C., & Collins, B. J. (2009). Resource dependence theory: A review. *Journal of Management*, *35*(6), 1404–1427. doi:10.1177/0149206309343469

Hilse, H., & Nicolai, A. T. (2004). Strategic learning in Germany's largest companies: Empirical evidence on the role of corporate universities within strategy processes. *Journal of Management Development*, *23*(4), 372–398. doi:10.1108/02621710410529811

Hitt, M. A., Ireland, R. D., & Hoskisson, R. E. (1997). *Strategic Management: Competitiveness and Globalization* (2nd ed.). St. Paul, MN: West Publishing Co.

Hmieleski, K. M., & Corbett, A. C. (2008). The contrasting interaction effects of improvisational behavior with entrepreneurial self-efficacy on new venture performance and entrepreneur work satisfaction. *Journal of Business Venturing*, *23*(4), 482–496. doi:10.1016/j.jbusvent.2007.04.002

Hochberg, Y. V., Ljungqvist, A., & Lu, Y. (2007). Whom you know matters: Venture capital networks and investment performance. *The Journal of Finance*, *62*(1), 251–301. doi:10.1111/j.1540-6261.2007.01207.x

Hoch, J. E. (2013). Shared leadership and innovation: The role of vertical leadership and employee integrity. *Journal of Business and Psychology*, *28*(2), 159–174. doi:10.100710869-012-9273-6

Hoegl, M., & Gemuenden, H. G. (2001). Teamwork quality and the success of innovative projects: A theoretical concept and empirical evidence. *Organization Science*, *12*(4), 435–449. doi:10.1287/orsc.12.4.435.10635

Hoffman, D. L., & Novak, T. P. (2009). Flow Online: Lessons Learned and Future Prospects. *Journal of Interactive Marketing*, *23*(1), 23–34. doi:10.1016/j.intmar.2008.10.003

Hoffmann, W. H., & Scholsser, R. (2001). Success factors of strategic alliances in small and medium-sized enterprises – An empirical survey. *Long Range Planning*, *34*(3), 357–381. doi:10.1016/S0024-6301(01)00041-3

Höge, T. (2009). When work strain transcends psychological boundaries: An inquiry into the relationship between time pressure, irritation, work–family conflict and psychosomatic complaints. *Stress and Health: Journal of the International Society for the Investigation of Stress*, *25*(1), 41–51. doi:10.1002mi.1226

Hollebeek, L. D., & Andreassen, T. W. (2018). The S-D logic-informed "hamburger" model of service innovation and its implications for engagement and value. *Journal of Services Marketing*, *32*(1), 1–7. doi:10.1108/JSM-11-2017-0389

Holmberg, S. R., & Cummings, J. L. (2009). Building successful strategic alliances. *Long Range Planning*, *42*(6), 164–193. doi:10.1016/j.lrp.2009.01.004

Holt-Lunstad, J., Smith, T., Baker, M., Harris, T., & Stephenson, D. (2015). Loneliness and social isolation as risk factors for mortality. *Perspectives on Psychological Science*, *10*(2), 227–237. doi:10.1177/1745691614568352 PMID:25910392

Ho, M. H.-W., Ghauri, P. N., & Kafouros, M. (2019). Knowledge Acquisition in International Strategic Alliances: The Role of Knowledge Ambiguity. *Management International Review*, *59*(3), 439–463. doi:10.100711575-019-00383-w

Hota, M., & Bartsch, F. (2019). Consumer socialization in childhood and adolescence: Impact of psychological development and family structure. *Journal of Business Research*, *105*, 11–20. doi:10.1016/j.jbusres.2019.07.035

Hovland, I. (2003). *Knowledge management and organisational learning: An international development perspective: An anotated bibliography*. London: Overseas Development Institute.

Howe, J. (2006). The Rise of Crowdsourcing. *Wired Magazine*, *14*(06), 1–5. doi:10.1086/599595

Hsieh, H. F., & Shannon, S. E. (2005). Three approaches to qualitative content analysis. *Qualitative Health Research*, *15*(9), 1277–1288. doi:10.1177/1049732305276687 PMID:16204405

Hu, Y. (2010). *Improving Chinese Expenditure Incentive Programs for Venture Capital Investment: A Comparative Study of Government Expenditure Supporting Policies of Venture Capital Investment in the United States, Canada and China* [Doctoral dissertation].

Huang, C. (2010). Internet use and psychological well-being: A meta-analysis. *Cyberpsychology, Behavior, and Social Networking*, *13*(3), 241–249. doi:10.1089/cyber.2009.0217 PMID:20557242

Huang, H.-C., Lai, M.-C., & Lo, K.-W. (2012). Do founders' own resources matter? The influence of business networks on start-up innovation and performance. *Technovation*, *32*(5), 316–327. doi:10.1016/j.technovation.2011.12.004

Huang, M., Bhattacherjee, A., & Wong, C.-S. (2018). Gatekeepers' innovative use of IT: An absorptive capacity model at the unit level. *Information & Management*, *55*(2), 235–244. doi:10.1016/j.im.2017.06.001

Huang, Y., Backman, S. J., & Backman, K. F. (2012). Exploring the impacts of involvement and flow experiences in Second Life on people's travel intentions. *Journal of Hospitality and Tourism Technology*, *3*(1), 4–23. doi:10.1108/17579881211206507

Huesmann, L. R. (1995). *Screen violence and real violence: Understanding the link*. Auckland, New Zealand: Media Aware.

Hunt, M. G., Marx, R., Lipson, C., & Young, J. (2018). No more FOMO: Limiting social media decreases loneliness and depression. *Journal of Social and Clinical Psychology*, *37*(10), 751–768. doi:10.1521/jscp.2018.37.10.751

Huotari, K., & Hamari, J. (2012). Definining Gamification - A Service Marketing Perspective. *Proceeding of the 16th International Academic MindTrek Conference* (pp. 17–22). Tampere: Mindtrek. 10.1145/2393132.2393137

Hut, J., & Molleman, E. (1998). Empowerment and team development. *Team Performance Management: An International Journal*, *4*(2), 53–66. doi:10.1108/13527599810214193

Huzinga, J. (1970). *Homo Ludens: A Study of the Play Element in Culture*. New York: J & J Harper.

Hyder, A. S., & Eriksson, L. T. (2005). Success is not enough: The spectacular rise and fall of a strategic alliance between two multinationals. *Industrial Marketing Management*, *34*(8), 783–796. doi:10.1016/j.indmarman.2004.12.008

IFC. (2013). *Small and medium enterprise finance: new findings, trends and G-20/global partnership for financial inclusion progress. International Finance Corporation (IFC) and Global Partnership for Financial Inclusion*. GPFI.

Ileri, Y. Y., & Soylu, Y. (2010). Bir Rekabet Üstünlüğü Araci Olarak Çeviklik Kavrami Ve Örgüt Yapisina Olasi Etkileri. *Selçuk Üniversitesi Sosyal Bilimler Meslek Yüksekokulu Dergisi*, *13*(1-2), 13–28.

Indulski, J.A., & Szubert, Z. (1997). System for analysing sickness absenteeism in Poland. *International Journal of Occupational Medicine and Environmental Health*, *10*, 159–166. PMID:9278128

Inglehart, R., & Abramson, P. R. (1994). Economic security and value change. *The American Political Science Review*, *88*(2), 336–354. doi:10.2307/2944708

Inkpen, A. C., & Tsang, E. W. (2005). Social capital, networks, and knowledge transfer. *Academy of Management Review*, *30*(1), 146–165. doi:10.5465/amr.2005.15281445

Inman, R. A., Sale, R. S., Green, K. W. Jr, & Whitten, D. (2011). Agile manufacturing: Relation to JIT, operational performance and firm performance. *Journal of Operations Management*, *29*(4), 343–355. doi:10.1016/j.jom.2010.06.001

Iqbal, Q., & Nawaz, R. (2019). Rife information pollution (infollution) and virtual organizations in industry 4.0: within reality causes and consequences. In Big Data and Knowledge Sharing in Virtual Organizations (pp. 117–135). Hershey, PA: IGI Global. doi:10.4018/978-1-5225-7519-1.ch005

Iqbal, Q., Ahmad, N. H., & Ahmad, B. (2018). Enhancing sustainable performance through job characteristics via workplace spirituality: A study on SMEs. *Journal of Science and Technology Policy Management*. doi:10.1108/JSTPM-02-2018-0022

Iqbal, Q., Hassan, S. H., & Ahmad, N. H. (2018). The assessment of perceived information pollution in banking sector: A scale development and validation study. *Business Information Review*, *35*(2), 68–76. doi:10.1177/0266382118772891

Iqbal, Q., Yang, S., Nawaz, R., & Iqbal, K. (2019). *A multi-dimensional construct of perceived information pollution in the era of rife infollution. VINE Journal of Information and Knowledge Management Systems*. doi:10.1108/VJIKMS-10-2018-0093

Ireland, R. D., Hitt, M. A., & Vaidyanath, D. (2002). Alliance Management as a source of competitive advantage. *Journal of Management*, *28*(3), 413–446. doi:10.1177/014920630202800308

Irving, M. J., Jan, S., Tong, A., Wong, G., Craig, J. C., Chadban, S., ... Howard, K. (2014). What factors influence people's decisions to register for organ donation? The results of a nominal group study. *Transplant International*, *27*(6), 617–624. doi:10.1111/tri.12307 PMID:24617320

Isaksson, A. (2006). *Studies on the venture capital process* [Doctoral dissertation]. Handelshögskolan vid Umeå universitet.

Isik, M., Timuroglu, K., & Aliyev, Y. (2015). The relationship between teamwork and organizational trust. *International Journal of Research in Business and Social Science*, *4*(1), 113-132.

Islam, J. U., Rahman, Z., & Hollebeek, L. D. (2018). Consumer engagement in online brand communities: A solicitation of congruity theory. *Internet Research*, *28*(1), 23–45. doi:10.1108/IntR-09-2016-0279

Islam, M. Z., Ahmed, S. M., Hasan, I., & Ahmed, S. U. (2011). Organizational culture and knowledge sharing: Empirical evidence from service organizations. *African Journal of Business Management*, *5*(14), 5900–5909.

Islam, M., Hossain, A. T., & Mia, L. (2018). Role of strategic alliance and innovation on organizational sustainability. *Benchmarking: An International Journal*, *25*(5), 1581–1596. doi:10.1108/BIJ-12-2016-0188

ITU. (2019). ITU World Telecommunication /ICT Indicators database. Retrieved from https://www.itu.int/en/ITU-D/Statistics/Pages/stat/default.aspx

Jackson, M. (2013). *The politics of storytelling: Variations on a theme by Hannah Arendt*. Copenhagen: Museum Tusculanum Press.

Jacobs, M., Droge, C., Vickery, S. K., & Calantone, R. (2011). Product and process modularity's effects on manufacturing agility and firm growth performance. *Journal of Product Innovation Management*, *28*(1), 123–137. doi:10.1111/j.1540-5885.2010.00785.x

Jansen, J. J., Van Den Bosch, F. A., & Volberda, H. W. (2006). Exploratory innovation, exploitative innovation, and performance: Effects of organizational antecedents and environmental moderators. *Management Science*, *52*(11), 1661–1674. doi:10.1287/mnsc.1060.0576

Jargalsaikhan, D (2019). SMEs are the reflection of the development. The Facto Gazette.

Jeng, L. A., & Wells, P. C. (2000). The determinants of venture capital funding: Evidence across countries. *Journal of Corporate Finance*, *6*(3), 241–289. doi:10.1016/S0929-1199(00)00003-1

Jennett, C., Cox, A. L., Cairns, P., Dhoparee, S., Epps, A., Tijs, T., & Walton, A. (2008). Measuring and defining the experience of immersion in games. *International Journal of Human-Computer Studies*, *66*(9), 641–661. doi:10.1016/j.ijhcs.2008.04.004

Jha, B., & Kumar, A. (2016). Employee engagement: A strategic tool to enhance performance. *DAWN: Journal for Contemporary Research in Management*, *3*(2), 21–29.

Jin, S. A. A. (2011). "I feel present. therefore, I experience flow:" A structural equation modeling approach to flow and presence in video games. *Journal of Broadcasting & Electronic Media*, *55*(1), 114–136. doi:10.1080/08838151.2011.546248

Jobe, L. E., & White, S. W. (2007). Loneliness, social relationships, and a broader autism phenotype in college students. *Personality and Individual Differences*, *42*(8), 1479–1489. doi:10.1016/j.paid.2006.10.021

Johnson, P. R., Heimann, V. L., & O'Neill, K. (2000). The wolf pack: Team dynamics for the 21st century. *Journal of Workplace Learning: Employee Counselling Today*, *12*(4), 159–164. doi:10.1108/13665620010332813

Johnson, S., McMillan, J., & Woodruff, C. (2002). Property rights and finance. *The American Economic Review*, *92*(5), 1335–1356. doi:10.1257/000282802762024539

Joia, L. A., & Malheiros, R. (2009). Strategic alliances and the intellectual capital of firms. *Journal of Intellectual Capital*, *10*(4), 539–558. doi:10.1108/14691930910996634

Joshi, A. M., & Nerkar, A. (2011). When do strategic alliances inhibit innovation by firms? Evidence from patent pools in the global optical disc industry. *Strategic Management Journal*, *32*(11), 1139–1160. doi:10.1002mj.929

Kabir, M. A., Sultana, F., Ibrahim, M., Islam, M. M., Rahman, M. M., Islam, M. J., & Bayen, S. (2018). Evaluation of stress, loneliness and depression among residential and non-residential students of Dhaka University: Case-control study. *Pharmaceutical Science and Technology*, *2*(1), 1–6. doi:10.11648/j.pst.20180201.11

Kagermann, H., Helbig, J., Hellinger, A., & Wahlster, W. (2013). *Recommendations for implementing the strategic initiative INDUSTRIE 4.0: Securing the future of German manufacturing industry; final report of the Industrie 4.0 Working Group*. Forschungsunion.

Kain, E. (2018). "PUBG" Tops 4 Million Players On Xbox One, Hands Out Free Stuff To Players. Forbes. Retrieved from https://www.forbes.com/sites/erikkain/2018/01/25/pubg-tops-4-million-players-on-xbox-one-hands-out-free-stuff-to-players/#5297c3d72621

Kaiser, K. M., & Westarp, C. (2010). Value Creation in the Private Equity: And Venture Capital Industry.

Kalafatis, S. P., Pollard, M., East, R., & Tsogas, M. H. (1999). Green marketing and Ajzen's theory of planned behaviour: A cross-market examination. *Journal of Consumer Marketing*, *16*(5), 441–460. doi:10.1108/07363769910289550

Kale, E., Aknar, A., & Başar, Ö. (2019). Absorptive capacity and firm performance: The mediating role of strategic agility. *International Journal of Hospitality Management*, *78*, 276–283. doi:10.1016/j.ijhm.2018.09.010

Kale, P., & Singh, H. (2009). Managing Strategic Alliances: What do we know now, and where do we go from here? *The Academy of Management Perspectives*, *23*(3), 45–62. doi:10.5465/amp.2009.43479263

Kale, P., Singh, H., & Perlmutter, H. (2000). Learning and protection of proprietary assets in strategic alliances: Building relational capital. *Strategic Management Journal*, *21*(3), 217–237. doi:10.1002/(SICI)1097-0266(200003)21:3<217::AID-SMJ95>3.0.CO;2-Y

Kalisch, B. J., & Lee, K. H. (2010). The impact of teamwork on missed nursing care. *Nursing Outlook*, *58*(5), 233–241. doi:10.1016/j.outlook.2010.06.004 PMID:20934078

Kalkavan, S., & Katrinli, A. (2014). The effects of managerial coaching behaviors on the employees' perception of job satisfaction, organisational commitment, and job performance: Case study on insurance industry in Turkey. *Procedia: Social and Behavioral Sciences*, *150*, 1137–1147. doi:10.1016/j.sbspro.2014.09.129

Kalra, R. (2012, March 5). Something in the Air. Retrieved from http://www.adweek.com/news/advertising-branding/something-air-138683

Kalshoven, K., Den Hartog, D. N., & de Hoogh, A. H. (2013). Ethical leadership and followers' helping and initiative: The role of demonstrated responsibility and job autonomy. *European Journal of Work and Organizational Psychology*, *22*(2), 165–181. doi:10.1080/1359432X.2011.640773

Kamal, K. S., Wahab, L. A., & Ahmad, A. C. (2004). Climate Design of the Traditional Malay House to Meet the Requirement of Modern Living. *Proceedings of the 38th Conference of Architectural Science Association (ANZAScA)*. Context of Architecture. Academic Press.

Kamaruddin, A. R., & Mokhlis, S. (2003). Consumer socialization, social structural factors and decision-making styles: A case study of adolescents in Malaysia. *International Journal of Consumer Studies*, *27*(2), 145–156. doi:10.1046/j.1470-6431.2003.00297.x

Kapetaniou, C., & Lee, S. H. (2019). Geographical proximity and open innovation of SMEs in Cyprus. *Small Business Economics*, *52*(1), 261–276. doi:10.100711187-018-0023-7

Kaplan, R. S., & Nortan, D. P. (1992). The balance scorecard-measures that drive performance. *Harvard Business Review*, *70*(1), 71–79. PMID:10119714

Kaplan, S. N., & Lerner, J. (2010). It ain't broke: The past, present, and future of venture capital. *Journal of Applied Corporate Finance*, *22*(2), 36–47. doi:10.1111/j.1745-6622.2010.00272.x

Kaplan, S. N., & Lerner, J. (2016). *Venture capital data: Opportunities and challenges (No. w22500)*. National Bureau of Economic Research. doi:10.3386/w22500

Kaplan, S. N., & Strömberg, P. (2003). Financial contracting theory meets the real world: An empirical analysis of venture capital contracts. *The Review of Economic Studies*, *70*(2), 281–315. doi:10.1111/1467-937X.00245

Kapp, K. M. (2012). *The gamification of learning and instruction: Game-based methods and strategies for training and education*. San Francisco, CA: Pfeiffer.

Karakose, T., Yirci, R., Uygun, H., & Ozdemir, T. Y. (2016). Relationship between high school students' Facebook addiction and loneliness status. *Eurasia Journal of Mathematics. Science and Technology Education*, *12*, 2419–2429. doi:10.12973/eurasia.2016.1557a

Kar-Purkayastha, I. (2010). An epidemic of loneliness. *Lancet*, *376*(9758), 2114–2115. doi:10.1016/S0140-6736(10)62190-3 PMID:21167590

Katja, B., & Ilpo, K. (2005). Co-experience: User experience as interaction. *CoDesign*, *1*, 1, 5–18. doi:10.1080/15710880412331289917

Katou, A. A., & Budhwar, P. (2015). Human resource management and organisational productivity: A systems approach based empirical analysis. *Journal of Organizational Effectiveness: People and Performance*, *2*(3), 244–266. doi:10.1108/JOEPP-06-2015-0021

Kauser, S., & Shaw, V. (2004). The influence of behavioural and organisational characteristics on the success of international strategic alliances. *International Marketing Review*, *21*(1), 17–52. doi:10.1108/02651330410522934

Kayes, D. C., & Burnett, G. (2006). G.: Team learning in organizations A review and integration. Proceedings of the OLKC 2006 Conference. Academic Press.

Kemal, A. R. (2006). Key Issues in Industrial Growth in Pakistan. *Lahore Journal of Economics, 11.*

Kempegowda, M. (2016). A study on employee retention measures and its impact on employee motivation. *Clear international journal of research in commerce & management, 7*(4).

Kenourgios, D., & Samitas, A. (2011). Equity market integration in emerging Balkan markets. *Research in International Business and Finance*, *25*(3), 296–307. doi:10.1016/j.ribaf.2011.02.004

Khan, A., & Turowski, K. (2016). A survey of current challenges in manufacturing industry and preparation for industry 4.0. *Paper presented at the Proceedings of the First International Scientific Conference "Intelligent Information Technologies for Industry" (IITI'16)*. Academic Press.

Khare, A. (2014). Consumers' susceptibility to interpersonal influence as a determining factor of ecologically conscious behaviour. *Marketing Intelligence & Planning*, *32*(1), 2–20. doi:10.1108/MIP-04-2013-0062

Kidd, P. T. (1994). *Agile Manufacturing "Forging new Frontiers." Addison-Wesley* Publication Company Inc.

Killeen, C. (1998). Loneliness: An epidemic in modern society. *Journal of Advanced Nursing*, *28*(4), 762–770. doi:10.1046/j.1365-2648.1998.00703.x PMID:9829664

Kim, W. C., & Mauborgne, R. A. (2014). *Blue ocean strategy, expanded edition: How to create uncontested market space and make the competition irrelevant*. Harvard business review Press.

Kim, E. Y., Ko, E., Kim, H., & Koh, C. E. (2008). Comparison of benefits of radio frequency identification: Implications for business strategic performance in the US and Korean retailers. *Industrial Marketing Management*, *37*(7), 797–806. doi:10.1016/j.indmarman.2008.01.007

Kindleberger, C. P. (1981). Dominance and leadership in the international economy: Exploitation, public goods, and free rides. *International Studies Quarterly*, *25*(2), 242–254. doi:10.2307/2600355

Kinjerski, V., & Skrypnek, B. J. (2008). The Promise of Spirit at WorkIncreasing Job Satisfaction and Organizational Commitment and Reducing Turnover and Absenteeism in Long-Term Care. *Journal of Gerontological Nursing*, *34*(10), 17–25. doi:10.3928/00989134-20081001-03 PMID:18942536

Kirkman, B. L., & Rosen, B. (1997). A model of work team empowerment. *Research in Organizational Change and Development*, *10*, 131–167.

Kirkman, B. L., & Rosen, B. (1999). Beyond self-management: Antecedents and consequences of team empowerment. *Academy of Management Journal*, *42*(1), 58–74.

Kirkman, B. L., & Rosen, B. (2000). Powering up teams. *Organizational Dynamics*, *28*(3), 48–66. doi:10.1016/S0090-2616(00)88449-1

Kirkman, B. L., Rosen, B., Tesluk, P. E., & Gibson, C. B. (2004). The impact of team empowerment on virtual team performance: The moderating role of face-to-face interaction. *Academy of Management Journal*, *47*(2), 175–192.

Kirkpatrick, D. L., & Kirkpatrick, J. D. (2006). *Evaluating Training Programs*. Tata McGraw-Hill Education.

Kitapch, H., Aydin, B., & Celiki, V. (2012). The Effects of Organizational Learning Capacity and Innovativeness on Financial Performance: An Empirical Study. *African Journal of Business Management*, *6*(6), 2332–2341. doi:10.5897/AJBM11.403

Kleiner, S., & Wallace, J. E. (2017). Oncologist burnout and compassion fatigue: Investigating time pressure at work as a predictor and the mediating role of work-family conflict. *BMC Health Services Research*, *17*(1), 639. doi:10.118612913-017-2581-9 PMID:28893255

Klöckner, C. A. (2013). A Comprehensive Model of the Psychology of Environmental Behaviour – a Meta-Analysis. *Global Environmental Change*, *23*(5), 1028–1038. doi:10.1016/j.gloenvcha.2013.05.014

Klus, M. A., Lohwasser, T. S., Holotiuk, F., & Moormann, J. (2019). Strategic Alliances between Banks and Fintechs for Digital Innovation: Motives to Collaborate and Types of Interaction. *The Journal of Entrepreneurial Finance*, *21*(1).

Kogut, B. (1991). Joint ventures and the option to expand and acquire. *Management Science*, *37*(1), 13–34. doi:10.1287/mnsc.37.1.19

Kohler, T. (2016). Corporate accelerators: Building bridges between corporations and startups. *Business Horizons*, *59*(3), 347–357. doi:10.1016/j.bushor.2016.01.008

Kohler, T., Fueller, J., Matzler, K., Stieger, D., & Füller. (2011). Co-Creation In Virtual Worlds: The Design Of The User Experience. *Management Information Systems Quarterly*, *35*(3), 773–788. doi:10.2307/23042808

Koka, B. R., & Prescott, J. E. (2008). Designing alliance networks: The influence of network position, environmental change, and strategy on firm performance. *Strategic Management Journal*, *29*(6), 639–661. doi:10.1002mj.679

Kolmakov, V. V., Polyakova, A. G., & Shalaev, V. S. (2015). An analysis of the impact of venture capital investment on economic growth and innovation: Evidence from the USA and Russia. *Ekonomski Anali/Economic Annals, 60*(207).

Komatsubara, T., Okimoto, T., & Tatsumi, K. I. (2017). Dynamics of integration in East Asian equity markets. *Journal of the Japanese and International Economies*, *45*, 37–50. doi:10.1016/j.jjie.2017.07.002

Kopfman, J. E., Smith, S. W., Yun, J. K. A., & Hodges, A. (1998). Affective and cognitive reactions to narrative versus statistical evidence organ donation messages. *Journal of Applied Communication Research, 26*(3), 279–300. doi:10.1080/00909889809365508

Kortum, S., & Lerner, J. (2001). Does venture capital spur innovation? In *Entrepreneurial inputs and outcomes: New studies of entrepreneurship in the United States* (pp. 1–44). Emerald Group Publishing Limited. doi:10.1016/S1048-4736(01)13003-1

Kotabe, M., & Scott Swan, K. (1995). The role of strategic alliances in high-technology new product development. *Strategic Management Journal, 16*(8), 621–636. doi:10.1002mj.4250160804

Kothari, C. R., & Garg, G. (2014). *Research methodology: Methods and techniques*. New Delhi: New Age International Publishers.

Kotler, P., Armstrong, G., Agnihotri, P. Y., & Haque, E. (2011). *Principles of Marketing-A south Asian Perspective*. New Delhi: Pearson Education.

Kotler, P., & Zaltman, G. (1971). Social Marketing: An Approach to Planned Social Change. *Journal of Marketing, 35*(3), 3–12. doi:10.1177/002224297103500302 PMID:12276120

Koza, M., & Lewin, A. (2000). Managing Partnerships and Strategic Alliances: Raising the Odds of Success. *European Management Journal, 18*(2), 146–151. doi:10.1016/S0263-2373(99)00086-9

Kozjek, T., & Ferjan, M. (2015). Organizational flexibility, employee security, and organizational efficiency–a case study of Slovenian public and private sector organizations. *Organizacija, 48*(1), 3–21. doi:10.1515/orga-2015-0001

Kraus, S., Klimas, P., Gast, J., & Stephan, T. (2019). Sleeping with competitors: Forms, antecedents and outcomes of coopetition of small and medium-sized craft beer breweries. *International Journal of Entrepreneurial Behaviour & Research, 25*(1), 50–66. doi:10.1108/IJEBR-09-2017-0356

Kräussl, R., & Krause, S. (2014). Has Europe been catching up? An industry level analysis of venture capital success over 1985–2009. *European Financial Management, 20*(1), 179–205. doi:10.1111/j.1468-036X.2013.12012.x

Kreiner, K., & Schultz, M. (1993). Informal collaboration in R & D. The formation of networks across organizations. *Organization Studies, 14*(2), 189–209. doi:10.1177/017084069301400202

Kross, E., Verduyn, P., Demiralp, E., Park, J., Lee, D., Lin, N., ... Ybarra, O. (2013). Facebook use predicts declines in subjective well-being in young adults. *PLoS One, 8*(8), e69841. doi:10.1371/journal.pone.0069841 PMID:23967061

Kubba, S. (2016). *Handbook of Green Building Design and Construction. LEEDS, BREEAM and Green Globes*. Butterworth-Heinemann.

Kumar, R., & Andersen, P. H. (2000). Inter firm diversity and the management of meaning in international strategic alliances. *International Business Review, 9*(2), 237–252. doi:10.1016/S0969-5931(99)00038-4

Kumar, S., & Bansal, L. K. (2008). The impact of mergers and acquisitions on corporate performance in India. *Management Decision, 46*(10), 1531–1543. doi:10.1108/00251740810920029

Kumkale, İ. (2016). Organization's Tool for Creating Competitive Advantage: Strategic Agility. *Balkan and Near Eastern Journal of Social Sciences, 2*(3), 118–124.

La Ferle, C., Edwards, S. M., & Lee, W. (2000). Teen's use of traditional media and the Internet. *Journal of Advertising Research, 40*(3), 55–65. doi:10.2501/JAR-40-3-55-65

Labov, W. (1997). Some further steps in narrative analysis. *Journal of Narrative and Life History*, *7*(1/4), 395–415. doi:10.1075/jnlh.7.49som

Labrecque, L. I., vor dem Esche, J., Mathwick, C., Novak, T. P., & Hofacker, C. F. (2013). Consumer power: Evolution in the digital age. *Journal of Interactive Marketing*, *27*(4), 257–269. doi:10.1016/j.intmar.2013.09.002

Lachance, M. J., & Legault, F. (2007). College students' consumer competence: Identifying the socialization sources. *Journal of Research for Consumers*, *13*, 1–21.

Lans, T., Blok, V., & Gulikers, J. (2015). Show me your network and I'll tell you who you are: Social competence and social capital of early-stage entrepreneurs. *Entrepreneurship and Regional Development*, *27*(7-8), 458–473. doi:10.1080/08985626.2015.1070537

Larsson, R., Bengtsson, L., Henriksson, K., & Sparks, J. (1998). The interorganizational learning dilemma: Collective knowledge development in strategic alliances. *Organization Science*, *9*(3), 285–305. doi:10.1287/orsc.9.3.285

Latham, G. P., & Ernst, C. T. (2006). Keys to motivating tomorrow's workforce. *Human Resource Management Review*, *16*(2), 181–198. doi:10.1016/j.hrmr.2006.03.014

Laurence, J. H., Naughton, J., & Harris, D. A. (1996). *Attrition Revisited: Identifying the Problem and Its Solutions*. Human resources research organization. doi:10.21236/ADA309172

Lawrence, P., & ul-Haq, R. (1998). Qualitative research into strategic alliances. *Qualitative Market Research*, *1*(1), 15–24. doi:10.1108/13522759810197569

Lechler, S., Canzaniello, A., & Hartmann, E. (2019). Assessment sharing intra-industry strategic alliances: Effects on sustainable supplier management within multi-tier supply chains. *International Journal of Production Economics*, 217, 1–52.

Lechner, C., & Dowling, M. (2003). Firm networks: External relationships as sources for the growth and competitiveness of entrepreneurial firms. *Entrepreneurship and Regional Development*, *15*(1), 1–26. doi:10.1080/08985620210159220

Lee, D., & Husk, K. K. (2014). Diversity in alliance portfolio and performance outcomes: A Meta-Analysis. *Journal of Management*, *43*(5), 1–26.

Lee, H. H., & Yang, T. T. (2015). Employee goal orientation, work unit goal orientation and employee creativity. *Creativity and Innovation Management*, *24*(4), 659–674. doi:10.1111/caim.12118

Lee, K. (2011). The green purchase behaviour of Hong Kong young consumers: The role of peer influence, local environmental involvement and concrete environmental knowledge. *Journal of International Consumer Marketing*, *23*(1), 21–44. doi:10.1080/08961530.2011.524575

Lee, K. S., Lim, G. H., & Tan, J. (2000). Feasibility pf strategic alliance as an entry strategy into markets dominated by major competitors. *Journal of Small Business and Enterprise Development*, *7*(1), 43–57. doi:10.1108/EUM0000000006804

Lee, S., Park, G., Yoon, B., & Park, J. (2010). Open innovation in SMEs—An intermediated network model. *Research Policy*, *39*(2), 290–300. doi:10.1016/j.respol.2009.12.009

Lengnick-Hall, C. A., & Beck, T. E. (2016). Resilience capacity and strategic agility: Prerequisites for thriving in a dynamic environment. In *Resilience Engineering Perspectives* (Vol. 2, pp. 61–92). CRC Press.

Lerner, J., & Tåg, J. (2013). Institutions and venture capital. *Industrial and Corporate Change*, *22*(1), 153–182. doi:10.1093/icc/dts050

Levenson, J., Shensa, A., Sidani, J., Colditz, J., & Primack, B. (2016). The Association between social media use and sleep disturbance among young adults. *Preventive Medicine*, *85*, 36–41. doi:10.1016/j.ypmed.2016.01.001 PMID:26791323

Lévi-Strauss, C. (1966). *The savage mind.* Chicago: The University of Chicago Press.

Levy, B. (1993). Obstacles to Developing Indigenous Small and Medium Enterprises: An Empirical Assessment. *The World Bank Economic Review*, *7*(1), 65–83. doi:10.1093/wber/7.1.65

Lewicki, R. J., & Bunker, B. B. (1996). Developing and maintaining trust in work relationships. *Trust in organizations: Frontiers of theory and research, 114*, 139.

Lewis, M. (2006). Customer acquisition promotions and customer asset value. *JMR, Journal of Marketing Research*, *43*(2), 195–203. doi:10.1509/jmkr.43.2.195

Liang, t. T. (1983). *Civilian manpower planning model for scientific and engineering personnel in the navy R&D centers.* Navy personnel research and development center.

Lichtenthaler, U. (2008). Open innovation in practice: An analysis of strategic approaches to technology transactions. *IEEE Transactions on Engineering Management*, *55*(1), 148–157. doi:10.1109/TEM.2007.912932

Lichtenthaler, U., & Lichtenthaler, E. (2009). A capability-based framework for open innovation: Complementing absorptive capacity. *Journal of Management Studies*, *46*(8), 1315–1338. doi:10.1111/j.1467-6486.2009.00854.x

Lifshitz-Assaf, H. (2018). Dismantling knowledge boundaries at NASA: The critical role of professional identity in open innovation. *Administrative Science Quarterly*, *63*(4), 746–782. doi:10.1177/0001839217747876 PMID:30443046

Li, K., Qiu, J., & Wang, J. (2019). Technology Conglomeration, Strategic Alliances, and Corporate Innovation. *Management Science*, 65(11), 1–26.

Li, L. P., Juric, B., & Brodie, R. J. (2017). Dynamic multi-actor engagement in networks: The case of United Breaks Guitars. *Journal of Service Theory and Practice*, *27*(4), 738–760. doi:10.1108/JSTP-04-2016-0066

Li, L. P., Juric, B., & Brodie, R. J. (2018). Actor engagement valence: Conceptual foundations, propositions and research directions. *Journal of Service Management*, *29*(3), 491–516. doi:10.1108/JOSM-08-2016-0235

Liles, P. R. (1977). *Sustaining the venture capital firm.* Management Analysis Center.

Lin, J. Y. (2008). The impact of the financial crisis on developing countries.

Lin, L. (2015). *Re-Engineering a Venture Capital Market: The Case of China* (Vol. 4). NUS Centre for Law & Business Working Paper 15.

Lin, C. L., & Tzeng, G. H. (2009). A value-created system of science (technology) park by using DEMETEL. *Expert Systems with Applications*, *36*(6), 9683–9697. doi:10.1016/j.eswa.2008.11.040

Lin, C.-J., & Wu, W.-W. (2008). A causal analytical method for group decision-making under fuzzy environment. *Expert Systems with Applications*, *34*(1), 205–213. doi:10.1016/j.eswa.2006.08.012

Lin, F.-J., & Lin, Y.-H. (2016). The effect of network relationship on the performance of SMEs. *Journal of Business Research*, *69*(5), 1780–1784. doi:10.1016/j.jbusres.2015.10.055

Lin, L., Sidani, J., Shensa, A., Radovic, A., Miller, E., Colditz, J., ... Primack, B. A. (2016). Association between social media use and depression among US young adults. *Depression and Anxiety*, *33*(4), 323–331. doi:10.1002/da.22466 PMID:26783723

Liou, J. J. H., Tzeng, G.-H., & Chang, H.-C. (2007). Airline safety measurement using a hybrid model. *Air Transport Management*, *13*(4), 243–249. doi:10.1016/j.jairtraman.2007.04.008

Liu, C. (2010). *A Comparison of the venture capitalists´ investment behavior pattern between China and America.*

Liu, R. R., & Zhang, W. (2010). Informational influence of online customer feedback: An empirical study. *Database Marketing & Customer Strategy Management*, *17*(2), 120–131. doi:10.1057/dbm.2010.11

Li, X., Chung, C., Goldsby, T. J., & Holsapple, C. W. (2008). A unified model of supply chain agility: The work-design perspective. *International Journal of Logistics Management*, *19*(3), 408–435. doi:10.1108/09574090810919224

Li-Ying, J. (2018). Dual boundary spanning: Toward a typology of outside-in open innovation in the Canadian context. *Canadian Journal of Administrative Sciences [Revue Canadienne des Sciences de l'Administration]*, *35*(3), 429-443.

Lkhagvasuren, T. (2014). Marketing of small and medium-sized enterprises in Mongolia. *International Journal of Technology Enhancements and Emerging Engineering Research*, *2*(5), 33–35.

Loch, A., Hilmi, I. N., Mazam, Z., Pillay, Y., & Choon, D. S. K. (2010). Differences in attitude towards cadaveric organ donation: Observations in a multiracial Malaysian society. *Hong Kong Journal of Emergency Medicine*, *17*(3), 236–243. doi:10.1177/102490791001700306

Lockett, A., Wright, M., Sapienza, H., & Pruthi, S. (2002). Venture capital investors, valuation and information: A comparative study of the US, Hong Kong, India and Singapore. *Venture Capital: An International Journal of Entrepreneurial Finance*, *4*(3), 237–252. doi:10.1080/13691060213715

López-Nicolás, C., & Meroño-Cerdán, Á. L. (2011). Strategic knowledge management, innovation and performance. *International Journal of Information Management*, *31*(6), 502–509. doi:10.1016/j.ijinfomgt.2011.02.003

Lu, A. S., Baranowski, T., Thompson, D., & Budday, R. (2012). Story Immersion of Videogames for Youth Health Promotion: A Review of Literature. *Games for Health Journal*, *1*(3), 199–204. doi:10.1089/g4h.2011.0012 PMID:24416639

Luczak, C., & Younkin, N. (2012). Net generation: A conceptual framework of the consumer socialization process. *Academy of Marketing Studies Journal*, *16*(2), 47–51.

Luo, X., & Deng, L. (2009). Do Birds of a Feather Flock Higher? The Effects of Partner Similarity on Innovation in Strategic Alliances in Knowledge-Intensive Industries. *Journal of Management Studies*, *46*(6), 1005–1030. doi:10.1111/j.1467-6486.2009.00842.x

Lusa, S., Häkkänen, M., Luukkonen, R., & Viikari-Juntura, E. (2002). Perceived physical work capacity, stress, sleep disturbance and occupational accidents among firefighters working during a strike. *Work and Stress*, *16*(3), 264–274. doi:10.1080/02678370210163301

Lwin, M. O., Williams, J. D., & Lan, L. L. (2002). Social marketing initiatives: National Kidney Foundation's organ donation programs in Singapore. *Journal of Public Policy & Marketing*, *21*(1), 66–77. doi:10.1509/jppm.21.1.66.17613

Lyotard, J. F. (1984). *The postmodern condition: A report on knowledge*. Minneapolis: University of Minnesota Press.

Lytovchenko, I. (2016). Corporate University As a Form of Employee Training and Development in American Companies. *Advanced Education*, 35. doi:10.20535/2410-8286.62280

Maccoby, E. E. (2006). Historical overview of socialization research and theory. In *J.E. Grusec & P.D. Hastings (Eds.), Handbook of Socialization: Theory and Research*. New York: Guilford Press.

Maddison, A. (1995). *Monitoring the world economy, 1820-1992* (p. 238). Paris: Development Centre of the Organisation for Economic Co-operation and Development.

Majdoub, J., & Mansour, W. (2014). Islamic equity market integration and volatility spillover between emerging and US stock markets. *The North American Journal of Economics and Finance*, *29*, 452–470. doi:10.1016/j.najef.2014.06.011

Majid, A., Yasir, M., & Yasir, M. (2017). Individual and work dynamics affecting the determinants of functional flexibility in SMEs: Evidence from Pakistan. *Journal of Entrepreneurship in Emerging Economies*, *9*(2), 144–160. doi:10.1108/JEEE-03-2016-0008

Makmor, T., Abdillah, N., Raja Noriza, R. A., Nurulhuda, M. S., Soo-Kun, L., & Kok-Peng, N. (2014). Low organ donation rate in Malaysia: A survey. *Journal of Health and Translational Medicine*, *17*(1), 14–17. doi:10.22452/jummec.vol17no1.3

Malay Mail. (2019). Survey shows majority support "no single-use plastic" campaign.

Malaysian Medical Council. (2006). Guideline on Organ Transplantation.

Malhotra, A. (2016). Impact of HRM Practices on Employee Attrition A Study of Selected BPOs in National Capital Region.

Manchanda, K., & Muralidharan, P. (2014, January). Crowdfunding: a new paradigm in start-up financing. *Proceedings of the Global Conference on Business & Finance Proceedings* (p. 369). Institute for Business & Finance Research.

Manigart, S., De Waele, K., Wright, M., Robbie, K., Desbrières, P., Sapienza, H. J., & Beekman, A. (2002). Determinants of required return in venture capital investments: A five-country study. *Journal of Business Venturing*, *17*(4), 291–312. doi:10.1016/S0883-9026(00)00067-7

Mansoor, S. & Ratna, R. (2014). Impact of Organizational Learning on Organizational Innovation.

Marshall, H. M. (2006). The use of multidimensional scaling in the assessment of college students' attitudes toward organ donation [PhD Thesis]. University of New York.

Marsick, V. J., & Watkins, K. E. (2003). Demonstrating the value of an organization's learning culture: The dimensions of the learning organization questionnaire. *Advances in Developing Human Resources*, *5*(2), 132–151.

Martin, R., Sunley, P., & Turner, D. (2002). Taking risks in regions: The geographical anatomy of Europe's emerging venture capital market. *Journal of Economic Geography*, *2*(2), 121–150. doi:10.1093/jeg/2.2.121

Mascarenhas, O. A. J., & Higby, M. A. (1993). Peer, parent and media influences in teen apparel shopping. *Journal of the Academy of Marketing Science*, *21*(1), 53–58. doi:10.1177/0092070393211007

Mason, C. M. (2009). Public policy support for the informal venture capital market in Europe: A critical review. *International Small Business Journal*, *27*(5), 536–556. doi:10.1177/0266242609338754

Massey, A. P., Montoya-Weiss, M. M., & Holcom, K. (2001). Re-engineering the customer relationship: leveraging knowledge assets at IBM. *Decision Support Systems, 32*(2), 155-170. Retrieved from http://www.sciencedirect.com/science/article/pii/S0167923601001087

Masthi, N. R. R. M., Pruthvi, S., & Mallekavn, P. (2017). A comparative study on social media addiction between public and private high school students of urban Bengaluru, India. *ASEAN Journal of Psychiatry*, *18*, 206–215.

Mavondo, F. T., Chimhanzi, J., & Stewart, J. (2005). Learning orientation and market orientation: Relationship with innovation, human resource practices and performance. *European Journal of Marketing*, *39*(11/12), 1235–1263. doi:10.1108/03090560510623244

Mayer, C., Schoors, K., & Yafeh, Y. (2005). Sources of funds and investment activities of venture capital funds: Evidence from Germany, Israel, Japan and the United Kingdom. *Journal of Corporate Finance*, *11*(3), 586–608. doi:10.1016/j.jcorpfin.2004.02.003

McCahery, J. A., & Vermeulen, E. P. (2010). Venture capital beyond the financial crisis: How corporate venturing boosts new entrepreneurial clusters (and assists governments in their innovation efforts). *Capital Markets Law Journal*, *5*(4), 471–500. doi:10.1093/cmlj/kmq018

McGlade, D., & Pierscionek, B. (2013). Can education alter attitudes, behaviour and knowledge about organ donation? A pretest-post-test study. *BMJ Open*, *3*(12), 1–7. doi:10.1136/bmjopen-2013-003961 PMID:24381257

McGrath, J. E. (1984). *Groups: Interaction and performance* (Vol. 14). Englewood Cliffs, NJ: Prentice-Hall.

McKee, R. (1997). *Story, substance, structure, style and the principle of screenwriting*. New York: HarperCollins.

McKim, B., & Hughes, A. M. (2001). How to measure consumer relationship management success. *Journal of Database Marketing*, *8*(3), 224–231.

McMillan, S. J., & Morrison, M. (2006). Coming of age with the Internet: A qualitative exploration of how the Internet has become an integral part of young people's lives. *New Media & Society*, *8*(1), 73–95. doi:10.1177/1461444806059871

McQuail, D. (2010). *McQuail's Mass Communication Theory*. London: Sage Publications.

Medvene, L., Nilsen, K., Smith, R., Ofei-Dodoo, S., DiLollo, A., Webster, N., ... Nance, A. (2015). Social networks and links to isolation and loneliness among elderly HCBS clients. *Aging & Mental Health*, *20*(5), 485–493. doi:10.1080/13607863.2015.1021751 PMID:25808879

Megginson, W. L. (2004). Toward a global model of venture capital? *Journal of Applied Corporate Finance*, *16*(1), 89–107. doi:10.1111/j.1745-6622.2004.tb00599.x

Mehran, H. (1995). Executive compensation structure, ownership, and firm performance. *Journal of Financial Economics*, *38*(2), 163–184. doi:10.1016/0304-405X(94)00809-F

Mei, N. S., Wai, C. W., & Ahamad, R. (2016). Environmental Awareness and Behaviour Index for Malaysia. *Procedia: Social and Behavioral Sciences*, *222*(07), 668–675. doi:10.1016/j.sbspro.2016.05.223

Meister, J. C. (1998). *Corporate Universities: Lessons in Building a World-Class Workforce*. New York: McGraw-Hill.

Memon, M. A., & Tahir, I. M. (2012). Performance analysis of manufacturing companies in Pakistan. *Business Management Dynamics*, *1*(7), 12–21.

Mensah, J. O., & Premaratne, G. (2018). Integration of ASEAN Banking Sector Stocks. *Journal of Asian Economics*, *59*, 48–60. doi:10.1016/j.asieco.2018.10.001

Metrick, A., & Yasuda, A. (2010). Venture capital and the finance of innovation. In A. Metrick and A. Yasuda (Eds.), Venture capital and the finance of innovation (2nd ed.) John Wiley and Sons, Inc.

Metrick, A., & Yasuda, A. (2011). *Venture Capital & The Finance of Innovation*. John Wiley & Sons.

Miles, R. E., & Snow, C. C. (1992). Causes of failure in network organization. *California Management Review*, *34*(4), 53–72. doi:10.2307/41166703

Miller, M. G., Fitzgerald, S. P., Murrell, K. L., Preston, J., & Ambekar, R. (2005). Appreciative inquiry in building a transcultural strategic alliance: The case of a biotech alliance between a US multinational and an Indian family business. *The Journal of Applied Behavioral Science*, *41*(1), 91–110. doi:10.1177/0021886304273060

Ministry of Finance. (2018). Pakistan economic survey 2017-18. Retrieved from http://www.finance.gov.pk/survey_1718.html

Miniwatts Marketing Group. (2018). Internet World Stats.

Miraz, M., Ali, M., Excell, P., & Picking, R. (2018). Internet of nano-things, things and everything: Future growth trends. *Future Internet, 10*(8), 68. doi:10.3390/fi10080068

Mirzaei Aliabadi, M., Aghaei, H., Kalatpour, O., Soltanian, A. R., & Nikravesh, A. (2018). Analysis of human and organizational factors that influence mining accidents based on Bayesian network. *International Journal of Occupational Safety and Ergonomics*, 1–8. doi:10.1080/10803548.2018.1455411 PMID:29560801

Mitchell, D. J., Kahn, B. E., & Knasko, S. C. (1995, September). There's something in the air: effects of congruent or incongruent ambient odor on consumer decision making. *Journal of Consumer Research, 22*(2), 229-238. Retrieved from http://www.jstor.org/stable/2489814

Mitchell, V.-W., Schlegelmilch, B. B., & Mone, S.-D. (2016). Why should I attend? The value of business networking events. *Industrial Marketing Management, 52*, 100–108. doi:10.1016/j.indmarman.2015.05.014

Mithas, S., Krishnan, M. S., & Fornell, C. (2005). Why do customer relationship management applications affect customer satisfaction? *Journal of Marketing, 69*(4), 201–209. doi:10.1509/jmkg.2005.69.4.201

Mohr, J., & Spekman, R. (1994). Characteristics of partnership success: Partnership attributes, communication behavior, and conflict resolution techniques. *Strategic Management Journal, 15*(2), 135–152. doi:10.1002mj.4250150205

Mol, M. J., & Birkinshaw, J. (2009). The sources of management innovation: When firms introduce new management practices. *Journal of Business Research, 62*(12), 1269–1280. doi:10.1016/j.jbusres.2009.01.001

Moore, E. S., & Bowman, G. D. (2006). Of friends and family: How do peers affect the development of intergenerational influences? *Advances in Consumer Research. Association for Consumer Research (U. S.), 33*, 536–542.

Moore, J. (2002). Running a Corporate University like a Business: a financial model. In M. Allen (Ed.), *The Corporate University Handbook: designing, managing and growing a successful program*. New York: Amacom.

Moore, J. N., Raymond, M. A., Mittelstaedt, J. D., & Tanner, J. F. Jr. (2002). Age and consumer socialization agent influences on adolescents' sexual knowledge, attitudes and behaviour: Implications for social marketing initiatives and public policy. *Journal of Public Policy & Marketing, 21*(1), 37–52. doi:10.1509/jppm.21.1.37.17612

Morschheuser, B., Hamari, J., Koivisto, J., & Maedche, A. (2017). Gamified crowdsourcing: Conceptualization, literature review, and future agenda. *International Journal of Human-Computer Studies, 106*(April), 26–43. doi:10.1016/j.ijhcs.2017.04.005

Moschis, G. P., & Churchill, G. A. Jr. (1978). Consumer socialization: A theoretical and empirical analysis. *JMR, Journal of Marketing Research, 15*(4), 599–609. doi:10.1177/002224377801500409

Mostafa, M. M. (2010). Altruistic, cognitive and attitudinal determinants of organ donation intention in Egypt: A social marketing perspective. *Health Marketing Quarterly, 27*(1), 97–115. doi:10.1080/07359680903519867 PMID:20155553

Muniz, A. M. J. Jr, & O'Guinn, T. C. (2001). Brand Community. *The Journal of Consumer Research, 27*(4), 412–432. doi:10.1086/319618

Murray, G. C. (1992). A challenging marketplace for venture capital. *Long Range Planning, 25*(6), 79–86. doi:10.1016/0024-6301(92)90173-Y

Murray-White, J. (2019). *What is Green Architecture? Retrieved from: Updated on 15 April 2019*. Sustainable Build.

Myers, S. P. (2013). Definition of Team Dynamics. Team Technology. Retrieved from http://www.teamtechnology.co.uk/team/dynamics/definition/

Myhrvold, N. (2010). The big idea: Funding eureka! *Harvard Business Review, 88*(2), 40–50.

Nadkarni, S., & Narayanan, V. K. (2007). Strategic schemas, strategic flexibility, and firm performance: The moderating role of industry clockspeed. *Strategic Management Journal, 28*(3), 243–270. doi:10.1002mj.576

Nam, J., Ekinci, Y., & Whyatt, G. (2011). Brand equity, brand loyalty and consumer satisfaction. *Annals of Tourism Research, 38*(3), 1009–1030. doi:10.1016/j.annals.2011.01.015

Nandonde, F. A. (2019). Building Commitment in Supplier–Retailer Relationship in Developing Economies: The Case of Tanzania. *FIIB Business Review, 8*(1), 39–50. doi:10.1177/2319714518821208

Narayandas, D. (1998). Measuring and managing the benefits of customer retention: An empirical investigation. *Journal of Service Research, 1*(2), 108–128. doi:10.1177/109467059800100202

Narula, R., & Hagedoorn, J. (1999). Innovating through strategic alliances: Moving towards international partnerships and contractual agreements. *Technovation, 19*(5), 283–294. doi:10.1016/S0166-4972(98)00127-8

National Architectural Accreditation Board. (2009). *Conditions for Accreditation*. Washington, DC: National Architectural Accrediting Board.

Nazni, N., Zaherawati, Z., Adnan, A., Mahazril, A. Y., Mohd Zool Hilmie, M. S., Mohd Shamsul, D., ... Kamarudin, N. (2012). Organ donation among Malaysian: The Malay dilemma toward social development. *Asian Social Science, 8*(10), 8–15.

Newell, F. (2000). *Loyalty.com: Customer Relationship Management in the New Era of Internet Marketing*. New York, NY: McGraw-Hill.

Nielsen, B. B., & Nielsen, S. (2009). Learning and Innovation in International Strategic Alliances: An Empirical Test of the Role of Trust and Tacitness. *Journal of Management Studies, 46*(6), 1031–1056. doi:10.1111/j.1467-6486.2009.00840.x

Ning, Y., Wang, W., & Yu, B. (2015). The driving forces of venture capital investments. *Small Business Economics, 44*(2), 315–344. doi:10.100711187-014-9591-3

Nixon, J. C., & Helms, M. M. (2002). Corporate universities vs higher education institutions. *Industrial and Commercial Training, 34*(4), 144–150. doi:10.1108/00197850210429129

Nor Hatina Shuib dan Mahani Ishak. (2019). Eksklusif: Malaysia negara terendah derma organ, BH Online. Retrieved from https://www.bharian.com.my/wanita/keluarga/2019/01/517319/eksklusif-malaysia-negara-terendah-derma-organ

Norazlina, A. A., Muhammad, N. A., & Zuramaznum, S. (2006). Donor's consent in organ procurement process in Malaysia: Issues and suggestions for solutions. Research Paper, Institut Penyelidikan, Pembangunan dan Pengkomersilan, Universiti Teknologi Mara.

Normawati, A., Faridah, I., Normah, M., & Chang, P. K. (2011). Pencarian Maklumat Pendermaan Organ dan Tisu dalam Pengikrar Penderma Organ: Satu Tinjauan Awal. *Malaysian Journal of Communication, 27*(1), 102–117.

Nowland, R., Necka, E., & Cacioppo, J. (2017). Loneliness and social internet use: Pathways to reconnection in a digital world? *Perspectives on Psychological Science, 13*(1), 70–87. doi:10.1177/1745691617713052 PMID:28937910

Obstfeld, M. (1994). Risk-taking, global diversification, and growth. *The American Economic Review, 84*(5), 1310–1329.

OECD. (2005). *The measurement of scientific and technological activities Oslo Manual. Guidelines for collecting and interpreting innovation data* (3rd ed.). Paris: OECD EUROSTAT.

OECD. (2016). Enhancing access to finance for micro, small and medium-sized enterprises in Mongolia.

Oh, J. R., & Park, C. H. (2011). Corporate Universities and Adult Education: Implications for Theory and Practice. *Adult Education Research Conference*. Retrieved from http://www.adulterc.org/Proceedings/2011/papers/rok_park.pdf

Ojha, D. (2008). *Impact of strategic agility on competitive capabilities and financial performance*. Clemson University.

Oliner, S. D., & Sichel, D. E. (2000). The resurgence of growth in the late 1990s: Is information technology the story? *The Journal of Economic Perspectives*, *14*(4), 3–22. doi:10.1257/jep.14.4.3

Oliver, R. L. (Special Issue 1999). Whence consumer loyalty? *Journal of Marketing, 63*, 33-44.

Oliver, R. L. (1997). Loyalty and profit: long term effects of satisfaction. In *Satisfaction: A Behavioural Perspective on the consumer*. New York, NY: McGraw-Hill.

Online, B. H. (2019). 80 peratus kes berpotensi derma organ ditolak ahli keluarga. Bharian. Retrieved from https://www.bharian.com.my/berita/nasional/2019/10/614745/80-peratus-kes-berpotensi-derma-organ-ditolak-ahli-keluarga

Örtenblad, A. (2001). On differences between organizational learning and learning organization. *The Learning Organization*, *8*(3), 125–133. doi:10.1108/09696470110391211

Osterloh, M., Rota, S., & von Wartburg, M. (2001). Open source-New rules in software development. University of Zurich, Institute of Organization and Administrative Science (IOU). Retrieved from http://www.iou.uzh.ch/orga/downloads/OpenSourceAoM.pdf

Owen-Smith, J., Cotton-Nessler, N. C., & Buhr, H. (2015). Network effects on organizational decision-making: Blended social mechanisms and IPO withdrawal. *Social Networks*, *41*, 1–17.

Oyedijo, A. (2012). Strategic agility and competitive performance in the Nigerian telecommunication industry: An empirical investigation. *American International Journal of Contemporary Research*, *2*(3), 227–237.

Ozdemir, U., & Tuncay, T. (2008). Correlates of loneliness among university students. *Child and Adolescent Psychiatry*, *2*, 1–6. doi:10.11866/1753-2000-2-29 PMID:18851744

Oztaysi, B., Sezgin, S., & Ozak, A. F. (2011). A measurement tool for consumer relationship management processes. *Industrial Management & Data Systems*, *111*(6), 943–960. doi:10.1108/02635571111144982

Padula, G., & Dagnino, G. B. (2007). Untangling the rise of coopetition: The intrusion of competition in a cooperative game structure. *International Studies of Management & Organization*, *37*(2), 32–52. doi:10.2753/IMO0020-8825370202

Pakistan Bureau of Statistics. (2006). Census of Manufacturing Industries (CMI) 2005-06. Retrieved from http://www.pbs.gov.pk/content/census-manufacturing-industries-cmi-2005-06

Paliszkiewicz, J. (2011). Inter-organizational trust: Conceptualization and measurement. *International Journal of Performance Measurement*, *1*(1), 15–28.

Palmer, A. J. (2010). Customer experience management: A critical review of an emerging idea. *Journal of Services Marketing*, *24*(3), 196–208. doi:10.1108/08876041011040604

Pandey, I. M. (1998). The process of developing venture capital in India. *Technovation*, *18*(4), 253–261. doi:10.1016/S0166-4972(98)00003-0

Papa, A., Dezi, L., Gregori, G. L., Mueller, J., & Miglietta, N. (2018). Improving innovation performance through knowledge acquisition: The moderating role of employee retention and human resource management practices. *Journal of Knowledge Management*. doi:10.1108/JKM-09-2017-0391

Park, S. H., Chen, R., & Gallagher, S. (2002). Firm resources as moderators of the relationship between market growth and strategic alliances in semiconductor start-ups. *Academy of Management Journal*, *45*(3), 527–545.

Parra-Lopez, E., Martínez-González, J.A., & Chinea-Martin, A. (2018). Drivers of the formation of e-loyalty towards tourism destinations. *European Journal of Management and Business Economics, 27*(1), 66 82. doi:. doi:10.1108/EJMBE-11-2017-0050

Patrucco, A. S., Pellizzoni, E., & Buganza, T. (2017). The design process of corporate universities: A stakeholder approach. *Journal of Workplace Learning*, *29*(4), 304–318. doi:10.1108/JWL-04-2016-0024

Payne, A. F., Storbacka, K., & Frow, P. (2008). Managing the co-creation of value. *Journal of the Academy of Marketing Science*, *36*(1), 83–96. doi:10.100711747-007-0070-0

Pentecôte, J. S., & Rondeau, F. (2015). Trade spillovers on output growth during the 2008 financial crisis. *Inter Economics*, *143*, 36–47. doi:10.1016/j.inteco.2015.04.003

Perumandla, S., & Kurisetti, P. (2018). Time - Varying Correlations, Causality, and Volatility Linkages of Indian Commodity and Equity Markets: Evidence from DCC - GARCH. *Indian Journal Of Finance*, *12*(9), 21–40. doi:10.17010/ijf/2018/v12i9/131558

Perumandla, S., & Kurisetti, P. (in press). An empirical examination of correlation dynamics between commodity and equity derivative indices: Evidence from India using DCC-GARCH models. *Afro Asian Journal of Finance and Accounting*.

Pervin, M. M., & Ferdowshi, N. (2016). Suicidal ideation in relation to depression, loneliness and hopelessness among university students. *Dhaka University Journal of Biological Sciences*, *25*(1), 57–64. doi:10.3329/dujbs.v25i1.28495

Peters, L., Raddats, C., & Robinson, W. (2017). Strategic B2B Customer Experience Management: The Importance of Outcomes-Based Measures. *Journal of Services Marketing*, *31*(2). doi:10.1108/JSM-10-2016-0350

Peterson, D. B., & Hicks, M. D. (1996). *Leader as coach*. Minneapolis: Personnel Decisions Inc.

Pett, T. L., & Dibrell, C. C. (2001). A process model of global strategic alliance formation. *Business Process Management Journal*, *7*(4), 349–364. doi:10.1108/14637150110401042

Petty, R., & Cacioppo, J. T. (1981). *Attitudes and Persuasion: Classic and contemporary approaches*. Iowa: Brown.

Pflanz, S. E., & Ogle, A. D. (2006). Job stress, depression, work performance, and perceptions of supervisors in military personnel. *Military Medicine*, *171*(9), 861–865. doi:10.7205/MILMED.171.9.861 PMID:17036607

Phillips, J. J., & Phillips, P. P. (2007). Next-Generation Evaluation: Searching for Value. In M. Allen (Ed.), *Next Generation of Corporate Universities*. San Francisco, CA: Pfeiffer.

Phillips, N., & Tracey, P. (2007). Opportunity recognition, entrepreneurial capabilities and bricolage: Connecting institutional theory and entrepreneurship in strategic organization. *Strategic Organization*, *5*(3), 313–320. doi:10.1177/1476127007079956

Phillips, P. C. B., & Perron, P. (1988). Testing for Unit Roots in Time Series Regression. *Biometrika*, *75*(2), 335–346. doi:10.1093/biomet/75.2.335

Pieterse, A. N., Van Knippenberg, D., Schippers, M., & Stam, D. (2010). Transformational and transactional leadership and innovative behavior: The moderating role of psychological empowerment. *Journal of Organizational Behavior*, *31*(4), 609–623. doi:10.1002/job.650

Piligrimiene, Z., Dovaliene, A., & Virvilaite, R. (2015). Consumer engagement in value co-creation: What kind of value it creates for company? *The Engineering Economist*, *26*(4), 452–460. doi:10.5755/j01.ee.26.4.12502

Pine, B. J. II, Peppers, D., & Rogers, M. (1995). Do you want to keep your customers forever? *Harvard Business Review, 73*(2), 103–114.

Pinto, M. B., & Mansfield, P. M. (2011). College students' attitudes toward the act of gambling: Influence from consumer socialization agents. *Services Marketing Quarterly, 32*(3), 210–227. doi:10.1080/15332969.2011.581968

Pitchbook. (2018). Annual European Venture Report. Retrieved from https://files.pitchbook.com/website/files/pdf/2018_Annual_European_Venture_Report.pdf

Pittman, M. (2018). Happiness, loneliness, and social media: perceived intimacy mediates the emotional benefits of platform use. *The Journal of Social media in Society*, 7, 164-176.

Plompen, M. (2005). *Innovative Corporate Learning: Excellent Management Development Practice in Europe*. New York: Palgrave Macmillan. doi:10.1057/9780230288799

Pollok, P., Lüttgens, D., & Piller, F. T. (2018). How firms develop capabilities for crowdsourcing to increase open innovation performance: The interplay between organizational roles and knowledge processes. *Journal of Product Innovation Management.*

Popa, I. L., Preda, G., & Boldea, M. (2010). A theoretical approach of the concept of innovation. *Managerial Challenges of the Contemporary Society. Proceedings*. Academic Press.

Popadić, M., Černe, M., & Milohnić, I. (2015). Organizational Ambidexterity, Exploration, Exploitation and Firms Innovation Performance. *Organizacija, 48*(2), 112–119. doi:10.1515/orga-2015-0006

Popa, S., Soto-Acosta, P., & Martinez-Conesa, I. (2017). Antecedents, moderators, and outcomes of innovation climate and open innovation: An empirical study in SMEs. *Technological Forecasting and Social Change, 118*, 134–142. doi:10.1016/j.techfore.2017.02.014

Premsankar, G., Di Francesco, M., & Taleb, T. (2018). Edge computing for the Internet of Things: A case study. *IEEE Internet of Things Journal, 5*(2), 1275–1284. doi:10.1109/JIOT.2018.2805263

Preqin. (2018). Venture Capital Deals and Exit Q4. Retrieved from https://docs.preqin.com/reports/Preqin-Venture-Capital-Deals-and-Exits-Q4-2018.pdf

Primack, B. A., Shensa, A., Sidani, J. E., Whaite, E. O., Lin, L. Y., Rosen, D., ... Miller, E. (2017). Social media use and perceived social isolation among young adults in the U.S. *American Journal of Preventive Medicine, 53*(1), 1–8. doi:10.1016/j.amepre.2017.01.010 PMID:28279545

Prince, C., & Stewart, J. (2002). Corporate universities – an analytical framework. *Journal of Management Development, 21*(10), 794–811. doi:10.1108/02621710210448057

Prugsamatz, R. (2010). Factors that influence organization learning sustainability in non-profit organizations. *The Learning Organization, 17*(3), 243–267. doi:10.1108/09696471011034937

Qiao, J. X. (2009). Corporate universities in China: Processes, issues and challenges. *Journal of Workplace Learning, 21*(2), 166–174. doi:10.1108/13665620910934852

Quick, B. L., Morgan, S. E., LaVoie, N. R., & Bosch, D. (2014). Grey's Anatomy viewing and organ donation attitude formation: Examining mediators bridging this relationship among African Americans, Caucasians and Latinos. *Communication Research, 41*(5), 690–716. doi:10.1177/0093650213475476

Rademakers, M. (2005). Corporate universities: Driving force of knowledge innovation. *Journal of Workplace Learning, 17*(1/2), 130–136. doi:10.1108/13665620510574513

Rademakers, M. (2014). *Corporate Universities: Drivers of the Learning Organization*. New York: Routledge. doi:10.4324/9780203073988

Rady, M. Y., McGregor, J. L., & Verheijde, J. L. (2012). Mass media campaigns and organ donation: Managing conflicting messages and interests. *Medicine, Health Care, and Philosophy*, *15*(2), 229–241. doi:10.100711019-011-9359-3 PMID:22020780

Ragheb, A., El-Shimy, H., & Ragheb, G. (2016). Green Architecture: A Concept of Sustainability. *Procedia: Social and Behavioral Sciences*, *216*, 778–787.

Ramaseshan, B., Wong, D., & Turner, B. (2011). Striking the right chord with organ donation, American Marketing Association, 22, 119.

Ramesh, S., & Gupta, A. (1995). *Venture capital and the Indian financial sector*. USA: Oxford University Press.

Rasch, R. H., & Tosi, H. L. (1992). Factors affecting software developers' performance: An integrated approach. *Management Information Systems Quarterly*, *16*(3), 395–413. doi:10.2307/249535

Rasmussen, E., Mosey, S., & Wright, M. (2015). The transformation of network ties to develop entrepreneurial competencies for university spin-offs. *Entrepreneurship and Regional Development*, *27*(7-8), 430–457. doi:10.1080/08985626.2015.1070536

Rather, A. R., Tehseen, S., & Parrey, H. S (2018). Promoting customer brand engagement and brand loyalty through customer brand identification and value congruity. *Spanish Journal of Marketing - ESIC, 22*(3), 319-337. doi:10.1108/SJME-06-2018-0030

Real, J. C., Roldán, J. L., & Leal, A. (2014). From entrepreneurial orientation and learning orientation to business performance: Analysing the mediating role of organizational learning and the moderating effects of organizational size. *British Journal of Management*, *25*(2), 186–208. doi:10.1111/j.1467-8551.2012.00848.x

Rehman, R. U., Zhang, J., & Ali, R. (2014). Firm Performance And Emerging Economies. *Journal of Applied Business Research*, *30*(3), 701. doi:10.19030/jabr.v30i3.8554

Remedios, R., & Boreham, N. (2004). Organisational learning and employees' intrinsic motivation. *Journal of Education and Work*, *17*(2), 219–235. doi:10.1080/1363908041000167741 9

Revees, B., & Read, J. L. (2009). *Total Engagement: Using Games and Virtual Worlds to Change the Way People Work and Businesses Compete*. Boston: Harvard Business School Press; doi:10.1016/j.chb.2010.03.035

Rezadoost, A., & Charvadeh, B. K. (2013). Studying narrative structure of one thousand and one night tales based on Labov's diamond- shaped model. *International Research Journal of Applied and Basic Sciences*, *4*(11), 3324–3327.

Rice, M. P., O'Connor, G. C., Leifer, R., McDermott, C. M., & Standish-Kuon, T. (2000). Corporate venture capital models for promoting radical innovation. *Journal of Marketing Theory and Practice*, *8*(3), 1–10. doi:10.1080/10696679.2000.11501870

Riessman, C. K. (2008). *Narrative Methods for Human Sciences*. London: Sage Publications.

Rinkesh. (2018). *What is Green Architecture?* Retrieved from https://www.conserve-energy-future.com/materials-used-and-principles-green-architecture.php

Riyanti, S., Hatta, M., Norhafizah, S., Balkish, M. N., Siti, Z. M., Hamizatul Akmal, A. H., & Normawati, A. (2014). Organ donation by sociodemographic characteristics in Malaysia. *Asian Social Science*, *10*(4), 264–272.

Roberts, N., & Grover, V. (2012). Investigating firm's customer agility and firm performance: The importance of aligning sense and respond capabilities. *Journal of Business Research*, *65*(5), 579–585. doi:10.1016/j.jbusres.2011.02.009

Robson, M. J., Katsikeas, C. S., Schlegelmilch, B. B., & Pramböck, B. (2018). Alliance capabilities, interpartner attributes, and performance outcomes in international strategic alliances. *Journal of World Business*, *54*(2), 137–153. doi:10.1016/j.jwb.2018.12.004

Rosenbusch, N., Brinckmann, J., & Bausch, A. (2011). Is innovation always beneficial? A meta-analysis of the relationship between innovation and performance in SMEs. *Journal of Business Venturing*, *26*(4), 441–457. doi:10.1016/j.jbusvent.2009.12.002

Roth, A. V. (1996). Achieving strategic agility through economies of knowledge. *Planning Review*, *24*(2), 30–36. doi:10.1108/eb054550

Rubio-Tamayo, J. L., Barrio, M. G., & García, F. G. (2017). Immersive Environments and Virtual Reality: Systematic Review and Advances in Communication, Interaction and Simulation. *Multimodal Technologies and Interaction*, *1*(4), 21. doi:10.3390/mti1040021

Rudtsch, V., Gausemeier, J., Gesing, J., Mittag, T., & Peter, S. (2014). Pattern-based Business Model Development for Cyber-Physical Production Systems. *Procedia CIRP*, *25*, 313–319. doi:10.1016/j.procir.2014.10.044

Ruhnka, J. C., & Young, J. E. (1991). Some hypotheses about risk in venture capital investing. *Journal of Business Venturing*, *6*(2), 115–133. doi:10.1016/0883-9026(91)90014-5

Ruiz Moreno, A., García Morales, V., & Lloréns Montes, F. J. (2005). Learning during the quality management process: Antecedents and effects in service firms. *Industrial Management & Data Systems*, *105*(8), 1001–1021. doi:10.1108/02635570510624419

Russell, D., Peplau, L. A., & Cutrona, C. E. (1980). The revised UCLA loneliness scale– concurrent and discriminate validity evidence. *Journal of Personality and Social Psychology*, *39*(3), 472–480. doi:10.1037/0022-3514.39.3.472 PMID:7431205

Russo, M., & Cesarami, M. (2017). Strategic alliances success factors: A literature review on alliance lifecycle. *International Journal of Business Administration*, *8*(3), 1–9. doi:10.5430/ijba.v8n3p1

Ryan, A. M. (2000). Peer groups as a context for the socialization of adolescents' motivation and education. *Educational Psychologist*, *35*, 2, 11–111. doi:10.1207/S15326985EP3502_4

Ryan, L., Prince, C., & Turner, P. (2015). The Changing and Developing Role of the Corporate University Post-Millennium. *Industry and Higher Education*, *29*(3), 167–174. doi:10.5367/ihe.2015.0256

Ryan, R. M., & Powelson, C. L. (1991). Autonomy and relatedness as fundamental to motivation and education. *Journal of Experimental Education*, *60*(1), 49–66. doi:10.1080/00220973.1991.10806579

Saadat, V., & Saadat, Z. (2016). Organizational learning as a key role of organizational success. *Procedia: Social and Behavioral Sciences*, *230*, 219–225. doi:10.1016/j.sbspro.2016.09.028

Saarijärvi, H., Kannan, P. K., & Kuusela, H. (2013). Value co-creation: Theoretical approaches and practical implications. *European Business Review*, *25*(1), 6–19. doi:10.1108/09555341311287718

Sahlman, W. A. (1990). The structure and governance of venture-capital organizations. *Journal of Financial Economics*, *27*(2), 473–521. doi:10.1016/0304-405X(90)90065-8

Sailer, M., Hense, J. U., Mayr, S. K., & Mandl, H. (2017). How gamification motivates: An experimental study of the effects of specific game design elements on psychological need satisfaction. *Computers in Human Behavior*, *69*, 371–380. doi:10.1016/j.chb.2016.12.033

Saini, S. (2016). *Role of Consumer Experience Management versus Consumer Relationship Management in Consumer Loyalty*. Unpublished Doctoral dissertation, Dr. B R Ambedkar National Institute of Technology.

Salavou, H., Baltas, G., & Lioukas, S. (2004). Organisational innovation in SMEs: The importance of strategic orientation and competitive structure. *European Journal of Marketing*, *38*(9/10), 1091–1112. doi:10.1108/03090560410548889

Salgın, B., Bayram, Ö. F., Akgün, A., & Agyekum, K. (2017). Sustainable Features of Vernacular Architecture: Housing of Eastern Black Sea Region as a Case Study. *Arts*, *6*(11), 1–14.

Santoro, G., Vrontis, D., Thrassou, A., & Dezi, L. (2018). The Internet of Things: Building a knowledge management system for open innovation and knowledge management capacity. *Technological Forecasting and Social Change*, *136*, 347–354. doi:10.1016/j.techfore.2017.02.034

Sanz-Valle, R., Naranjo-Valencia, J. C., Jiménez-Jiménez, D., & Perez-Caballero, L. (2011). Linking organizational learning with technical innovation and organizational culture. *Journal of Knowledge Management*, *15*(6), 997–1015. doi:10.1108/13673271111179334

Sapienza, H. J. (1992). When do venture capitalists add value? *Journal of Business Venturing*, *7*(1), 9–27. doi:10.1016/0883-9026(92)90032-M

Sapsed, J., Grantham, A., & DeFillippi, R. (2007). A bridge over troubled waters: Bridging organisations and entrepreneurial opportunities in emerging sectors. *Research Policy*, *36*(9), 1314–1334. doi:10.1016/j.respol.2007.05.003

Saran, A., & Guo, C. (2005). Competing in the global marketplace: The case of India and China. *Business Horizons*, *48*(2), 135–142. doi:10.1016/j.bushor.2004.10.010

Sarasvathy, S. D., Dew, N., Read, S., & Wiltbank, R. (2008). Designing organizations that design environments: Lessons from entrepreneurial expertise. *Organization Studies*, *29*(3), 331–350. doi:10.1177/0170840607088017

Satrijono, W., Djawahir, K. M., & Sugiarsono, J. (Eds.). (2017). *Indonesia's Best Practices of Corporate University*. Jakarta: PLN Corporate University.

Scharp, K. M., Seiter, J. S., & Curran, T. (2019). Learning supportive communication through an adaptation of the board game Apples to Apples. *Communication Teacher*, *33*(1), 5–10. doi:10.1080/17404622.2018.1530798

Schau, J. H., Muniz, A. M. Jr, & Arnould, E. J. (2009). How brand communities create value. *Journal of Marketing*, *73*(September), 30–51. doi:10.1509/jmkg.73.5.30

Scherer, F. M. (2000). The size distribution of profits from innovation. In *The Economics and Econometrics of Innovation* (pp. 473–494). Boston, MA: Springer. doi:10.1007/978-1-4757-3194-1_19

Schertler, A. (2001). *Venture capital in Europe's common market: a quantitative description* (No. 1087). Kiel working paper.

Schiffer, M., & Weder, B. (2001). Firm size and the business environment: Worldwide survey results. International Finance Cooperation.

Schmidt, N., & Sermat, V. (1983). Measuring loneliness in different relationships. *Journal of Personality and Social Psychology*, *44*(5), 1038–1047. doi:10.1037/0022-3514.44.5.1038

Schmitt, B. H. (1999). *Experiential Marketing: How to Get Consumers to SENSE FEEL THINK ACT and RELATE to Your Company and Brands*. New York, NY: The Free Press.

Schönhals, A., Hepp, T., & Gipp, B. (2018). Design Thinking using the Blockchain: Enable Traceability of Intellectual Property in Problem-Solving Processes for Open Innovation. *Paper presented at the 1st Workshop on Cryptocurrencies and Blockchains for Distributed Systems*. Academic Press. 10.1145/3211933.3211952

Schultz, D., & Schultz, H. (2004). *IMC the next generation: Five steps for delivering value and measuring financial returns*. USA: McGraw Hill.

Scott, G., Boyle, E., Czerniawska, K., & Courtney, A. (2018). Posting photos on Facebook: The impact of narcissism, social anxiety, loneliness, and shyness. *Personality and Individual Differences*, *133*, 67–72. doi:10.1016/j.paid.2016.12.039

Scuotto, V., Del Giudice, M., Bresciani, S., & Meissner, D. (2017). Knowledge-driven preferences in informal inbound open innovation modes. An explorative view on small to medium enterprises. *Journal of Knowledge Management*, *21*(3), 640–655. doi:10.1108/JKM-10-2016-0465

Semrau, T., Ambos, T., & Kraus, S. (2016). Entrepreneurial orientation and SME performance across societal cultures: An international study. *Journal of Business Research*, *69*(5), 1928–1932. doi:10.1016/j.jbusres.2015.10.082

Sengupta, S. (2006). The New World of Private Equity. *Business World*, 16-17.

Senyard, J., Baker, T., & Davidsson, P. (2009). Entrepreneurial bricolage: Towards systematic empirical testing. *Frontiers of Entrepreneurship Research*, *29*(5), 5.

Senyard, J., Baker, T., Steffens, P., & Davidsson, P. (2014). Bricolage as a path to innovativeness for resource-constrained new firms. *Journal of Product Innovation Management*, *31*(2), 211–230. doi:10.1111/jpim.12091

Sharkie, R. (2005). Perceptions of trust as a determinant of the level of knowledge sharing. *International Journal of Knowledge, Culture and Change Management*, *4*(a), 1795-1801.

Sharma, R. K. (2003). Understanding organizational learning through knowledge management. *Journal of Information & Knowledge Management*, *2*(4), 343–352. doi:10.1142/S021964920300053X

Sherehiy, B., Karwowski, W., & Layer, J. K. (2007). A review of enterprise agility: Concepts, frameworks, and attributes. *International Journal of Industrial Ergonomics*, *37*(5), 445–460. doi:10.1016/j.ergon.2007.01.007

Shettar, M., Karkal, R., Kakunje, A., Mendonsa, R. D., & Chandran, V. V. M. (2017). Facebook addiction and loneliness in the post-graduate students of a university in southern India. *The International Journal of Social Psychiatry*, *63*(4), 325–329. doi:10.1177/0020764017705895 PMID:28504040

Shieh, J. I., Wu, H. H., & Huang, K. K. (2010). A DEMATEL method in identifying key success factors of hospital service quality. *Knowledge-Based Systems*, *23*(3), 277–282. doi:10.1016/j.knosys.2010.01.013

Shim, S. (1996). Adolescent consumer decision-making styles: The consumer socialization perspectives. *Psychology and Marketing*, *13*(6), 547–569. doi:10.1002/(SICI)1520-6793(199609)13:6<547::AID-MAR2>3.0.CO;2-8

Shin, H., Lee, J.-N., Kim, D., & Rhim, H. (2015). Strategic agility of Korean small and medium enterprises and its influence on operational and firm performance. *International Journal of Production Economics*, *168*, 181–196. doi:10.1016/j.ijpe.2015.06.015

Shi, W., & Prescott, J. E. (2011). Sequence patterns of firms' acquisition and alliance behaviour and their performance implications. *Journal of Management Studies*, *48*(5), 1044–1070. doi:10.1111/j.1467-6486.2010.00953.x

Shockley-Zabalak, P., & Morely, D. D. (2000). *Trust in Organizations*. San Francisco, CA: IABC Research Foundation.

Siegrist, J. (1996). Adverse health effects of high-effort/low-reward conditions. *Journal of Occupational Health Psychology, 1*(1), 27–41. doi:10.1037/1076-8998.1.1.27 PMID:9547031

Sieradzan, P., & Sobańska, K. (2004). Risk of venture capitale / private equity investments. Basic conditions for the construction of the private equity fund's portfolio Part 1. Studies and Works Collegium of Management and Finance / Warsaw School of Economics, (52), 62-75.

Simonin, B. L. (1999). Ambiguity and the process of knowledge transfer in strategic alliances. *Strategic Management Journal, 20*(7), 595–623. doi:10.1002/(SICI)1097-0266(199907)20:7<595::AID-SMJ47>3.0.CO;2-5

Simonin, B. L. (1999). Transfer of marketing know-how in international strategic alliances: An empirical investigation of the role and antecedents of knowledge ambiguity. *Journal of International Business Studies, 30*(3), 463–490. doi:10.1057/palgrave.jibs.8490079

Singh, J., & Saini, S. (2016). Managing consumer loyalty through acquisition, retention and experience efforts: an empirical study on service consumers in India. *Vision- The Journal of Business Perspective, 20*(3), 121-134.

Singh, I. A. K. (2014). Role of Interpersonal communication in Organizational effectiveness. *International Journal of Research in Management and Business Studies, 1*(4), 36–39.

Sirois, B. C., Sears, S. F. Jr, & Marhefka, S. (2005). Do new drivers equal new donors? An examination of factors influencing organ donation attitudes and behaviors in adolescents. *Journal of Behavioral Medicine, 28*(1), 201–212. doi:10.100710865-005-3669-y PMID:15957575

Sivadas, E., & Dwyer, F. R. (2000). An examination of organizational factors influencing new product success in internal and alliance-based processes. *Journal of Marketing, 64*(1), 31–49. doi:10.1509/jmkg.64.1.31.17985

Slater, M. D., & Rouner, D. (2002). Entertainment-education and elaboration likelihood: Understanding the processing of narrative persuasion. *Communication Theory, 2*(12), 173–191.

Smith, K. (2019). *126 Amazing Social Media Statistics and Facts.* Retrieved from https://www.brandwatch.com/blog/amazing-social-media-statistics-and-facts/#section-10

Smith, K. G., Smith, K. A., Olian, J. D., Sims, H. P., O'Bannon, D. P., & Scully, J. A. (1994). Top management team demography and process: The role of social integration and communication. *Administrative Science Quarterly, 39*(3), 412–438. doi:10.2307/2393297

Smith, M., & Chang, C. (2010). Improving customer outcomes through the implementation of customer relationship mangement evidence from Taiwan. *Asian Review of Accounting, 18*(3), 260–285. doi:10.1108/13217341011089658

Smith, P. A. (2012). The importance of organizational learning for organizational sustainability. *The Learning Organization, 19*(1), 4–10. doi:10.1108/09696471211199285

Soh, P. C. H., Charlton, J. P., & Chew, K. W. (2014). *The influence of parental and peer attachment on Internet usage motives and addiction.* Peer reviewed Journal on the Internet, 19, 7.

Soltani, I., Damirchi, Q. V., & Darban, M. Z. (2011). Organizational culture and organizational innovation. *Arabian journal of business and management review, 4.*

Song, G., Min, S., Lee, S., & Seo, Y. (2017). The effects of network reliance on opportunity recognition: A moderated mediation model of knowledge acquisition and entrepreneurial orientation. *Technological Forecasting and Social Change, 117*, 98–107. doi:10.1016/j.techfore.2017.01.004

Song, H., Zmyslinski-Seelig, A., Kim, J., Drent, A., Victor, A., Omori, K., & Allen, M. (2014). Does Facebook make you lonely?: A meta analysis. *Computers in Human Behavior, 36*, 446–452. doi:10.1016/j.chb.2014.04.011

Spangenberg, E. R., Sprott, D. E., Grohmann, B., & Tracy, D. L. (2006). Gender-congruent ambient scent influences on approach and avoidance behaviors in a retail store. *Journal of Business Research, 59*(November), 1281–1287. doi:10.1016/j.jbusres.2006.08.006

Spekman, R. E., Forbes, T. M., Isabella, L. A., & MacAvoy, T. C. (1998). Alliance management: A view from the past and a look to the future. *Journal of Management Studies, 35*(6), 747–772. doi:10.1111/1467-6486.00118

Spreitzer, G. M. (1995). Psychological empowerment in the workplace: Dimensions, measurement, and validation. *Academy of Management Journal, 38*(5), 1442–1465.

Srai, J. S., Kumar, M., Graham, G., Phillips, W., Tooze, J., Ford, S., ... Tiwari, M. K. (2016). Distributed manufacturing: Scope, challenges and opportunities. *International Journal of Production Research, 54*(23), 6917–6935. doi:10.1080/00207543.2016.1192302

Šramová, B. (2017). *Children's Consumer Behavior*. Consumer Behavior - Practice Oriented Perspectives.

Srivastava, A., Bartol, K. M., & Locke, E. A. (2006). Empowering leadership in management teams: Effects on knowledge sharing, efficacy, and performance. *Academy of Management Journal, 49*(6), 1239–1251. doi:10.5465/amj.2006.23478718

Stasser, G., Stewart, D. D., & Wittenbaum, G. M. (1995). Expert roles and information exchange during discussion: The importance of knowing who knows what. *Journal of Experimental Social Psychology, 31*(3), 244–265. doi:10.1006/jesp.1995.1012

Stasser, G., Vaughan, S. I., & Stewart, D. D. (2000). Pooling unshared information: The benefits of knowing how access to information is distributed among group members. *Organizational Behavior and Human Decision Processes, 82*(1), 102–116. doi:10.1006/obhd.2000.2890

Statista, (2019). Global digital population as of July 2019 (in millions). Retrieved from https://www.statista.com/statistics/617136/digital-population-worldwide/

Steers, M., Wickham, R., & Acitelli, L. (2014). Seeing everyone else's highlight reels: How Facebook usage is linked to depressive symptoms. *Journal of Social and Clinical Psychology, 33*(8), 701–731. doi:10.1521/jscp.2014.33.8.701

Steptoe, A., Shankar, A., Demakakos, P., & Wardle, J. (2013). Social isolation, loneliness, and all-cause mortality in older men and women. *Proceedings of the National Academy of Sciences of the United States of America, 110*(15), 5797–5801. doi:10.1073/pnas.1219686110 PMID:23530191

Storbacka, K., Brodie, R. J., Böhmann, T., Maglio, P. P., & Nenonen, S. (2016). Actor engagement as a microfoundation for value co-creation. *Journal of Business Research, 69*(8), 3008–3017. doi:10.1016/j.jbusres.2016.02.034

Storey, D. J. (1985). The Problems Facing New Firms. *Journal of Management Studies, 22*(3), 327–345. doi:10.1111/j.1467-6486.1985.tb00079.x

Subhash, K. B. (2004). Globalisation and Indian venture capital system.

Sun, Y., Liu, J., & Ding, Y. (2019). Analysis of the relationship between open innovation, knowledge management capability and dual innovation. *Technology Analysis and Strategic Management*, 1–14. doi:10.1080/09537325.2019.1632431

Sussman, S. W., & Siegal, W. S. (2003). Informational influence in organizations: An integrated approach to knowledge adoption. *Information Systems Research, 14*(1), 47–65. doi:10.1287/isre.14.1.47.14767

Swafford, P. M., Ghosh, S., & Murthy, N. (2006a). The antecedents of supply chain agility of a firm: Scale development and model testing. *Journal of Operations Management, 24*(2), 170–188. doi:10.1016/j.jom.2005.05.002

Swafford, P. M., Ghosh, S., & Murthy, N. N. (2006b). A framework for assessing value chain agility. *International Journal of Operations & Production Management, 26*(2), 118–140. doi:10.1108/01443570610641639

Swenson, R. R., & Prelow, H. M. (2005). Ethnic identity, self-esteem and perceived efficacy as mediators of the relation of supportive parenting to psychosocial outcomes among urban adolescents. *Journal of Adolescence, 28*(4), 465–477. doi:10.1016/j.adolescence.2004.09.005 PMID:16022882

Swift, P. E., & Hwang, A. (2013). The impact of affective and cognitive trust on knowledge sharing and organizational learning. *The Learning Organization, 20*(1), 20–37. doi:10.1108/09696471311288500

Tahir-Kheli, S., & Biringer, K. L. (2001). *Reducing risk in South Asia: Managing India-Pakistan tensions.* Cooperative Monitoring Center, Sandia National Laboratories.

Taiwo, M., Ayodeji, A., & Yusuf, B. (2012). Impact of small and medium enterprises on economic growth and development. *American Journal of Business and Management, 1*(1), 18–22. doi:10.11634/21679606170644

Tallon, P. P., & Pinsonneault, A. (2011). Competing perspectives on the link between strategic information technology alignment and organizational agility: Insights from a mediation model. *Management Information Systems Quarterly, 35*(2), 463–486. doi:10.2307/23044052

Tang, A., Tietze, F., & Molloy, J. (2019). Openness in Intellectual Property Strategies of Synthetic Biology Start-ups. *Paper presented at the 2019 IEEE Technology & Engineering Management Conference (TEMSCON).* IEEE Press. 10.1109/TEMSCON.2019.8813740

Tao, F., Qi, Q., Liu, A., & Kusiak, A. (2018). Data-driven smart manufacturing. *Journal of Manufacturing Systems, 48*, 157–169. doi:10.1016/j.jmsy.2018.01.006

Tartaglione, A. M., Cavacece, Y., Russo, G., & Granata, G. (2019). A Systematic Mapping Study on Customer Loyalty and Brand Management, *Administrative Sciences, MDPI. Open Access Journal, 9*(1), 1–21.

Teker, D., Teker, S., & Teraman, Ö. (2016). Venture capital markets: A cross country analysis. *Procedia Economics and Finance, 38*, 213–218. doi:10.1016/S2212-5671(16)30192-7

Teng, K. T., Yen, S. H., Chua, S. Y., & Lean, H. H. (2013). Financial market integration of ASEAN-5 with China and India. In *Emerging markets and the global economy: A handbook* (pp. 825-859). Elsevier.

Teo, H. H., & Wang, X. (2005). Organizational Learning Capacity and Attitude toward Complex Technological Innovations: An Empirical Study. *International Science, 57*(2), 264–279.

Teoh, A. P., Lee, K. Y., & Muthuveloo, R. (2017). The impact of enterprise risk management, strategic agility, and quality of internal audit function on firm performance. *International Review of Management and Marketing, 7*(1), 222–229.

Thacker, S., & Kaul, M. (2017). The Evolution of Private Equity in India. *Ct. Uncourt, 4*, 5.

Tharion, W. J., Lieberman, H. R., Montain, S. J., Young, A. J., Baker-Fulco, C. J., DeLany, J. P., & Hoyt, R. W. (2005). Energy requirements of military personnel. *Appetite, 44*(1), 47–65. doi:10.1016/j.appet.2003.11.010 PMID:15604033

The Star. (2017). Malaysia has the most Wazers.

The Star. (2018). Make green growth a priority. Retrieved from https://www.thestar.com.my/news/nation/2018/10/19/make-green-growth-a-priority-we-must-focus-on-making-sustainable-development-a-mainstream-goal/

Thiruchelvan, S. (2017). *Working Environment and Its Influence on Employees' Performance: A Case Study of an Oils and Gas Vendor Company in Malaysia* [Doctoral dissertation]. UTAR.

Thomas, J. S. (2001). A methodology for linking customer acquisition to customer retention, *Journal of Marketing Research, 38*(2), 262-268. Retrieved from http://www.jstor.org/stable/1558629

Thomas, R. (2009). The Principles of Green Building Design,

Thomas, J. B., & Trevino, L. K. (1993). Information processing in strategic alliance building: A multiple-case approach. *Journal of Management Studies, 30*(5), 779–814. doi:10.1111/j.1467-6486.1993.tb00326.x

Thompson, G. (2000). Unfulfilled Prophecy: The Evolution of Corporate Colleges. *The Journal of Higher Education, 71*(3), 322–341. Retrieved from http://www.jstor.org/stable/2943781

Thornton, S. C., Henneberg, S. C., & Naudé, P. (2014). Conceptualizing and validating organizational networking as a second-order formative construct. *Industrial Marketing Management, 43*(6), 951–966. doi:10.1016/j.indmarman.2014.05.001

Tian Xie, F., & Johnston, W. J. (2004). Strategic alliances: Incorporating the impact of e-business technological innovations. *Journal of Business and Industrial Marketing, 19*(3), 208–222. doi:10.1108/08858620410531342

Tidd, J., & Bessant, J. R. (2018). *Managing innovation: integrating technological, market and organizational change*. John Wiley & Sons.

Tippins, M. J., & Sohi, R. S. (2003). IT competency and firm performance: Is organizational learning a missing link? *Strategic Management Journal, 24*(8), 745–761. doi:10.1002mj.337

Tipu, S. A. A., & Fantazy, K. A. (2014). Supply chain strategy, flexibility, and performance: A comparative study of SMEs in Pakistan and Canada. *International Journal of Logistics Management, 25*(2), 399–416. doi:10.1108/IJLM-01-2013-0001

Tobin, S., Vanman, E., Verreynne, M., & Saeri, A. (2014). Threats to belonging on Facebook: Lurking and ostracism. *Social Influence, 10*(1), 31–42. doi:10.1080/15534510.2014.893924

Todeva, E., & Knoke, D. (2005). Strategic alliances & model of collaboration. *Management Decision, 43*(1), 1–22. doi:10.1108/00251740510572533

Toma, A., Secundo, G., & Passiante, G. (2018). Open innovation and intellectual property strategies: Empirical evidence from a bio-pharmaceutical case study. *Business Process Management Journal, 24*(2), 501–516. doi:10.1108/BPMJ-11-2016-0230

Toma, C. (2013). Feeling better but doing worse: Effects of Facebook self-presentation on implicit self-esteem and cognitive task performance. *Media Psychology, 16*(2), 199–220. doi:10.1080/15213269.2012.762189

Topu, F. B., Reisoğlu, İ., Yılmaz, T. K., & Göktaş, Y. (2018). Information retention's relationships with flow, presence and engagement in guided 3D virtual environments. *Education and Information Technologies, 23*(4), 1621–1637. doi:10.100710639-017-9683-1

Trott, P., & Hartmann, D. (2009). Why 'open innovation' is old wine in new bottles. *International Journal of Innovation Management, 13*(04), 715–736. doi:10.1142/S1363919609002509

Tuul, O., & Bing, S. (2019). Employee training in small and medium-sized enterprises in Mongolia. *Proceedings of the Mongolian Academy of Sciences, 59*(1), 75–84.

Tyson, S., & Ward, P. (2004). The Use of 360 Degree Feedback Technique in the Evaluation of Management Development. *Management Learning, 35*(2), 205–223. doi:10.1177/1350507604043025

Tzannatos, Z. (1999). Women and labor market changes in the global economy: Growth helps, inequalities hurt and public policy matters. *World Development*, *27*(3), 551–569. doi:10.1016/S0305-750X(98)00156-9

Tzeng, G. H., Chiang, C. H., & Li, C. W. (2007). Evaluating intertwined effects in e-learning programs: A novel hybrid MCDM model based on factor analysis and DEMATEL. *Expert Systems with Applications*, *32*(4), 1028–1044. doi:10.1016/j.eswa.2006.02.004

U.S. Environmental Protection Agency. (2016). Green Building. Retrieved from https://archive.epa.gov/greenbuilding/web/html/

Umberson, D. K., & Montez, J. (2010). Social relationships and health: A flashpoint for health policy. *Journal of Health and Social Behavior*, *51*(1 suppl), S54–S66. doi:10.1177/0022146510383501 PMID:20943583

Upadhyay, P., & Paul, M. (2019). The linkage between knowledge management practices and organization based projects for better learning outcome. *International Journal of Educational Management*, *33*(1), 166–178. doi:10.1108/IJEM-04-2018-0133

Urbain, C., Gonzalez, C., & Le Gall-Ely, M. (2013). What does the future hold for giving? An approach using the social representations of generation Y. *International Journal of Nonprofit and Voluntary Sector Marketing*, *18*(3), 159–171. doi:10.1002/nvsm.1448

Uzniene, R. (2010). Media – Agents of socialization. *Regional Formation and Development Studies*, *3*(8), 231–239.

Uzuegbunam, I., Ofem, B., & Nambisan, S. (2019). Do Corporate Investors Affect Entrepreneurs' IP Portfolio? Entrepreneurial Finance and Intellectual Property in New Firms. *Entrepreneurship Theory and Practice*, *43*(4), 673–696. doi:10.1177/1042258717738247

Valtorta, N., & Hanratty, B. (2012). Loneliness, isolation and the health of older adults: Do we need a new research agenda? *Journal of the Royal Society of Medicine*, *105*(12), 518–522. doi:10.1258/jrsm.2012.120128 PMID:23288086

Valtorta, N., Kanaan, M., Gilbody, S., & Hanratty, B. (2016). Loneliness, social isolation and social relationships: What are we measuring? A novel framework for classifying and comparing tools. *BMJ Open*, *6*(4), e010799. doi:10.1136/bmjopen-2015-010799 PMID:27091822

Van de Vrande, V., De Jong, J. P., Vanhaverbeke, W., & De Rochemont, M. (2009). Open innovation in SMEs: Trends, motives and management challenges. *Technovation*, *29*(6-7), 423–437. doi:10.1016/j.technovation.2008.10.001

van den Broek, J., Boselie, P., & Paauwe, J. (2018). Cooperative innovation through a talent management pool: A qualitative study on coopetition in healthcare. *European Management Journal*, *36*(1), 135–144. doi:10.1016/j.emj.2017.03.012

van der Klauw, M., Hengel, K. O., Roozeboom, M. B., Koppes, L. L., & Venema, A. (2016). Occupational accidents in the Netherlands: Incidence, mental harm, and their relationship with psychosocial factors at work. *International Journal of Injury Control and Safety Promotion*, *23*(1), 79–84. doi:10.1080/17457300.2014.966119 PMID:25531558

Varadarajan, P. R., & Cunningham, M. H. (1995). Strategic alliances: A synthesis of conceptual foundations. *Journal of the Academy of Marketing Science*, *23*(4), 282–296. doi:10.1177/009207039502300408

Vargo, S. L., & Lusch, R. F. (2004). Evolving to a New Dominant Logic for Marketing. *Journal of Marketing*, *68*(1), 1–17. doi:10.1509/jmkg.68.1.1.24036

Vargo, S. L., & Lusch, R. F. (2008). Service-dominant logic: Continuing the evolution. *Journal of the Academy of Marketing Science*, *36*(1), 1–10. doi:10.100711747-007-0069-6

Vargo, S. L., & Lusch, R. F. (2016). Institutions and axioms: An extension and update of service-dominant logic. *Journal of the Academy of Marketing Science*, *44*(1), 5–23. doi:10.100711747-015-0456-3

Varra, E., Kleymann, B., & Seristo, H. (2004). Strategies as discursive constructions: The case of airline alliances. *Journal of Management Studies*, *41*(1), 1–35. doi:10.1111/j.1467-6486.2004.00419.x

Vega-Vazquez, M., Revilla-Camacho, M. Á., Cossío-Silva, F. J., & Cossı, F. J. (2013). The value co-creation process as a determinant of customer satisfaction. *Management Decision*, *51*(10), 1945–1953. doi:10.1108/MD-04-2013-0227

Verhoef, P. C., & Donkers, B. (2001). Predicting customer potential value: An application in the insurance industry. *Decision Support Systems*, *32*(2), 189–199. doi:10.1016/S0167-9236(01)00110-5

Vickery, S., Droge, C., Setia, P., & Sambamurthy, V. (2010). Supply chain information technologies and organisational initiatives: Complementary versus independent effects on agility and firm performance. *International Journal of Production Research*, *48*(23), 7025–7042. doi:10.1080/00207540903348353

Victor, C., Scambler, S., & Bond, J. (2009). *The social world of older people: Understanding Loneliness and Social Isolation in Later Life*. Maidenhead: Open University Press.

Vinayak, H. V., Thompson, F., & Tonby, O. (2014). Understanding ASEAN: Seven things you need to know. *growth*, *2000*, 13.

Virk, N., & Javed, F. (2017). European equity market integration and joint relationship of conditional volatility and correlations. *Journal of International Money and Finance*, *71*, 53–77. doi:10.1016/j.jimonfin.2016.10.007

Volberda, H. W. (1996). Toward the flexible form: How to remain vital in hypercompetitive environments. *Organization Science*, *7*(4), 359–374. doi:10.1287/orsc.7.4.359

Von Zedtwitz, M. (2004). Managing foreign R&D laboratories in China. *Research Management*, *34*(4), 439–452.

Vygotsky, L. S. (1978). *The mind in society: The development of higher psychological processes*. Cambridge, MA: Harvard University Press.

Wallin, M. W., & Krogh, G. (2010). Organizing for open innovation: Focus on the integration of knowledge. *Organizational Dynamics*, *39*(2), 145–154. doi:10.1016/j.orgdyn.2010.01.010

Walter, A., Auer, M., & Ritter, T. (2006). The impact of network capabilities and entrepreneurial orientation on university spin-off performance. *Journal of Business Venturing*, *21*(4), 541–567. doi:10.1016/j.jbusvent.2005.02.005

Wälti, S. (2011). Stock market synchronization and monetary integration. *Journal of International Money and Finance*, *30*(1), 96–110. doi:10.1016/j.jimonfin.2010.07.004

Walton, J. (2005). Would the real corporate university please stand up? *Journal of European Industrial Training*, *29*(1), 7–20. doi:10.1108/03090590510576181

Walton, R. E., Allen, C., & Gaffney, M. (1987). *Innovating to compete: Lessons for diffusing and managing change in the workplace*. Jossey-Bass.

Wang, A. (2009). Government's role as public venture capitalist in high-technology small and medium sized enterprises.

Wang, M., Zhang, W., & Wang, X. (2008). The principle of family determination in organ donation: The application of Confucian Ethics. *HEC Forum*, *20*(2), 183–196. doi:10.100710730-008-9069-2 PMID:18642090

Waze. (2018). Data, Driven. Retrieved from https://www.waze.com/brands/drivers/

Weber, R. P. (1990). Basic content analysis. Thousand Oaks, CA: Sage. doi:10.4135/9781412983488

Weber, Y., & Tarba, S. Y. (2014). Strategic Agility: A State of the Art Introduction to the Special Section on Strategic Agility. *California Management Review, 56*(3), 5–12. doi:10.1525/cmr.2014.56.3.5

Weeks, D., Michela, J., Peplau, L., & Bragg, M. (1980). Relation between loneliness and depression: A structural equation analysis. *Journal of Personality and Social Psychology, 39*(6), 1238–1244. doi:10.1037/h0077709 PMID:7205551

Weiss, R. S. (1973). *Loneliness: The experience of emotional and social isolation London.* MIT Press. Retrieved from https://mitpress.mit.edu/ books/loneliness

Weiss, A. M. (1985). Absenteeism and Wages. *Economics Letters, 19*(3), 277–279. doi:10.1016/0165-1765(85)90036-9

Wemmer, F., Emrich, E., & Koenigstorfer, J. (2016). The impact of coopetition-based open innovation on performance in nonprofit sports clubs. *European Sport Management Quarterly, 16*(3), 341–363. doi:10.1080/16184742.2016.1164735

Wentzel, K. R. (1991). Social competence at school: Relations between social responsibility and academic achievement. *Review of Educational Research, 61*(1), 1–24. doi:10.3102/00346543061001001

Wentzel, K. R., & Looney, L. (2006). Socialization in school settings. In *J.E. Grusec & P.D. Hastings (Eds.), Handbook of Socialization: Theory and Research.* New York: Guilford Press.

Werbach, K., & Hunter, D. (2012). *For win: How game thinking can revolutionize your business.* Philadelphia: Wharton Digital Press.

Wernerfelt, B. (1984). A resource-based view of the firm. *Strategic Management Journal, 5*(2), 171–180. doi:10.1002mj.4250050207

West, J., & Gallagher, S. (2006). Challenges of open innovation: The paradox of firm investment in open-source software. *Research Management, 36*(3), 319–331.

Whipple, J. M., & Frankel, R. (2000). Strategic alliance success factor. *The Journal of Supply Chain Management, 36*(2), 21–28. doi:10.1111/j.1745-493X.2000.tb00248.x

Wilson, D. (2005). *New perspectives in applied communication.* Ibadan: Sterlin-Hordan Publishers.

Wiltshire, T. J., Steffensen, S. V., & Fiore, S. M. (2019). Multiscale movement coordination dynamics in collaborative team problem solving. *Applied Ergonomics, 79*, 143–151. doi:10.1016/j.apergo.2018.07.007 PMID:30103905

Wiman, A. R. (1983). Parental influence and children's responses to television advertising. *Journal of Advertising, 12*(1), 12–18. doi:10.1080/00913367.1983.10672825

Wines, J. (2019). Green Architecture. Encyclopedia Britannia. Retrieved from https://www.britannica.com/art/green-architecture

Winkel, V. M., Wichers, M., Collip, D., Jacobs, N., Derom, C., & Thiery, E. (2017). Unraveling the role of loneliness in depression: The relationship between daily life experience and behavior. *Psychiatry, 80*, 104–117. doi:10.1080/00332747.2016.1256143

Wiseman, H., Mayseless, O., & Sharabany, R. (2006). Why are they lonely? Perceived quality of early relationships with parents, attachment, personality predispositions and loneliness in first-year university students. *Personality and Individual Differences, 40*(2), 237–248. doi:10.1016/j.paid.2005.05.015

Wiwchar, M., Walker, D., & Marsh, R. (2020). *Protecting Your Digital Health Intellectual Property: Fundamentals of Intellectual Property and How It Applies to Software, Hardware and Business Processes. In Digital Health Entrepreneurship* (pp. 103–118). Springer.

Wolniczak, I., Cáceres-DelAguila, J., Palma-Ardiles, G., Arroyo, K., Solís-Visscher, R., Paredes-Yauri, S., ... Bernabe-Ortiz, A. (2013). Association between Facebook dependence and poor sleep quality: A study in a sample of undergraduate students in Peru. *PLoS One*, *8*(3), e59087. doi:10.1371/journal.pone.0059087 PMID:23554978

Wong, A. (2011). The Evolution of the Venture Capital Market in China: Current Trends in Venture Capital Financing Strategies and Investment Preferences. *Journal of Investing*, *20*(4), 16–26. doi:10.3905/joi.2011.20.4.016

Wong, L. P. (2010). Information needs, preferred educational messages and channel of delivery, and opinion on strategies to promote organ donation: A multicultural perspectives. *Singapore Medical Journal*, *51*(10), 790–795. PMID:21103814

Wood, P. (1998). The rise of consultancy and the prospect for regions. *Paper presented at the 38th Congress of the European Regional Science Association*. Academic Press.

Wood, C. (2013). Loneliness: A silent epidemic. *Independent Nurse*, *6*. doi:10.12968/indn.2013.17.6.99114

WorkforceSkills Preparedness Report. (2016). PayScale.

World Bank. (2019). *Doing Business 2019* (16th ed.). World Bank Group.

Wright, M. L. (2005). On the gains from international financial integration. *Economics Letters*, *87*(3), 379–386. doi:10.1016/j.econlet.2005.01.007

Wu, S. P.-J., Straub, D. W., & Liang, T.-P. (2015). How information technology governance mechanisms and strategic alignment influence organizational performance: Insights from a matched survey of business and IT managers. *Management Information Systems Quarterly*, *39*(2), 497–518. doi:10.25300/MISQ/2015/39.2.10

Wu, W.-W., & Lee, Y.-T. (2007). Developing global managers' competencies using the fuzzy DEMATEL method. *Expert Systems with Applications*, *32*(2), 499–507. doi:10.1016/j.eswa.2005.12.005

Xia, N., & Li, H. (2018). Loneliness, social isolation, and cardiovascular health. *Antioxidants & Redox Signalling*, *28*(9), 837–851. doi:10.1089/ars.2017.7312 PMID:28903579

Yadav, R., Shukla, N. K., & Gupta, R. (2019). All-Digital Phase Locked Loop (ADPLL) as an Intellectual Property (IP) Core for an Application-specified Integrated Circuit (ASIC) Product: A Survey. *Journal of VLSI Design Tools & Technology*, *5*(3), 12–17.

Yamakawa, Y., Yang, H., & Lin, Z. J. (2011). Exploration versus exploitation in alliance portfolio: Performance implications of organizational, strategic, and environmental fit. *Research Policy*, *40*(2), 287–296. doi:10.1016/j.respol.2010.10.006

Yang, Y. P. O., Shieh, H. M., Leu, J. D., & Tzeng, G. H. (2008). A novel hybrid MCDM model combined with DEMATEL and ANP with applications. *International journal of operations research, 5*(3), 160-168.

Yang, Y. P., Shieh, H. M., Leu, J. D., & Tzeng, G. H. (2008). A novel hybrid MCDM model combined with DEMATEL and ANP with applications. *International Journal Operational Research*, *5*(3), 160–168.

Yavich, R., Davidivutch, N., & Frenkel, Z. (2019). Social media and loneliness - forever connected? *Higher Education Studies*, *9*(2), 10–21. doi:10.5539/hes.v9n2p10

Yoo, Y., & Kanawattanachai, P. (2001). Developments of transactive memory systems and collective mind in virtual teams. *The International Journal of Organizational Analysis*, *9*(2), 187–208. doi:10.1108/eb028933

Yousaf, Z., & Majid, A. (2016). Strategic performance through inter-firm networks: Strategic alignment and moderating role of environmental dynamism. *World Journal of Entrepreneurship, Management and Sustainable Development*, *12*(4), 282–298. doi:10.1108/WJEMSD-03-2016-0015

Yousaf, Z., & Majid, A. (2018). Organizational network and strategic business performance: Does organizational flexibility and entrepreneurial orientation really matter? *Journal of Organizational Change Management*, *31*(2), 268–285. doi:10.1108/JOCM-12-2016-0298

Yousfi, A., Saidi, R., & Dey, A. K. (2016). Variability patterns for business processes in BPMN. *Information Systems and e-Business Management*, *14*(3), 443–467. doi:10.100710257-015-0290-7

Yuan, L. J. (1987). *The Malay House: Rediscovering Malaysia's Indigenous Shelter System*. Penang: Institute Masyarakat.

Yunus, A., & Landau, E. (2019), Youth Now Defined as Those Between 15 to 30. New Straits Times. Retrieved from https://www.nst.com.my/news/nation/2019/07/501288/youth-now-defined-those-between-15-and-30

Yu, R., & Tseng, G. H. (2006). A soft computing method for multi-criteria decision making with dependence and feedback. *Applied Mathematics and Computation*, *180*(1), 63–75. doi:10.1016/j.amc.2005.11.163

Zacharakis, A. L., McMullen, J. S., & Shepherd, D. A. (2007). Venture capitalists' decision policies across three countries: An institutional theory perspective. *Journal of International Business Studies*, *38*(5), 691–708. doi:10.1057/palgrave.jibs.8400291

Zachary, W. B., Zachary, W. B., & Kuzuhara, L. W. (2005). *Organizational behavior: Integrated models and applications*. South-Western Pub.

Zaheer, A., & Bell, G. G. (2005). Benefiting from network position: Firm capabilities, structural holes, and performance. *Strategic Management Journal*, *26*(9), 809–825. doi:10.1002mj.482

Zahra, S. A., Sapienza, H. J., & Davidsson, P. (2006). Entrepreneurship and dynamic capabilities: A review, model and research agenda. *Journal of Management Studies*, *43*(4), 917–955. doi:10.1111/j.1467-6486.2006.00616.x

Zeithaml, V. A., Berry, L. L., & Parasuraman, A. (1996). The behavioral consequences of service quality. *Journal of Marketing, 60*(2), 31-46. Retrieved from http://www.jstor.org/stable/1251929

Zelikowsky, M., Hui, M., Karigo, T., Choe, A., Yang, B., Blanco, M., ... Anderson, D. J. (2018). The neuropeptide controls a distributed brain state induced by chronic social isolation stress. *Cell*, *173*(5), 1265–1279. doi:10.1016/j.cell.2018.03.037 PMID:29775595

Zeng, F. (2004). *Venture capital investments in China*. Santa Monica, CA: Rand Corp.

Zeng, Y., Zhang, J., He, K., & Cheng, L. (2019). Who cares what parents think or do? Observational learning and experience-based learning through communication in rice farmers' willingness to adopt sustainable agricultural technologies in Hubei Province, China. *Environmental Science and Pollution Research International*, 26(12), 1–15. PMID:30850982

Zhang, B., Wardrop, R., Rau, P. R., & Gray, M. (2015). Moving mainstream: Benchmarking the European alternative finance market. *Journal of Financial Perspectives*, *3*(3).

Zhang, Y., & Mayes, D. G. (2018). The performance of governmental venture capital firms: A life cycle perspective and evidence from China. *Pacific-Basin Finance Journal*, *48*, 162–185. doi:10.1016/j.pacfin.2018.02.002

Zhao, X.-y., Frese, M., & Giardini, A. (2010). Business owners' network size and business growth in China: The role of comprehensive social competency. *Entrepreneurship and Regional Development*, *22*(7-8), 675–705. doi:10.1080/08985620903171376

Zhao, Y., & Zhu, Q. (2014). Effects of extrinsic and intrinsic motivation on participation in crowdsourcing contest. *Online Information Review*, *38*(7), 896–917. doi:10.1108/OIR-08-2014-0188

Zheng, H., Li, D., & Hou, W. (2011). Task Design, Motivation, and Participation in Crowdsourcing Contests. *International Journal of Electronic Commerce*, *15*(4), 57–88. doi:10.2753/JEC1086-4415150402

Zheng, S., Zhang, W., & Du, J. (2011). Knowledge-based dynamic capabilities and innovation in networked environments. *Journal of Knowledge Management*, *15*(6), 1035–1051. doi:10.1108/13673271111179352

Zider, B. (1998). How venture capital works. *Harvard Business Review*, *76*(6), 131–139. PMID:10187243

Zineldin, M. (2006). The royalty of loyalty: CRM quality and retention. *Journal of Consumer Marketing*, *23*(7), 430–437. doi:10.1108/07363760610712975

Zineldin, M., & Bredenlow, T. (2003). Strategic alliance: synergies and challenges: A case of strategic outsourcing relationships "SOUR". *International Journal of Physical Distribution & Logistics Management*, *33*(5), 449–464. doi:10.1108/09600030310482004

Zolkepli, I. A., & Kamarulzaman, Y. (2015). Social Media Adoption: The Role of Media Needs and Innovation Characteristics. *Computers in Human Behavior*, *43*, 189–209. doi:10.1016/j.chb.2014.10.050

Zolkiewski, J., Story, V., Burton, J., Chan, P., Gomes, A., Hunter-Jones, P., ... Robinson, W. (2017). Strategic B2B customer experience management: The importance of outcomes-based measures. *Journal of Services Marketing*, *31*(2), 172–184.

# About the Contributors

**Patricia Ordóñez de Pablos** is a professor in the Department of Business Administration in the Faculty of Business and Economics at The University of Oviedo (Spain). She completed her education in The London School of Economics (UK). Her teaching and research interests focus on the areas of strategic management, knowledge management, organizational learning, human resource management, intellectual capital, information technologies, with special interest in Asia (Bhutan, China, Laos, Myanmar). She is Editor-in-Chief of the International Journal of Learning and Intellectual Capital (IJLIC) and International Journal of Asian Business and Information Management (IJABIM), respectively. She has edited books for IGI Global, Routledge, and Springer.

**Xi Zhang** is a professor of College of management and Economics, Tianjin University of China. His research include knowledge management and big data analytics, with several publications in international journals such as JMIS, IEEE TEM, CHB, JKM.

**Kwok Tai Chui** received the B.Eng. degree and Ph.D. degree from City University of Hong Kong. He had industry experience as Senior Data Scientist in Internet of Things (IoT) company. He joined the Department of Technology, School of Science and Technology, at The Open University of Hong Kong as Research Assistant Professor. He was the recipient of 2nd Prize Award (Postgraduate Category) of 2014 IEEE Region 10 Student Paper Contest. Also, he received Best Paper Award in IEEE The International Conference on Consumer Electronics-China, in both 2014 and 2015. He has more than 40 research publications: edited books, book chapters, journal papers, and conference papers. His research interests include computational intelligence, wireless communication, pattern recognition, healthcare, machine learning algorithms and optimization.

***

**Noor Hazlina Ahmad**, Ph.D., is a Professor and Dean at the School of Management USM. She joined the university after completing her Ph.D. at the University of Adelaide, Australia. She teaches entrepreneurship and organizational behavior related subjects. She also supervises Ph.D. and MA students in the field of entrepreneurship and SMEs. Currently, she serves as the President of the Asian Academy of Management and sits on the editorial board of the Asian Academy of Management Journal. She has served as a visiting professor and external assessor at the Management & Science University. She is also actively involved with consultation programs and served as a facilitator for community, entrepreneurship and leadership projects. Her research interests are in the areas of Entrepreneurship, SMEs and Organisational Behaviour.

**Elbegjargal Enkh-Amgalan** is an Economist of the Research Division at the Bank of Mongolia. Her research interests focus on microeconomics, international economics and survey designing.

**Widi Galih Anggara** is an employee at the Financial Education and Training Agency which is also known as the Ministry of Finance Corporate University - Ministry of Finance of the Republic of Indonesia since 2010. He received his master's degree in management science from the School of Business and Management, Bandung Institute of Technology. He actively follows corporate learning issues in Indonesia, particularly corporate university.

**Khuslen Batmunkh** is an Economist of the Central bank of Mongolia. His research interests include monetary policy transmission mechanism, economic connectedness and business cycle. His research has been published in peer reviewed international journals such as Journal of Applied Econometrics and Central Bank Review journal. He is also the one author of BBQ add-in software and published the quest blog about the business cycle in EViews.

**Som Bhattacharyya** is a professor on the Strategic Management faculty at National Institute of Industrial Engineering (NITIE), Mumbai, India. He has published many research papers and teaching cases in international journals of repute. He was awarded Emerald Literati Awards for Excellence: Outstanding Paper Award – 2011 and Emerald Literati Awards for Excellence: Highly Commended Paper Award – 2014. For teaching, he received NITIE best teacher award 2014 and Business School Award for Best Professor in Strategic Management in India in 2014.

**Izzal Asnira Zolkepli** is best known for her work on convergence and adoption of communication technologies; and the proliferation of digital, social media and networked media pertaining to user engagement (i.e. behavioral response). Her current research projects are on user engagement (voluntary behavior) which look at crowdsourcing as well as user engagement on adopting/using mobile apps. Dr. Izzal is currently serving as a senior lecturer at Universiti Sains Malaysia, a research-intensive public university that is recognised as the top 50 university under 50 years in the world by QS World Ranking. Dr. Izzal is lecturing marketing communication subjects (persuasion communication, marketing management, and international advertising) that are offered in the School of Communication.

**Cheah Wen Kit** is currently a PhD candidate in Universiti Sains Malaysia and has served in various universities and colleges in Malaysia. Trained in various discipline of marketing communication - including digital marketing, advertising, and broadcasting, he obtained his academic qualification in the field of psychology and communication. His research interest is in Internet user experience and behaviour in the age of Industry 4.0. He is currently developing an integrated communication model of value co-creation by incorporating gamification technology.

**Boon Cheong Chew** obtained his first degree-Bachelor of Engineering (Civil) from Faculty of Civil Engineering University of Technology Malaysia (UTM) as well as his master's degree-Master of Technology Management. He spent four-year time working at one of the Malaysian Best Managed Companies, YTL Corporation, Bhd., under one of the divisions-YTL Cement Bhd. In 2007, he obtained a scholarship (funded by the Minister of Higher Education) and went for further study to the University of Edinburgh Scotland. He obtained his PHD on 26 June 2012.

**Mohammad Ashraful Ferdous Chowdhury** is the Associate Professor of the Department of Business Administration, Shahjalal University of Science & Technology (SUST), Sylhet, Bangladesh. He is a senior researcher graduated with a bachelor's degree in finance and MBA in finance & banking from Shahjalal University of Science & Technology (SUST), Bangladesh. He also holds PhD. and MSc degree in Islamic Finance from International Centre for Education in Islamic Finance (INCEIF), The Global University of Islamic Finance, a post-graduate university specializing in Islamic Finance, established by the Central Bank of Malaysia. He has managed to publish a number of articles in refereed international and local journals published by Inderscience, Emerald, Springer, and other publishers. His main research interests are in the area of Islamic capital market, Islamic banking, sustainable development, ethical and socially responsible investments. Mohammad Ashraful Ferdous Chowdury is a financial economist with consulting, research and teaching experiences in development finance, money and capital markets, Islamic finance, corporate finance, investments.

**Ajay Dheer** served in the Army Medical Corps, Indian Army and superannuated in September 2018. His field of interest is combat medicine, with a specialty in high altitudes. He is a PhD scholar presently, at the Sikkim Manipal Institute of Medical Sciences.

**Gan-Ochir Doojav** is the Chief Economist of the Bank of Mongolia and an Associate Professor of the National University of Mongolia. He earned his master and Ph.D. degree in Economics from the Australian National University. He is the author of A Small Open Economy Modelling: A Bayesian DSGE Approach and has published many articles in international peer-reviewed journals.

**Hary Febriansyah** was a material physician in 2001. He received his bachelor's degree from Mathematics and Natural Sciences at Brawijaya University Malang-East Java. His special expertise is in mathematics especially integral and differential calculus and material science. Continuing his studies, he graduated with a master's degree in 2004 from Management and Industrial Engineering (TMI-ITB). He decided to pursue his Master's in TMI-ITB, in contrast with his bachelor's studies because he wanted to enrich and enlarge his knowledge from two sides; natural science and industrial management. He chose to do this since he was interested in solving practical and academic problems in the real world with a holistic and balanced view. He has gained professional experience working in February 2004 as a human resources and operation manager in a general consultant company in Palembang, South of Sumatera-Indonesia. There he had several responsibilities, including overseeing the treatment of employees and monitoring the company's policy in this area to ensure that it was effective and efficient. He was also a consultant in the human resources management area for private and national-state companies during this time. However, in August 2005, he decided to switch careers becoming an academic in a management and industrial engineering department in a private university. His responsibilities included lecturing and being head of the management laboratory until June 2006. In July 2006, he moved to School of Business and Management ITB, Bandung-West Java as a teaching and research assistant, at the same time preparing for his doctoral degree in human resources management. After some deliberation he chose to pursue his doctoral degree in School of Management, Innsbruck University-Austria, after receiving a recommendation from ASEA-UNINET and scholarships from the Directorate General of Higher Education, the Ministry of Education and the Culture of Indonesia and Institute Technology Bandung.

**Munkhbayar Gantumur** is an Economist of the Research Division at the Bank of Mongolia. His research interests include microeconomics and computational economics.

**Waseem Ul Hameed** completed his Ph.D. in finance and banking from Universiti Utara Malaysia. His research interest areas are women empowerment, microfinance, and open innovation. He has a number of publications in high indexed journals. Currently, he is serving the Islamia University of Bahawalpur as a lecturer. Email: expert_waseem@yahoo.com; waseemulhameed@gmail.com

**Muhammad Imran** has completed PhD (Management) from Universiti Utara Malaysia. He served as assistant professor in Iqra University, Karachi, Pakistan. Currently, he is working as an assistant professor in department of management sciences, The Islamia University of Bahawalpur, Pakistan. He has published many research papers in various SSCI, SECI, ESCI, Scopus index journals. He is serving as a reviewer of many journals as well.

**Jawad Iqbal** is working as an Associate Professor & Chairman of the Department of Management Sciences at The Islamia University of Bahawalpur. He completed his doctoral degree from the Universiti Teknologi Malaysia (UTM) in 2013. His doctoral research was completed under the supervision of Prof. Dr. Amran Bin Md. Rasli in the area of knowledge-based research collaboration for economic transformation. He has published 24 research articles in highly reputed local and international journals. He has organized different national and international conferences. He also has worked as a chief editor of two research journals.

**Qaisar Iqbal** is currently pursuing his Ph.D. in organizational behavior at School of Management, Universiti Sains Malaysia. He has published many research papers in international and national journals including VINE Journal of Information and Knowledge management Systems, Business Information Review, Global Business Review, Journal of Science and Technology Policy Management, Administrative Sciences, Global IGI, World Journal of Science, Technology and Sustainable Development, and The Lahore Journal of Business. Currently, he is also the editorial board member of Probe-Business & Management, Insight-Information, Probe-Accounting, auditing, and taxation, International Journal of Economics, Finance, and Management and Journal of Business Administration Research. He has also served his university as USM Student Ambassador and President-PhD Club at School of Management. He had visited as Universiti Muhammadiyah Yogyakarta, Indonesia and Minjiang University, China as an invited Guest Speaker in 2018 and 2019. Previously, he has served as Manager Operations in the banking industry for five years. Professionally, he has been awarded the SPOT Award and the President Award as well.

**Saifuddin Isa** is a Master of Sciences in the field of Technology Management (Innovation). He is a researcher at Universiti Teknikal Malaysia Melaka (UTeM), Faculty of Technology Management and Technopreneurship. His main area of expertise is quality management, operation management, smart manufacturing, and Industrial Revolution 4.0. His research focuses on quality tools and techniques of Industrial Revolution 4.0, particularly in the Smart Manufacturing context.

**Md. Aminul Islam** is a faculty at the Department of Media Studies and Journalism at the University of Liberal Arts Bangladesh (ULAB). Before joining the ULAB, he was a lecturer at the Department of Journalism, Communication and Media Studies, Varendra University. He did masters degree in mass communication and journalism at the University of Rajshahi. He completed a post-graduate diploma in journalism (New Media) from Asian College of Journalism, Chennai, India. Currently, he is working in the fields of cyberpsychology, social interaction and mental health, social network analysis, human behavior, and media psychology. He can be reached at aminul.islam@ulab.edu.bd/aminul.vu@gmail.com.

**Ajeya Jha** is a Professor at the Department of Management Studies at Sikkim Manipal Institute of Technology (SMIT) since February 2000. In addition to having post-graduate and doctoral degrees in Business Administration, he possesses graduation and post-graduation in pharmaceutical sciences also. He is a fellow of the International Society of Management. He has a rich and diverse industrial experience of 15 years (includes launching two green-field projects), teaching 21 years and has been with NGO for 8 months. He is the founder Chief coordinator of Entrepreneurship Development Cell (EDC) at SMIT established with support from Department of Science and Technology, Government of India.

**Bezon Kumar** is now serving as a Lecturer in Economics at Rabindra University, Bangladesh. He started his career first at Varendra University as a Lecturer in Economics just after passing the master's in economics from the University of Rajshahi. Besides teaching, he also works as the Director of BK School of Research. His expertise belongs to behavioural and psychological issues, sustainable development, climate change, poverty, welfare and so on. He has several international peer-review journal articles and currently he is engaged in three research projects. He can be reached at bezon.kumar3@gmail.com.

**Padma Kurisetti** has been working as Associate Professor in School of Management, National Institute of Technology-Warangal. She has been teaching for more than 20 years. Her teaching and research interests include economics, finance and social issues. She obtained her doctoral degree from IIT Kharagpur. She published several articles in national and international journals.

**Davaajargal Luvsannyam** is Director of Research Division at the Bank of Mongolia. He holds master and PhD degree in Economics from the Australian National University. His research has been published in peer reviewed international journals such as Journal of Applied Econometrics and International economic journal.

**Pushpavalli Maniam** is a lecturer of the Faculty of Communication and Creative Industries at the Tunku Abdul Rahman University College and currently pursuing PhD in Communication at University of Science Malaysia. Her areas of interest are in the communication and social media studies and her main research area is crowdsourcing. For the past ten years, she has been teaching communication and media subjects.

**Mohd Syaiful** graduated with his first bachelor's degree of Human Resources Management from University Utara Malaysia (UUM) as well as his master's degree of Business Administration from UUM. Subsequently, he obtained his Doctor of Philosophy in Quality and Technology Manufacturing from Strathclyde University, School of Engineering, Design Manufacture Engineering Management, United Kingdom. He worked as a Business Development Executive at Era Riang and Human Resource Officer at ONKYO Electric (Malaysia) Private Limited.

**Hassan Mujtaba Nawaz Saleem** has been involved in research and teaching since year 1998 soon after completing his MBA degree. During his last 20 years he has been able to work on the diverse set of assignments that involved undergraduate/postgraduate level teaching and research where had chance to work with the multidisciplinary teams of researchers/consultants nationally/internationally.

**Tsolmon Otgonbat** is an Economist of Research Division at the Bank of Mongolia. His research interests focus on the areas of effectiveness of central bank intervention, banking sector fragility, economic connectedness, government expenditure and survey designing.

**Swamy Perumandla** completed his Ph.D. from National Institute of Technology, Warangal. Currently, he is working as an Assistant Professor at MITS School of Business, Madanapalle Institute of Technology and Science. His teaching and research interests include financial management, financial econometrics, international finance, equity, and commodity derivative markets.

**Saibal Kumar Saha** (UGC NET Qualified) holds a first class Master's degree in MBA and a first class Bachelor's Degree in Electronics and Communication Engineering. He has 8+ years of experience and has worked in MNCs like Cognizant Technology Solutions and Tata Aig Life. He has served National Institute of Technology - Silchar, Jyotirmoy School of Business – Kolkata, University of Technology and Management - Shillong and is presently working as Assistant Professor at Sikkim Manipal Institute of Technology - Majitar, India.

**Shivani Saini** received her PhD in area of marketing from Dr. B. R. Ambedkar National Institute of Technology, Jalandhar, India. She is an educationist and has six years of teaching experience in various engineering and management institutes including Dr. B R Ambedkar National institute of Technology, Jalandhar and Lovely Professional University. She has published numerous articles in referred journals and conferences. Her research interests include customer experience management, consumer loyalty, services marketing, and sustainable marketing.

**Jagwinder Singh** is working as an Associate Professor and Head of the Department of Humanities and Management Dr B R Ambedkar NIT Jalandhar, India. He has 21 years of academic experience. His areas of research are consumer behavior, social marketing, consumer loyalty and educator development. He has contributed articles to reputed journals in the areas of business management.

**Bilguunzul Sodnomdarjaa** is an Economist of the Research Division of the Bank of Mongolia. She holds her master's degree in Finance from University of London (UK). Her research interest focus on financial market, household finance and micro data.

**Bilguun Sukhbaatar** is a Senior economist of the Research division at the Bank of Mongolia. He earned his master's degree in Economics from the Charles University (Czech Republic). His research interest focus on survey designing and Bayesian DSGE models.

**Heru Kurnianto Tjahjono** is a Professor at the Department of Management, Faculty of Economics and Business, Universitas Muhammadiyah Yogyakarta. He received his Ph.D. in Organisational Behaviour from the Universitas Gadjah Mada, 2008. His research interests include organizational justice, organizational behavior, and human resource management

**Izzal Asnira Zolkepli** is best known for her work on convergence and adoption of communication technologies; and the proliferation of digital, social media and networked media pertaining to user engagement (i.e. behavioral response). Her current research projects are on user engagement (voluntary behavior) which look at crowdsourcing as well as user engagement on adopting/using mobile apps. Dr. Izzal is currently serving as a senior lecturer at Universiti Sains Malaysia, a research-intensive public university that is recognised as the top 50 university under 50 years in the world by QS World Ranking. Dr. Izzal is teaching marketing communication subjects—persuasion communication, marketing management, and international advertising—that are offered in the School of Communication.

# Index

www.ingramcontent.com/pod-product-compliance
Lightning Source LLC
LaVergne TN
LVHW060410300926
849162LV00006B/1387